RF Systems, Components, and Circuits Handbook

For a complete listing of the *Artech House Microwave Library*,
turn to the back of this book.

RF Systems, Components, and Circuits Handbook

Ferril Losee

Artech House
Boston • London

Library of Congress Cataloging-in-Publication Data
Losee, Ferril.
  RF systems, components, and circuits handbook / Ferril Losee.
    p.   cm.
  Includes bibliographical references and index.
  ISBN 0-89006-933-6 (alk. paper)
  1. Radio circuits—Design and construction—Handbooks, manuals, etc.
  I. Title.
  TK6560.L65   1997
  621.384—dc21   97-13330
                                                                              CIP

British Library Cataloguing in Publication Data
Losee, Ferril
  RF systems, components, and circuits handbook
  1. Electronic circuit design
  I. Title.
  621.3'815

  ISBN 0-89006-933-6

Cover design by Jennifer L. Stuart

© 1997 ARTECH HOUSE, INC.
685 Canton Street
Norwood, MA 02062

All rights reserved. Printed and bound in the United States of America. No part of this book may be reproduced or utilized in any form or by any means, electronic or mechanical, including photocopying, recording, or by any information storage and retrieval system, without permission in writing from the publisher.
  All terms mentioned in this book that are known to be trademarks or service marks have been appropriately capitalized. Artech House cannot attest to the accuracy of this information. Use of a term in this book should not be regarded as affecting the validity of any trademark or service mark.

International Standard Book Number: 0-89006-933-6
Library of Congress Catalog Card Number: 97-13330

10 9 8 7 6 5 4 3 2 1

Contents

Preface			xvii
Part I	RF Systems		1
Chapter 1	Introduction to Communication Systems		3
	1.1	Overview	3
	1.2	Background Information and Terminology	3
		1.2.1 Units and Conversion Information	3
		1.2.2 Frequency Bands for Communication Systems	4
		1.2.3 The Use of Decibels	4
	1.3	Introduction to Communication Services and Systems	6
		1.3.1 Frequency Allocations and FCC Regulations	6
		1.3.2 Aeronautical Mobile Service	7
		1.3.3 Aeronautical Mobile Satellite Service	9
		1.3.4 Amateur Service	9
		1.3.5 Broadcasting Service	9
		1.3.6 Broadcasting-Satellite Service	12
		1.3.7 Citizens Band Radio	12
		1.3.8 Earth Exploration-Satellite Service	12
		1.3.9 Fixed Service	12
		1.3.10 Fixed-Satellite Service	19
		1.3.11 Intersatellite Service	19
		1.3.12 Land Mobile-Satellite Service	20
		1.3.13 Maritime Mobile Service	20

		1.3.14	Maritime Mobile-Satellite Service	20
		1.3.15	Meteorological Aids Service	20
		1.3.16	Meteorological-Satellite Service	20
		1.3.17	Mobile Service	21
		1.3.18	Mobile Telephone Service	22
		1.3.19	Mobile-Satellite Service	25
		1.3.20	Space Operation Service	25
		1.3.21	Space Research Service	25
		1.3.22	Standard Frequency and Time Signal Service	26
		1.3.23	Telephone Service	26
	1.4	Transmission Media		27
		1.4.1	Open-Wire Lines	28
		1.4.2	Twisted-Pair Lines	28
		1.4.3	Coaxial Cable	28
		1.4.4	Microwave Terrestrial Radio (Microwave Relay)	28
		1.4.5	Communication Satellites	29
		1.4.6	Fiber Optics Cable Communication Systems	30
		1.4.7	Submarine Cable Communication Systems	31
Chapter 2		Introduction to Radar, Navigation, and Other RF Systems		35
	2.1	Radar Systems		35
		2.1.1	Basic Concepts	35
		2.1.2	Radar Frequencies	36
		2.1.3	Types of Radars	40
	2.2	Navigation Systems		50
		2.2.1	Omega System Navigational Aids	50
		2.2.2	Loran Navigational Aids	52
		2.2.3	Radio Beacons and Airborne Direction Finders	52
		2.2.4	VHF Omnidirectional Range System	55
		2.2.5	Distance-Measuring Equipment	56
		2.2.6	Instrument Landing System	56
		2.2.7	Tactical Air Navigation	59
		2.2.8	Microwave Landing System	60
		2.2.9	Air Traffic Control Radar Beacon System	60
		2.2.10	Transit	60
		2.2.11	NAVSTAR/Global Positioning System	60
	2.3	Electronic Warfare Systems		63
		2.3.1	Communication Intelligence Gathering Systems	63
		2.3.2	Electronic Intelligence Systems	64
		2.3.3	Electronic Countermeasure Systems for Communications	64
		2.3.4	Electronic Countermeasure Systems for Radars	64
	2.4	Radio Astronomy Systems		68

	2.5	Radiometer Systems		68
	2.6	Microwave Heating Systems		69
	References			70
Chapter 3	Radio Frequency Propagation			71
	3.1	Antennas		71
		3.1.1	Transmit Antennas	71
		3.1.2	Receive Antenna Gain and Capture Area	72
	3.2	Electromagnetic Waves, Fields, and Power Density		73
	3.3	RF Electrical Field Waveforms and Vector Addition		75
	3.4	Free-Space Path Loss		75
	3.5	Excess Path Loss and Atmospheric Attenuation		76
		3.5.1	Atmospheric Absorption	76
		3.5.2	Attenuation Produced by Rain, Snow, and Fog	78
	3.6	Atmospheric Refraction		79
	3.7	Diffraction of Radio Waves		80
	3.8	Multipath		82
	3.9	Ionospheric Propagation		84
	3.10	Ground-Wave Propagation		86
	3.11	Scatter Propagation		87
	3.12	Fiber Optic Cable Propagation		88
	3.13	Radar Cross-Section of Targets		89
	3.14	Equations for Calculating Propagation Performance for Communication Systems		90
		3.14.1	Example 1: HF Ionospheric Reflection Communication System	91
		3.14.2	Example 2: VHF Base Station to Mobile Unit Communication System	91
		3.14.3	Example 3: Microwave Uplink to Satellite Relay Located at Geostationary Orbit	92
	3.15	Equations for Calculating Propagation Performance for Radar Systems		92
		3.15.1	Example 4: L-Band Aircraft Surveillance Radar	93
		3.15.2	Example 5: X-Band Airborne Multiple-Function Radar	94
	References			94
Chapter 4	RF Noise and Link Analysis			95
	4.1	Concepts of RF Noise and Signal-to-Noise Ratio		95
	4.2	Noise Power, Noise Temperature, and Noise Figure		97
	4.3	Multiple-Stage Systems With Noise		98
	4.4	Types of Noise		100
		4.4.1	Atmospheric Noise	100
		4.4.2	Galactic Noise	101

	4.4.3	Solar Noise	103
	4.4.4	Ground Noise	104
	4.4.5	Man-Made Noise and Interference	105
4.5	Signal-to-Noise Improvement by Use of Integration		105
4.6	Signal-to-Noise Ratio		107
4.7	Communication System Link Analysis		109
4.8	Radar System Link Analysis		111
4.9	Performance Calculations for Radar Systems With Electronic Countermeasures		112
References			114

Chapter 5 Modulation Techniques — 115

5.1	Pulsed Continuous-Wave Signals	115
5.2	Conventional Amplitude Modulation	116
5.3	Double Sideband Suppressed Carrier Modulation	118
5.4	Vestigial Sideband Modulation	120
5.5	Single-Sideband Modulation	120
5.6	Standard Frequency Modulation	121
5.7	Modulation for Telemetry	124
5.8	Combination Communication and Range-Measurement Systems	125
5.9	Modulation for Radar	126
	5.9.1 Pulsed CW Modulation	126
	5.9.2 High-Power Impulse Generators and Ultra-Wideband, High-Power Microwave Generators	128
	5.9.3 Chirp Pulse Modulation	129
	5.9.4 Phase Code Modulated Pulse Modulation	129
	5.9.5 Continuous-Wave Modulation	130
	5.9.6 Frequency-Modulated CW Modulation	131
5.10	Single-Channel Transmitter System	131
5.11	Frequency Division Multiplex Transmitter System	133
5.12	Sample Circuits and Analog-to-Digital Converter Concepts	134
5.13	Time Division Multiplex Transmitter System With Pulse Code Modulation	135
5.14	Two-State Modulation Types for Binary Signals	137
	5.14.1 On-Off or Two-State Amplitude Keying	137
	5.14.2 Frequency Shift Keying	137
	5.14.3 Binary Phase-Shift Keying	138
5.15	Four-State and Eight-State Phase-Shift Keying	138
5.16	Sixteen Phase-State Keying (16-PSK)	139

	5.17	Sixteen Amplitude-Phase Keying	140
	References		141
	Selected Bibliography		141

Chapter 6 RF Amplifiers, Oscillators, Frequency Multipliers, and Mixers 143

6.1	Amplifiers		143
	6.1.1	Front-End Low-Noise RF Amplifiers for Receivers	145
	6.1.2	IF Amplifiers	151
	6.1.3	Audio and Other LF Amplifiers	154
	6.1.4	Transmitter RF Amplifier Chains	154
	6.1.5	Transmitter RF Power Amplifiers for Communication Systems	154
	6.1.6	RF Power Amplifiers and Oscillators for Radars, Navigation, and Electronic Countermeasure Applications	157
6.2	Oscillators and Frequency Synthesizers		159
	6.2.1	Transistor Feedback Oscillators	159
	6.2.2	Negative Resistance Two-Terminal Oscillators	161
	6.2.3	Frequency Synthesizers	162
6.3	Frequency Multipliers		163
	6.3.1	Varactor Diode Frequency Multipliers	164
	6.3.2	Step-Recovery Diode Frequency Multipliers	164
	6.3.3	Transistor Multipliers	165
6.4	Mixers		167
	6.4.1	Diode Mixers	167
	6.4.2	Transistor Mixers	173
References			175
Selected Bibliography			175

Chapter 7 Modulators and Demodulators 177

7.1	Modulators		177
	7.1.1	Modulators for Conventional Amplitude Modulation	177
	7.1.2	Modulators for Double-Sideband Modulation	180
	7.1.3	Vestigial-Sideband Modulators	181
	7.1.4	Modulators for Single-Sideband Modulation	183
	7.1.5	Modulators for Frequency-Division Multiplex	186
	7.1.6	Modulators for Standard Frequency Modulation	186
	7.1.7	Modulators for Frequency-Shift Keying	189
	7.1.8	Modulators for Phase-Shift Keying	189
	7.1.9	Modulators for Pulse Code Modulation Time-Division Multiplex Modulation	192
	7.1.10	Time-Division Multiple Access	194

	7.2	Demodulators or Detectors	195
		7.2.1 Amplitude Modulation Detectors	195
		7.2.2 Product Detectors	196
		7.2.3 Frequency Modulation Detector Concepts	199
		7.2.4 Phase Detectors	202
	References		203
Chapter 8	Older Communication Systems		205
	8.1	HF Communication System Using Single-Sideband Modulation	205
	8.2	VHF or UHF Ground-to-Air Communication System Using Either Amplitude Modulation or Narrowband FM	208
	8.3	Frequency Modulation Broadcast Systems	211
	8.4	Microwave Relay Systems	214
	8.5	Satellite Relay Communication Systems	219
	8.6	Satellite Relay Earth Stations	223
	References		225
Chapter 9	Current and Future Commercial Communication Systems		227
	9.1	Business and Personal Communication Systems	227
		9.1.1 49-MHz Cordless Telephones	227
		9.1.2 900-MHz Cordless Telephones	229
		9.1.3 Pager Systems	230
		9.1.4 Citizens Band Radios	231
		9.1.5 VHF and UHF FM Business and Personal Two-Way Radio	232
	9.2	Cellular Telephone Systems	232
		9.2.1 The Concept of Spatial Frequency Reuse	232
		9.2.2 Propagation Characteristics of Cellular Telephone Systems	234
		9.2.3 Advance Mobile Phone Service	236
		9.2.4 Narrowband Advanced Mobile Phone Service	246
		9.2.5 North American Digital Cellular Telephone Systems	246
		9.2.6 PDC Japan Digital Cellular System	251
		9.2.7 GSM Europe System Digital Cellular System	251
		9.2.8 TIA IS-95 North America Digital Cellular System	252
	9.3	GPS Receivers	254
	9.4	Fiber Optic Communication Systems	256
	References		260
Chapter 10	Radar Systems		261
	10.1	Radar Cross-Section of Targets	261

Contents | xi

	10.2	Radar Clutter	263
	10.3	Radar Measurements	267
		10.3.1 Range Measurements	267
		10.3.2 Velocity Measurement Using CW Radar	270
		10.3.3 Velocity Measurements Using FMCW Radar	272
		10.3.4 Velocity Measurements Using Pulse Type Radar	272
		10.3.5 Angle Measurements for Radars	273
	10.4	Continuous-Wave Radar Systems	274
		10.4.1 Continuous-Wave Radars for Target Velocity Measurement	274
		10.4.2 Frequency-Modulated Continuous-Wave Radars	276
	10.5	Moving-Target Indicator and Pulse Doppler Radars	278
	10.6	Signal Processing for Moving-Target Indicator Radars	281
		10.6.1 Delay-Line Cancelers	281
		10.6.2 Digital Signal Processing	283
		10.6.3 MTI From a Moving Platform	284
	10.7	Tracking Radars	285
		10.7.1 Monopulse Tracking Radars	285
		10.7.2 Tracking in Range Using Sequential Gating	286
	References		287

Part II   RF Components and Circuits   289

Chapter 11   Transmission Lines and Transmission Line Devices   291

11.1	Two-Wire Transmission Lines	291
11.2	Coaxial Transmission Lines	292
11.3	Coaxial Cable Connectors	295
11.4	Microstrip Transmission Lines	296
11.5	Stripline Transmission Lines	298
11.6	Characteristics of Transmission Lines	300
	11.6.1 Wave Velocity on Transmission Lines	300
	11.6.2 Reflection Coefficients	301
	11.6.3 Standing-Wave Ratio	301
11.7	The Smith Chart	304
	11.7.1 Impedance and Admittance Coordinates	304
	11.7.2 Voltage Standing-Wave Ratio Circles	306
	11.7.3 Reflection Coefficients	306
	11.7.4 Examples Using the Smith Chart	306
11.8	Impedance Matching Using the Smith Chart	310
	11.8.1 Impedance Matching With a Quarter-Wave Transformer	310
	11.8.2 Impedance Matching With a Short-Circuited Stub	313
11.9	Coaxial Terminations	315
11.10	Coaxial Directional Couplers	316

11.11	Baluns	317
11.12	Two-Wire Transmission Line Impedance Transformer	318
11.13	Stripline and Microstrip Circuits	319
	11.13.1 Shunt Stub DC Returns	319
	11.13.2 Branch Line 90-Degree Hybrid Couplers	319
	11.13.3 Stripline or Microstrip Rat Race Hybrid Coupler	320
	11.13.4 Split Inline Hybrid Dividers and Combiners	322
	11.13.5 Quarter-Wave Coupled-Line Directional Couplers	323
	11.13.6 90-Degree Coupled-Line Hybrid Coupler	323
	11.13.7 Stripline Lowpass Filters	325
	11.13.8 Stripline Highpass Filters	326
	11.13.9 Stripline Bandpass Filters	327
11.14	Ferrite Circulators and Isolators	329
11.15	Coaxial Electromechanical Switches	331
11.16	Pin Diode Switches	333
11.17	Sparkgap Switches for Lightning Protection	334
References		335
Selected Bibliography		335

Chapter 12 Waveguides and Waveguide-Related Components 337

12.1	Introduction to Waveguides	337
12.2	Rectangular Waveguides	338
12.3	Higher Order Modes in Rectangular Waveguides	346
12.4	Launching the TE_{10} Mode Using a Coaxial Line Input	347
12.5	Characteristic Wave Impedance for Waveguides	349
12.6	Other Types of Waveguides	350
	12.6.1 Ridged Waveguides	350
	12.6.2 Circular Waveguides	350
12.7	Waveguide Hardware	352
	12.7.1 Waveguide Flanges	352
	12.7.2 Rotary Joints	352
	12.7.3 Tapered Transition Sections of Waveguides	353
	12.7.4 Flexible Waveguides	353
	12.7.5 Waveguide Accessories	353
12.8	Waveguide Hybrid Junctions	354
12.9	Waveguide Impedance Matching	356
12.10	Waveguide Resistive Loads and Attenuators	357
12.11	Waveguide Directional Couplers	357
12.12	Waveguide Ferrite Isolators, Circulators, and Switches	358
12.13	Waveguide Detectors and Mixers	361
12.14	Gas-Tube Switches	361
12.15	Duplexers	362

	12.16	Cavity Resonators	365
	References		368
	Selected Bibliography		368
Chapter 13	Antennas		369
	13.1	Monopole Antennas	371
		13.1.1 Thin-Wire Monopole Antennas	371
		13.1.2 Wideband Monopoles	372
		13.1.3 Impedance of Monopole Antennas	372
		13.1.4 Large-Size Monopole Antennas	373
		13.1.5 Electrically Small Monopole Antennas	375
	13.2	Dipole Antennas	377
		13.2.1 Thin-Wire Dipole Antennas	377
		13.2.2 Other Types of Dipole Antennas	378
		13.2.3 Dipole Impedance	378
		13.2.4 Dipole Current Distribution and Antenna Patterns for Different L/λ Ratios	381
		13.2.5 Turnstile Antenna	384
	13.3	Yagi-Uda Antennas	385
	13.4	Sleeve Antennas	386
		13.4.1 Sleeve Monopoles	386
		13.4.2 Sleeve Dipoles	387
	13.5	Loop Antennas	387
		13.5.1 Air-Core Loop Antennas	388
		13.5.2 Ferrite-Core Loop Antennas	389
	13.6	Helical Antennas	390
	13.7	Spiral Antennas	391
		13.7.1 Equiangular Spiral Antennas	392
		13.7.2 Archimedean Spiral Antennas	393
		13.7.3 Conical Spiral Antennas	394
	13.8	Log-Periodic Antennas	394
		13.8.1 Log-Periodic Dipole Array	394
		13.8.2 Trapezoidal-Toothed Log-Periodic Antennas	398
		13.8.3 Triangular-Toothed Log-Periodic Antennas	400
	13.9	Slot Antennas	401
		13.9.1 Open-Slot Antennas	401
		13.9.2 Cavity-Backed Rectangular-Slot Antennas	402
		13.9.3 Waveguide-Fed Slot Antennas	402
	13.10	Notch Antennas	405
	13.11	Horn Antennas	405
	13.12	Lens Antennas	408
		13.12.1 Dielectric Lens Antenna	408
		13.12.2 Luneburg Lens Antenna	409
		13.12.3 Metallic-Plate Lens Antenna	409

	13.13	Antenna Arrays	410
		13.13.1 End-Fire Line Antenna Arrays	410
		13.13.2 Broadside Line Antenna Arrays	412
	13.14	Planar Arrays	414
	13.15	Scanning Methods	415
		13.15.1 Mechanically Scanned Arrays	415
		13.15.2 Arrays With Space Feeds	418
	13.16	Flat-Plate Reflector Type Antennas	418
		13.16.1 Half-Wave Dipole Antennas With Reflectors	419
		13.16.2 Corner Reflector Antennas	419
	13.17	Parabolic Reflector Antennas	420
	References		423
Chapter 14		Lumped Constant Components and Circuits	425
	14.1	Conductors and Skin Effect	425
	14.2	RF Resistors	426
	14.3	Inductors and Inductive Reactance	427
	14.4	RF Chokes	431
	14.5	Capacitors and Capacitive Reactance	431
	14.6	Series Resonant RLC Circuits	435
	14.7	Parallel Resonant RLC Circuits	436
	14.8	Complex Resonant Circuits	438
	14.9	The Use of the Smith Chart for Circuit Analysis	441
	14.10	S-Parameters	443
	14.11	Impedance Matching Using LC Circuits	444
	14.12	Impedance-Matching Design Using the Smith Chart	446
	14.13	LC Filters	452
	References		455
Chapter 15		RF Transformer Devices and Circuits	457
	15.1	Conventional Transformers	457
	15.2	Magnetic Core Material for RF Transformers	460
	15.3	Tuned Transformers	462
	15.4	High-Frequency Wideband Conventional RF Transformers	464
	15.5	Transmission-Line Transformers	468
	15.6	Power Combiners and Splitters	471
	References		472
Chapter 16		Piezoelectric, Ferrimagnetic, and Acoustic Devices and Circuits	473
	16.1	Quartz Crystal Resonators and Oscillators	473
	16.2	Monolithic Crystal Filters	477
	16.3	Ceramic Filters	477

	16.4	Dielectric Resonant Oscillators	479
		16.4.1 Dielectric Resonator Description and Parameters	479
		16.4.2 Coupling Between a Dielectric Resonator and a Microstrip Line	482
		16.4.3 Mechanical and Electrical Tuning of Dielectric Resonators	482
		16.4.4 Examples of Dielectrically Stabilized Oscillators	482
	16.5	YIG Resonators and Filters	484
		16.5.1 Ferrimagnetic Resonance in Yttrium Iron Garnet Crystals	484
		16.5.2 YIG Bandpass Filters	486
		16.5.3 YIG-Tuned Oscillators	486
	16.6	Surface Acoustic Wave Delay Lines	486
		16.6.1 Nondispersive Delay Lines	486
		16.6.2 Tapped Delay Lines	489
		16.6.3 Dispersive Delay Lines	489
	16.7	Surface Acoustic Wave Delay Line Oscillators	490
	16.8	Bulk Acoustic Wave Delay Lines	493
	References		497
Chapter 17	Semiconductor Diodes and Their Circuits		499
	17.1	Semiconductor Materials	499
	17.2	"Ordinary" Junction Diodes	500
	17.3	Zener Diodes	500
	17.4	Schottky-Barrier Diodes	501
	17.5	Pin Diodes	501
	17.6	Varactor Diodes	502
	17.7	Step-Recovery Diodes	504
	17.8	Microwave Tunnel Diodes and Circuits	504
	17.9	Microwave Gunn Diodes and Circuits	505
	17.10	Microwave Impatt Diodes	507
	17.11	Semiconductor IR Laser Diodes	508
	17.12	Light-Emitting Diodes	510
	17.13	IR Photodiodes	511
	References		512
Chapter 18	Bipolar and Field-Effect Transistors and Their Circuits		513
	18.1	Bipolar Junction Transistors	513
	18.2	BJT Amplifier Configurations	515
		18.2.1 Common-Emitter Amplifier	515
		18.2.2 Common-Base Amplifier	522
		18.2.3 Common-Collector Amplifier	524
	18.3	Field Effect Transistors and Circuits	526
		18.3.1 Junction Field-Effect Transistors	526

		18.3.2	Metal-Semiconductor Field-Effect Transistors	529
		18.3.3	Metal-Oxide Semiconductor Field Effect Transistors	530
	18.4		Comparison of FET and BJT Amplifiers	532
	18.5		High Electron Mobility Transistors and Heterojunction Bipolar Transistors	534
		18.5.1	High Electron Mobility Transistors	534
		18.5.2	Heterojunction Bipolar Transistors	535
	18.6		DC Bias Circuits for BJT Amplifiers	535
	18.7		Bias Circuits for FET Amplifiers	538
	18.8		Stability With BJT and FET Amplifiers	539
	18.9		Impedance Matching	539
	18.10		Design Methods With S-Parameters	541
	18.11		Manufacturers' Data Sheets for Transistors	543
	References			547
Chapter 19			High-Power Vacuum Tube Amplifiers and Oscillators	549
	19.1		Grid Tubes	549
		19.1.1	Triode, Tetrode, and Pentode Vacuum Tubes	549
		19.1.2	Grid-Type Vacuum Tube Amplifiers and Modulated Amplifiers	551
		19.1.3	The Use of Cavities as Resonators for High-Power, High-Frequency Triodes and Tetrodes	557
	19.2		Microwave Tubes and Circuits	558
		19.2.1	Introduction to Microwave Tubes	558
		19.2.2	Multiple-Cavity Klystron Amplifiers	559
		19.2.3	Helix-Type Traveling-Wave Tube Amplifiers	562
		19.2.4	Coupled-Cavity Traveling-Wave Tube Amplifiers	566
		19.2.5	Conventional Magnetrons	566
		19.2.6	Coaxial Cavity Magnetron	567
		19.2.7	Amplitrons	568
		19.2.8	Gyrotron Oscillators and Amplifiers	569
		19.2.9	Circuit Configurations for Microwave Tubes	571
	References			572
About the Author				573
Index				575

Preface

I wrote this book from my perspective as both a university educator and an RF engineer. I have spent 22 years as an RF engineer working in industry, 18 years as a university professor, and 3 years as a consultant and author.

While working in industry, I came to appreciate the need for a handbook or self-teaching text that would cover the total field of RF systems, components, and circuits. I frequently heard individuals say they wished they had an up-to-date, easy-to-understand handbook to help them learn more about RF. In many cases, these persons had specialized in computers or other technical areas while in college and had very little RF training. In other cases, they simply were out of date or had forgotten the things they had learned while in college.

As a university professor, I often wished a textbook was available that covered the total RF field and that was at a technical level that beginners could use in a one- or two-semester course. No such text was available. I believe other professors and schools would also like such a book.

There have been rapid changes in the RF field in recent years. Many of our most important RF systems did not even exist 10 years ago, for example, cellular telephones, fiber optic systems, digital satellite TV, and GPS systems. Thus, there is a need for an up-to-date text that covers new subjects and materials, as well as important older RF systems and devices.

It was that set of needs that prompted me to write this handbook. I tried to make the text easy to read and understand. I believe that a good technical document should have about one figure for each page of text (the total number of figures in this handbook is 348). I assumed that the readers would have a fairly limited technical background and wrote the text accordingly.

This handbook is in two parts. Part I, which comprises Chapters 1 through 10, covers RF systems. Chapter 1 provides background information and terminology, followed by an introduction to communication services and systems. A final section introduces transmission media for telephones.

Chapter 2 provides a similar simplified introduction to radar, navigation, ECM, and other RF systems. Chapter 3 discusses RF propagation. Chapter 4 presents information about RF noise and signal-to-noise ratio concepts. It also shows how to conduct link analysis and determine the performance of communication and radar systems.

Chapter 5 discusses modulation techniques and performance characteristics. It introduces some of the newer modulation concepts, as well as discussing more conventional systems. Chapter 6 presents concepts and design information for RF amplifiers, oscillators, frequency multipliers, and mixers. Chapter 7 discusses modulators and demodulators. Chapter 8 presents several examples of some of the older types of communication systems that are still important.

Chapter 9 presents current and future commercial communication systems. It includes a section devoted to cellular telephone systems. Chapter 10 presents concepts and design information for radar systems. These systems are important for both military and civilian applications.

Part II, Chapters 11 through 19, covers RF components and circuits. Chapter 11 discusses transmission lines and transmission line–related devices. Chapter 12 presents information about waveguides and related devices. Chapter 13 presents concepts and design information for antennas. Chapter 14 presents useful information about RF lumped constant components and circuits. This chapter discusses resistors, inductors, and capacitors. Applications discussed include series resonant circuits, parallel resonant circuits, impedance matching networks, and L-C filters. Chapter 15 discusses RF transformer devices and circuits. Topics include conventional transformers, core material for radio frequency (RF) transformers, intermediate frequency (IF) amplifier transformers, high-frequency wideband conventional transformers, transmission-line transformers, and power combiners and splitters.

Chapter 16 presents concepts and design information for piezoelectric, ferrimagnetic, and acoustic wave devices and circuits. Topics include quartz crystals, quartz crystal oscillators, quartz crystal filters, ceramic filters, dielectric resonant oscillators, yttrium iron garnet (YIG) filters, YIG oscillators, surface acoustic wave (SAW) delay lines, and bulk acoustic wave (BAW) delay lines.

Chapter 17 presents information about semiconductor diodes and circuits. Topics include semiconductor materials, ordinary junction diodes, zener diodes, Schottky-barrier diodes, PIN diodes, varactor diodes, step-recovery diodes, microwave tunnel diodes, microwave Gunn diodes, microwave IMPATT diodes, IR laser diodes used for fiber optics systems, and IR photo diodes used for fiber optics systems.

Chapter 18 discusses transistors and transistor circuits. Topics include bipolar junction transistors and circuits, field-effect transistors and circuits, HEMTs, HBTs, dc bias circuits, impedance matching circuits, design methods using S-parameters, and manufacturers' data sheets for transistors.

Chapter 19 discusses grid-type vacuum tubes and circuits. Tube types include high-power triodes, tetrodes, and pentodes. Both lumped constant-type tube circuits and cavity resonator-type tube circuits are discussed.

Chapter 19 also discusses microwave tubes and circuits. Types of microwave tubes discussed include multiple cavity klystrons, helix-type traveling wave tubes, coupled-cavity traveling wave tubes, conventional-type magnetrons, coaxial cavity magnetrons, amplitrons, and gyrotrons. I sincerely hope that you will enjoy this text and find it interesting reading as well as a valuable teaching aid. I further hope it will meet your needs in learning about modern RF systems, components, and circuits.

PART I
RF Systems

CHAPTER 1

Introduction to Communication Systems

1.1 OVERVIEW

The term communication systems refers to any type of radio frequency (RF) or optical frequency system in which the main objective is to transfer information from one point to another. The means of communication may be by land lines, underground cables, underwater cables, ground-wave propagation, free-space propagation, tropospheric scatter propagation, ionospheric reflection propagation, ground-to-ground microwave relay, ground-to-ground fiber optics relay, ground-to-satellite-to-ground relay, and other systems. This chapter discusses the functions of many of those systems, frequency assignments, and the ways in which signals propagate. In some cases, the mode of modulation is indicated. Very little detail is provided about the hardware and components involved (that information is discussed in later chapters). This chapter also provides some very limited background information that may be helpful to the reader.

1.2 BACKGROUND INFORMATION AND TERMINOLOGY

1.2.1 Units and Conversion Information

A wavelength is the distance an electromagnetic wave travels in one RF cycle. That distance depends on the speed of the wave in the medium involved. In the case of free space, the speed of the RF electromagnetic wave is the speed of light, which is approximately $3 \cdot 10^8$ meters per second. The equation for wavelength with free-space propagation is

$$\lambda = c/f$$

where

λ = wavelength in meters

c = speed of light in meters per second = $3 \cdot 10^8$ m/s

f = frequency in hertz (1 Hz = 1 cycle per second)

In media other than free space, the RF energy travels at a speed less than the speed of light, and the wavelength at a given frequency is smaller. The ratio of the wavelength in free space to the wavelength in a medium other than free space is equal to the square root of the relative dielectric constant, permittivity, of the medium.

Table 1.1 shows units and conversion information for frequency and wavelength.

1.2.2 Frequency Bands for Communication Systems

Table 1.2 shows frequency band names and frequency coverage information for communication systems. Corresponding free-space wavelengths also are shown.

1.2.3 The Use of Decibels

Decibels (dBs) are used to indicate ratios of power in a logarithmic fashion. As such, a ratio expressed in decibels is simply 10 times the logarithm to the base 10 of the numeric power ratio. The following are examples.

1.2.3.1 Loss

If the output power from a filter is one-half the input power, the insertion loss for the filter is $10 \log_{10} 2 = 3$ dB.

Table 1.1
Common Frequency and Wavelength Units and Conversion

Frequency		
1 hertz	= 1 Hz	= 1 cycle per second
1 kilohertz	= 1 kHz	= 10^3 Hz
1 megahertz	= 1 MHz	= 10^6 Hz
1 gigahertz	= 1 GHz	= 10^9 Hz
1 terahertz	= 1 THz	= 10^{12} Hz
Wavelength		
1 centimeter	= 1 cm	= 10^{-2} m
1 millimeter	= 1 mm	= 10^{-3} m
1 micron (or micrometer)	= 1 μ	= $1 \cdot 10^{-6}$ m
1 angstrom	= 1 A	= $1 \cdot 10^{-10}$ m

Table 1.2
Frequency Bands for Communication Systems

Frequency	Wavelength	Frequency Band
3–30 kHz	10^5–10^4 m	VLF (very low frequency)
30–300 kHz	10^4–10^3 m	LF (low frequency)
0.3–3 MHz	10^3–10^2 m	MF (medium frequency)
3–30 MHz	10^2–10 m	HF (high frequency)
30–300 MHz	10–1 m	VHF (very high frequency)
0.3–3 GHz	1–0.1 m	UHF (ultra high frequency)
3–30 GHz	10–1 cm	SHF (super high frequency)
30–300 GHz	1–0.1 cm	EHF (extremely high frequency)
0.3–3 THz	1–0.1 mm	Band 12
1–417 THz	300–0.72 mm	Infrared
417–789 THz	0.72–0.38 mm	Visible light
789 to $5 \cdot 10^6$ THz	0.38 to $6 \cdot 10^{-5}$ mm	Ultraviolet
$3 \cdot 10^4$ to $3 \cdot 10^8$ THz	100 to $1 \cdot 10^{-2}$ A	X-rays
$>3 \cdot 10^7$ THz	<0.1 A	Gamma rays

If the output power from a cable is one-quarter the input power, the insertion loss is $10 \log_{10} 4 = 6$ dB.

If the received power for a communication link is 1 μW (microwatt) and the transmitted power is 1 kW, the propagation loss for the link is $10 \log_{10}(1 \text{ μW} / 1 \text{ kW}) = 10 \log_{10}(10^{-6} \text{ W} / 10^3 \text{ W}) = 10 \log_{10}(10^9) = 90$ dB.

1.2.3.2 Amplifier Gain

If the power output of an amplifier is 50 times the input power, the gain of the amplifier is $10 \log_{10} 50 = 17$ dB.

1.2.3.3 Antenna Gain

If the power density in a given direction is 40 times what it would be if the antenna had been an isotropic radiator (uniform in all directions), the gain of the antenna with respect to isotropic would be $10 \log_{10} 40 = 16$ dBi, where i indicates isotropic. Sometimes antenna gain is given with respect to that of a dipole antenna. In that case, if the power density in a given direction is two times what it would be if the antenna had been a dipole antenna, the gain of the antenna would be $10 \log_{10} 2 = 3$ dBd (where d indicates dipole). That would be about 5 dBi, since the gain of a dipole with respect to isotropic is about 1.8 dBi.

1.2.3.4 Power Level

If the transmitted power level is 300 W, the power level expressed in decibel format would be $10 \log_{10} 300 = 25$ dBW, where W stands for watts. Equivalently the power can be expressed as $10 \log_{10} (30{,}000 \text{ mW}) = 55$ dBm, where m stands for milliwatts. It is emphasized that dBm and dBW are ratios relative to 1 mW and 1W, respectively.

If the receive power is $8 \cdot 10^{-9}$ W, or 8 nW (nanowatts), the received power expressed in dBW is $10 \log_{10}(8 \cdot 10^{-9}) = 10 \log_{10}(8) + 10 \log_{10}(10^{-9}) = 10(0.9) + 10(-9) = 9 - 90 = -81$ dBW.

1.2.3.5 A Number

If the number in an equation is 4π, that number expressed in decibels is $10 \log_{10}(4\pi) = 11$.

If the wavelength in an equation is 0.01m, that wavelength expressed in decibels is $10 \log_{10}(0.01) = -20$ dB.

1.2.3.6 Voltage Ratios

When expressing a voltage ratio in decibels, we express the ratio as a power ratio. Power is voltage squared divided by resistance. Thus, the voltage ratio $100/10$ expressed in decibels is $20 \log_{10}(100/10) = 20$ dB.

1.2.3.7 Using Decibels in an Equation

When using decibels in an equation, we add decibels for multiplication and subtract decibels for division. Thus, $\log_{10}(a \cdot b) = \log_{10} a + \log_{10} b$, and $\log_{10}(a/b) = \log_{10} a - \log_{10} b$. We multiply decibels by a number when we want to raise to a power. Thus, $\log_{10}(m)^n = n \log_{10}(m)$.

1.3 INTRODUCTION TO COMMUNICATION SERVICES AND SYSTEMS

This section briefly introduces types of RF communication services and systems. The different types of communication systems are presented in alphabetical order. More details for many of these systems and associated subsystems are provided in subsequent chapters. Much of the material presented in this section is quoted or adapted from [1].

1.3.1 Frequency Allocations and FCC Regulations

The frequency allocations for the different communication services are provided on a worldwide basis by the International Telecommunication Union (ITU). The allocations for Region 2, which includes North and South America, have been adopted by the U.S. Federal Communications Commission (FCC), which has responsibility for RF spectrum management within the United States. Individual frequency assignments or authorizations must be requested by the user and subsequently approved by the FCC before the user may transmit at the requested frequencies. Other FCC regulations also must be followed, such as the maximum transmitted power levels, out-of-band harmonics, and spurious signal levels. Most of the ITU and FCC frequency allocations can be obtained from Van Valkenburg [1] and other similar sources. Some of the wording that follows is derived from the definitions of services provided by the ITU as listed in [1].

1.3.2 Aeronautical Mobile Service

Aeronautical mobile service is a mobile service between aeronautical stations and ground stations or between aircraft stations. Emergency position-indicating radio beacon stations also may participate in this service on designated distress and emergency frequencies. For purposes of frequency allocation, aeronautical mobile service is divided into two parts: aeronautical mobile (R) service and aeronautical mobile (OR) service. Aeronautical mobile (R) service is reserved for communications relating to safety and regularity of flight, primarily along national or international civil air routes. Frequency allocations for this service include HF, VHF, and UHF frequencies. Allocated HF frequencies are in the frequency range of 2.85 MHz to 23.35 MHz. Those frequencies are used for long-range communication, such as communication between land-based stations and aircraft flying over the Atlantic or Pacific Ocean. The propagation mode is ionospheric reflection propagation, sometimes referred to as sky wave propagation. The mechanism involved is actually refraction rather than reflection, but it is common to think of the result as reflection. Long-range communication is made possible by this mechanism because the ionosphere is so high above the Earth. Different frequencies are used for different times of day because of the changes in maximum usable frequency for this propagation mode as a function of time of day, season of the year, path, and other factors. This mode of operation is illustrated in Figure 1.1(a).

Figure 1.1 shows transceivers connected to antennas. A transceiver is a combination transmitter and receiver system that permits the operator to select either transmit or receive operation and thereby achieve two-way communication with a single unit. Separate input and output (I/O) devices are connected to the transceivers to facilitate required input and output functions. Examples of input systems are microphones, keyboards, telephone interfaces, and controls. Examples of output systems are speakers, headphones, printers, telephone interfaces, recorders, meters, and displays.

Most of the aircraft that fly only over land use only VHF or UHF frequencies. VHF frequencies used are in the range of 118 MHz to 137 MHz. That band normally

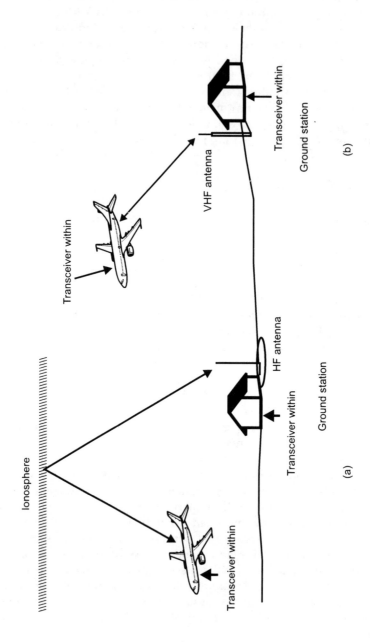

Figure 1.1 Types of aeronautical mobile service: (a) aeronautical mobile service using HF ionospheric reflection propagation and (b) aeronautical mobile service using VHF or UHF free-space propagation.

is used for amplitude modulation (AM) signals. UHF frequencies used are in the range of 225 MHz to 400 MHz. That band normally is used for frequency modulation (FM) signals. The propagation mode in each case is free-space propagation, sometimes called space wave. The space wave represents the energy that travels from the transmitter to the receiver antenna in the Earth's troposphere, that is, the portion of the Earth's atmosphere in the first 10 mi adjacent to the Earth's surface (Figure 1.1(b)). The range capability with free-space propagation is radio horizon limited and for airborne systems is typically less than 200 mi, depending on aircraft altitude and the surrounding topology.

Aeronautical mobile (OR) service is used primarily outside national or international civil air routes. Otherwise, it is the same as aeronautical mobile (R) service.

1.3.3 Aeronautical Mobile Satellite Service

Aeronautical mobile satellite service is a mobile service in which mobile earth stations are located on aircraft. Survival craft stations and emergency position-indicating radio beacon stations also may participate in this service. The frequency allocations for the service are 1545–1555 MHz for the space-to-Earth link, and two bands at 1646.5–1656.5 MHz and 1660–1660.5 MHz for the Earth-to-space link. At the time of this writing (1997), U.S. commercial airlines were not using this service. It is anticipated that that will change because of the many advantages of satellite communications and the many problems with using HF ionospheric reflection communication for communication with flights over the oceans. Those problems include crowded spectrum and possible radio blackouts due to ionospheric disturbances. Figure 1.2 illustrates the aeronautical mobile satellite service.

1.3.4 Amateur Service

Amateur service is a radiocommunciation service for the purpose of self-training, enjoyment, and technical investigations carried out by amateurs. Amateurs are defined as duly authorized persons interested in radio technique solely with a personal aim and without pecuniary interest. Frequency allocations for this service include MF, HF, VHF, UHF, SHF, and EHF frequencies.

Amateur radio is an enjoyable hobby for many individuals. Most of the radio amateur activity is in the HF band, where the propagation mode is long-range ionospheric reflection communication. Examples of frequency assignments available to amateurs in the HF band are 3.5–3.75 MHz, 7.0–7.3 MHz, 14.0–14.25 MHz, 18.069–18.168 MHz, 21.0–21.450 MHz, 24.89–24.99 MHz, and 28.0–21.45 MHz.

1.3.5 Broadcasting Service

Broadcasting service is a radio communication service in which the transmissions are intended for direct reception by the general public. This service includes sound

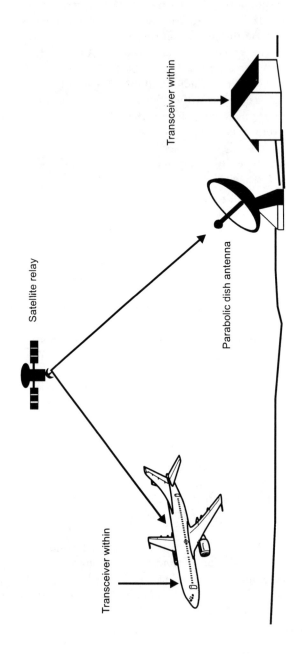

Figure 1.2 Aeronautical mobile satellite service.

transmissions, television transmissions, and other types of transmission. Frequency allocations for this service include MF, HF, VHF, UHF, and higher frequencies.

The MF frequencies are used for standard AM broadcasting. The mode of propagation in this case is normally ground wave during day time. Coverage range in this mode is usually less than 100 mi. Ground wave is sometimes called surface wave. It can exist when the transmitting and receiving antennas are close to the surface of the Earth and are vertically polarized. Ground wave is supported at its lower edge by the presence of the ground. Charges move in the Earth and constitute a current with loss as the wave moves. It is effective only at the lower frequencies because losses become too great at the higher frequencies.

At night, it is possible to have much longer range capability at MF frequencies by means of E-layer ionospheric reflection propagation. That mode, however, is not depended on for MF broadcast service.

The HF frequencies are used for broadcasting over very long ranges. The mode of propagation is ionospheric reflection propagation. Only narrowband operation is possible.

VHF frequencies are used for FM radio broadcasting and television broadcasting. FM radio broadcasting uses frequencies in the 88 to 108 MHz band. VHF television broadcasting uses the 54–72 MHz band, the 76–88 MHz band, and the 174–216 MHz band. UHF television broadcasting uses the 470–608 MHz band and the 614–890 MHz band.

FM radio and television broadcasting is illustrated in Figure 1.3. Here we see that a single high-power transmitter with a wide-angle antenna can provide signals to a large number of receivers. Coverage range can be 50 mi or more, depending on antenna height, transmitter power, and surrounding topology.

The normal mode of propagation for VHF and UHF broadcasting is free-space propagation. Diffraction also is often involved where the line-of-sight path is

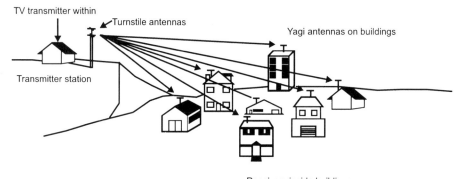

Figure 1.3 Illustration of VHF and UHF broadcast services.

blocked by trees, buildings, or hills. Often there is attenuation as the waves pass through trees and other types of natural and unnatural structures. These propagation modes are discussed in more detail in Chapter 3.

1.3.6 Broadcasting-Satellite Service

Broadcasting-satellite service is a radio communication service in which signals transmitted or retransmitted by space stations are intended for direct reception by the general public. In broadcasting-satellite service, the term direct reception includes both individual reception and community reception. Broadcasting-satellite service is illustrated in Figure 1.4.

Frequency allocations for broadcasting-satellite service include 2520 to 2670 MHz, 3.7 to 4.2 GHz, 12.2 to 12.7 GHz, and 17.3 to 17.7 GHz.

1.3.7 Citizens Band Radio

The 26.96–27.23 MHz frequency band is allocated for citizens who are not licensed amateurs but who desire to use radio transmitters and receivers for pleasure or business. These systems often are used by truck drivers and others who travel the nation's highways to communicate with each other. CB radio may also include ground stations for ground-to-mobile or ground-to-ground communication.

1.3.8 Earth Exploration-Satellite Service

Earth exploration-satellite service is a radio communication service between Earth stations and one or more space stations or between space stations. With this service, information relating to the characteristics of the Earth and its natural phenomena can be obtained from active or passive sensors on Earth satellites. Similar information is collected from airborne or earth-based platforms. Such information can be distributed to Earth stations within the system concerned. Platform interrogation may be included in this service.

This service may include feeder links necessary for its operation. Frequency allocations for this service include UHF, SHF, and higher frequencies.

1.3.9 Fixed Service

Fixed service is a radio communication service between specific fixed points. By fixed points, we mean ground-based communication units or stations.

There are many types of ground-to-ground communication systems and many frequency allocations. The military uses VLF and LF frequencies for long-range

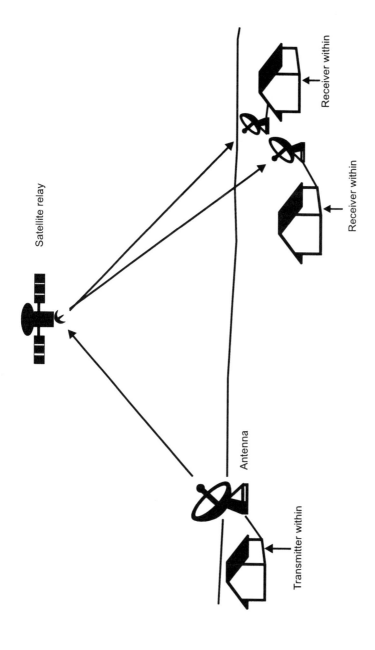

Figure 1.4 Broadcasting-satellite service.

ground-wave communications. MF frequencies are used for shorter range ground-wave communications. In this mode, the RF signal travels along the surface of the earth with the electrical-field (E-field) vertical. The lower the frequency, the lower the loss will be due to currents in the Earth. At VLF the communication range could be as large as 2,000 mi, whereas at MF, the range typically would be less than 100 mi.

HF frequencies are used mainly for long-range communication using ionospheric reflection propagation. As pointed out earlier, many different frequency allocations are needed over the HF band for ionospheric reflection propagation because the optimum usable frequency changes widely as a function of time of day, season of the year, and path length. Very long range is possible in this mode because the ionosphere is so high above the surface of the Earth. With HF it is common to operate with ranges of 2,000 to 4,000 mi. In fact, it is not uncommon to operate at ranges exceeding 8,000 mi. More details about this mode of propagation are provided in Chapter 3.

VHF and UHF frequencies are used extensively for short-range communication between fixed points on the ground. Transceivers may be located in base stations, or they may be portable, hand-held transceivers. The mode of propagation is normally free-space propagation. Frequently line-of-sight propagation is not possible due to blockage by buildings, trees, small hills, or other objects. In such cases, communication may be possible with increased propagation loss by diffraction around or over the blocking object. The lower the frequency, the lower the excess loss will be. Propagation of this type is illustrated in Figure 1.5.

UHF and SHF frequencies also are frequently used for long-range communication between specified fixed points. The long range is made possible by the use of microwave relay communication. A system of this kind is shown in Figure 1.6. Such systems are used for transmitting telephone, television, and data over large distances. They ordinarily use high-gain antennas mounted on towers. The towers and associated stations often are located on hills or mountaintops to extend the line-of-sight distance between stations.

Simultaneous two-way operation is possible by using two antennas and two circulators. The signal from the A-direction antenna is received, processed, and retransmitted by the B-direction antenna. The signal from the B-direction antenna is received, processed, and retransmitted by the A-direction antenna.

Most of the microwave relay links operate in the 4- and 6-GHz region, although there are systems that operate at other frequencies. The 4-GHz band has an authorized frequency range of 3.7 to 4.2 GHz. The 6-GHz band has an authorized range of 5.925 to 6.425 GHz.

Microwave relay systems usually are low-power systems with typical power levels less than 10W. They usually are wideband systems with many multiplexed channels. Diversity reception often is used, in which two or more receiver antennas are mounted one above the other and separate receivers used for each with appropriate diversity combiners. That helps reduce fading problems caused by multipath. Some

Introduction to Communication Systems | 15

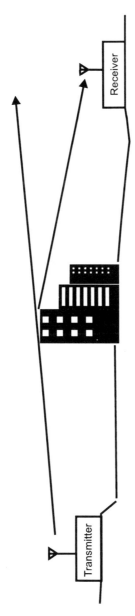

Figure 1.5 Fixed service using VHF free-space propagation with diffraction.

16 | RF Systems, Components, and Circuits Handbook

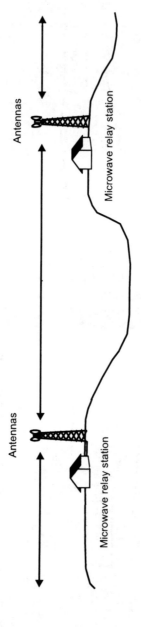

Figure 1.6 Two sections of a long-range microwave relay communication system.

of the material presented in the following paragraphs is quoted or adapted from [2].

Figure 1.7 is a simplified diagram of a tropospheric scatter communication system. The troposphere is the region of the atmosphere just below the ionosphere (within about 15 km of the ground). In this region of the upper atmosphere, it is possible to have low-level forward scattering from nonuniform atmospheric conditions, including changes in the dielectric properties of the atmosphere. Sometimes these reflections are caused by ionized gas produced by meteors entering the atmosphere. In that case, a mode of communication known as meteor burst communication is possible.

The frequency range used for tropospheric scatter is the UHF band (300 to 3000 MHz). The best frequencies and those most often used are centered on 900 and 2000 MHz. The 790–960 MHz band is the most common frequency range.

Tropospheric scatter is recognized as a very reliable method of over-the-horizon communication. It is not affected by the abnormal solar phenomena that afflict HF ionospheric reflection propagation. Its disadvantage is the fact that the scattered signal is very weak, making it necessary to use high transmitter power, high transmitter-antenna gain, high receiver-antenna gain, low noise receivers, and narrow bandwidth.

The troposcatter link terminal is similar to a microwave link terminal. The main differences lie in the higher output powers and the lower receiver noise figures in troposcatter links. Typical output powers are 1 to 10 kW, but transmitter powers as high as 100 kW have been used for broadband links.

Diversity is nearly always used in troposcatter links. Diversity reception is the case where signals from more than one receiver are combined to form a single receive signal. The types of diversity used for tropospheric scatter communication include space diversity, in which two or more spaced antennas are used; frequency diversity, in which two or more frequencies are used simultaneously; polarization diversity, in which two or more polarizations are used for the antennas; and quadruple diversity, in which a combination of these are used.

A high percentage of troposcatter links are single-span links. A single span over inaccessible terrain is likely to have a communication range of 300 to 1,000 km. On the other hand, a link designed to provide communications for a group of islands, such as in the Caribbean, Indonesia, or the Philippines, will have several spans with access at each point.

Antenna sizes vary with required link span, bandwidth, and antenna gain. Parabolic dish antenna diameters of 15m may be used for broadband links. Even larger antennas have been used for long scatter relay paths and large bandwidth. Thus, they may be even larger than satellite earth station antennas.

A typical broadband link may carry 192 two-way voice channels. Capabilities in excess of five times that number are available.

Tropospheric scatter systems are costly because of the high power and large antenna requirements, so they are used only where special considerations so dictate.

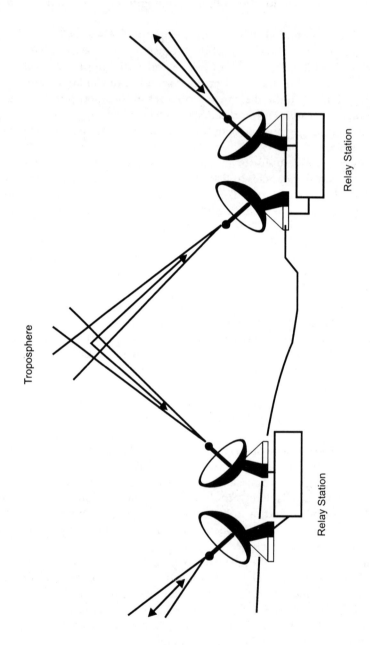

Figure 1.7 Fixed service using tropospheric scatter propagation.

1.3.10 Fixed-Satellite Service

Fixed-satellite service is a radio communication service between Earth stations at given positions when one or more satellites are used. The given position may be a specified fixed point or any fixed point within specified areas. In some cases, the service includes satellite-to-satellite links, which also may be operated in the intersatellite service. The fixed-satellite service also may include feeder links for other space radio communication services.

The frequency allocations for fixed-satellite service include UHF, SHF, and higher frequencies. The earliest frequency bands used for satellite transmission of telephone were in the 4-GHz and 6-GHz bands, used for terrestrial microwave transmission. The 4-GHz band (3.7 to 4.2 GHz) was used for the downlink, and the 6-GHz band (5.925 to 6.425 GHz) was used for the uplink. Some newer satellites operate in Ku-band, utilizing 11 GHz (10.95 to 11.2 GHz and 11.45 to 11.7 GHz) for the downlink and 14 GHz (14.0 to 14.5 GHz) for the uplink. A third frequency range at Ka-band, 17 and 30 GHz, also has been authorized for broadcasting-satellite service.

Generally the frequency separation between the uplink frequency and the downlink frequency is about 2 GHz. That separation is needed to permit simultaneous reception and transmission by the satellite repeaters.

Most of the fixed-satellite services use geostationary orbit satellites with orbit altitudes of about 35,800 km and equatorial orbit planes. At that altitude, the orbital period is 24 hours, so the satellites appear stationary to observers on the surface of the Earth. Satellite communication systems of this type typically use high-gain, parabolic dish antennas at the ground stations. They also use high transmitter power for the ground-based transmitter.

One communication satellite of this type is INTELSAT V. All communication satellites launched before INTELSAT V use the 5.925–6.425 GHz frequency range for the uplink and the 3.7–4.2 GHz frequency range for the downlink. INTELSAT V uses those same frequencies and in addition uses a second set of frequencies: 14.0–14.5 GHz for the uplink and both 10.95–11.20 GHz and 11.45–11.70 GHz for the corresponding downlink. The use of the 14/11 GHz range significantly increases the available system capacity when compared with earlier satellites [2].

1.3.11 Intersatellite Service

Intersatellite service is a radio communication service that provides links between artificial Earth satellites. Frequency allocations for this service include SHF and higher frequencies.

1.3.12 Land Mobile-Satellite Service

Land mobile-satellite service is a mobile satellite service in which mobile Earth stations are located on land. Frequency allocations for this service include VHF and UHF frequencies.

1.3.13 Maritime Mobile Service

Maritime mobile service is a mobile service between coast stations and ship stations, from ship station to ship station, or between associated on-board communication stations. Survival craft stations and emergency position-indicating radio beacon stations also may participate in this service. Frequency allocations include VLF, LF, MF, HF, and VHF frequencies.

The mode of propagation for VLF, LF, and MF frequencies is ground-wave propagation. The mode of propagation at HF frequencies is ionospheric reflection propagation. At VHF the mode of propagation is free-space propagation.

1.3.14 Maritime Mobile-Satellite Service

Maritime mobile-satellite service is a mobile-satellite service in which mobile Earth stations are located on board ships. Survival craft stations and emergency position-indicating radio beacon stations also may participate in this service. Frequency allocations for this service include the following:

- Space to Earth (downlink): 1530–1544 MHz;
- Earth to space (uplink): 1626.5–1645.4 MHz.

1.3.15 Meteorological Aids Service

Meteorological aids service is a radio communication service used for meteorological, including hydrological, observations and exploration. Frequency allocations for this service include HF, VHF, UHF, SHF, and higher frequencies. Examples of this type of system are radiosondes.

1.3.16 Meteorological-Satellite Service

Meteorological-satellite service is an Earth exploration-satellite service for meteorological purposes. Frequency allocations for this service include VHF, UHF, and SHF frequencies. The satellites often contain television-type cameras and means for sending visual images of the clouds and Earth features below the satellites to the ground-based receiving systems. These weather-mapping satellites are of great value in the prediction of weather conditions in all parts of the world.

1.3.17 Mobile Service

Mobile service is radio communication service between mobile and land stations or between mobile stations. Frequency allocations for this service include LF, MF, HF, VHF, UHF, and SHF. Mobile communication systems include public safety systems, industrial systems, land transportation systems, broadcast remote pickup, and television pickup. Public safety systems include police, fire, highway, forestry, ambulance, and emergency services. Industrial systems include power, petroleum, pipeline, forest products, factories, builders, ranchers, motion picture, press relay, and radio-controlled appliance repair services. Land transportation systems include taxis, trucks, buses, and railroads.

Most of these examples of mobile communication systems operate in either the VHF or the low UHF frequency bands. Typical operating frequencies are less than 470 MHz. Most of the systems use narrowband FM modulation [2].

The following frequency allocations are for land transportation (taxis, trucks, buses, railroads):

- 30.56–32.00 MHz;
- 33.00–33.01 MHz;
- 43.68–44.61 MHz;
- 150.8–150.98 MHz;
- 152.24–152.48 MHz;
- 157.45–157.74 MHz;
- 159.48–161.575 MHz;
- 452.0–453.0 MHz;
- 457.0–458.0 MHz.

The following frequency allocations apply to public safety systems, including police, fire, highway, forestry, and emergency services:

- 1.605–1.750 MHz;
- 2.107–2.170 MHz;
- 2.194–2.495 MHz;
- 2.505–2.850 MHz;
- 3.155–3.400 MHz;
- 30.56–32.00 MHz;
- 33.01–33.11 MHz;
- 37.01–37.42 MHz;
- 37.88–38.00 MHz;

- 39.00–40.00 MHz;
- 42.00–42.95 MHz;
- 44.61–46.60 MHz;
- 47.00–47.69 MHz;
- 150.98–151.49 MHz;
- 153.7325–154.46 MHz;
- 154.6275–156.25 MHz;
- 158.7–159.48 MHz;
- 162.0–172.4 MHz;
- 453.0–454.0 MHz;
- 458.0–459.0 MHz.

1.3.18 Mobile Telephone Service

Much of the material presented in this subsection is quoted or adapted from [3–5]. Another important example of mobile communication systems is cellular mobile telephone systems. Cellular mobile telephone service is a high-capacity system that provides direct-dial telephone service to automobiles and other forms of portable telephone by using two-way radio transmission in the 800–900 MHz band.

A first-generation cellular telephone system uses analog (FM and frequency shift keying, or FSK) modulation. This system was used exclusively in the United States and Canada prior to the advent of digital cellular systems. This first-generation system is known as advanced mobile phone service (AMPS). It is still in use as the primary system in less heavily populated areas for users without digital end instruments and for users whose end instruments cannot operate on the standard of their selected system.

The second generation cellular telephone system is a time-division multiple access (TDMA) system. It has many advanced features that permit it to be a higher performance system than the first-generation system. The details of design and operation for both systems are presented in Chapter 9.

A first-generation, or AMPS, cellular telephone system is illustrated in Figure 1.8.

An AMPS cellular telephone system consists of a switching system, a number of cell sites or base stations, a data-link network, enhancers and converters, and mobile subscribers' units. The transmission frequency band for the base stations is 869 to 894 MHz; for the mobile stations, it is 824 to 849 MHz. Corresponding wavelengths are 0.35–0.34m and 0.36–0.35m. The antennas used for the cellular telephone transceivers often are 5/8 wave monopole or whip antennas and thus can have dimensions of about 22 cm (approximately 8.7 in). Hand-held systems may use 1/4 wavelength monopole antennas with dimensions of only about 9 cm (approximately 3.5 in) or less.

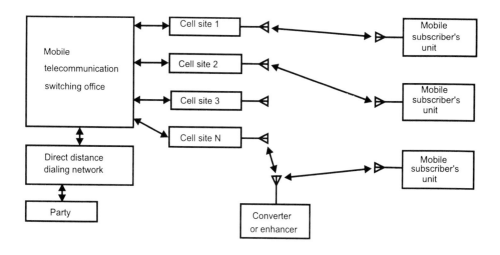

Figure 1.8 Cellular telecommunications system. (*After:* [3].)

The mobile units are two-way systems with 45 kHz between transmission and reception frequencies and with 25 kHz spacing between channels. The total number of channels is 832.

The coverage radius per cell site is 2 to 20 km. The type of modulation is FM with a frequency deviation of ±12 kHz. The control signal type is FSK with a deviation of ±8 kHz. The data transmission rate is 10 Kbps.

The cell site consists of an antenna with a height of about 30m (100 ft) to 45m (150 ft), a controller, and a number of transceivers. The controller is used to handle the call process between the switching system and the mobile units via a setup channel.

The mobile switching office (MTSO) is a special-purpose switch that connects the call between mobile units and the landline telephone. The function of this system is to assign the voice channel to each cell, perform the handoffs and service features, and monitor the calling information for billing.

The data link network carries the data between the cell sites and the MTSO. The data link networks may use wire lines, microwave links, or optical fibers.

The enhancer is a repeater with amplification by which the signal range can be extended. Converters can be used to change the frequency of the cellular signal to either microwave frequency or optical frequency.

The mobile subscriber's units are a transceiver, a logic unit, and a control unit. Mobile units are mounted in vehicles while portable units are carried by the user.

Figure 1.9 illustrates the concepts of frequency reuse and cell divisions in the service area. A key idea is to use sufficiently low power that the coverage for a given set of frequencies is limited to a small cell or small area within the larger service

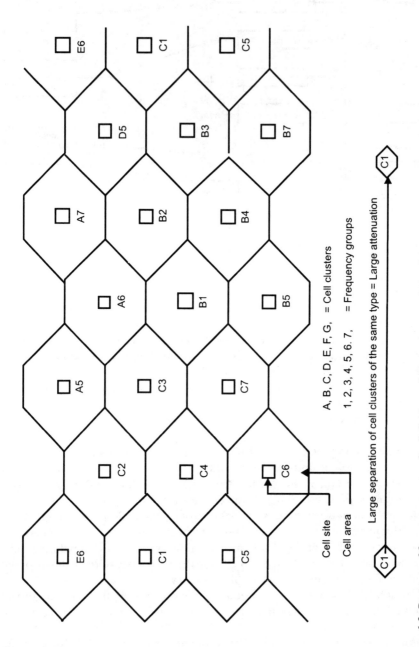

Figure 1.9 Concepts of frequency reuse and cell divisions of the service area. (*After:* [6].)

area. That same set of frequencies can then be used in other cells that are separated by a substantial distance or multiple other cell diameters.

In Figure 1.9, cells are organized into clusters, with seven cells per cluster. The radio channels are allocated across the seven cells. The clusters are then repeated to cover the whole geographic area served by the system. Although the cells are depicted as hexagons, their actual shape is irregular and depends on the terrain and the radio propagation.

Not all cells of a service area are the same size. In areas of high-density use, the cell size may be very small. That makes more channels available for use. As we move to areas of low-density use, the cell size may be made larger.

The same concept of frequency reuse and cell operation is also used for the second-generation digital cellular telephone systems. More detail about that type of communication system is presented later.

Code division multiple access (CDMA) can be viewed as a third-generation modulation type for cellular telecommunications systems. CDMA soon may become the standard, replacing TDMA. Chapter 9 discusses in more detail cellular telephone systems and the expected developments in that important communication area.

1.3.19 Mobile-Satellite Service

Mobile-satellite service is a radio communication service between mobile Earth stations and one or more space stations, between space stations used by this service, or between mobile Earth stations by means of one or more space stations. This service also may include feeder links necessary for its operation. Frequency allocations for mobile-satellite service include UHF, SHF, and higher frequencies.

1.3.20 Space Operation Service

Space operation service is a radio communication service concerned exclusively with the operation of spacecraft, in particular space tracking, space telemetry, and space telecommand. Those functions normally are provided within the service in which the space station is operating. Frequency allocations for this service include VHF and UHF frequencies.

1.3.21 Space Research Service

Space research service is a radio communication service in which spacecraft or other objects in space are used for scientific or technological research purposes. Frequency allocations for this service include VHF, UHF, SHF, and higher frequencies.

1.3.22 Standard Frequency and Time Signal Service

Standard frequency and time signal service is a radio communication service for scientific, technical, and other purposes, providing the transmission of specified frequencies, time signals, or both with stated high precision, intended for general reception. The exact frequencies transmitted are 2.5 MHz, 5.0 MHz, 10.0 MHz, 15.0 MHz, and 20.0 MHz.

1.3.23 Telephone Service

Much of the material presented in this subsection is quoted or adapted from [5–7]. The most important communication service that we have is the telephone. It is estimated that on the average each person in the United States and Canada uses the telephone eight times per day. Telephones are used not only for voice communication but also for personal computer communications via modems, Internet access, fax transmissions, and other communication services.

We can now make calls to almost any place in the world and have essentially the same quality of voice reception as if we were calling someone in a neighboring town. Calling rates are low and applications are many. Telecommunication is the key component of the information age and the services that form the basis of our information economy.

Figure 1.10 illustrates the main elements of a telephone system. On the left are the customer location and a number of telephone sets. The sets are connected by wire lines to a protector block. A single twisted pair of small-gauge wire known as the local loop connects each customer telephone system to a central office. Local loops from many customers are combined in a cable. A typical cable has 1,400 pairs of copper wires carried in a single plastic-sheathed cable of about 3 in in diameter.

The central offices or central switching centers in different communities are connected to each other by so-called trunk lines. These lines may be telephone wire lines, or they may be some type of RF communication system. Trunk lines

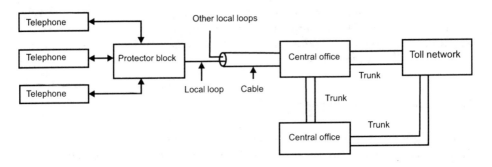

Figure 1.10 Main elements of a telephone system. (*After:* [6].)

include single transmitter and receiver microwave communication links, microwave relay systems, satellite relay systems, and cable systems.

Toll networks are provided for long-distance telephone, which is provided by such companies as AT&T, MCI, and Sprint rather than by the local telephone company. Links provided by these companies include microwave relay systems, satellite relay systems, land-based coax cable systems, undersea coax cable systems, and fiber optic cable systems. (Each of these types of links is discussed later.)

It is common to send many signals over each single wire or other channel so that the total number of voice channels is much greater than the number of wires or channels. This is done by a process known as multiplexing. Advanced techniques are used so that a single transmission medium can carry thousands of combined signals.

Two main approaches have been used for multiplexing telephone signals: frequency-division multiplexing and time-division multiplexing. Frequency-division multiplexing is an analog, or A-type, system, while time-division multiplexing is a digital, or D-type, system. The earlier systems used for telephone were mainly frequency multiplex systems, but the age of analog multiplexing is now over. Digital multiplexing has proved superior in nearly all respects and is now quite widespread. In the late 1980s, AT&T replaced nearly all the analog multiplexing in use on its long-distance network with digital multiplexing systems; MCI followed in the early 1990s. Each of these multiplexing methods is discussed in detail in Chapter 5.

The centralized switching systems or telephone exchanges are complex systems that involve large amounts of equipment. The function of the exchange is to interconnect wire lines so as to permit a call to be established correctly. If both the calling and the called subscribers are connected to the same exchange, it merely has to interconnect them. If the wanted subscriber is connected to some other exchange, the call from the calling subscriber must be routed correctly so it will reach the wanted number.

With modern exchange systems, interconnections are made by the exchange processor or computer; as a result, the space occupied by the exchange is much smaller than was required with earlier systems. The earlier systems used large numbers of relays in what is referred to as a crossbar exchange. Computers are also used to keep track of the calls made by each subscriber and to provide billing information.

1.4 TRANSMISSION MEDIA

The remainder of this chapter briefly discusses the transmission media used by telephone systems. Much of the material presented in this section is quoted or adapted from [7].

1.4.1 Open-Wire Lines

Open-wire lines consist of uninsulated bare copper wires strung on poles with insulators and separated by about 1 ft. In a telephone system, direct current is supplied to the telephone set through the two wires entering the set from the central office or switching center. A dc voltage of about 45V normally is used for communication with 75V dc used for ringing. Open-wire lines are still found in some rural areas.

1.4.2 Twisted-Pair Lines

A twisted pair consists of two individually insulated copper wires twisted together with a full twist about every 2 to 6 in. Twisting the wires reduces interference to the lines from outside sources. Small-gauge wire such as 26 gauges (0.016-in diameter) to 19 gauge (0.036-in diameter) typically are used. Many twisted pairs are combined into a single cable. Paired cable can be strung on poles, buried underground, or installed in a conduit. Paired cable is used primarily for the local loop and is also utilized between local exchange central offices. Since the telephonic speech signal has a required passband extending from approximately 300 Hz to 3300 Hz, the bandwidth requirement for a twisted pair is small.

1.4.3 Coaxial Cable

A large number of one-way voice circuits can be multiplexed together on a single coax cable. A number of 3/8-in or smaller diameter cables typically are combined into a larger cable with an outside plastic, vinyl, or rubber coating. For example, the cable used with the AT&T L5 system has 11 pairs of coaxial cables enclosed in a single outer jacket. The characteristics and performance of the L5 system are reported to be as follows:

- *Service date:* 1978;
- *Technology:* Integrated circuits;
- *Repeater spacing:* 1.0 mi;
- *Capacity per 3/8-inch coax pairs:* 13,200 voice circuits;
- *Capacity per group (10 working pairs):* 132,000 voice circuits.

1.4.4 Microwave Terrestrial Radio (Microwave Relay)

Microwave relay was once the backbone of the long-distance network and still is a fairly important technology. This type of system was discussed briefly in Section 1.3.9. As pointed out there, this type of system uses radio towers for antennas spaced about 26 mi apart along the route. A minimum system includes two antennas, one aimed in each of the two directions, with waveguide connected between the antennas and relay circuits. The most common types of antennas used are either horn antennas or parabolic dish reflector antennas, which are wideband antennas that may be dual polarized, permitting operation in each of two polarizations. That permits the capacity of the system to be doubled, compared to a system with a single polarization.

Two microwave bands currently in use are 3.7–4.2 GHz (called the 4-GHz band) and 5.925–6.425 GHz (termed the 6-GHz band). Notice that the width of each band is 500 MHz. Each band is further subdivided into a number of channels. Channel widths are 20 MHz for the 4-GHz band and 30 MHz for the 6-GHz band.

As an indication of the performance capability of a microwave relay system for telephone service using frequency multiplex, consider the AR6A radio system. This system, introduced in 1981 by AT&T, used single-sideband, suppressed-carrier, microwave transmission for the 6-GHz band. Six thousand voice channels could be placed in each radio channel. Thus, the seven active channels permitted 42,000 simultaneous two-way voice circuits. The AR6A was used in addition to the existing TD system, which used the 4-GHz band. The TD system had a capacity of 19,800 circuits. Combining the two systems, the total capacity, using the same antennas, was 61,800 two-way voice circuits.

In the early 1980s, AT&T introduced digital radio systems that used time-division multiplexing. Systems introduced included the DR6-30 system, the DR11-40 system, the DR6-30-135 system, and the DR4-20-90 system. The DR6-30 operates in the 6-GHz band with channels that have 30-MHz bandwidth. The total capacity is 9,408 two-way digital voice circuits. The DR11-40 operates in the 11-GHz band with 40-MHz channels. The total capacity is 13,444 two-way digital voice channels. The DR6-135 system operates in the 6-GHz band and uses 64 quadrature amplitude modulation (QAM). It has a capacity of 14,112 two-way digital voice circuits. The DR11-40-135 system operates in the 11-GHz band and has a total system capacity of 20,160 two-way digital voice circuits.

The time-division multiplexed signal is transmitted using QAM, a technique that is a combination of phase and amplitude modulation. A 16-QAM has 16 possible states and represents 4 bits per baud. That involves 4 phase states and 4 amplitude states. A 64-QAM has 64 possible states and represents 6 bits per baud. That involves 8 amplitude states and 8 phase states.

1.4.5 Communication Satellites

Fixed-satellite service was discussed briefly in Section 1.3.10. Satellite relay systems are used extensively for relaying telephone signals for long-distance telephone systems. The earliest frequency bands used for satellite transmission were the same 4-GHz and 6-GHz bands used for terrestrial microwave transmission. The 4-GHz band (3.7 to 4.2 GHz) was used for the downlink, and the 6-GHz band (5.925 to 6.425 GHz) for the uplink. Those two bands taken together are called the C-band (as in C-band radar). Some newer communication satellites use the radar Ku-band operating at 11 GHz (10.95 to 11.2 GHz plus 11.45 to 11.7 GHz) for the downlink and 14 GHz (14.0 to 14.5 GHz) for the uplink. International agreements also have authorized the use of a third band, the Ka-band, with operation at 17 and 30 GHz.

The typical spectrum width of the radio channel served by one transponder is 36 MHz. The modulation type is FM. A single transponder can be used for one color television signal, 1,200 voice circuits, or digital data at a rate of 50 Mbps. The total width of each half of the C-band is 500 MHz. Horizontal and vertical polarization of the radio signal are used to double the capacity. The result is that 24 channels are available for use, which provides for 12 two-way transponder pairs. Many satellites can be used with separations adequate to permit coverage of a single satellite per antenna beam.

1.4.6 Fiber Optics Cable Communication Systems

Figure 1.11 shows a simplified block diagram for a fiber optic cable communication system. This system uses a laser diode transmitter to transmit infrared (IR) waves to a distant optical receiver. As in the case of RF transmitters, the coherent IR waves are modulated by the incoming audio, digital, or video message. They are then sent through the fiber optic cable to either an optical repeater or to the end receiving system, depending on the range involved.

One type of optical fiber is a small-diameter cylindrical section of extremely pure glass with typical core diameters in the range of 2 to 10 μm. The optical fiber is covered with an outside cladding of glass of a slightly different chemical composition and a different refractive index. This type of fiber is known as a step-index fiber. The outside cladding of glass is subsequently covered with a protective covering.

The step-index fibers may be used for multimode or single-mode propagation. Single-mode propagation yields the best performance and lowest attenuation. Single-mode propagation in a step-index fiber is similar to single-mode propagation in a circular waveguide.

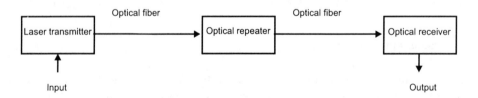

Figure 1.11 Fiber optics cable communication system.

Some earlier optical fiber cables were made with glass having a gradually changing index of refraction. These optical fibers are known as graded-index fibers. They are the easiest to make and were therefore developed first. However, because these larger diameter fibers use multimode propagation and have greater attenuation per unit length than the step-index fiber, their use is greatly diminishing [2].

In 1975 the lowest attenuation available for fiber optic cable was about 3.2 dB/km. By 1983 the lowest attenuation available for fiber optic cable was about 0.25 dB/km, a remarkable improvement. Hence, a 100-km fiber optic cable would have an attenuation of only 25 dB instead of 320 dB, thus significantly reducing the number of in-line amplifiers required to periodically boost the transmitted signal level so as to maintain an acceptable signal-to-noise ratio.

The receiver converts the modulated optical signal from the fiber optic cable to an electrical signal using a photodiode or a phototransistor. That is followed by an amplifier, which increases the level of the converted signal. The output of the amplifier feeds a detector that demodulates the electrical signal. The detected signal is then fed to the output system.

This type of cable communication system has advantages of wide bandwidth capability, security, and relatively low loss. Repeaters are required along the lines for very long range communication, as in cross-country telephone or television.

The following capacity information for fiber optic cables is taken from [6]. The increase in the capacity of commercially available optical fiber communication systems is astonishing. AT&T's first fiber system, FT3, was introduced in 1979 and used graded-index fiber at 45 Mbps per fiber. Seventy-two fiber pairs were placed together in a single cable, and 66 were active. The total route capacity of the system was 44,352 two-way digital voice circuits. The FT3C system was introduced soon afterward, with each fiber operating at 100 Mbps and carrying 1,344 digital voice circuits. Using cable with 66 active pairs, the system had a maximum route capacity of 88,704 two-way digital voice circuits. Repeaters were located every 4 mi along the route.

In 1984, AT&T introduced the FTX-180 system, with each single-mode fiber operating at 180 Mbps and carrying 2,688 digital voice circuits. A 400-Mbps system introduced in 1986 had each single-mode fiber carrying 6,048 digital voice circuits and repeaters spaced every 20 mi. Fiber optic systems operating at 2,000 Mbps are now standard. The capacities of today's fiber systems are mind boggling. A 2,000-Mbps system has 31,250 digital voice circuits, or 40 television signals, per single-mode fiber.

1.4.7 Submarine Cable Communication Systems

Submarine cable transmission lines also can be used for long-distance telephone communication. Such system use power boosters spaced along the path at appropriate intervals to compensate for line losses experienced by the electrical signals. A system of this type is shown in Figure 1.12.

32 | RF Systems, Components, and Circuits Handbook

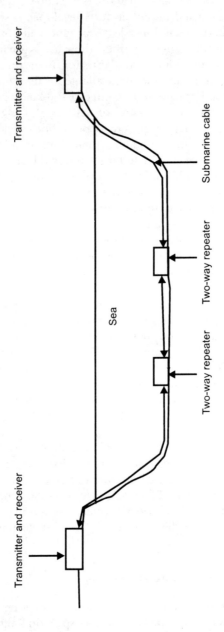

Figure 1.12 Submarine cable communication system.

Submarine cable systems have been around for a long time and have greatly improved in capacity with time. The first installed system had a design capacity of only 36 two-way voice circuits. That system, which was named TAT-1, had a service date of 1958. It used vacuum tube technology with repeaters spaced at intervals of 44 mi. The first system that used transistor technology was the TAT-5, having an in-service date of 1970. It had a design capacity of 845 two-way voice circuits. Repeaters for the TAT-5 system were spaced 12 mi apart. The TAT-7, with a service date of 1983, also used transistor technology. It had a design capacity of 4,200 two-way voice circuits. Repeaters for that system were spaced only 6 mi apart.

The first transatlantic submarine cable system using fiber optics was the TAT-8, with a service date of 1988. It utilized three fiber pairs, diode lasers, and integrated circuit technology. It has a design capacity of 8,000 two-way digital voice circuits. That capacity can be extended using a technique called time-assignment speech interpolation, or TASI. TASI uses the silent intervals in speech conversations to carry signals from other speech conversations. The maximum capacity for the TAT-8 system using TASI was 40,000 voice circuits. Repeaters for the TAT-8 system are spaced 41 mi apart.

The TAT-9 system has a service date of 1991. It is a fiber optic system with a maximum capacity of 80,000 two-way voice circuits. Repeaters are spaced every 75 mi.

The large capacity and reliability of optical fiber make it the transmission medium choice. The cost of international telephone calls will continue to decrease as capacity continues to increase [7].

References

[1] Van Valkenburg, Mac E., ed., *Reference Data for Engineers*, 8th ed., SAMS, Carmel, IN: Prentice Hall Computer Publishing, 1993, Chap. 1, "Frequency Data" by Frederick Matos.
[2] Kennedy, George, *Electronic Communication Systems*, 3rd ed., New York: McGraw-Hill, 1985, Chap. 15.
[3] Van Valkenburg, op. cit., Chap. 46, "Cellular Telecommunication Systems" by William C.Y. Lee.
[4] Keiser, Bernhard E., and Eugene Strange, *Digital Telephony and Network Integration*, 2nd ed., New York: Van Nostrand Reinhold, 1995, Appendix C.
[5] Noll, A. Michael, *Introduction to Telephones and Telephone Systems*, 2nd ed., Boston: Artech House, 1991, Chap. 6.
[6] Ibid., Chap. 1.
[7] Ibid., Chap. 3.

CHAPTER 2

Introduction to Radar, Navigation, and Other RF Systems

The first section of this chapter is an introduction to radar systems, followed by an introduction to navigation systems. Other RF systems that are briefly discussed in this chapter include electronic warfare (EW) systems, radio astronomy systems, radiometer systems, and microwave heating systems.

2.1 RADAR SYSTEMS

2.1.1 Basic Concepts

The term radar refers to any type of radio frequency or optical frequency system in which the main objective is to gain information regarding remote objects or targets by means of reflections of transmitted signals. Radar is an acronym for radio detection and ranging. Types of information that can be obtained by radar include range or distance to targets, azimuth and elevation angles to targets, radar cross-section (RCS) of targets, and radial velocity of targets with respect to the radar. Some radars measure all those parameters, while others measure only some of them.

Radars are of two general types: pulse radar and continuous wave (CW) radar. With pulse radar, the range to the target is measured by the time delay between the short transmitted RF pulses and the received (reflected) RF pulses. Angles are measured by using highly directional antennas and target-scanning methods. Radar cross-section of the target is determined by measuring the amplitude of the return signal. Velocity is measured by Doppler frequency measurements.

Three main types of RF pulses are used with pulse radar: pulsed CW, in which the frequency and phase remain constant during the short pulse period; pulsed

FM, in which the frequency is swept in a linear fashion from f_1 to f_2 during the pulse period (chirp modulation); and pulsed phase code modulation (PCM), in which two-state coded phase modulation is used during the pulse period. The latter two methods of modulation are referred to as pulse compression modulations. A matched filter is used in the receiver to compress the received pulse to a much shorter duration pulse than the transmitted pulse. That permits improved resolution and improved range accuracy compared to the pulsed CW modulation.

With CW radar, range is measured by using either frequency or phase modulation of the CW signal and measuring the delay time between the transmitted waveform and the received, reflected waveform. Frequency-modulated CW is referred to as FMCW. Angles, radar cross-section, and velocity are measured in the same manner as in pulse radar.

2.1.2 Radar Frequencies

Radars operate over a wide range of frequencies. Standard radar frequency bands and ITU allocations for radar are given in Table 2.1 [1].

Each frequency region has its own set of characteristics that make it better than other frequencies for certain applications. The following discussion of characteristics is based in part on information presented in [2].

2.1.2.1 MF Radar (0.3–3.0 MHz)

With MF radar, the wavelength is in the range of 1,000–100m. With long wavelengths, such as MF radar utilizes, a significant portion of the radiated energy can be propagated by diffraction beyond the horizon, sometimes called the ground wave or surface wave. The lower the frequency, the lower will be the diffraction loss. The main advantage for this mode of propagation is its over-the-horizon capability. The disadvantages include requirements for very large size antennas, large clutter levels, high ambient noise levels, and crowded electromagnetic spectrum. The result is that MF radar is not attractive for most radar applications.

2.1.2.2 HF Radar (3.0–30.0 MHz)

With HF radar, the wavelength is in the range of 100–10m. At HF frequencies, it is also possible to provide over-the-horizon radar with somewhat larger attenuation of the ground wave. It is also possible to provide very long range radar using ionospheric reflection. The military has used both bistatic and monostatic radars of this kind for detecting targets such as missiles in flight. Bistatic means that the transmitter and receiver antennas are not at the same location. In some bistatic systems, the antennas may be hundreds of miles apart. Monostatic means that the

Table 2.1
Standard Radar Frequency Bands

Band Designation	Nominal Frequency Range	ITU Allocated Radar Frequency Bands for Region 2
MF	0.3–3 MHz	1.85–2 MHz
HF	3–30 MHz	
VHF	30–300 MHz	138–144 MHz
		216–225 MHz
UHF	300–1,000 MHz	420–450 MHz
		890–942 MHz
L	1.0–2.0 GHz	1.215–1.4 GHz
S	2.0–4.0 GHz	2.3–2.5 GHz
		2.7–3.7 GHz
C	4.0–8.0 GHz	5.25–5.925 GHz
X	8.0–12.0 GHz	8.5–10.68 GHz
K_u	12.0–18.0 GHz	13.4–14.0 GHz
		15.7–17.7 GHz
K	18.0–26.5 GHz	24.05–24.25 GHz
K_a	26.5–40.0 GHz	33.4–36.0 GHz
V	40.0–75.0 GHz	59.0–64.0 GHz
W	75.0–110.0 GHz	76.0–81.0 GHz
		92.0–100.0 GHz
Millimeter (mm) wave	110–300 GHz	126–142 GHz
		144–149 GHz
		231–235 GHz
		238–248 GHz

Source: [1].

transmitter and receiver antennas are at the same location. In many cases, monostatic radar uses the same antenna for both transmit and receive functions.

The disadvantages of HF radar again include requirements for very large size antennas, large clutter levels, high ambient noise levels, and crowded electromagnetic spectrum. The result is that HF radar, like MF radar, is not attractive for most radar applications.

2.1.2.3 VHF Radar (30.0–300.0 MHz)

With VHF radar, the wavelength is in the range of 10–1m. VHF has been used for long-range intercontinental ballistic missile (ICBM) early-warning radars and for satellite and aircraft surveillance radars. Such radars use very large antennas and high transmitted power. Advantages for VHF radars include very long range capability for targets such as missile warheads and small satellites. External noise levels are much lower at VHF than at HF and lower frequencies. They are not as low as at higher frequencies, such as L-, S-, and C-bands. VHF radars are not subject to unwanted weather echoes and atmospheric attenuation. (A weather echo is a radar reflection or return from clouds, rain, or snow.) Disadvantages of VHF radars include the very large size antennas needed for decent angle resolution.

2.1.2.4 UHF Radar (300.0–1,000.0 MHz)

With UHF radar, the wavelength is in the range of 1.0–0.3m. UHF is a good frequency band for reliable, long-range surveillance radar. It is free from weather effects and has good moving target indicator (MTI) capability. MTI capability is the ability to detect moving targets in a background of stationary clutter or ground return on the basis of Doppler frequency shift. Chapter 10 presents concepts and design information for radars, including detailed information about MTI systems and performance. External noise levels for UHF radar are small, and the required antenna size for good target angle resolution is not too great. Application of the UHF band for radar is limited by the wide spectrum allotted to UHF television.

2.1.2.5 L-Band Radar (1.0–2.0 GHz)

With L-band radar, the wavelength is in the range of 30–15 cm. This frequency band is popular in the United States for aircraft surveillance radars. It has the advantages of improved angle resolution and low external noise. It is free from weather effects and atmospheric attenuation and has good MTI capability.

2.1.2.6 S-Band Radar (2.0–4.0 GHz)

With S-band radar, the wavelength is in the range of 15–7.5 cm. S-band radars usually are not used for long-range surveillance but are used for precise target location and tracking. Good angular resolution can be achieved with reasonable antenna size. External noise levels are low, and the band is free from weather effects and atmospheric attenuation. S-band is a good compromise frequency for medium-range aircraft detection and tracking radars when a single radar must be used for both functions.

2.1.2.7 C-Band Radar (4.0–8.0 GHz)

With C-band radar, the wavelength is in the range of 7.5–3.75 cm. C-band has been successfully used for moderate range surveillance applications in which precision information is necessary, as in the case of ship-navigation radar. It is also the frequency used for many precision long-range instrumentation radars, as might be used for accurate tracking of missiles. Relatively long range military weapon control radars also operate in this band.

2.1.2.8 X-Band Radar (8.0–12.0 GHz)

With X-band radar, the wavelength is in the range of 3.75–2.5 cm. This is a popular frequency band for military weapon control and for commercial applications. Civil marine radar, airborne weather-avoidance radar, and Doppler navigation radars are found at X-band. X-band radar antennas are fairly small and are favored where mobility and light weight are important. X-band permits large bandwidth capability and short pulses. Small-beam-angle antennas are also practical with small-size antennas.

At X-band there are significant weather effects and atmospheric attenuation. That makes it possible to use X-band for weather-detection radars. The two-way low-altitude atmospheric attenuation at 10.0 GHz is about 0.024 dB/km.

2.1.2.9 K_u-Band Radar (12.0–18.0 GHz)

With K_u-band radar, the wavelength is in the range of 2.5–1.67 cm. K_u-band radars have the advantages of good resolution in both angle and range. High power is difficult to achieve, and the antennas are small. There is increased atmospheric attenuation and higher external noise. In the past, there have been less sensitive receivers. That is changing, however, with the recent development of pHemt devices as low-noise amplifiers. Noise figures equal to or less than 1.5 dB are now possible without cooling at K_u-band. Even lower noise figures are possible with cooling. Chapter 4 presents information about RF noise and defines noise figures.

These factors result in a relatively short range for K_u-band radar. Limitations due to rain clutter and attenuation are increasingly serious at the higher frequencies. The two-way low-altitude atmospheric attenuation is about 0.055 dB/km at 15.0 GHz.

2.1.2.10 K-Band Radar (18.0–27.0 GHz)

With K-band radar, the wavelength is in the range of 1.67–1.11 cm. The resonance frequency for water vapor is 22.2 GHz. At that frequency, the two-way low-altitude atmospheric absorption is about 0.3 dB/km. High power is difficult to achieve with the use of solid-state devices, and the antennas are small. K-band radars have the advantages of small-size antennas and good resolution in both angle and range.

2.1.2.11 Ka-Band Radar (26.5–40.0 GHz)

With K_a-band radar, the wavelength is in the range of 1.11-0.75 cm. The two-way atmospheric attenuation for K_a-band radars is about 0.14 dB/km at 35 GHz. High power is difficult to achieve, and the antennas are small. K_a-band radars have the advantages of good resolution in both angle and range.

2.1.2.12 V-Band Radar (40.0–75.0 GHz)

With V-band radar, the wavelength is in the range of 7.5–4.0 mm. That part of the frequency spectrum offers wide bandwidth and narrow antenna beams from small-aperture antennas. It suffers the same limitations as the K-band, only more so. Oxygen molecules have resonance at 60 GHz, which is in the center of the V-band. Near that frequency, the two-way atmospheric attenuation due to absorption is about 35 dB/km [3], which is clearly an absorption band. There are military radar applications where this very high attenuation is of value, for example, with preventing long-range detection of the V-band radars.

2.1.2.13 W-Band Radar (75–110 GHz)

With W-band radar, the wavelength is in the range of 4.0–2.73 mm. That part of the frequency spectrum again offers wide bandwidth and narrow antenna beams from very small aperture antennas. It suffers the same limitations as the K_a-band, only more so. The two-way atmospheric attenuation at 95 GHz is about 0.8 dB/km [3].

2.1.2.14 Millimeter-Wave Radar (110–300 GHz)

With millimeter-wave radar, the wavelength is in the range of 2.73–1.0 mm. That part of the frequency spectrum offers wide bandwidth and narrow antenna beams from very small aperture antennas. It suffers the same limitations as the W-band, only more so. The two-way atmospheric attenuation at 140 GHz is about 1.0 dB/km. At 240 GHz, the two-way atmospheric attenuation is 15 dB/km [3].

2.1.3 Types of Radars

The following paragraphs present information about some of the more important types of radars. More details and design information are provided about each radar in later chapters. Details about the performance of radars are presented in Chapters 4 and 10.

2.1.3.1 Surveillance Radars

Many ground-based surveillance radars are used throughout the world for the purpose of controlling air traffic flying to and from airports. That ability to detect and track is very important for the safety of the aircraft. A system of this type is illustrated in Figure 2.1.

Surveillance radar typically uses L-band frequencies with a rotating dish antenna that has a fan-shaped beam. The azimuth beam angle is small, and the elevation beam is fairly large. It detects aircraft within line of sight, measuring their range, azimuth angle, and radar cross-section. The surveillance radar system used for aircraft surveillance may also include a secondary surveillance radar that works with an information friend or foe (IFF) system or beacon on the aircraft. The IFF unit is used to identify the aircraft and to receive status information from the aircraft. The secondary surveillance radar operates at a frequency of 1.030 GHz for the ground-to-air link (interrogation) and 1.090 GHz for the air-to-ground link (reply). A fan-shaped antenna beam is used here also [4].

Another radar that may be used along with the ground-based surveillance radar is a height-finding radar, which determines the elevation angle to the target. In that case, the radar antenna has a fan-shaped beam with the elevation beam angle much smaller than the azimuth beam angle. A nodding height finder moves the beam up and down in a nodding motion to locate the elevation angle for the target [5].

A common type of display used for surveillance radars is a cathode ray tube (CRT) plan position indicator (PPI), a map-like display. As the radar antenna scans in azimuth, the PPI display shows reflected signals received as bright spots on the circular screen. The degree of brightness is an indication of the radar cross-section of the target. The distance from the center of the display is an indication of the range to the target. The angle of the spot is an indication of the azimuth angle to the target. The persistence of the phosphor used with the CRT allows the bright spots to remain in view for a period of time longer than the azimuth scan period.

If a secondary surveillance radar is used with a beacon on-board the aircraft, the identification of the aircraft can be indicated on the PPI map along with other important information about the aircraft. Possible information provided includes altitude, speed, and direction of flight. If a height-finder radar is also used along with the azimuth surveillance radar, the altitude of the aircraft can also be indicated on the PPI map.

A number of other radar indicators and nonpolar displays are used with radars. Such indicators include the A-scope display, which shows the signal intensity as a function of range for a given azimuth angle; the B-scope display, which shows target range as a function of azimuth angle; and the C-scope, which shows the elevation of targets as a function of azimuth angle [6].

2.1.3.2 Airport High-Resolution Vehicular Monitoring Radar

Aircraft and ground vehicular traffic at large airports frequently is monitored by means of high-resolution radars. High-resolution radar has the capability to produce

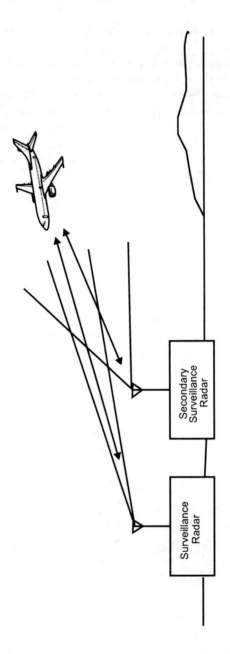

Figure 2.1 Ground-based surveillance radar.

a PPI map of the airport with aircraft and ground vehicles clearly indicated in relation to other fixed elements of the airport, which is important for the safe operation of an airport. Requirements for this radar include use of short RF pulses for good range resolution and use of short wavelengths to provide very small antenna beam angles for small-angle resolution.

2.1.3.3 Speed Measurement Radar

CW Doppler radar is widely used by the police to measure the speed of automobiles and other vehicles. Speed is measured by the Doppler frequency shift of the return or scattered signals from cars and other vehicles. These radars typically are small, solid-state microwave systems that may be either vehicle-mounted or hand-held. They typically use a transmitter known as a Gunn diode oscillator.

Doppler frequency measurements also are made by other types of radars and often are used to reject unwanted clutter signals.

The two-way radar Doppler frequency is given by the following equation:

$$f_d = 2vf/c \qquad (2.1)$$

where
- v = relative velocity of moving vehicle
- f = frequency transmitted
- c = speed of electromagnetic propagation in air (about 3×10^8 m/s)

If the target is moving toward the radar, the Doppler frequency is positive and the received signal is higher than the transmit frequency. If the target is moving away from the radar, the Doppler frequency is negative, and the received signal is lower than the transmit frequency.

For example, at 10 GHz and a relative velocity of 300 m/s, the Doppler frequency is

$$f_d = 2vf/c = 2 \times 300 \times 10^{10} / (3 \times 10^8) = 20 \text{ kHz}$$

A second type of speed-measurement radar that is finding increasing use is a laser speed detector. The same Doppler frequency may be used to determine vehicle velocity with respect to the radar. The difference is that we are here dealing with IR frequencies rather than standard radar frequencies.

2.1.3.4 Ground-Based Weather Radar

Ground-based X-band or higher frequency weather radars are used to measure the location of clouds, rain, and snowstorms. They provide the necessary information to permit weather maps to be produced showing the locations of storms and the intensity of rainfall and snowfall. Doppler radar is typically used to provide that information. Often these radars are located on mountaintops to permit long-range detection capability.

2.1.3.5 Airborne Weather-Avoidance Radar

Weather-avoidance radar is used on commercial airlines and other civilian and military aircraft to identify regions of precipitation or clouds to the pilot. Such radar is important for the safety of the aircraft and the comfort of the passengers. A system of this type is shown in Figure 2.2.

Weather-avoidance radar can measure the range and the angle to clouds, snow, or rain. It also can indicate the rate of precipitation. X-band or higher frequency radars are needed to provide good reflection characteristics for clouds and precipitation.

2.1.3.6 Radar Altimeters

Radar altimeters, which are located on many aircraft, are used to accurately measure the altitude of aircraft. The radar type used typically is an FMCW radar and is very simple and low cost. The frequency band 4.2–4.4 is reserved for radar altimeters. Figure 2.3 shows a radar system of this type. Such radars are discussed in detail in Chapter 10.

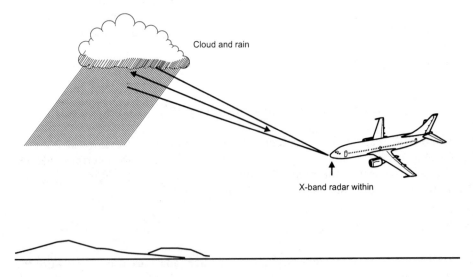

Figure 2.2 Airborne weather-avoidance radar.

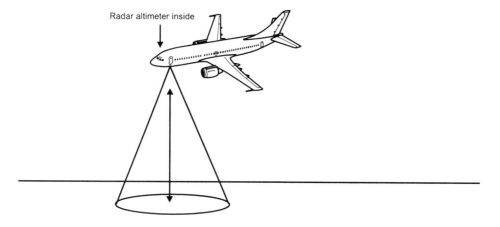

Figure 2.3 Airborne radar altimeter.

2.1.3.7 Airborne Doppler Navigation Radars

Figure 2.4 shows an airborne Doppler navigation radar system. A system of this type may use three or four separate radar beams. In a four-beam system, two beams point down and forward, one on either side of the flight path, and two beams point rearward, one on either side of the flight path. The frequencies of the return signals have a Doppler shift that is determined by the motion of the aircraft with respect to the ground. By comparing the Doppler frequency shift for each beam, it is possible to measure accurately the aircraft's direction and rate of travel.

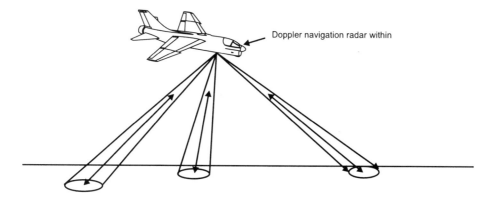

Figure 2.4 Doppler navigation radar.

2.1.3.8 Ship-Based Search and Surveillance Radars

Ship-based search and surveillance radars warn of potential collision with other ships or land objects. They also are used to detect navigation buoys.

2.1.3.9 Shore-Based Search and Surveillance Radars

Shore-based radars of moderately high resolution are used for the surveillance of harbors as an aid to navigation. A shore-based navigation or tracking radar may be used to measure the range and angle to a ship or other target and the radar cross-sections of the targets. It must be able to do that in the presence of large clutter return from the water.

2.1.3.10 Space Applications of Radar

Space vehicles use radar for rendezvous and docking and for tracking other man-made objects in space. Satellite-based radars also have been used for remote sensing of the Earth, planets, and other space objects. Space-based radar can accurately measure the range to targets and the azimuth and elevation angles. It can also measure the radar cross-section of targets and the relative velocity of the targets with respect to the radar.

2.1.3.11 Ground-Based Radars for Locating and Tracking Missiles and Satellites

Some of the largest ground-based radars are used for detection and tracking of missiles and satellites. These systems often use parabolic dish antennas with dish diameters of 60–90 ft. A radar of this type may operate at S- or C-band with pulse powers of the order of 1 MW. Such systems sometime use a pulse compression modulation known as chirp to improve the range resolution and tracking accuracy. The term chirp refers to a type of modulation in which the frequency of the pulse is changed during the pulse period in a linear sweep.

2.1.3.12 Airborne Multiple-Function Radar

Military airborne radar uses advanced multiple-function radars. Functional capability includes weather detection, fire control, navigation, and terrain avoidance. Figure 2.5 illustrates radar of this type. Terrain-avoidance capability is especially important for military aircraft, which often must fly at very high speeds very near to the ground.

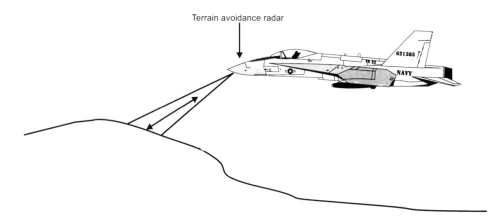

Figure 2.5 Airborne terrain-avoidance radar.

2.1.3.13 Airborne Terrain-Following Radar

Cruise missiles and other vehicles may use terrain-following radar. Such radar measures the altitude of the vehicle with respect to the ground using a small-angle beam for high angular resolution. The information is processed to obtain a map of the terrain directly under the cruise missile. The map then is compared to a stored map in the vehicle's computer to determine where the vehicle is at any given time. Errors in flight path thus can be detected and the necessary corrections made to allow the vehicle to fly to the desired target.

2.1.3.14 Airborne Side-Looking Radar

Figure 2.6 shows a side and a top view for an airborne side-looking radar. Airborne side-looking radar may be used for high-resolution radar ground mapping. It is possible to take advantage of the forward motion of the aircraft to produce what is known as a synthetic aperture antenna of very large size. This allows the effective azimuth beamwidth to be very small. The radar pulse width also is made very small to permit the range resolution to be small. This combination of small-range resolution and small-azimuth beamwidth permits the needed high resolution for a good radar map of the area to the side of the aircraft path. These radar-mapping systems have the advantage over optical systems that they can "look" through clouds, dust, and other cover. They have the disadvantage that their resolution is not as good as that of optical systems.

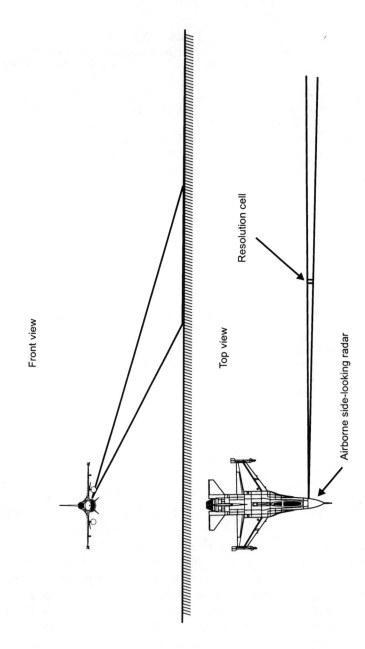

Figure 2.6 Airborne side-looking radar.

2.1.3.15 Ground-Based Military Radar for Aircraft and Missile Defense

Ground-based military radar is used to detect and track aircraft and missiles. It is also used for fire control and pointing of weapons against such threats. These radars normally have pulse-Doppler or MTI capability.

2.1.3.16 Ground-Based Military Radar for Ground-Based Targets

Ground-based military radar is also used to locate moving targets on the ground, such as tanks and trucks. The radar must be able to see the targets in the presence of much larger return from the ground (referred to as ground clutter). These radars also are used to locate gun or mortar emplacements and for fire control and tracking.

2.1.3.17 Ship-Based Military Radar for Ships, Aircraft, and Missile Defense

Ship-based military radar is used to detect and track ships, aircraft, and missiles. It also is used for fire control and pointing of weapons against such threats. One of the U.S. Navy's most demanding tasks for radar is the detection and tracking of low-flying missiles that travel only a few feet above the water. The radar must be able to see the missiles in the presence of very much larger return from the water (referred to as sea clutter).

2.1.3.18 Ground-Based Radars for ICBM Defense

Experimental ground-based military radars have been developed and tested for ICBM defense systems. Types of radar for that function include long-range VHF early-warning radars, UHF and L-band acquisition radars, and C-band target-tracking radars. The early-warning radars typically are very large. Such a radar would use one or more fan beams. Acquisition radars also are large but not as large as the VHF radars. These radars can provide the needed information to permit the target-tracking radars to lock on to targets. Target-tracking radars can be fairly small because of the short wavelengths used. A typical tracking radar antenna beam angle would be 1.0–degree azimuth by 1.0–degree elevation.

2.1.3.19 Radar Cross-Section Test Ranges

Another area of radar application is instrumentation radars for RCS test ranges. Examples are static and dynamic radar cross-section measurement ranges for aircraft and missile targets. A static radar cross-section test range is one in which the targets being measured do not move during measurements but are mounted on special low-reflection support structures. A dynamic radar cross-section test range is one in which the targets being measured are not mounted on the ground but fly past the instrumentation radars.

Figure 2.7 illustrates an outdoor static radar cross-section range. An example of such a range is the RATSCAT main range located at the White Sands Missile Range. At that facility, many radars are used to cover the full frequency range from VHF through Ku-band. The radars are located about half a mile from the target. The surface between radars and target is made flat and coated with asphalt. The target under test is mounted on a support that is designed to have very low backscatter reflection characteristics. A below-ground control system is used to adjust the height and orientation angles of the target.

2.1.3.20 Infrared and Optical Laser Radars (LASARS)

Small infrared lasers are used extensively for surveying instruments. Such radars provide precision measurements of range and angle to corner reflector targets mounted on hand-held or other supports. Laser radars also are used for a number of short-range military applications. These lasers can measure range, angle, and optical RCS in much the same way that the lower frequency conventional radars do.

2.1.3.21 Laser Target Designators

Laser target designators for guided missiles and so-called smart bombs are also a type of radar. This type of radar can be viewed as a bistatic radar, in which separate locations are used for transmitter and receiver. The laser target designator may be located in an aircraft. This part of the system is the transmitter. The missile contains the IR receiver part of the bistatic radar and the guidance system. Such a system can have very high accuracy because of the small size of the IR spot that is placed on the target and the very high angular accuracy of the IR receiving system.

2.2 NAVIGATION SYSTEMS

Many different types of navigation systems are used by aircraft, ships, and land vehicles. Some of the systems use land-based signal sources and measuring equipment. One of the more important, newer navigational systems uses satellites. In the future, we should expect many new types of navigation systems and greatly increased use of those systems. Some of the more important navigation systems are discussed in the following paragraphs. Part of the information is based on [7].

2.2.1 Omega System Navigational Aids

The Omega system is a worldwide VLF navigation system used for marine and en route air navigation. The system uses pulsed-CW transmitting stations, which

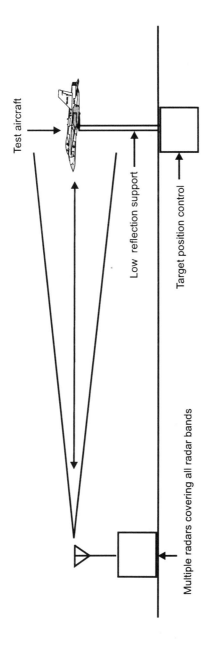

Figure 2.7 Instrumentation radars for RCS test range.

sequentially transmit long, precisely timed pulses at four frequencies: 10.2 kHz, 11.3 kHz, 11.05 kHz, and 13.6 kHz. Position information is obtained by measuring the relative phase difference between the received signals. Accuracies of 2–4 nautical miles are achieved in most of the coverage areas by this system. The operation of an Omega system is shown in Figure 2.8.

2.2.2 Loran Navigational Aids

Loran-C, a long-range hyperbolic radio navigation system, is illustrated in Figure 2.9. The system transmits synchronized, phase-coded pulses from a master station and two or more secondary stations at 100 kHz. The transmitting stations form a chain characterized by a group repetition interval (GRI) in which the pulses are repeated. The term chain means a network of stations operating as a group. The GRI starts with the master station transmitting eight pulses, each spaced 1 millisecond (ms) apart, followed by a ninth pulse 2 ms later. The master station transmission is followed, after a prescribed coding delay, by transmissions from each of the secondary stations in the chain, each transmitting eight pulses at 1-ms intervals. Phase coding is used to differentiate the master pulses from those of the secondaries. The pulse spacing and phase code allow the ground wave to be differentiated from the sky wave. Currently there are 22 chains throughout the world, each with a different GRI.

A typical Loran-C receiver makes use of a microprocessor for signal processing, navigation computation, and control. The accuracy for this system is about 0.25 mi absolute and less than 100 ft relative. The Loran-A system uses the same basic principal but operates at MF frequencies. Pulses 45 ms in duration are transmitted 20–34 times per second by the master station and repeated by slave stations about 300 mi away. An aircraft, using CRT display, notes the difference in time of arrival and computes its position by use of two or more master-slave pairs. Lines of constant-time-difference are hyperbolic, with the stations as foci. Pairs are distinguished by frequency—1,850 kHz, 1,900 kHz, and 1,950 kHz—and by slight variations in pulse repetition frequency (PRF).

2.2.3 Radio Beacons and Airborne Direction Finders

Radio beacons are nondirectional transmitters that operate in the LF and MF bands. Frequencies used are 190–415 kHz and 510 kHz-535 kHz. A radio beacon system is illustrated in Figure 2.10. A radio direction finder is used to measure the relative bearing to the transmitter with respect to the heading of an aircraft or marine vessel. The direction finder is sometimes referred to as an airborne direction finder (ADF). Angular accuracy is in the range of ±3 to ±10 degrees. The beacons transmit either a coded or a modulated CW signal for station identification. The radio direction finder may use a rotating loop antenna.

Introduction to Radar, Navigation, and Other RF Systems | 53

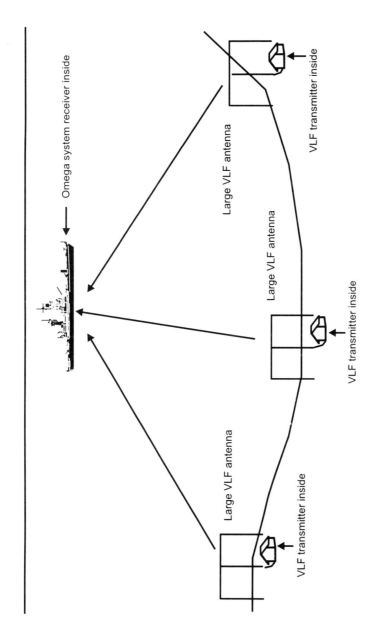

Figure 2.8 Omega system navigational aid.

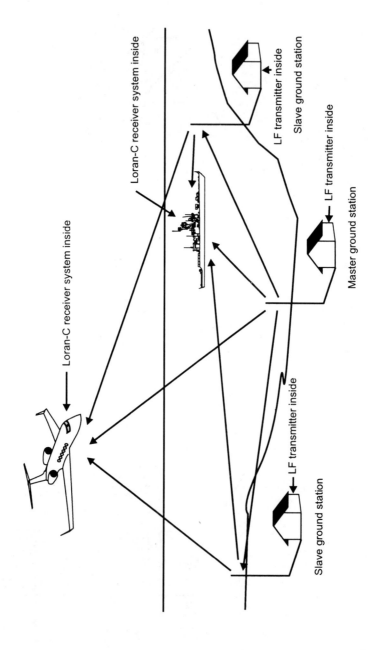

Figure 2.9 Loran-C system navigational aid.

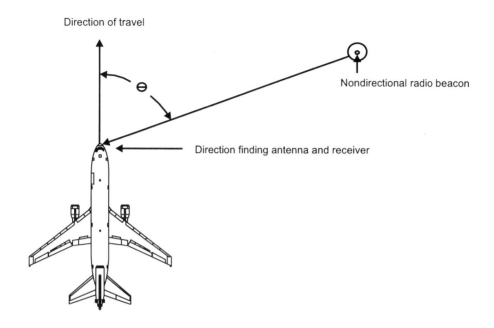

Figure 2.10 Radio beacons and ADF navigational aids.

2.2.4 VHF Omnidirectional Range System

A VHF omnidirectional range (VOR) navigational aid transmits CW signals on one of 20 assigned channels in the 108–112 MHz band and 60 channels in the 112–118 MHz band, with 100 kHz channel separation. A 30-Hz reference signal is transmitted by an omnidirectional antenna. A frequency modulation of ±480 Hz on a 9,960-Hz subcarrier is transmitted along with a carrier from a rotating antenna with a horizontal cardioid pattern. The cardioid antenna pattern rotates at a 30-Hz rate, allowing the airborne receiver to determine its bearing from the station as a function of phase between the reference and the rotating signal. A VOR navigation system is illustrated in Figure 2.11.

A VOR system is essentially a line-of-sight system since it uses VHF. With the aircraft at 5,000 ft, the range is approximately 100 nautical miles. Above 20,000 ft altitude, the range is approximately 200 nautical miles. Typical *en route* VOR stations are rated at 50W. The accuracy of the ground station is better than ±1.4 degrees when properly calibrated. Distance-measuring equipment (DME) often is collocated with the VOR station to provide ranging information. In the United States and other countries, tactical air navigation (TACAN) installations usually are collocated with VOR systems to provide a navigation system that is utilized primarily by the military.

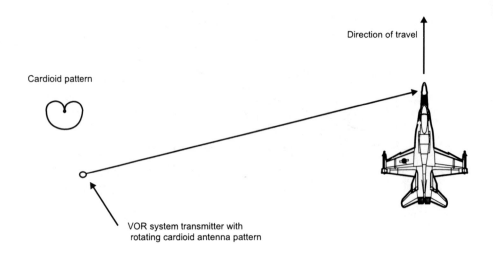

Figure 2.11 VOR system navigational aid.

2.2.5 Distance-Measuring Equipment

A typical DME system is illustrated in Figure 2.12. The airborne equipment (interrogator) generates a pulse signal that is recognized by the ground equipment (transponder), which then transmits a reply that is identified by the tracking circuit in the interrogator. The distance is computed by measuring the total round-trip time of the interrogation, the reply, and the fixed delay introduced by the ground transponder.

The airborne transponder transmits about 30 pulse pairs per second on one of the 126 allocated channels between 1,025 and 1,150 MHz. The ground transponder replies on one of the paired channels in the 962–1,024 MHz band or in the 1,151–1,213 MHz band. A DME and a collocated VOR constitute the International Civil Aviation Organization (ICAO) standard rho-theta system. The ICAO is a United Nations agency that formulates standards and recommended practices, including navigational aids, for all civil aviation. The term rho-theta is a generic term for navigation systems that derive position by measurement of distance and bearing from a single station, that is, by the use of a geometry involving circular coordinates.

2.2.6 Instrument Landing System

A typical instrument landing system (ILS) navigational aid is illustrated in Figure 2.13. Currently the ILS operating in the 108–112 MHz frequency band is the primary worldwide, ICAO-approved, precision landing system.

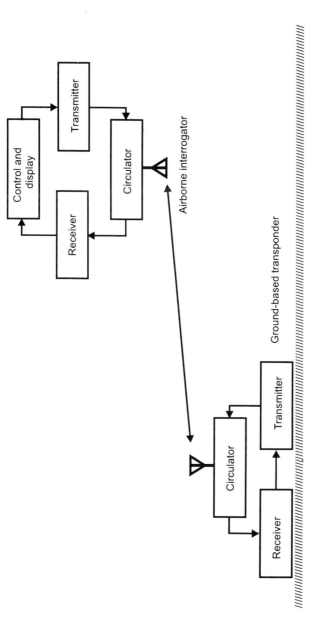

Figure 2.12 DME navigational aid.

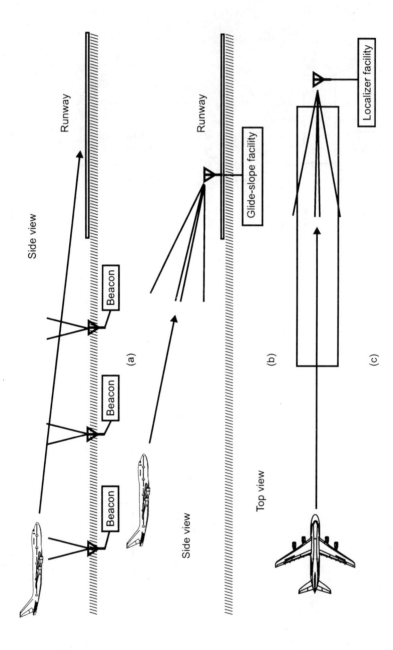

Figure 2.13 ILS: (a) fan-shaped markers along course; (b) glide-slope indicator; and (c) azimuth guidance.

An ILS normally consists of two or three marker beacons, a localizer, and a glide slope indicator to provide both vertical and horizontal guidance information. The localizer, operating in the 108–112 MHz band, normally is located 1,000 ft beyond the stop end of the runway. The glide slope normally is positioned 1,000 ft after the approach end of the runway and operates in the 328.6–335.4 MHz band. Marker beacons operating along the extension of the runway centerline at 75 MHz are used to indicate decision height points for the approach or distances to the threshold of the runway.

Azimuth guidance provided by the localizer is accomplished by use of a 90-Hz-modulated left-hand antenna pattern and a 150-Hz-modulated right-hand antenna pattern as viewed from the aircraft on an approach. A 90-Hz signal detected by the aircraft receiver causes the course deviation indicator (CDI) to deviate to the right. A 150-Hz signal drives the CDI vertical to the left when the aircraft is right of the centerline course. When the aircraft is on the centerline, the CDI vertical needle is centered. The ILS localizer system provides a total of 40 channels. Each channel is paired with a possible glide-slope channel.

Vertical guidance is provided by the glide-slope facility, which is normally located to the side of the approach end of the runway. A total of 40 channels is provided in the 328.6–335.4 MHz band. The carrier radiated in the antenna pattern below the glide slope is amplitude modulated with a 150-Hz signal. The antenna pattern above the glide slope produces a signal with the 90-Hz amplitude modulation. When the approaching aircraft is on the glide slope, the CDI horizontal glide slope needle is centered. If the approaching aircraft is either too high or too low, that is indicated by the CDI glide slope needle.

The marker-beacon facilities along the course provide vertical fan markers to mark the key locations along the approach. The inner marker is normally at the runway threshold. The middle marker is about 3,500 ft from the runway threshold. The outer marker usually is about 5 mi from the runway threshold.

2.2.7 Tactical Air Navigation

The TACAN system provides both omnidirectional bearing and distance-measuring capability. The antenna system provides a rotating cardioid antenna pattern plus a rotating nine-lobed pattern. The antenna system rotates at a 15-Hz rate producing a 15-Hz course bearing and a 135-Hz (9 × 15 Hz) fine bearing in the aircraft. Reference signals are transmitted by coded pulse trains to provide the phase reference. Bearing is obtained by the airborne receiver by comparing the 15-Hz and the 135-Hz sine waves with the reference pulse groups.

The TACAN system operates in the 960–1,215 MHz band with 1-MHz channel separations. The higher frequency used by TACAN permits the use of smaller antennas than are used by VOR. The multilobe principle improves the accuracy.

2.2.8 Microwave Landing System

The microwave landing system (MLS) is the ICAO-approved replacement for the current ILS system. The system is designed to meet the full range of user operational requirements for the year 2000 and beyond. The MLS system is based on the time-referenced scanning beams, referenced to the runway, that enable the airborne unit to determine the precise azimuth angle and elevation angle. Azimuth and elevation angle functions are provided by 200 channels in the 5,000–5,250 MHz band. Range information for the MLS system is provided by DMEs operating in the 960–1,215 MHz band. An option is included in the signal format to permit a special-purpose system operating in the 15,400–15,700 MHz band.

2.2.9 Air Traffic Control Radar Beacon System

The air traffic control radar beacon system (ATCRBS), illustrated in Figure 2.14, is a ground-based radar system that operates with a beacon in the aircraft. The ground based-interrogator transmits at 1,030 MHz. The system uses a fan-shaped antenna beam that is wide in elevation but narrow in azimuth. The rotation or scan rate in azimuth is 5 Hz en route and 2.5 Hz for terminal areas. The interrogator transmits approximately 400 pulse pairs per second and receives replies from aircraft transponders that are within the beam of the antenna pattern.

The airborne transponder replies at 1,090 MHz with one of the 4,096 pulse codes available. The decoded replies are displayed on the surveillance radar PPI along with primary radar returns. An omnidirectional pulse pattern is also radiated from the ground to suppress unwanted sidelobe replies. This system is often referred to as a secondary surveillance radar.

2.2.10 Transit

Transit is a satellite navigational system operated by the U.S. Navy. It consists of four or more satellites in approximately 600-nautical-mile polar orbits. The satellites broadcast ephemeris information continuously at 150 MHz and 400 MHz. A ship-based receiver measures successive Doppler shifts of the signal as the satellite approaches or passes the user. The geographical position of the receiver is then calculated from the satellite position information and the Doppler measurements.

Coverage is worldwide, but not continuous. The period between updates of navigation information can be as short as 1 hour but as long as 8 hours, depending on latitude. The transit system is to be replaced by the global positioning system (GPS) in the 1990s.

2.2.11 NAVSTAR/Global Positioning System

The GPS, illustrated in Figure 2.15, is a worldwide satellite navigation system developed by the U.S. Department of Defense. The complete system includes 24 satellites,

Introduction to Radar, Navigation, and Other RF Systems | 61

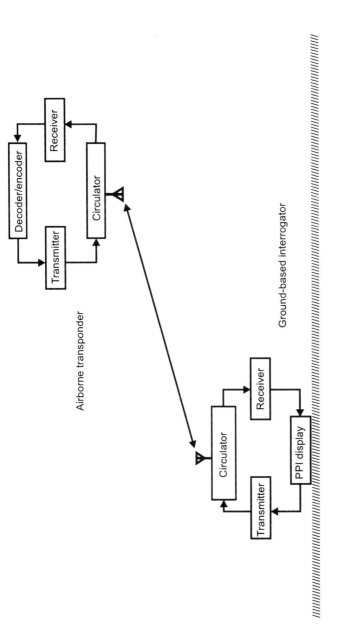

Figure 2.14 ATCRBS.

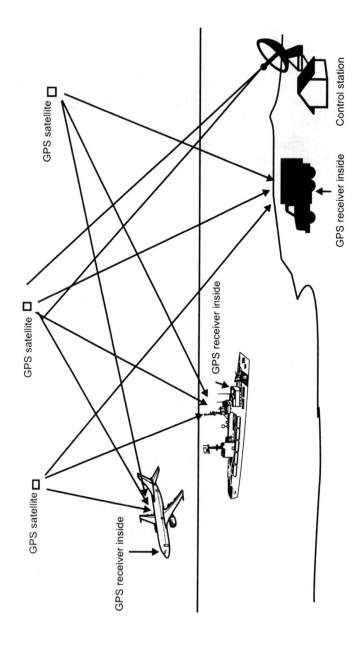

Figure 2.15 NAVSTAR/GPS.

a master control station, a monitor station, an uplink antenna, and multiple user segments. User segments may be ground-based, ship-based, or airborne.

The successful operation of the GPS depends on an accurate knowledge of the position of each satellite as a function of time. A unique ephemeris data table transmitted by each satellite is periodically updated by the master control station. The user's position relative to the satellites is determined by processing signals received from at least four satellites to solve for the time-of-arrival difference. A time correction then relates the satellite system to Earth coordinates.

The satellite signals are transmitted at 1,575.42 MHz and 1,227.6 MHz. The use of two frequencies permits corrections for ionospheric delays in propagation. The signals are modulated with pseudorandom noise code in phase quadrature. The signals also are continuously modulated with the navigation data bit stream at 50 bps. The codes and the modulation allow identification of the satellites and measurement of the transit time through measurement of the phase shift required to match the codes. Two pseudorandom noise codes are used, one operating at 10.23 Mbps and one at 1.023 Mbps.

The accuracy of a position fix varies with the capability of the user's equipment and with the user-to-satellite geometry. Typical accuracies are about 15m for a single measurement. Accuracies for differential measurements typically are in the range of 1–3m. GPS navigational aids are discussed in more detail in Chapter 9.

2.3 ELECTRONIC WARFARE SYSTEMS

The military uses a number of different types of EW systems. This section briefly discusses some of those systems.

2.3.1 Communication Intelligence Gathering Systems

There are two main types of specialized electronic receive-only systems used by the military to gain intelligence. One type is communication intelligence gathering systems, or COMINT systems. Possible platforms or locations for these systems include satellites, aircraft, ships, mobile ground stations, and fixed ground stations. In some cases, the COMINT equipment is portable or hand held. RF equipment used by COMINT systems includes special antennas, receivers, signal-processing equipment, display equipment, and recording equipment. Sometimes it is possible to receive from the main antenna beam of the enemy transmitter. In many other cases, it is necessary to receive from sidelobes of the antenna or from reflections of the transmitted signal. Because COMINT system equipment usually is highly classified, it cannot be discussed in detail in this text.

2.3.2 Electronic Intelligence Systems

A second type of intelligence-gathering system is the electronic intelligence (ELINT) system. This type of system is designed to gather intelligence about the location and characteristics of electronic equipment used by an enemy. Possible platforms or locations for ELINT systems include satellites, aircraft, ships, mobile ground stations, and fixed ground stations. RF equipment used by an ELINT system includes special antennas, receivers, signal-processing equipment, display equipment, spectrum analyzers, and recording equipment. Because ELINT systems usually are classified, they are not discussed in detail in this text.

2.3.3 Electronic Countermeasure Systems for Communications

Electronic countermeasures (ECM) for communications are an important EW system for the military. The types of jammers used in an ELINT system typically are main-beam or sidelobe noise jammers. The goal is to increase the noise level for enemy communication systems to the point that they are no longer effective. Both narrowband adaptive spot noise jammers and wideband barrage noise jammers can be used. Jammer platforms may include airborne platforms, ground-based vehicular platforms, and ship-based platforms.

Figure 2.16 illustrates communication jamming for a ground-to-air communication system. The ground station is assumed to use a monopole-type antenna with 360 degrees of azimuth coverage by the main beam. A noise jammer thus can jam the ground-based receiver through the main beam. The jammer may use either wideband barrage noise jamming or narrowband spot noise jamming, depending on available intelligence for the jammer. An adaptive spot jammer uses a collocated receiver and processing system to determine the correct frequency band for the noise jamming.

As an electronic counter-countermeasure (ECCM), the military communication system may use fast-frequency hop or other spread-spectrum techniques to spread the spectrum of the communication signal over a large bandwidth. The goal is to force the jammer to use only wideband barrage jamming, which may not be effective.

2.3.4 Electronic Countermeasure Systems for Radars

ECM for radars is also an important electronic warfare system for the military. The types of jammers used include adaptive spot noise jammers, wideband barrage noise jammers, and deception jammers. Types of jammers used include sidelobe jammers and main-beam jammers. Jammer platforms may be airborne, ground-based, or ship-based.

Introduction to Radar, Navigation, and Other RF Systems | 65

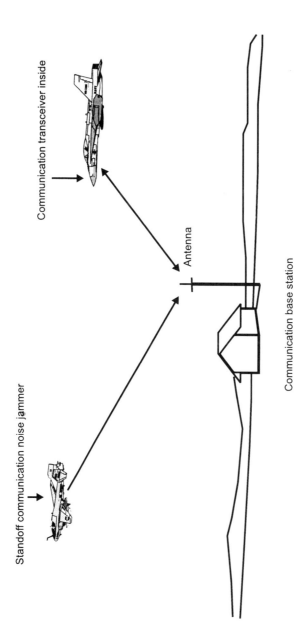

Figure 2.16 ECM systems for communications.

2.3.4.1 Sidelobe Noise Jammers

Airborne sidelobe noise jammers include on-board jammers, standoff jammers, and precursor jammers (Figure 2.17). With on-board jammers, the jammer is located on the same platform as the target being defended. With standoff jammers, the jammers are located on one or more separate platforms that are at greater distances from the radar than the target being defended. With precursor or stand-in jammers, the jammers are located on one or more separate platforms that are at much shorter distances from the radar than the target being defended. Precursor jammers have a large range advantage over the standoff jammers but are more vulnerable to attack. One advantage of precursor and standoff sidelobe jammers is that locating the jammer in no way locates the real target. These jammer systems can also simultaneously defend any number of targets.

There will be some range at which the signal from the target can be seen by the radar with sidelobe noise jamming. That range is referred to as the burn-through range. The goal of the ECM system is to minimize that range to protect the aircraft. With standoff sidelobe jamming systems, doing that normally requires high power, high antenna gain, and small jamming noise bandwidths. On-board sidelobe jammers often are used when it is not possible to use precursor or standoff jammers. An on-board sidelobe jammer typically reduces the jammer power level when illuminated by the main beam of the radar and transmits full power when illuminated by the sidelobes of the radar.

2.3.4.2 Main-Beam Noise Jammers

Main-beam noise jammers normally are located on the platform being defended. They operate into the main antenna beam of the radar. Thus, they are able to provide much higher jamming power levels at the input to the radar receiver than sidelobe jammers having equal power outputs. In theory, therefore, they should be able to prevent range measurements to very short distances. The main weakness of main-beam noise jammers is that if two or more widely spaced radars are used to view the same platform, it is possible to locate the jammers and, thus, the main target being defended by triangulation.

2.3.4.3 Main-beam Repeater or Deception Jammers

Repeater jammers are used to generate false targets for radars [8]. These jammers can be effective with much smaller jamming power than noise jammers. A true repeater is one that retransmits the same kind of signal that the target would reflect or reradiate. A transponder repeater plays back a stored replica of the radar signal after it is initially triggered by reception of the radar signal. The transmitted signal may be made to resemble the radar signal from a real target as closely as possible.

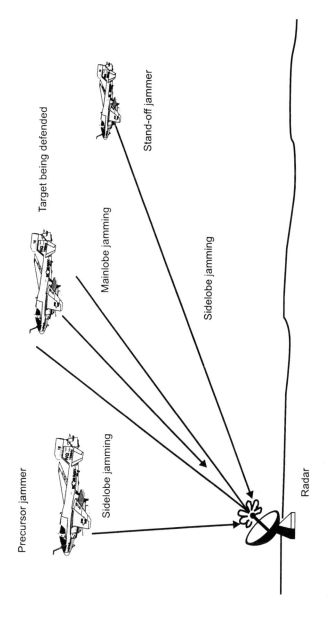

Figure 2.17 Types of airborne sidelobe jammers.

In other cases, the transmitted signal may have noise added. The transponder can be programmed to remain silent when illuminated by the main radar beam and to transmit only when illuminated by the sidelobes, thereby creating spurious targets on the radar display at directions other than that of the true target. A range-gate stealer is a repeater jammer whose function is to cause a tracking radar to break lock on the target. The range-gate stealer operates by initially transmitting a single pulse in synchronism with each pulse received by the radar, thereby strengthening the target echo. The repeater slowly shifts the timing of its own pulse transmission through a time delay or memory loop to cause an apparent change in the target range. If the simulated radar return is larger than the return from the real target, the radar tracking system will lock on to the false target and break track on the real target. Thus, the countermeasure technique is referred to as range-gate pulloff (RGPO).

Repeaters (especially transponder types) that generate false Doppler-shifted return signals also can be used to break lock, referred to as velocity-gate pulloff (VGPO). Both RGPO and VGPO can be combined in the most sophisticated systems.

Much of the information about ECM, especially ECCM, is classified secret or higher and is therefore not presented in this text.

2.4 RADIO ASTRONOMY SYSTEMS

Radio astronomy systems are used to detect stars or galaxies on the basis of radiated RF signals. To be effective, these systems must use very large antennas that provide very small beam angles and very low noise receivers. They must be erected in locations where external noise levels are as small as possible.

One such radio astronomy system is the very large array (VLA) radio astronomy site located in western New Mexico. This facility is an interferometer that uses two arms or lines of very large parabolic dish antennas distributed along each of the arms. The arms are orthogonal to each other. This facility uses aperture synthesis techniques to map the sky.

The receivers for radio astronomy systems use cryogenically cooled front-end amplifiers for very low noise capability. They typically use liquid helium as a refrigerant with a compressor for converting the gas to a liquid. (More details about low-noise receivers are presented in later chapters.) The system includes signal-processing systems, display systems, and data-processing and recording systems.

The ITU and the FCC frequency allocations for radio astronomy service are listed in Table 2.2.

2.5 RADIOMETER SYSTEMS

A millimeter-wave radiometer is a low-noise, high-gain receiver that senses thermal radiation at its antenna as it scans targets and background. A very high gain, very

Table 2.2
FCC Frequency Allocations for Radio Astronomy Service

Low Band	High Band
13.36–13.41 MHz	10.6–10.7 GHz
22.55–26.10 MHz	15.35–15.4 GHz
73.00–74.60 MHz	23.6–14.0 GHz
322.0–328.6 MHz	42.5–43.5 GHz
406.1–410.0 MHz	86.0–92.0 GHz
608.0–614.0 MHz	164.0–168.0 GHz
1,400–1,427 MHz	182.0–185.0 GHz
1,660–1,668.4 MHz	217.0–231.0 GHz
2,690–2,700 MHz	265.0–275.0 GHz

small beamwidth antenna is used to generate the needed high resolution. Because of the very small wavelength involved, such an antenna can be quite small.

Microwave radiometers have been in use for many years for laboratory instrumentation, meteorological research, and radio astronomy. We now have the ability to produce high-performance millimeter-wave radiometers as well. Such radiometers can be used to detect stationary or moving metallic targets such as tanks, trucks, or other vehicles even when hidden by dust, clouds, fog, or other climatic conditions.

A perfect reflector reflects the "cold" sky. At millimeter-wave frequencies, a metal target has an emissivity of approximately 0, while the emissivity of the Earth background is nearly 1. Ground background noise temperature is normally at or near 290°K.

The effective sky temperature depends on the weather, the frequency, and the viewing angle. The largest temperature differentials (colder sky) occur at the lower frequencies and higher viewing angles. The higher frequencies, however, permit the utilization of smaller beam angles and therefore less interception of background radiation within the beam when centered on the metallic target. Details of military applications are classified.

2.6 MICROWAVE HEATING SYSTEMS

Microwaves can be used for the heating of foods and other products. The home microwave oven is a familiar example. Larger commercial microwave ovens are used by food-processing industries. In each case, a magnetron oscillator operating at a frequency of about 2,450 MHz generates the microwave radiation that enters the oven. There, the energy is absorbed by the food, with the resulting conversion of electromagnetic energy to thermal energy. (The magnetron oscillator is discussed in detail later in this text.)

The fact that microwave energy can be used to cook food should alert us to the fact that RF radiation can be dangerous. RF energy is particularly damaging to the male testes and to the eyes—it can cause sterility and blindness if the energy level is great enough. For that and other reasons, it is common practice to provide a fence or a screen that limits access to the area near radars or other high-power RF transmitters when they are operating.

References

[1] Barton, David K., *Modern Radar System Analysis*, Boston: Artech House, 1988, p. 6.
[2] Skolnik, Merrill I., ed., *Radar Handbook*, New York: McGraw-Hill, 1970, Chap. 1.
[3] Barton, op. cit., pp. 278–279.
[4] Skolnik, Merrill I., ed., *Radar Handbook*, New York: McGraw-Hill, 1970, Chap. 38, "Beacons" by James Ashley.
[5] Ibid., Chap. 22, "Radar Height Finding" by Burton P. Brown and James S. Perry.
[6] Ibid., Chap. 6, "Radar Indicators and Displays" by Alton A. Berg.
[7] Van Valkenburg, Mac E., ed., *Reference Data for Engineers*, 8th ed., SAMS, Carmel, IN: Prentice Hall Computer Publishing, 1993, Chap. 37, "Radio Navigation Aids" by E. A. Robinson.
[8] Skolnik, Merrill I., *Introduction to Radar Systems*, 1st ed., New York: McGraw-Hill, 1962, pp. 559–567.

CHAPTER 3

Radio Frequency Propagation

This chapter provides information about RF propagation, including the mechanisms by which electrical signals are converted to electromagnetic waves at the transmitter, the characteristics of those generated electromagnetic waves, the way that the electromagnetic waves move from one point to another, the loss mechanisms involved in the movement, and the mechanisms by which electromagnetic waves are converted back to electrical signals at the receiver.

3.1 ANTENNAS

3.1.1 Transmit Antennas

A transmitter antenna is a transducer that transforms electrical power delivered to the antenna into RF electromagnetic radiation. If the radiation from an antenna is uniform in all directions, it is called an isotropic antenna. Such an antenna is not possible in practice, but it is convenient to use as a reference.

The theoretical isotropic antenna is taken as the reference, and the gain of the antenna in a given direction is a measure of how the power level in that direction compares with that which would exist if the isotropic antenna had been present. That gain can be either less or greater than 1. Expressed in decibels, it can be either positive or negative, since the logarithm of 1 is zero. For example, a typical tracking radar might use a parabolic dish antenna that produces a 1-degree azimuth by 1-degree elevation main beam. There are 41,300 square degrees in a spherical solid angle. The directional gain in that main beam then would be about 41,300, or 46 dBi, where dBi is decibels with respect to isotropic. At angles other than the main beam, there will be sidelobes. A typical power gain in the sidelobe directions might be 0 dBi or less.

The power gain of the antenna is less than the directional gain because of antenna losses and radiation through the sidelobes. In a typical case, the antenna with a 1-degree by 1-degree beam will have a power gain of about 27,000, or 44 dBi.

The typical relationship between antenna gain and beam angle is as follows:

$$G = 27{,}000/(\theta_{az} \cdot \theta_{el}) \tag{3.1}$$

where

G = numeric antenna gain
θ_{az} = the -3 dB azimuth beam angle (degrees)
θ_{el} = the -3 dB elevation beam angle (degrees)

Beam angles are measured at the half power (-3 dB) points on the beam. The antenna gain expressed in decibels with respect to isotropic (dBi) is $10 \log_{10} G$.

For an example of the use of (3.1), assume the case of a 2.0-degree azimuth by 10-degree elevation fan beam. The gain would be

$$G = 27{,}000/(\theta_{az} \cdot \theta_{el})$$
$$G = 27{,}000/(2 \cdot 10) = 1{,}350 = 31.3 \text{ dBi}$$

The effective radiated power (ERP) is defined as the product of the antenna gain and the radiated power. For example, if the numeric antenna gain is 10 and the radiated power is 200W,

$$\text{ERP} = 200 \cdot 10 = 2{,}000\text{W, or 33 dBW, or 63 dBm}$$

As discussed in Chapter 1, dBW is decibels with respect to 1W, while dBm is decibels with respect to 1 mW.

3.1.2 Receive Antenna Gain and Capture Area

The antenna gain of an antenna when it is used as a receiver antenna is the same as the antenna gain when it is used as a transmitter antenna, that is, an antenna is bilateral. The effective capture area of the receiver antenna is the antenna gain times the area of an ideal isotropic antenna for the frequency of interest. The effective capture area of a receiver antenna is given by (3.2).

$$A_e = G_r \lambda^2 / 4\pi \tag{3.2}$$

where

A_e = effective capture area of an antenna (square meters)

λ = wavelength (meters)

G_r = receive directional antenna gain

For an example of the use of (3.2), assume an antenna with a gain of 13 dBi operating at a wavelength of 2m. The effective capture area of the antenna would be

$$A_e = G_r \lambda^2 / 4\pi$$
$$A_e = 20 \times 4 / 4\pi = 6.37 \text{ m}^2$$

In the case of aperture-type antennas, such as horn antennas and parabolic dish reflector antennas, the effective capture area of the antenna is typically about half the actual aperture area. The reason is that the aperture is not illuminated uniformly by the feed system. For example, if we have a parabolic dish antenna with a diameter of 60 ft (18.3m), the aperture area is about 263 m^2, and the effective capture area is about half that, or 132 m^2.

3.2 ELECTROMAGNETIC WAVES, FIELDS, AND POWER DENSITY

Figure 3.1 illustrates an electromagnetic wave traveling in free space. The wave is transverse electric and transverse magnetic, that is, a transverse electromagnetic (TEM) wave in which the E-field is at right angles to the magnetic field, and both are at right angles to the direction of propagation. Also, there is no electric or magnetic field in the direction of propagation, only transverse to the direction of propagation, hence the term transverse electromagnetic. Figure 3.1 shows a front view and a side view of the wave.

The essential properties of a radiated RF electromagnetic wave are wavelength, frequency, intensity, direction of travel, and direction of polarization. The wavelength is the distance the wave travels in one complete RF cycle period. The frequency is the number of cycles in a second. The intensity of the wave is the field strength given in volts per meter or the power density of the wave in watts per square meter. Performance calculations usually deal with the power density.

The direction of polarization is the direction of the free-space electrical (E) field. If the direction of polarization is horizontal with respect to the surface of the Earth, the wave is said to have horizontal polarization. If the free-space E-field is vertical with respect to the surface of the earth, the wave is said to have vertical polarization. These polarizations are considered linear.

It is also possible to have elliptical or circular polarization. Such waves have a rotating E-vector. They may be assumed to be composed of two linearly polarized

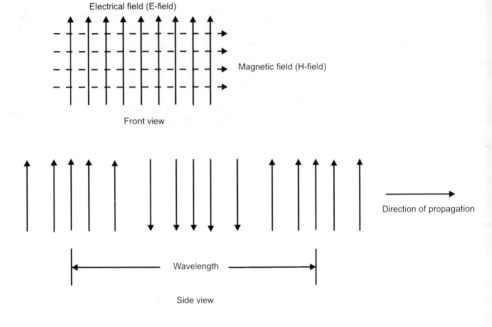

Figure 3.1 Illustration of a TEM wave.

waves that have orthogonal polarizations, that is, their E-field vectors differ by 90 degrees. In those cases, the polarization vector rotates in either the left or the right direction as the wave propagates, depending on whether the phase shift between the two linear polarization components is +90 degrees or −90 degrees.

The power density of a TEM wave at a point in space can be expressed in terms of its E-field and the characteristic impedance of free space as follows:

$$P_1 = E^2 / Z_o \qquad (3.3)$$

where

P_1 = power density (watts per square meter)

E = E-field strength (volts per meter)

Z_o = characteristic impedance of free space = 377Ω

The magnetic field strength, H, has units of amperes per meter. Alternatively, the power density in watts per square meter equals 377 H².

For free-space propagation, the electric and magnetic field intensities decrease directly with range from the transmitter. Hence, the power density decreases as the square of the range. Thus, if point B is four times the distance from the transmitter as point A, the E-field intensity at point B will be one-fourth the E-field intensity at point A, and the power density at point B will be one-sixteenth the power density at point A.

The equation for power density in terms of range for free-space propagation is as follows:

$$P_1 = P_T G_T / 4\pi R^2 \tag{3.4}$$

where

P_1 = power density at a distance R
P_T = transmitter power
G_T = transmitter antenna gain
R = range

As an example of E-field strength and power density for free-space propagation, assume a transmitter power output of 100W and an antenna gain of 10. The power density at a range of 100 km (10^5 m) thus would be

$$P_1 = P_T G_T / 4\pi R^2$$
$$P_1 = 100 \times 10 / (4\pi \times 10^{10}) = 7.96 \times 10^{-9} \text{ W/m}^2$$

The RMS E-field strength would be

$$E = (Z_o P_1)^{0.5}$$
$$E = (377 \times 8 \times 10^{-9})^{0.5} = 1.73 \times 10^{-3} \text{ V/m} \tag{3.5}$$

3.3 RF ELECTRICAL FIELD WAVEFORMS AND VECTOR ADDITION

It is common to have two or more E-fields added. Examples are the addition of fields from the individual elements of a phased array antenna and the addition (or subtraction) of the direct and reflected waves in the case of multipath propagation. Because E-fields have both magnitude and phase, they are vector quantities and must be added or subtracted by standard vector addition and subtraction techniques.

3.4 FREE-SPACE PATH LOSS

The term free-space path loss refers to the spreading loss for a radiated signal between a transmitter antenna and a receiver antenna with gain equal to 1 (0 dBi) for both transmit and receiver antennas. It is given by

$$L_P = (4\pi)^2 R^2/\lambda^2 \tag{3.6}$$

where

L_P = Free-space path loss (numeric)
R = range (meters)
λ = wavelength (meters)

Free-space path loss increases as the frequency is increased because of the smaller capture area of the antenna. It also increases rapidly with increased range because of the range squared proportionality. For a fixed frequency, the one-way free-space path loss, as experienced by communication systems, increases 6 dB per octave (range doubling), or 20 dB per decade. The two-way free-space path loss, as experienced by radar systems, increases 12 dB per range doubling, or 40 dB per decade, since round-trip loss is proportional to R^4. Twelve dB is calculated from $10\log_{10}(2^4)$.

3.5 EXCESS PATH LOSS AND ATMOSPHERIC ATTENUATION

The radiated signal sees a number of other losses in addition to free-space path loss. The sum of those losses is called the excess path loss. Examples of such losses are atmospheric attenuation, diffraction loss, multipath loss, ground-wave loss, ionosphere refraction loss, and scatter propagation loss. Excess path losses are discussed in the following paragraphs. Some of the information presented is adapted from [1].

3.5.1 Atmospheric Absorption

As electromagnetic waves pass through the atmosphere, they are attenuated by atmospheric absorption. At the lower frequencies, the additional attenuation is very small and propagation losses are nearly the same as for free-space propagation. At X-band and higher frequencies, atmospheric absorption loss becomes important and can become large. That is especially true at the resonance frequencies for water vapor and oxygen. The first resonance frequency for water vapor is about 22 GHz, while the first resonance frequency for oxygen is approximately 60 GHz. There are other resonance frequencies above 100 GHz for both water vapor and oxygen.

Figure 3.2 shows a plot of two-way (radar case) atmospheric attenuation coefficients versus frequency for clear atmospheric conditions, which includes oxygen plus water vapor, at sea level. Table 3.1 lists the attenuation coefficients in the so-called atmospheric windows and at the major peaks of Figure 3.2. It also lists attenuation coefficients at the center frequencies of the standard radar bands [1].

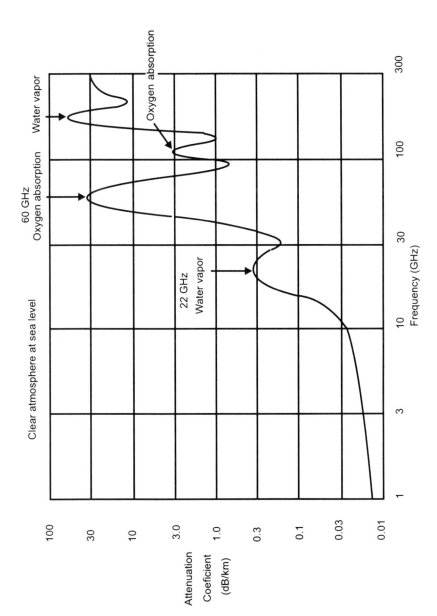

Figure 3.2 Two-way atmospheric attenuation coefficients versus frequency. (*After:* [2].)

Table 3.1
Attenuation Coefficients

Radar Band	Approximate Center Frequency (GHz)	Attenuation Coefficient (dB/km)
L	1.3	0.012
S	3.0	0.015
C	5.5	0.017
X	10	0.024
K_u	15	0.055
K	22	0.3
K_a	35	0.14
V	60	35.0
W	95	0.8
mm	140	1.0
mm	240	15

In the case of a communication system, a one-way path is involved. The atmospheric attenuation, therefore, is one-half the two-way value shown in the figure when expressed in decibels. For example, if the two-way atmospheric attenuation is computed as 26 dB for a given path, the one-way atmospheric attenuation would be 13 dB.

The attenuation coefficients given in Figure 3.2 are for sea-level atmospheric conditions. At higher elevations, the loss is less because of lower oxygen and water vapor densities. For example, at an altitude of 30,000 feet, the two-way atmospheric attenuation coefficient at 10 GHz is reduced to about 0.003 dB/km. That compares with 0.024 dB/km at sea level. A ground-based, X-band radar that is looking up at a 10-degree elevation angle and a range of 100 km will see an atmospheric loss of only about 0.9 dB, compared to the 2.4-dB loss at sea level for a low elevation angle.

3.5.2 Attenuation Produced by Rain, Snow, and Fog

Figure 3.3 shows a plot of attenuation coefficients for rain as a function of frequency [2]. The attenuation coefficients for one-way communication will be one-half the two-way value when expressed in decibels.

At L-band (1.3 GHz typical), only the most intense rain can cause a doubling of the clear-atmosphere attenuation coefficient. Such rainfall cannot exist over any substantial path length. At S-band (3 GHz typical), the clear-atmosphere value is doubled by rain at 10 mm/h. Higher frequencies are much more affected by rain.

The attenuation produced by ice particles in the atmosphere, whether occurring as hail, snow, sleet, or ice-crystal clouds, is much less than that caused

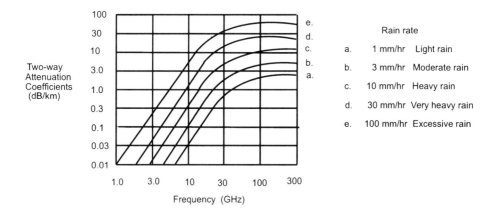

Figure 3.3 Two-way rain attenuation coefficients versus frequency and rate. (*After:* [2].)

by rain of an equivalent rate of precipitation. Fog also may produce attenuation at the higher frequencies. Attenuation from heavy fog is comparable to that of moderate rain.

3.6 ATMOSPHERIC REFRACTION

Another important effect of the atmosphere on electromagnetic wave propagation is refraction. As pointed out earlier, the velocity of the electromagnetic wave is inversely proportional to the square root of the dielectric constant of the medium. Because the atmosphere is not of constant density as altitude varies and because the composition changes as a function of altitude, the propagation velocity also differs with altitude. A transmitted wave does not propagate in a straight line but follows a curved path. One consequence of that bending is that greater range is possible for communication and radar systems without the inherent horizon limitations.

If the bending of the ray path for the waves is just right, the bending just matches the curvature of the Earth, and there are no radio horizon limitations. Normally, however, the rate of bending is not that great. Even so, the horizon limit is extended significantly beyond that of straight-line propagation.

With the linear constant-gradient model and the so-called SBF method, the range capability is determined by plotting the ray path as a straight line and assuming that the radius of the earth is increased. With the so-called standard atmosphere, the earth's radius is assumed to be four-thirds times the true radius. It should be emphasized that this "four-thirds earth" rule is only an approximation.

For an example of radio horizon extension using the four-thirds earth rule, assume an aircraft flying at 40,000 ft. With no refraction and 0-degree elevation angle at the Earth's surface, the radio horizon would be at a range of 245 mi. That is based on an Earth radius of 3,960 mi, or a circumference of 24,881 mi. With the "four-thirds Earth" model, the Earth "radius" is 5,283 mi. The radio horizon in that case would 283 mi.

Under special cases, it is possible to produce ducting of the electromagnetic wave. At some elevated region above the Earth, the dielectric gradient becomes so large that the wave is refracted back to the Earth. There it is reflected, and the wave travels back to the strong refracting layer, where it is again refracted down. By that process, the wave may travel great distances without radio horizon limitations. Ducting is most common over water, but it may also appear over land areas.

3.7 DIFFRACTION OF RADIO WAVES

Figure 3.4 illustrates diffraction of radio waves. The term diffraction refers to the process by which an electromagnetic wave is bent in its path by a material edge. An example is a VHF or UHF communication system in which there are buildings that block the path of the electromagnetic wave. A fraction of the energy in the wave is bent around or over the buildings and may be received by a receiver that does not have optical line of sight back to the transmitter. There is loss in the diffraction process, but the signal may nevertheless be strong enough for good reception.

Figure 3.5 shows the approximate one-way attenuation for knife-edge diffraction [3]. The two-way attenuation, as would be experienced by radar, would be twice the one-way value expressed in decibels. The propagation factor or added attenuation for knife-edge diffraction depends on the vertical distance, h, between the top of the obstacle and the direct path between the radar and the target, and on the wavelength. In Figure 3.5, it is shown as a function of the diffraction parameter, v, which is calculated as follows:

$$v = [(2h\theta)/\lambda]^{0.5} \qquad (3.7)$$

where θ = angle and λ = wavelength.

For h/d_1 and h/d_2, each small the approximate value of v is given by (3.8).

$$v = h[(2/\lambda)(1/d_1 + 1/d_2)]^{0.5} \qquad (3.8)$$

Positive v implies a blocked path, and negative v implies that the line-of-sight path clears the obstacle.

For an example of the use Figure 3.5 in predicting attenuation due to diffraction, assume the case of a communication system operating at a frequency of 1 GHz. Other assumptions are as follows:

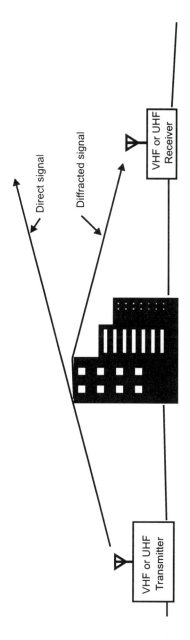

Figure 3.4 Example of diffraction with a communication system.

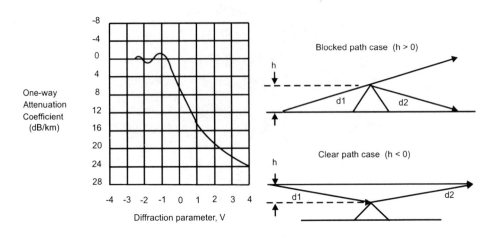

Figure 3.5 Attenuation for knife-edge diffraction. (*After:* [3].)

$h = 100$m
$d_1 = 20,000$m
$d_2 = 40,000$m
$\lambda = 0.3$m (1 GHz)

Using (3.8),

$$v = h[(2/\lambda)(1/d_1 + 1/d_2)]^{0.5}$$
$$v = 100[(2/0.3)(1/20,000 + 1/40,000)]^{0.5} = 2.23$$

From Figure 3.5 with v = 2.23, the one-way attenuation due to refraction is 20 dB. If the frequency had been 100 MHz instead of 1 GHz, the value of v would be 0.71. The attenuation in that case would be about 12 dB. Thus, we see that the diffraction loss is much lower at the lower frequencies than at the higher frequencies.

3.8 MULTIPATH

Another important property of electromagnetic waves is that they reflect from the Earth's surface and other objects as well as travel directly to an antenna or target. The reflected signal adds to the direct signal, producing either a larger signal than the direct signal at the receiver antenna (constructive interference) or a smaller signal (destructive interference), depending on the relative phase angles of the two signals. That vector addition of a direct signal and a reflected signal is referred to as multipath. Multipath usually is considered undesirable because of the substantial reduction of signal in the nulls. In a few cases, it can be considered desirable because of the added signal in the peaks.

Figure 3.6 illustrates the concepts of multipath peaks extending beyond the free-space pattern and nulls extending below that pattern. In the limit where the direct and reflected signals are exactly the same amplitude, the peaks could extend 12 dB beyond the free-space pattern for radar and 6 dB for communication. The nulls in this case would extend to zero signal. In the typical case, the signals are not equal, peaks are not so great, and null is not so deep.

In the case of communication systems in which large elevation angles are used, there may be lobing at high elevation angles as well as at low angles. With radar and highly directional communication systems, multipath is important only at the lower elevation angles where the Earth is included in the main beam of the antenna. The surface reflection coefficient is the product of three factors: the Fresnel reflection coefficient, the specular scattering coefficient for a rough surface, and the coefficient for vegetative absorption. On reflection, the phase shift for horizontal polarization is 180 degrees, independent of grazing angles. For vertical polarization, the phase shift on reflection is 180 degrees for grazing angles less than the Brewster angle and near zero degrees for angles greater than the Brewster angle. Thus, for near-zero elevation angle and multipath, we should expect a null. How deep the null is depends on the magnitude of the reflection coefficient.

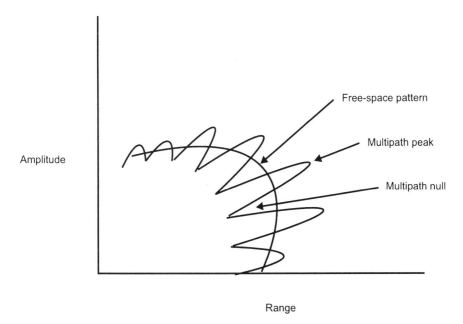

Figure 3.6 Coverage diagram with multipath lobbing showing constructive and destructive interference as a function of elevation angle. (*After:* [4].)

The Brewster angle is the angle that corresponds to the minimum reflection coefficient. That angle depends on the frequency and the type of reflecting surface. For example, for seawater the Brewster angle at 3 GHz is about 6 degrees. It is only about 1 degree for seawater and 100 MHz.

The scattering coefficient for a rough surface is much less than unity, with the exact value depending on the grazing angle and the root mean square (rms) deviation from a smooth surface. The same is true for a surface covered with vegetation. A layer of vegetation covering the surface absorbs much of the incident radiation and scatters the remainder in an irregular pattern [4].

Multipath can be very important in determining the propagation characteristics for both communication and radar systems. That is equally true over land and over water. It is most important when a surface is flat and free from vegetation.

3.9 IONOSPHERIC PROPAGATION

Some of the material presented in this and subsequent sections is quoted or adapted from [5,6].

Important terms in talking about ionospheric propagation or HF communication are critical frequency, maximum usable frequency (MUF), and lowest usable frequency (LUF). The critical frequency is that frequency for a given layer that reflects a vertically incident wave. Frequencies higher than the critical frequency pass through the layer at vertical incidence. The MUF is the critical frequency times the secant of the angle of incidence at the reflecting layer times a correction factor for earth curvature. The MUF varies for each layer with local time of day, season, latitude, and throughout the 11-year sunspot cycle. In addition, ionization is subject to frequent abnormal variations, such as those caused by solar flares.

Ionospheric losses are a minimum near the MUF and increase rapidly for lower frequencies during daylight. An optimum working frequency is selected below the MUF to provide some margin for variations. The LUF is that frequency below which ionospheric absorption and radio noise levels make required radiated power impractical. At altitudes between about 50 km and 400 km, the atmosphere may be ionized by energy from the sun. The so-called D-layer extends from altitudes of about 50 to 90 km and exists only during daylight hours. The D-layer reflects VLF and LF waves, absorbs MF waves, and weakens HF waves through partial absorption. VHF and higher frequency waves pass through the D-layer with very little attenuation.

The E-layer exists at a height of about 110 km and may be present both day and night. It provides reflections for both MF and HF energy and small attenuation for higher frequencies.

The F_1-layer exists at a height of about 175 to 250 km. It exists only during the daylight hours. It provides reflection of HF waves and low attenuation of higher frequency waves.

The F_2-layer is at heights of about 250 to 400 km. This layer is the principal reflecting region for long-distance HF communication. It is present both day and night. Above about 100 MHz, the attenuation caused by the F_2 and the other ionized layers is less than about 1 dB and may be neglected.

It is possible to achieve very long range communication using the ionosphere as a reflecting medium because the ionosphere is so high. The signal travels from the ground-based transmitter antenna to the ionospheric layer where the path is bent downward. Although it is really refraction rather than reflection that is involved, it is common to speak of the process as reflection. The downward directed wave then travels to the ground, where it is received. The process constitutes single-hop propagation. For two-hop propagation, the wave is reflected from the ground back to the ionosphere, where it is again refracted downward to the ground, where it is received. Three-hop (or more) propagation also is possible. By these means, it often is possible to communicate many thousands of miles.

Figure 3.7 illustrates HF ionospheric reflection propagation between Washington, D.C., and Chicago and between Washington, D.C., and San Francisco. In the case of communication between Washington and Chicago, single-hop paths are possible using either E-layer reflection or F_2-layer reflection. In the case of communication between Washington and San Francisco, a two-hop path is possible using F_2-layer reflection.

In evaluating the performance of HF communication systems, three main losses must be taken into account: the free-space path loss, the loss in the ionosphere, and the loss through reflection on land or sea. With more than one hop, there will be losses each time the signal passes through the ionosphere and each time

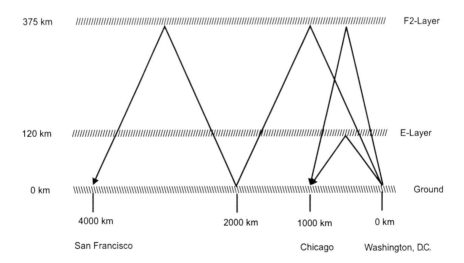

Figure 3.7 Example reflection paths for ionospheric propagation.

reflection takes place. Typical losses in the ionosphere range from as low as 2 dB per pass to as high as 30 dB, depending on the time of day and the frequency used. Losses due to reflection are typically less than 4 dB per reflection, depending on the frequency and the land type.

HF communication using the ionosphere as a means of reflection has only limited bandwidth capability. It has high noise levels and large losses. It has problems of multipath-induced fading unless operation is near the MUF. HF communication is not reliable over paths near the magnetic poles and has been largely replaced by satellite communication and other means for long-distance communication [5,6].

3.10 GROUND-WAVE PROPAGATION

Another type of propagation that is important at the lower frequencies is ground-wave propagation [6]. The ground wave glides over the surface of the Earth and is vertically polarized. Any horizontal component of the E-field in contact with the earth is short-circuited by the Earth and therefore is quickly attenuated. The ground wave induces charges in the Earth that travel with the wave and so constitute a current. As the ground wave passes over the surface of the Earth, it is weakened as a result of energy absorbed by the Earth. Losses are the lowest at the lower frequencies and increase with frequency. Ground-wave propagation is normally not used above HF frequencies because the attenuation is too great. Ground-wave losses are much lower over seawater than over land [6]. For example, with good ground, a frequency of 150 kHz, and a range of 1,000 miles, the ground-wave loss is about 30.5 dB greater than free-space loss. In the case of seawater and that same range and frequency, the ground-wave loss is about 22.5 dB greater than free-space loss. That is a difference of 8.0 dB. As another example, with good ground, a frequency of 500 kHz, and a range of 200 miles, the ground-wave loss is about 20 dB greater than free-space loss. In the case of seawater and the same range and frequency, the ground-wave loss is about 6 dB greater than free-space loss. That is a difference of 14 dB.

Ground-wave VLF and LF frequencies are used for high survivability, high value, military command links, and navigation systems. Such systems have range capabilities on the order of 1,000 to 2,500 mi. Ground waves also are used for standard AM broadcasting (535 to 1,605 kHz), where the range capability is approximately 60 to 100 mi. Ground-wave propagation also is used by the military at low HF frequencies where the range capability is 50 to 80 mi. Most ground-wave communication systems involve very large wavelengths. For example, at 30 kHz, the free-space wavelength is 10,000m. At 300 kHz, the free-space wavelength is 1,000m. At 3 MHz the free-space wavelength is 100m. At the higher frequencies, it is typical for a vertically polarized monopole antenna to be on the order of a quarter-wave high for good efficiency. At that height, the antenna would be resonant. That, however, would be impractical at 30 kHz, where the antenna height would be 2,500m. The same is true at 300 kHz, where the required antenna height would be 250m. For those long wavelength systems, it is necessary to use electrically small antennas for transmitter antennas rather than quarter-wave monopoles. Such antennas require tuning and often use capacitive loading. With such tuning, the antennas can be made as small as 0.1 wavelength high or less. However, tuned antennas typically have low efficiency because of losses in the tuning systems.

In the case of receiver antennas for long wavelength signals, it is not necessary to have the best efficiency. That is due to the fact that the limiting noise for these systems is external to the receiver and is received by the antenna. Making a better receiver antenna increases the received signal, but it also increases the noise to the same degree. The signal-to-noise ratio thus is not improved.

Again, it must be emphasized that ground-wave propagation requires vertical polarization. Therefore, the antennas used must provide vertical polarization. Details about antennas for all bands and modes of propagation are discussed in Chapter 13.

3.11 SCATTER PROPAGATION

Three modes of scatter propagation have been used for communication: tropospheric scatter, ionospheric scatter, and meteor-burst scatter. Weak but reliable fields are propagated several hundred miles beyond the horizon in the VHF, UHF, and SHF bands by those scatter modes. High-gain directional antennas and high power are used for tropospheric scatter and ionospheric scatter because of the weakness of the signals.

In the case of tropospheric scatter, the signal is reflected by regions high above the Earth, where changes in the dielectric constant of the atmosphere take place. In a similar way, reflections can take place at those higher frequencies from the ionosphere where there are changes in the ionization density. Meteor burst scattering takes place when there is a meteor that enters the atmosphere, creating an ionized trail. Signals may be scattered from that trail as long as it exists. This type of system normally operates at VHF frequencies.

Figure 3.8 illustrates the concept of propagation via tropospheric scatter [5]. On the left of the diagram is a transmitter station with a high-power transmitter feeding a high-gain parabolic dish antenna directed at the sky. On the right is a receiver station with a high-gain parabolic dish receiver antenna. The two antenna beams are directed so there is a common volume high in the atmosphere for the two antenna main beams.

Although tropospheric scatter is subject to fading with little signal scattered forward, it nevertheless forms a reliable method of over-the-horizon communication. It is not affected by the abnormal phenomena that afflict HF ionospheric propagation. Path lengths typically are 300 to 500 km. Operating frequencies typically are centered around 900, 2,000, and 5,000 MHz. Losses include free-space

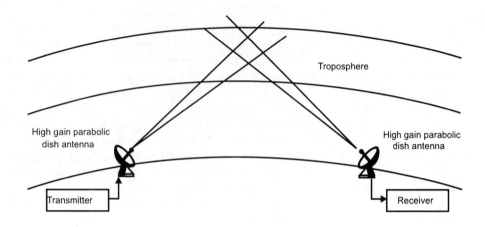

Figure 3.8 Tropospheric scatter propagation.

path loss plus scatter loss in the troposphere. Scatter loss may be as high as 60 to 90 dB. High transmitting power, extremely good receiver sensitivity, and high antenna gain obviously are needed [5].

3.12 FIBER OPTIC CABLE PROPAGATION

In recent years, fiber optic communication has become an important mode of communication. Fiber optic cables have many advantages over other types of cable, including much larger bandwidth capability, smaller size and weight, reduced loss per unit length, and lower cost.

Many applications are now being found for fiber optic cables in addition to long-range communication links. Examples are fiber optic delay lines for radars and other RF systems, fiber optic control lines, and fiber optic lines used in the medical field for viewing inside the human body.

An optical fiber is a piece of highly pure glass of very small diameter. The fiber has an outside cladding of glass similar to the fiber itself but, because of a slightly different chemical composition, with a different refractive index. This type of optical fiber is known as a step-index fiber. Other types of optical fibers are possible, but they have higher attenuation than the step-index fiber.

Figure 3.9 shows attenuation in typical modern fiber optic cables as a function of wavelength. The majority of early commercial fiber optic links operated at a wavelength of about 0.85 mm. Attenuation at that wavelength is about 2.5 dB/km. Most of the fiber optic links now being installed operate in a second window, at 1.3-mm wavelength. Attenuation at that wavelength is about 0.4 dB/km. It is likely that future fiber optic systems will operate in the third window at a wavelength of about 1.6 mm to permit attenuations as low as 0.25 dB/km [7]. That lower attenua-

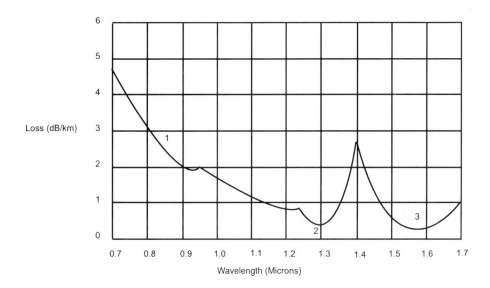

Figure 3.9 Attenuation in low-loss optic fibers. (*Source:* [7].)

tion permits reduced numbers of repeaters in long-distance links, higher reliability due to reduction in parts, and reduction in system cost.

3.13 RADAR CROSS-SECTION OF TARGETS

The RCS of a real target is the cross-section that the target would have if the signal is scattered uniformly in all directions. A physically small target can have a very large RCS if the signal reaching the target is scattered back largely in the direction of the radar rather than uniformly in all directions. A corner reflector is an example of a target of this kind. On the other hand, a very large target can have a very small RCS if very little of the scattered signal is in the direction of the radar.

Figure 3.10 illustrates the concepts of propagation for radars using backscatter from targets. The figure shows that the signal experiences two-way spreading loss, two-way excess path loss, and reflection by an ideal spherical metallic target having the same back-scatter level as the real target but with uniform radiation in all directions. The radar is shown with separate transmitter and receiver antennas rather than a single antenna, as is normally used simply for ease in showing the loss functions. The radar transmitter has a transmitter antenna gain, G_T, which is normally quite high and which provides a fan or pencil beam, depending on the application. The receiver antenna has an effective capture area, A_e, that is equal to $G_R \lambda^2 / 4\pi$, where G_R is the receiver antenna gain and λ is the wavelength.

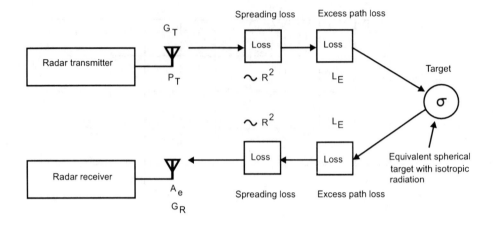

Figure 3.10 Concepts of propagation for radar with an equivalent ideal spherical target with RCS σ.

The radar may also see back-scatter signals from the ground or the sea and from weather such as rain or clouds. A more detailed discussion of RCS of targets and clutter for radar is presented in Chapter 10.

3.14 EQUATIONS FOR CALCULATING PROPAGATION PERFORMANCE FOR COMMUNICATION SYSTEMS

The received power at the input to the receiver is given by (3.9).

$$P_r = P_T G_T G_R \lambda^2 / [(4\pi)^2 R^2 L_E L_{CT} L_{CR}] \tag{3.9}$$

where

P_T = transmitter output power
G_T = transmitter antenna gain
G_R = receiver antenna gain
λ = wavelength
R = range (same units as wavelength)
L_E = excess path loss, numeric
L_{CT} = cable loss between transmitter and antenna
L_{CR} = cable loss between receiver and antenna

The following examples use (3.9) to determine the performance of communication systems.

3.14.1 Example 1: HF Ionospheric Reflection Communication System

Assumptions:

P_T = transmitter output power = 1,000W
G_T = transmitter antenna gain = 2 dBi
G_R = receiver antenna gain = 2 dBi
λ = wavelength = 30m
R = range = 2,000 km = 2×10^6 m
L_E = excess path loss = 15 dB
L_{CT} = cable loss between transmitter and antenna = 2 dB
L_{CR} = cable loss between receiver and antenna = 2 dB

Expressed in decibels, (3.9) is

$$P_r = P_T + G_T + G_R + 20 \log \lambda - 20 \log(4\pi) - 20 \log(R) \\ - L_E - L_{CT} - L_{CR} \qquad (3.10)$$

Substituting values:

$$P_r = 30 + 2 + 2 + 29.5 - 22 - 126 - 15 - 2 - 2$$
$$P_r = -103.5 \text{ dBW} = -73.5 \text{ dBm} = 4.47 \times 10^{-11} \text{W}$$

3.14.2 Example 2: VHF Base Station to Mobile Unit Communication System

Assumptions:

P_T = transmitter output power = 100W
G_T = transmitter antenna gain = 3 dBi
G_R = receiver antenna gain = 0 dBi
λ = wavelength = 2m
R = range = 20 km
L_E = excess path loss = 20 dB
L_{CT} = cable loss between transmitter and antenna = 2 dB
L_{CR} = cable loss between receiver and antenna = 1 dB

Substituting values into (3.10):

$$P_r = 20 + 3 + 0 + 6 - 22 - 86 - 20 - 2 - 1$$
$$P_r = -102 \text{ dBW} = -72 \text{ dBm} = 6.3 \times 10^{-11} \text{W}$$

3.14.3 Example 3: Microwave Uplink to Satellite Relay Located at Geostationary Orbit

Assumptions:

P_T = transmitter output power = 10,000W
G_T = transmitter antenna gain = 44 dBi
G_R = receiver antenna gain = 24 dBi
λ = wavelength = 0.05m
R = range = 40,000 km
L_E = excess path loss = 1 dB
L_{CT} = cable loss between transmitter and antenna = 2 dB
L_{CR} = cable loss between receiver and antenna = 1 dB

Substituting values into (3.10):

$$P_r = 40 + 44 + 24 - 26 - 22 - 152 - 1 - 2 - 1$$
$$P_r = -96 \text{ dBW} = -66 \text{ dBm} = 2.5 \times 10^{-10} \text{W}$$

3.15 EQUATIONS FOR CALCULATING PROPAGATION PERFORMANCE FOR RADAR SYSTEMS

The received power at the input to the radar receiver is given by (3.11).

$$P_r = P_T G_T G_R \lambda^2 \sigma / [(4\pi)^3 R^4 L_{E1} L_{E2} L_{GT} L_{GR}] \tag{3.11}$$

where

P_T = transmitter output power
G_T = transmitter antenna gain
G_R = receiver antenna gain
λ = wavelength
σ = target RCS, same units as range and wavelength (i.e., meters squared)
R = range
L_{E1} = excess path loss for transmitted signal
L_{E2} = excess path loss for reflected signal
L_{GT} = waveguide or cable loss between transmitter and antenna
L_{GR} = waveguide or cable loss between receiver and antenna

The following examples use (3.11) to determine the performance of radar systems.

3.15.1 Example 4: L-Band Aircraft Surveillance Radar

Assumptions:

P_T = transmitter output power = 100 kW
G_T = transmitter antenna gain = 33 dBi
G_R = receiver antenna gain = 33 dBi
λ = wavelength = 0.23 m
σ = target RCS = 10 m^2
R = range = 100 km
L_{E1} = excess path loss for transmitted signal = 2 dB
L_{E2} = excess path loss for back-scatter signal = 2 dB
L_{GT} = waveguide loss between transmitter and antenna = 2 dB
L_{GR} = waveguide loss between receiver and antenna = 2 dB

Expressed in decibels, (3.11) is

$$P_r = P_T + G_T + G_R + 20 \log \lambda + 10 \log \sigma - 30 \log(4\pi)$$
$$- 40 \log(R) - L_{E1} - L_{E2} - L_{GT} - L_{GR} \qquad (3.12)$$

Substituting values:

$$P_r = 50 + 33 + 33 - 12.8 + 10 - 33 - 200 - 2 - 2 - 2 - 2$$
$$P_r = -127.8 \text{ dBW} = -97.8 \text{ dBm} = 1.6 \times 10^{-13} \text{W}$$

3.15.2 Example 5: X-Band Airborne Multiple-Function Radar

Assumptions:

P_T = transmitter output power = 100 kW
G_T = transmitter antenna gain = 38 dBi
G_R = receiver antenna gain = 38 dBi
λ = wavelength = 0.033m
σ = target RCS = 10 m^2
R = range = 100 km
L_{E1} = excess path loss for transmitted signal = 2 dB
L_{E2} = excess path loss for back-scatter signal = 2 dB
L_{GT} = waveguide loss between transmitter and antenna = 2 dB
L_{GR} = waveguide loss between receiver and antenna = 2 dB

Substituting values into (3.12):

$$P_r = 50 + 38 + 38 - 29.6 + 10 - 33 - 200 - 2 - 2 - 2 - 2$$
$$P_r = -134.6 \text{ dBW} = -104.6 \text{ dBm} = 3.5 \times 10^{-14} \text{W}$$

References

[1] Barton, David K., *Modern Radar System Analysis*, Boston: Artech House, 1988, Chap. 6.
[2] Ibid, p 283.
[3] Ibid., p 300.
[4] Ibid., pp 288–296.
[5] Kennedy, George, *Electronic Communication Systems*, 3rd ed., New York: McGraw-Hill, 1985, pp. 221–231.
[6] Van Valkenburg, Mac E., ed., *Reference Data for Engineers*, 8th ed., SAMS, Carmel, IN: Prentice Hall Computer Publishing, 1993, pp. 33-3 to 33-12, "Electromagnetic-Wave Propagation" by Douglass D. Crombic.
[7] Kennedy, op. cit, pp. 545–548.

CHAPTER 4

RF Noise and Link Analysis

4.1 CONCEPTS OF RF NOISE AND SIGNAL-TO-NOISE RATIO

Noise limits the performance of all receivers. To be effective, the received signal must be substantially greater than the RF noise. Figure 4.1 lists noise sources, including internal noise sources, where the noise is generated by the receiver itself, and external noise sources, in which the noise is received by the receiver antenna. External noise sources include both natural and man-made sources. Man-made noise sources include industrial noise, signals from radar or communication transmitters that provide interference, and jamming sources. Natural noise sources include atmospheric noise such as lightning storms, galactic noise, solar noise, atmospheric-loss noise, and hot-Earth noise. At the lower frequencies (VLF, LF, MF, and HF frequencies), external noise received by the receiver antenna from either a natural or a man-made source generally is much greater than internal noise and therefore is the limiting factor in system performance. At VHF and higher frequencies, internally generated noise generally is greater than external noise received by the antenna and therefore is the limiting factor in system performance.

The intended signal is transmitted by a signal transmitter and is received by the receiver antenna. At the receiver, the signal is added to the external noise, also received by the receiver antenna, and to the internal noise generated by the receiver. If the combined noise is much smaller than the signal, it will not significantly degrade the detected signal. However, if the noise is large compared to the desired signal, detection of the signal may not be possible.

Figure 4.2 illustrates detection of a signal in the presence of noise for a detected radar pulse. The assumption is made that both the signal and the noise have been filtered by a bandpass filter having a near-optimum bandwidth for detection of the radar signal. Thus, the frequency components for the noise are the same as those

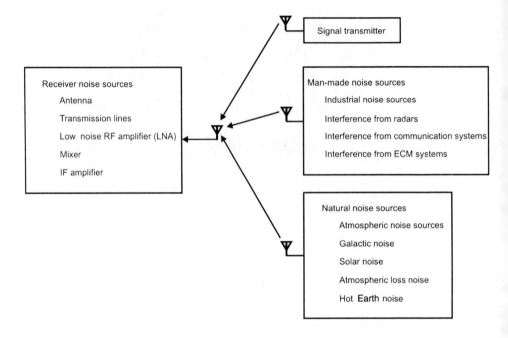

Figure 4.1 Noise sources.

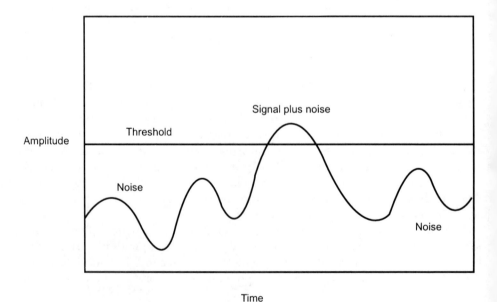

Figure 4.2 Detection of a radar signal in the presence of noise. (*After:* [3].)

for the signal. Without filtering, the noise would have a wide range of frequency components with details, depending on the type of noise present.

Figure 4.2 is an amplitude-versus-time plot. Often a threshold is set as shown to permit a decision of whether a signal is present. The threshold level is set sufficiently above the noise level that the false alarm rate will be acceptable. Signals greater than that threshold level are classed as true signals. It cannot be set too high, however, or the probability of detection may be too low.

Power requirements are such to permit an acceptable probability of detection of the signal. A typical required minimum signal-to-noise ratio for radar after integration is about +15 dB. A typical required minimum signal-to-noise ratio for a cellular telephone system might be +18 dB. Broadcast communication systems for entertainment would require +30 dB or higher signal-to-noise ratios.

4.2 NOISE POWER, NOISE TEMPERATURE, AND NOISE FIGURE

The information in this chapter is based on [1–4].

Noise power is given by

$$P_n = kTB \qquad (4.1)$$

where

P_n = available noise power (watts)

k = Boltzmann's constant = $1.38 \cdot 10^{-23}$ J/K

T = noise temperature (kelvin)

B = effective receiver noise bandwidth (hertz)

For example, the noise power corresponding to the reference temperature $T_0 = 290$K and a bandwidth of 1 Hz would be

$$P_n = 1.38 \cdot 10^{-23} \cdot 290 \cdot 1 = 4 \cdot 10^{-21} \text{W}$$

Expressed in decibels, that would be

$$P_n = -204 \text{ dBW/Hz} = -174 \text{ dBm/Hz}$$

If a system had a noise temperature of 5,000K and an effective noise bandwidth of 1 MHz, the output noise power would be

$$P_n = 1.38 \cdot 10^{-23} \cdot 5{,}000 \cdot 10^6 = 6.9 \cdot 10^{-14} \text{W}$$

Expressed in decibels, that would be −131.6 dBW or −101.6 dBm.

Another important expression is the noise power expressed in terms of the reference temperature, T_0, and a noise figure F_n, where F_n is the ratio of the real noise temperature to the reference temperature. The expression is

$$P_n = kT_0 BF_n \tag{4.2}$$

where

P_n = available noise power (watts)
k = Boltzmann's constant = $1.38 \cdot 10^{-23}$ J/K
T_0 = reference noise temperature (290K)
B = effective receiver noise bandwidth (hertz)
F_n = noise figure

As an example of the use of (4.2), assume that a system has a noise figure of 7 dB or 5.1 numeric and an effective noise bandwidth of 25 kHz. Substituting values into (4.2), the noise power would be

$$P_n = 1.38 \cdot 10^{-23} \cdot 290 \cdot 25 \cdot 10^3 \cdot 5 = 5 \cdot 10^{-16} \text{W}$$

Expressed in decibels, we have

$$P_n = 10 \log_{10}(kT_0) + 10 \log_{10}(B) + F_n$$
$$P_n = -204 + 44 + 7 = -153 \text{ dBW} = -123 \text{ dBm}$$

4.3 MULTIPLE-STAGE SYSTEMS WITH NOISE

The information in this section is based on [2,3]. It is common practice to use noise temperatures in performing noise calculations for complex systems involving noise stages and gain or loss stages in series. Figure 4.3 illustrates such a system. In the figure, T_1, the antenna-noise temperature, represents the received noise from outside sources. T_2 is the thermal temperature of the transmission line system, usually assumed to be 290K. G_2 is the gain of the transmission line system. That

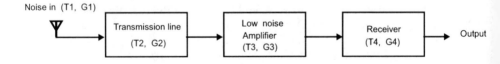

Figure 4.3 Multiple stage system with noise.

gain is less than 1 and is actually a loss. T_3 and G_3 are the noise temperature and the gain of the low noise amplifier (LNA), a stage that may not be used in some systems. T_4 and G_4 are the noise temperature and the gain of the receiver.

It is common practice to provide the noise characteristics of LNAs and receivers in terms of noise figures, which are converted using (4.3).

$$T_e = (F_n - 1) T_0 \qquad (4.3)$$

where

T_e = noise temperature (kelvin)
F_n = noise figure (numeric)
T_0 = reference temperature (290K)

For example, if the noise figure of a receiver is given as 7 dB, we first would convert to a numeric equivalent, which for this case would be 5.0. Then we would calculate the noise temperature as follows:

$$T_e = (5 - 1) \cdot 290 = 1{,}160 \text{K}$$

To convert a noise temperature to a noise figure, we use (4.4), a simple conversion from (4.3):

$$F_n = 1 + T_e/T_0 \qquad (4.4)$$

For example, if the noise temperature of a system is 3,000K, the noise figure for the system would be

$$F_n = 1 + 3{,}000/290 = 11.3 = 10.5 \text{ dB}$$

The effective noise temperature of a receiver consisting of a number of networks in cascade is given by (4.5):

$$T_e = T_1 + T_2/G_1 + T_3/G_1 G_2 + \ldots \qquad (4.5)$$

where T_i and G_i are the effective noise temperature and gain of the ith network. Both T_i and G_i are real numbers, not decibels.

An alternative equation for noise figure for a number of networks in cascade is given by (4.6):

$$F = F_1 + (F_2 - 1)/G_1 + (F_3 - 1)/G_1 G_2 + (F_4 - 1)/G_1 G_2 G_3 \qquad (4.6)$$

where F_i is the noise figure, numeric.

For an example of the use of (4.5) for the system shown in Figure 4.3, assume the following:

$T_1 = 500K \qquad G_1 = 1$
$T_2 = 290K \qquad G_2 = 0.3$
$T_3 = 200K \qquad G_3 = 100$
$T_4 = 1,200K \qquad G_4 = 10^{10}$

Then,

$T_e = T_1 + T_2/G_1 + T_3/G_1 G_2 + \ldots$
$T_e = 500 + 290/1 + 200/(.3 \cdot 1) + 1,200/(0.3 \cdot 1 \cdot 100)$
$T_e = 500 + 290 + 667 + 40 = 1,497K$

The system noise figure would be found using (4.4), as follows:

$F_n = 1 + T_e/T_0$
$F_n = 1 + 1,497/290 = 6.2 = 7.9$ dB

4.4 TYPES OF NOISE

The information in this section is based on [1,4]. The types of external noise that are important to RF systems include the following:

- Atmospheric noise;
- Galactic noise;
- Solar noise;
- Ground noise;
- Man-made noise.

4.4.1 Atmospheric Noise

Figure 4.4 shows typical effective antenna noise figures, in decibels, above $kT_0 B$ for frequencies below 100 MHz and locations in the central region of the United States. Below 30 MHz, the strongest source of noise is atmospheric noise, generated mostly by lightning discharge in thunderstorms. The noise level depends on the frequency, the time of day, weather, the season of the year, and geographical location. Atmospheric noise is greatest at the lowest frequencies and decreases with increasing frequency. It propagates over the Earth in the same fashion as ordinary radio waves of the same frequencies. Thus, at HF and lower frequencies, it is not only the lightning storms in close proximity that cause the interference or noise, but storms any place in the world where propagation paths are possible.

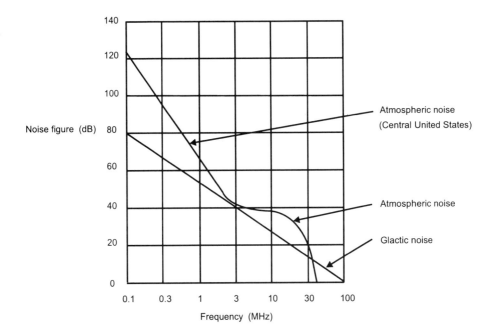

Figure 4.4 RF antenna noise figure below 100 MHz. (*After:* [1].)

Atmospheric noise becomes less severe at frequencies above about 30 MHz because of two factors. First, the higher frequencies are limited to line-of-sight propagation. Second, the mechanism generating the noise is such that very little of it is created in the VHF range and higher.

The region with the largest atmospheric lightning noise for the western hemisphere is the Caribbean near Panama. The median noise level for that area is about 85 dB above kT_0B at 1.0 MHz. The central and eastern regions of the United States have median values of noise power of about 70 dB above kT_0B at 1 MHz. The average atmospheric noise level for the western portion of the United States is substantially less, being about 60 dB in Utah and 50 dB in California at 1 MHz. The atmospheric noise levels decrease rapidly as we move northward toward the pole and as we move out over the Pacific and Atlantic oceans. The peak atmospheric noise can exceed the median values of average noise power by 10 dB or more. Atmospheric noise is lower than that in the summer by about 10 dB over much of the frequency range.

4.4.2 Galactic Noise

Figure 4.4 shows that above about 30 MHz and less than 100 MHz galactic noise is the largest type of natural noise. It is similar in magnitude to man-made noise in a quiet location.

Figure 4.5 shows typical antenna temperatures for frequencies above 100 MHz. Galactic noise is the largest natural noise component between 100 and 400 MHz. Above 400 MHz, other noise components dominate. Those other components are solar noise, atmospheric-loss noise, and hot-Earth noise. The plot in Figure 4.5 assumes that the Sun and the Earth are seen only through antenna sidelobes. If the earth is viewed with the main beam, the antenna noise temperature due to the hot Earth could be as high as 300K. The solar noise temperature could be much greater.

Galactic noise is most intense in the galactic plane and reaches a maximum in the direction of the galactic center. That center appears to be about 3 degrees in diameter. Antenna noise temperature resulting from galactic noise may be greater than 18,000K for a narrow-beam antenna pointed at the galactic center in the region of 100 MHz and less than 3K above about 1,000 MHz. Stronger than average noise radiation is received from a broader region within 30 degrees of the galactic plane. The minimum galactic noise antenna temperature that is measured when pointing at the galactic poles is about 500K at 100 MHz. The geometric mean of these maximum and minimum temperatures is 3,000K at 100 MHz.

For an example of galactic antenna noise temperature for directional antennas, assume the use of an antenna with 30 dBi antenna gain (beam angle 5 degrees

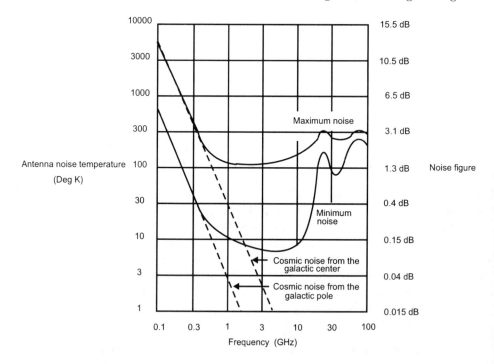

Figure 4.5 Radio frequency antenna noise temperature versus frequency above 100 MHz. (*After:* [4].)

diameter) operating at 100 MHz and pointing at the galactic center (3 degrees diameter). The fraction of the beam occupied by the source would be about 0.36. The noise temperature of the galactic center at 100 MHz is assumed at 18,000K, and the atmospheric loss is assumed to be 1 (0 dB). The antenna temperature due to galactic noise from only the galactic center may be found using (4.7).

$$T_a = a_C T_C / L_A \qquad (4.7)$$

where

T_a = antenna noise temperature (kelvin)
a_C = fraction of beam filled by noise source
T_C = temperature at galactic center (kelvin)
L_A = Atmospheric loss (numeric)

Using those assumptions:

$$T_a = .36 \cdot 18,000/1 = 6,480K$$

If the antenna beam angle is less than 3 degrees (the approximate diameter of the galactic center) and the antenna points at the galactic center, the antenna temperature will be no greater than 18,000K. If the antenna beam angle is much larger than 3 degrees and points at the galactic center, it will see other parts of the galaxy in addition to the center.

If a high-gain (highly directional) antenna points at the galactic pole, it will see a noise temperature of about 500 degrees at 100 MHz. In no case will the antenna see less than the temperature of the galactic pole.

It is clear from Figure 4.5 that galactic noise usually is the strongest natural noise between 100 MHz and about 400 MHz. Other types of natural noise become important above 400 MHz and typically produce the maximum and minimum temperatures as shown.

4.4.3 Solar Noise

The Sun is a powerful noise source. If a directional antenna is pointed at the Sun, it will see a large antenna noise temperature. The Sun also can contribute appreciably to antenna noise through sidelobes of the antenna. During high levels of Sun spot activity, noise temperatures from 100 to 10,000 times greater than those of the quiet sun may be observed for periods of seconds in what is called solar bursts, followed by levels about 10 times the quiet level lasting for several hours.

At microwave frequencies, the Sun's noise diameter is one-half degree. The Sun's effective noise temperature seen by an antenna of gain G_s is given by (4.8) and (4.9).

$$T_a = a_S T_S / L_A \qquad (4.8)$$

$$T_a = 4.75 \cdot 10^{-6}\, G_s T_S / L_A \qquad (4.9)$$

where

T_a = antenna noise temperature

a_S = fraction of the beam filled by the sun

T_S = noise temperature of the sun

L_A = atmospheric loss (numeric)

G_s = antenna gain

Equation 4.9 can be derived from (4.8) as follows: The directional gain of the antenna is $G = 41{,}300/$(beam angle squared). The solid angle covered by the antennas is thus $41{,}300/G_S$.

The solid angle covered by the sun is 0.196 square degrees. Thus, the fraction of the beam filled by the Sun is

$$0.196/(41{,}300/G_S) = 4.75 \cdot 10^{-6} = a_s$$

The value for a_s is used in (4.8) to yield (4.9).

For example, if the frequency is 1,000 MHz and there are quiet sun conditions, the noise temperature of the sun is about $(2 \cdot 10^5)$K. If we assume an antenna gain in the direction of the sun of 31 dBi and an atmospheric loss of 1 dB, the antenna noise temperature will be

$$T_a = 4.75 \cdot 10^{-6}\, G_s T_S / L_A$$
$$T_a = 4.75 \cdot 10^{-6} \cdot 1{,}259 \cdot 2 \cdot 10^5 / 1.26$$
$$T_a = 949.3\text{K}$$

4.4.4 Ground Noise

The Earth is a radiator of electromagnetic noise. The thermal temperature of the Earth is typically about 290K. In radar systems and directional communication systems, the Earth will be viewed mainly through the sidelobes of the antennas. The average sidelobe antenna gain typically is about −10 dBi. A rough estimate of antenna noise temperature in that case is 29K. The maximum possible ground antenna noise temperature is about 290K for directional antennas pointing at the ground.

4.4.5 Man-Made Noise and Interference

Man-made noise is due chiefly to electric motors, neon signs, power lines, and ignition systems located within a few hundred yards of the receiving antenna. In addition, there may be radiation from hundreds of communication and radar systems that may interfere with reception. The interference may include signals on assigned frequencies, harmonics of those signals, unwanted spurs (mixing products of signals generated within transmitters), subharmonics, and leakage from low-frequency oscillators used in multiplier chains.

Propagation of man-made noise and interference may be by means of power lines, by ground wave, and by any of the other propagation modes discussed in Chapter 3. The amplitude of man-made noise decreases with increasing frequency and varies considerably with location. Generally this type of noise is assumed to decrease with frequency as shown in the (4.10):

$$T_a = T_{100}(100/f_{MHz})^{2.5} \tag{4.10}$$

where T_{100} is the man-made noise temperature at 100 MHz.

For an example of the use of (4.10), assume an operating frequency of 400 MHz and a man-made noise temperature at 100 MHz of 300,000K (F_a = 30.2 dB above kT_0B). The calculated antenna noise temperature would be

$$T_a = T_{100}(100/f_{MHz})^{2.5}$$
$$T_a = 300,000 \cdot (100/400)^{2.5} = 9,375K$$

The temperature is about 15.2 dB above kT_0B.

To determine the man-made noise at a given site, it is necessary to make noise measurements, which will differ for time of day, season of the year, and direction. The man-made noise and interference signal typically exceed natural noise in the frequency range of 10 to 1,000 MHz.

4.5 SIGNAL-TO-NOISE IMPROVEMENT BY USE OF INTEGRATION

The information in this section is based on [3].

Figure 4.6 shows the signal-to-noise ratio improvement that is possible with the use of different signal processing techniques. Three cases are given in the figure. The greatest improvement results from coherent integration, where the term coherent refers to having phase reference and equal phase. That is referred

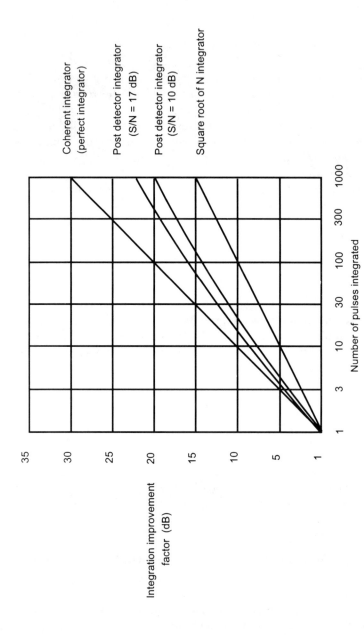

Figure 4.6 Integration improvement factor versus number of pulses integrated. (*After:* [3].)

to as a "perfect" integrator, because the processing improvement is equal to the number of pulses integrated. Coherent integration systems use synchronous detectors or phase detectors. The integration improvement factor (IIF), expressed in decibels, for this type of integration is

$$\text{IIF} = 10 \log_{10} N$$

where N = number of pulses integrated.

The next level of improvement is for noncoherent integration. The type of detectors used in this case may be simple envelope detectors for AM. Here the improvement is less than perfect and is a function of the signal-to-noise ratio or the probability of detection.

The lowest level of improvement is the case where the integration improvement is proportional to the square root of the number of pulses integrated. An example of this type of integration is an operator viewing a CRT screen, where the IIF is

$$\text{IIF} = 10 \log_{10}(N^{0.5}) = 5 \log_{10} N$$

where N = number of pulses integrated.

Integration often is used with radar systems: many pulses are integrated to improve the probability of detection and to improve tracking accuracy. For example, assume that a radar integrates 100 pulses. If the radar uses coherent integration, the integration improvement factor, as given by Figure 4.6, is 20 dB. If the single-pulse signal-to-noise ratio is 17 dB and postdetector integration is used, the improvement factor as shown by Figure 4.6 is about 16 dB. If the single-pulse signal-to-noise ratio is 10 dB, the postdetector IIF is about 14 dB. If the integrator is a square root of the N integrator, the improvement factor, as shown by Figure 4.6, is only 10 dB.

4.6 SIGNAL-TO-NOISE RATIO

The information in this section is based on [3]. The required signal-to-noise ratio for a radar system can be found in Figure 4.7. This figure shows the probability of detection, also called the probability of intercept (POI), for a sinusoidal signal in the presence of noise as a function of the signal-to-noise (power) ratio and the probability of false alarm. For an example of the use of Figure 4.7, assume that we want a probability of detection of 0.99 or greater and a probability of false alarm of no more than 10^{-7}. From Figure 4.7, we see that the required signal-to-noise ratio without integration, that is, postprocessing (single-pulse detection), is about 15 dB. for another example, assume a probability of false alarm of 10^{-6} and a probability of detection of 0.9. From Figure 4.7, the required signal-to-noise ratio is about 13.3 dB, again without any postprocessing.

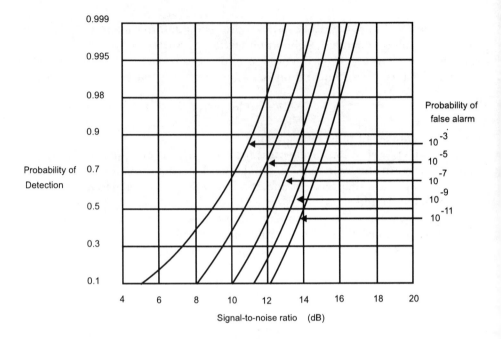

Figure 4.7 Required signal-to-noise ratio versus probability of detection and probability of false alarm. (*After:* [3].)

The term false alarm means that the system indicates the presence of a signal when in reality it is merely noise. An interpretation of the false-alarm probability is as follows: On the average, there will be one false decision out of n_f possible decisions in the false-alarm time interval T_{fa}. The total number of decisions, n_f, in T_{fa} is equal to the number of range intervals per pulse period times the number of pulses per second (pulse repetition rate, or PRR) times the false-alarm time. Thus, the number of possible decisions is T_{fa}/τ, where τ is the pulse width.

A generally employed assumption is that the product of the receiver bandwidth and the pulse width ($B\tau$) is equal to 1. Errors in that approximation are not serious in practice because of the exponential relationship between the threshold and the probability of false alarm.

Using those assumptions, the probability of false alarm is

$$P_{fa} = 1/n_f = 1/T_{fa}B \tag{4.11}$$

where

n_f = total number of decisions

T_{fa} = false-alarm time interval

B = receiver bandwidth

For an example of the use of (4.11), assume a radar with a pulse width of 1 μs and a bandwidth of 1 MHz. If we want a false alarm of no more than once an hour (3,600 sec), the false-alarm probability is calculated to be $1/(3,600 \cdot 10^6) = 2.8 \cdot 10^{-10}$.

If n pulses are processed so as to improve detection, the number of independent decisions in the time T_{fa} will be reduced by a factor of n. As a result, the equation for probability of false alarm becomes

$$P_{fa} = n/n_f = n/T_{fa}B \qquad (4.12)$$

where the terms are the same as for (4.11) and n is the number of pulses.

For example, to compute the probability of false alarm with integration, assume the following:

$B = 0.5$ MHz

$T_{fa} = 1,000$ sec

$n = 10$

$P_{fa} = 10/(1,000 \cdot 0.5 \cdot 10^6) = 2 \cdot 10^{-8}$

The required signal-to-noise ratio for a communication system depends on the type of modulation used and the type of application. For AM and FM communication, it generally is sufficient to have a signal-to-noise ratio of 20 dB or greater. For broadcast of music and video, it usually is necessary to have a somewhat higher signal-to-noise ratio. A typical requirement might be 30 dB or more. These requirements are for analog signals. With digital signals, such as the newer DSS and DBS systems, the signal-to-noise requirement can be much less (perhaps as low as 15 dB).

4.7 COMMUNICATION SYSTEM LINK ANALYSIS

The information in this section is based on [2,3].

A communication system link analysis is a set of calculations that shows the signal-to-noise ratio that is achieved at the receiver output based on a given set of assumptions for a communication system. This section shows the equations used in the communication system link analysis and an example link analysis for a typical communication system.

The equation for received power at the receiver antenna output is as follows:

$$P_R = P_T G_T G_R \lambda^2 / [(4\pi)^2 R^2 L_{El}] \qquad (4.13)$$

where

P_R = receiver antenna power output
P_T = transmitter output power at the antenna input
G_T = transmitter antenna gain
G_R = receiver antenna gain
λ = wavelength (meters)
R = range (meters)
L_{E1} = one-way excess propagation loss (>1)

The equation for the receiver system noise power is as follows:

$$N = kT_0 BF_n \tag{4.14}$$

where

$kT_0 = -174$ dBm/Hz $= 4 \cdot 10^{-18}$ mW/Hz $= 4 \cdot 10^{-21}$ W/Hz
B = receiver bandwidth (hertz)
F_n = receiver system noise figure (or noise factor) (numeric)

The equation for the signal-to-noise ratio is thus

$$S/N = P_T G_T G_R \lambda^2 / [(4\pi)^2 R^2 L_{E1} kT_0 BF_n] \tag{4.15}$$

where terms are as defined for (4.13) and (4.14).

For an example of the use of (4.15), assume the following:

P_T = transmitter output power = 100W = 50 dBm
G_T = transmit antenna gain = 3 dBi
G_R = receive antenna gain = 3 dBi
λ = wavelength = 0.1m = $-$10 dB
R = range = 100 km = 50 dB
L_{E1} = receive excess propagation loss = 6 dB
kT_0 = reference noise level = $-$174 dBm
B = receiver bandwidth (20 kHz) = 43 dB
F_n = receiver system noise figure = 6 dB

Using those values and (4.15) expressed in decibels, we have

$$S/N = P_T + G_T + G_R + \lambda^2 - (4\pi)^2 - R^2 - L_{E1} - kT_0 - B - F_n$$
$$S/N = 50 + 3 + 3 - 20 - 22 - 100 - 6 + 174 - 43 - 6 = 33 \text{ dB}$$

4.8 RADAR SYSTEM LINK ANALYSIS

The information in this section is based on [3,4].

A radar system link analysis is a set of calculations that shows the signal-to-noise ratio that is achieved at the receiver output based on a given set of assumptions for a radar system. This section shows equations that are used in the radar system link analysis and an example link analysis for a typical radar system.

The equation for received power at the receiver antenna output is as follows:

$$P_R = P_T G_T G_R \lambda^2 \sigma / [(4\pi)^3 R^4 L_{E1} L_{E2}] \tag{4.16}$$

where

P_R = receiver antenna power output
P_T = transmitter output power
G_T = transmitter antenna gain
G_R = receiver antenna gain
λ = wavelength (meters)
σ = RCS (square meters)
R = range (meters)
L_{E1} = transmit excess propagation loss
L_{E2} = receive excess propagation loss

The equation for signal-to-noise ratio with integration gain, G_i, included is as follows:

$$S/N = P_T G_T G_R \lambda^2 \sigma G_i / [(4\pi)^3 R^4 L_{E1} L_{E2} kT_0 BF_n] \tag{4.17}$$

where terms are as previously defined.

Assumptions for the radar system are as follows:

P_T = transmitter output power (100 kW)		80 dBm
G_T = transmitter antenna gain		40 dBi
G_R = receiver antenna gain		40 dBi
λ = wavelength (0.063m)		−12 dB
σ = RCS (1.6 m^2)		2 dB
G_i = integration signal-to-noise gain		8 dB
R = range (100 km)		50 dB
L_{E1} = transmit excess propagation loss		6 dB
L_{E2} = receive excess propagation loss		6 dB
kT_0 = reference noise level		−174 dBm
B = receiver bandwidth (100 kHz)		50 dB
F_n = receiver system noise figure		6 dB

Using those values and (4.16) expressed in decibels, we have

$$S/N = P_T + G_T + G_R + \lambda^2 + \sigma + G_i - (4\pi)^3 - R^4 - L_{E1} - L_{E2} - kT_0 - B - F_n$$
$$S/N = 80 + 40 + 40 - 24 + 3 + 8 - 33 - 200 - 6 - 6 + 174 - 50 - 6$$
$$S/N = 20 \text{ dB}$$

We see that the signal-to-noise ratio changes by 12 dB per octave or 40 dB per decade in range. It is directly proportional to the RCS of the target.

4.9 PERFORMANCE CALCULATIONS FOR RADAR SYSTEMS WITH ELECTRONIC COUNTERMEASURES

When ECM is present with a radar, the jamming signal may be much greater than the noise signal. For that case, a different equation must be used to calculate system performance. In the equation, the signal due to jamming is substituted for $kT_0 BF_n$. The jamming signal is as follows:

$$J = P_J G_J G_{RJ} \lambda^2 B_R / [(4\pi)^2 (R_J)^2 L_J B_J] \tag{4.18}$$

where

P_J = jammer transmitter power
G_J = jammer transmitter antenna gain
G_{RJ} = receiver antenna gain for jammer signal
λ = wavelength
B_R = receiver bandwidth
R_J = jammer range to receiver
L_J = jammer loss in addition to free-space loss between the jammer transmitter and the receiver
B_J = jammer signal bandwidth

The equation for the signal-to-jamming ratio for radar with integration of a number of pulses is as follows:

$$S/J = P_T G_T G_R \sigma G_i (R_J)^2 L_J B_J / [(4\pi)R^4 L_{E1} L_{E2} P_J G_J G_{RJ} B_R] \tag{4.19}$$

where the terms are as defined previously.

For an example of the use of (4.18) for radar jamming, use the following assumptions.

Radar assumptions:
P_T = transmitter output power		80 dBm
G_T = transmitter antenna gain		40 dBi
G_R = receiver antenna gain		40 dBi
σ = RCS		2 dB
L_{E1} = transmit excess propagation loss		6 dB
L_{E2} = receive excess propagation loss		6 dB
B = receiver bandwidth		50 dB
G_i = integration gain of signal-to-noise ratio		8 dB

Jammer assumptions:
P_J = jammer output power		60 dBm
G_J = jammer antenna gain		10 dBi
G_{RJ} = receiver antenna gain seen by jammer		−3 dBi
L_J = jammer excess propagation loss		6 dB
B_J = jammer noise bandwidth (200 MHz)		83 dB
R_J = jammer range (100 km)		50 dB

Expressing (4.18) in decibels and substituting values, we have

$$S/J = P_T + G_T + G_R + \sigma + G_i + (R_J)^2 + L_J + B_J - (4\pi) - R^4 - L_{E1} - L_{E2} - P_J - G_J - G_{RJ} - B_R$$
$$S/J = 80 + 40 + 40 + 3 + 8 + 100 + 6 + 83 - 33 - R^4 - 6 - 6 - 60 - 10 + 3 - 50$$
$$S/J = 198 - R^4$$

If the radar range is 100 km, the S/J ratio is $198 - 200 = -2$ dB. If it is 30 km, the signal-to-jamming ratio is $198 - 179 = 19$ dB.

References

[1] Van Valkenburg, Mac E., ed., *Reference Data for Engineers*, 8th ed., SAMS, Carmel, IN: Prentice Hall Computer Publishing, 1993, Chapter 34, "Radio Noise and Interference" by George W. Swenson, Jr., and A. Richard Thompson.
[2] Kennedy, George, *Electronic Communication Systems*, 3rd ed., New York: McGraw-Hill, 1985, Chap. 2.
[3] Skolnik, Merrill I., *Introduction to Radar Systems*, 2nd ed., New York: McGraw-Hill, 1980, pp. 15–33.
[4] Barton, David K., *Modern Radar System Analysis*, Boston: Artech House, 1988, p. 13–18.

CHAPTER 5

Modulation Techniques

Modulation techniques can be either analog or digital. An analog modulation scheme is one in which the carrier signal is modulated or changed in amplitude, frequency, or phase in a way that is directly proportional to the amplitude of the modulating signal. This type of system is continuously variable. A digital modulation system is one in which the signal to be sent is first sampled on a periodic basis and converted to a digital signal using an analog-to-digital (A-to-D) converter. The digitized signal then is used to modulate the carrier signal in amplitude, frequency, phase, or a combination thereof. Both modulation techniques are used for communications and are discussed in this chapter, as are modulation types for radar.

5.1 PULSED CONTINUOUS-WAVE SIGNALS

Figure 5.1 shows the waveform and frequency spectrum for a pulsed CW signal. This type of modulation is used for CW telegraphy, pulsed radar, and many other pulse-type systems. The frequency spectrum for rectangular-shaped RF pulses in a pulse train is a $(\sin x)/x$-type voltage spectrum or a $[(\sin x)/x]^2$ power spectrum. The main spectral lobe (maximum amplitude lobe) is centered at the carrier frequency. It has a width at the nulls of two times the reciprocal of the minimum RF pulse width. For example, if the RF pulse width is 1 μs, the width of the main spectrum lobe at the nulls is 2 MHz. Each of the sidelobes has half that width between nulls. The first sidelobe is reduced in amplitude by about 13.5 dB from the main spectral lobe. Higher-order sidelobes are progressively lower, falling off at a rate of about 20 dB per decade. The second and third sidelobes are down from the main lobe peak by about 17.9 dB and 20.8 dB, respectively.

Figure 5.1 shows only the envelope of the spectral lines, which are separated by the pulse repetition frequency. Thus, if the pulse repetition frequency is 1,000 Hz, there would be spectral lines every 1,000 Hz. The required −3 dB bandwidth for a pulsed CW signal as used for radar is approximately the reciprocal of

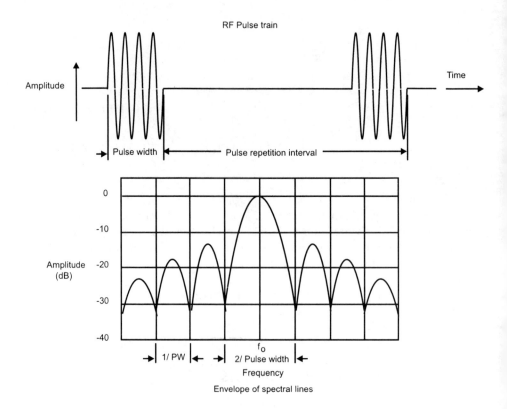

Figure 5.1 Waveform and frequency spectrum for a pulsed CW signal.

the pulse width. Thus, if the RF pulse width is 1 μs, the required bandwidth is about 1 MHz.

Figure 5.1 also can be used to show the spectrum for pulsed phase modulation and phase code modulation. In phase code modulation, the pulse width is the shortest phase change interval.

5.2 CONVENTIONAL AMPLITUDE MODULATION

Some of the material presented in this and the following four sections is quoted or adapted from [1].

Figure 5.2 shows the waveform and the frequency spectrum for conventional AM as used for standard AM radio. The modulating signal in the figure is a single sine wave with 100% modulation. The RF frequency is constant, and the amplitude is made proportional to the amplitude of the information being sent.

The carrier amplitude may be indicated as V_c and the modulating signal amplitude as V_m. The ratio V_m/V_c is the modulation index (M) and is a number

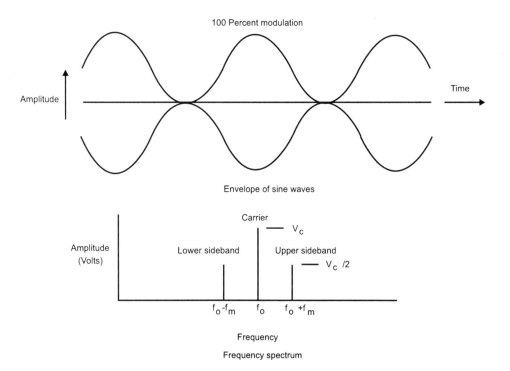

Figure 5.2 Waveform and frequency spectrum for a conventional AM signal with single sine wave modulation and 100% modulation.

between 0 and 1. This ratio often is expressed as a percentage modulation varying between 0% and 100%. This is an example of analog communication in which signals are continuously variable rather than being digital or on-off in nature.

The relative power in the sidebands and the carrier depends on the percentage of modulation. If we had 100% modulation, the voltage of each sideband for single sine wave modulation would be one-half the voltage of the carrier. The power in each sideband therefore would be one-fourth the carrier power, and the total sideband power would be one-half the carrier power. The maximum power in the sidebands is thus only one-third the total power transmitted. Since it is only the sideband power that carries information, this is very poor efficiency compared to FM and some other modulation methods.

Figure 5.3 shows conventional amplitude modulation with 50% modulation (M = 0.5). Again, the modulating signal is a single sine wave. The sideband voltage

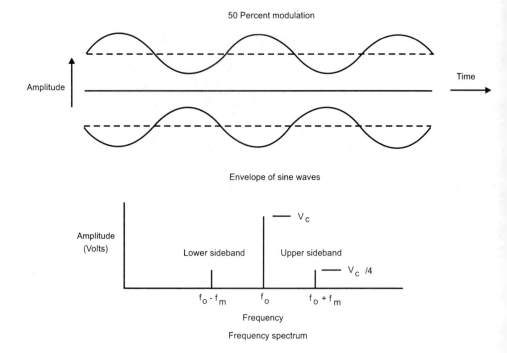

Figure 5.3 Waveform and frequency spectrum for a conventional AM signal with single sine wave modulation input and 50% modulation.

amplitude for each sideband is one-fourth the amplitude of the carrier. The sideband power for each sideband is one-sixteenth the power of the carrier, and the total sideband power is one-eighth the carrier power.

If we had only 10% modulation, the voltage of each sideband for single sine wave modulation would be only one-twentieth the voltage of the carrier, and the power in each sideband would be only one-four hundredth the power in the carrier. The efficiency in this case would be very low.

The required bandwidth for conventional AM is typically twice the highest modulating frequency that we wish to pass. For example, with telephone and many voice communication radio systems, it usually is adequate to pass only a 3-kHz (300 to 3,300 Hz) bandwidth. The two sidebands on each side of the carrier thus would have a maximum width of 3 kHz and the total bandwidth would be 6.6 kHz.

5.3 DOUBLE SIDEBAND SUPPRESSED CARRIER MODULATION

Figure 5.4 shows the case of double-sideband (DSB) suppressed carrier modulation for a single sine wave modulation signal. The frequency spectrum thus shows only a lower sideband and an upper sideband but no carrier line.

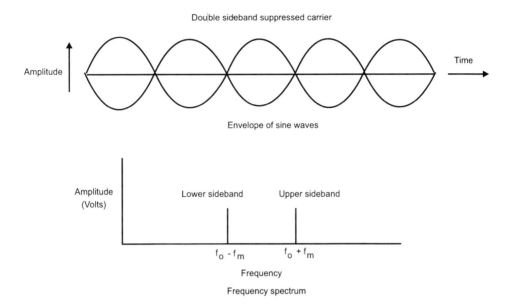

Figure 5.4 Waveform and frequency spectrum for a DSB suppressed carrier modulation (AM) signal with single sine wave modulation.

The DSB signal usually is produced by a modulator known as a balanced modulator, which rejects the carrier frequency component. The operation of this circuit is as follows. Assume that the carrier input signal is sinusoidal and appears at the local oscillator (LO) input of the modulator as $\cos \omega_c t$ and that the baseband input signal to the modulator is $\cos \omega_m t$. The modulator performs a multiplication of the input signals, and the output is

$$[\cos \omega_c t][\cos \omega_m t] = 1/2[\cos (\omega_c + \omega_m) t + \cos(\omega_c - \omega_m) t] \qquad (5.1)$$

We see that the modulator has generated the sum and difference frequencies. It should be pointed out that that is the same circuit or device that could be used for a mixer. The balanced mixer has as one of its inputs an LO signal and as the other signal a signal that we want to frequency convert. The mixer generates sum and difference frequencies as indicated by (5.1). A filter is used to select which of the two output frequencies we want to pass. If the output selected is the sum frequency, we say that the system is a frequency up-converter. If the output selected is the difference frequency, we say that the system is a frequency down-converter. (More details about mixers are presented in Chapter 6.)

5.4 VESTIGIAL SIDEBAND MODULATION

Vestigial-sideband (VSB) modulation may be derived from a conventional AM signal by filtering out part of one of the two sidebands as well as part of the carrier. Figure 5.5 shows the frequency spectrum for VSB as used for television video transmission. We see that 1.25 MHz of the lower sideband is transmitted with about 0.75 MHz of the lower sideband passed undiminished. This makes sure that the lowest frequencies in the wanted upper sideband are not distorted in phase by the VSB filter. By using this method, 3.0 MHz of the needed frequency spectrum is saved. Thus, instead of requiring 9-MHz bandwidth, as would be the case for AM, it requires only 6.0 MHz [1].

5.5 SINGLE-SIDEBAND MODULATION

One of the more important types of analog modulation used for communications is single-sideband (SSB) modulation. This type of amplitude modulation is much more efficient than standard AM because it is not necessary to transmit the full carrier. It also uses only half the frequency bandwidth.

Figure 5.6 shows the spectra for two types of SSB modulation. For simplicity, a single sine wave modulating signal is assumed. Figure 5.6(a) shows an SSB reduced carrier. In this case, sufficient carrier is transmitted to permit the receiver to have a carrier reference on which to lock. The receiver then generates the necessary carrier signal for demodulation of the SSB signals.

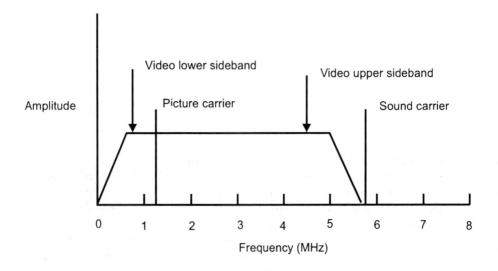

Figure 5.5 Spectrum for transmitted video signals for television. (*After:* [1].)

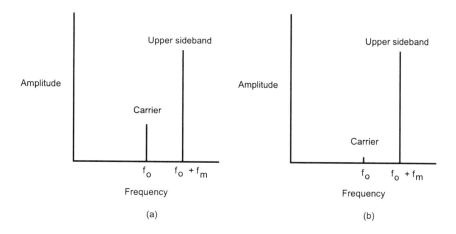

Figure 5.6 SSB modulation: (a) SSB reduced carrier frequency spectrum and (b) SSB suppressed carrier frequency spectrum.

Figure 5.6(b) shows the second type of SSB modulation, SSB suppressed carrier. In this case, there is essentially no carrier signal for the receiver to lock on.

In some cases, SSB modulation can be produced with a balanced modulator and a bandpass filter. The balanced modulator eliminates the carrier; the filter must eliminate nearly all the signals on one side of the carrier. In some cases, that is done easily. In other cases, adequate filtering is difficult to realize. That is especially true for the case of low baseband frequencies that are very close to the carrier. An alternative approach for SSB modulation is use of a phase cancellation method. Both methods are discussed in detail in Chapter 7.

5.6 STANDARD FREQUENCY MODULATION

With standard FM, the waveform amplitude is constant, and the RF frequency is made proportional to the amplitude of the information being sent. This type of modulation is used for voice for television, broadcast FM radio, and many other types of communication systems. The frequency spectrum for FM is more complex than that for AM. In the case of a simple single sine wave modulation signal, the FM voltage is given by

$$v = A \sin(w_c t + m_f \sin \omega_m t) \tag{5.2}$$

where

A = magnitude

$\omega_c = 2\pi \times$ carrier frequency

$\omega_m = 2\pi \times$ modulating frequency

t = time

m_f = modulation index for FM

= frequency deviation / modulating frequency

Note that we have the case of a sine of a sine. This requires Bessel functions of the first kind for solution. An expansion of (5.2) using Bessel functions is shown as (5.3).

$$v = A\{J_0(m_f)\sin \omega_c t$$
$$+ J_1(m_f)[\sin(\omega_c + \omega_m)t - \sin(\omega_c - \omega_m)t]$$
$$+ J_2(m_f)[\sin(\omega_c + 2\omega_m)t - \sin(\omega_c - 2\omega_m)t]$$
$$+ J_3(m_f)[\sin(\omega_c + 3\omega_m)t - \sin(\omega_c - 3\omega_m)t] \ldots\} \quad (5.3)$$

It can be seen that the output consists of a carrier and an apparently infinite number of pairs of sidelobes, each preceded by a J coefficient. These are Bessel functions of the first kind and of the order denoted by the subscript, with the argument m_f. The spacing between spectral lines is the modulation frequency. Thus, if the modulation frequency is 3 kHz, the spacing between spectrum lines is 3 kHz. Fortunately, only a limited number of spectral lines are large enough in amplitude to be important. It is important to note that the carrier term is not always the largest; in fact, it can be zero amplitude.

Table 5.1 shows Bessel functions of the first kind as a function of the modulation index, m_f. The number of lines on either side of the carrier that are significant are shown by this plot as well as the magnitude and sign (phase) of the components.

For an example of the use of Table 5.1, assume a narrowband FM system such as used for many types of voice communication. The maximum deviation is assumed to be 5 kHz, and the modulation frequency is assumed to be 1 kHz. The FM modulation index is the deviation divided by the modulation frequency, so the modulation index for this example is 5.0. From Table 5.1, we see that there are six spectrum lines with significant amplitudes on either side of the carrier. Because each line is spaced by the modulation frequency of 1 kHz, the required bandwidth to pass all spectrum lines shown is 12 kHz.

Some of the terms in Table 5.1 are negative and some are positive. A negative sign indicates a change of phase by 180 degrees. That is the case for the carrier and the first sidebands for the example in the preceding paragraph. The rule of thumb known as *Carson's rule* states that the approximate bandwidth required to pass an FM signal is twice the sum of the deviation and the highest modulating

Table 5.1
Bessel Functions of the First Kind

x	J0	J1	J2	J3	Order (n) J4	J5	J6	J7	J8	J9
0.00	1.00									
0.25	0.98	0.12								
0.5	0.94	0.24	0.03							
1.0	0.77	0.44	0.03							
1.5	0.51	0.56	0.23	0.01						
2.0	0.22	0.58	0.35	0.13	0.03					
2.5	−0.05	0.50	0.45	0.22	0.07	0.02				
3.0	−0.26	0.34	0.49	0.31	0.13	0.04	0.01			
4.0	−0.40	−0.07	0.36	0.43	0.28	0.13	0.05	0.02		
5.0	−0.18	−0.33	0.05	0.36	0.39	0.26	0.13	0.05	0.02	
6.0	0.15	−0.28	−0.24	0.11	0.36	0.36	0.25	0.13	0.06	0.02
7.0	0.30	0.00	−0.30	−0.17	0.16	0.35	0.34	0.23	0.13	0.06
8.0	0.17	0.23	−0.11	−0.29	−0.10	0.19	0.34	0.32	0.22	0.13
9.0	−0.09	0.24	0.14	−0.18	−0.27	−0.06	0.20	0.33	0.30	0.21

frequency. Using the previous example and Carson's rule, the required bandwidth to pass this FM wave is 2 × (5 + 1) = 12 kHz, the same answer that we obtained using the Bessel plot.

Figure 5.7 shows plots of the FM frequency spectrum for modulation index values of 0.5, 2.5, and 5.0 with amplitudes determined by Table 5.1 [1]. The modulation frequency is assumed constant at 1 kHz for those plots.

Wideband frequency modulation is used for FM broadcast and television. Maximum permissible deviation is 75 kHz for FM broadcast. The permissible deviation for the sound accompanying television transmissions is 25 kHz in the United States and 50 kHz in Europe. The modulation frequency range for FM broadcast is 30 Hz to 15 kHz. Thus, the maximum modulation index ranges from 5 to 2,500 in entertainment broadcasting.

Narrowband FM with modulation bandwidths of 3 kHz or less is used by the so-called FM mobile communication services, which include police, ambulance, taxi cabs, fire departments, radio-controlled appliance repair services, ground-to-air communication services, and short-range ship-to-shore services. The higher audio frequencies (above 3 kHz) are attenuated as they are in most long-distance telephone systems, but the resulting speech quality is still perfectly adequate. Maximum deviations of 5 to 10 kHz are permitted, and channel spacing is not much greater than for AM broadcasting (of the order of 15 to 30 kHz) [1].

Preemphasis and deemphasis are used with all FM transmissions. Noise has a greater effect on the higher modulating frequencies than on the lower ones. If the higher audio frequency signals are artificially boosted in amplitude at the

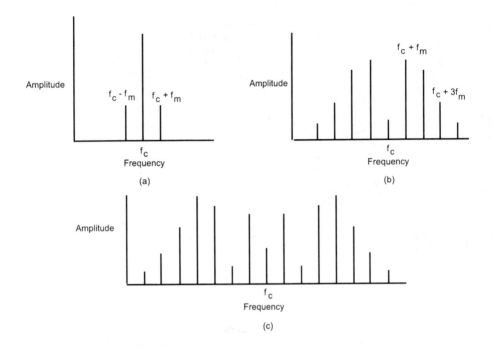

Figure 5.7 Example frequency spectrura for FM: (a) spectrum for a modulation index of 0.5; (b) spectrum for a modulation index of 2.5; and (c) spectrum for a modulation index of 5.0.

transmitter and correspondingly cut in amplitude at the receiver, an improvement in noise immunity can be expected [1]. This boosting in amplitude of the higher frequencies is called preemphasis. The corresponding reduction in amplitude of the higher frequencies at the receiver is called deemphasis.

Other FM spectra of interest are those for frequency sweep modulation (chirp modulation), as used for radar. They have complex multiple-line spectra with frequency components spaced by the pulse repetition frequency and a spectral width somewhat greater than two times the frequency deviation.

5.7 MODULATION FOR TELEMETRY

Some of the following material is adapted from [2].

Telemetry, as the name suggests, consists of performing and reporting measurements on distant objects. With radio telemetry, there are always a number of channels of information that must be transmitted in parallel. Thus, either frequency division multiplexing (FDM) or time division multiplexing (TDM) is used. With FDM, a number of separate frequency channels are used in parallel to transmit the various channels of data. With TDM, the data alternates, using a single channel.

First data channel 1 uses the single-frequency channel, then data channel 2 uses the frequency channel, then data channel 3 uses the frequency channel, and so on, until all data channels have been sampled. Then the process repeats.

In many telemetry systems, there is a need for both narrowband channels and relatively wideband channels. The mixture is achieved by a process known as subcommutating. Subcommutating consists of taking one of the wideband channels and subdividing it into several narrowband ones. For example, a system might use FDM in general and TDM for the subcommuted channels.

A large number of different types of telemetering systems have been used in the past. The FM/FM FDM system is used quite widely in the United States and is extended to pulse amplitude modulation (PAM)/FM/FM when subcommutating is required. It has the advantages of reliability and flexibility but requires greater bandwidth and carrier strength than the purely pulse systems. Systems such as pulse position modulation (PPM)/AM, pulse width modulation (PWM)/FM, PAM/FM, and pulse code modulation (PCM)/FM all have been used in certain applications, with varying advantages for each.

In PPM, the amplitude and the width of the pulse are kept constant, while the position of each pulse is varied with respect to a reference pulse. The position delay is proportional to the amplitude of the sampled signal. In PWM, the amplitude and the starting time for each pulse are fixed, but the width of each pulse is made proportional to the amplitude of the sampled signal. In PAM, the pulse width and the starting time are constant, and the amplitude of the pulse is made proportional to the amplitude of the sampled signal. In PCM, the total amplitude range that the signal may occupy is divided into a number of standard levels. The amplitude level of the sampled signal is transmitted in a binary code. For example, if 128 levels are used, it would take seven pulses of 1s or 0s to represent the amplitude of the sampled signal.

5.8 COMBINATION COMMUNICATION AND RANGE-MEASUREMENT SYSTEMS

Phase-shift keying and high-bit-rate pseudorandom code modulation sometimes are used in systems in which both communication and range measurement are required. Examples are the Position Location and Reporting System (PLARS) and the Joint Tactical Information Distribution System (JTIDS) military communication systems. Very short duration, minimum phase-change intervals, and complex pseudorandom codes are used for modulation, producing a spread spectrum signal in which the energy is spread over very large bandwidths. The use of spread spectrum provides reduced vulnerability to jamming and interference by forcing a potential jammer to spread its energy over a large bandwidth. While the communication or radar system using the spread spectrum modulation also has its energy spread over a large bandwidth, it has the advantage that coherent integration is used to provide an effective narrow bandwidth.

Spread spectrum modulation using binary phase-shift keying (BPSK) or quaternary phase-shift keying (QPSK) with direct sequence pseudorandom code modulation also is used in some communication systems in which range measurement is not necessary. For those cases, a noise-like spectrum is produced that has antijam features and low observable signal capability.

As mentioned earlier in this chapter, simple phase-shift modulation that does not use wave shaping has a $[\sin x/x]^2$ spectrum with the first sidelobes down 13.5 dB and a rolloff rate of approximately 6 dB per octave (20 dB per decade) of frequency. The null-to-null main lobe frequency spectrum width is twice the clock frequency. Thus, with a code clock rate of 100 MHz, the null-to-null main lobe frequency spectrum width is 200 MHz. The −3 dB spectral bandwidth is 0.88 times the code clock frequency. For a code clock rate of 100 MHz, the −3 dB spectral bandwidth would be 88 MHz.

In the case of JTIDS, it was determined to be necessary to provide a rolloff rate faster than 6 dB per octave and lower sidelobes due to interference with other communication systems. The solution was to use a type of phase-shift keying called minimum shift keying (MSK). With this type of phase-shift keying, the phase is not changed abruptly at each phase change position in the code but is changed over a period of time. The result is a somewhat narrower main lobe (1.5 times code frequency for null-to-null main lobe width rather than 2.0 times code frequency) and a falloff rate of 12 dB per octave (40 dB per decade).

Spread spectrum phase-shift modulation may also find important applications for radar. It is one way to achieve high range resolution with long pulses, thereby providing large average power from a peak power-limited system.

5.9 MODULATION FOR RADAR

5.9.1 Pulsed CW Modulation

The most common waveform for radar is pulsed CW modulation. For example, a radar might use 1-μs duration pulses of RF at a pulse repetition frequency (PRF) of 1,000 pulses per second. The pulse repetition interval (PRI) would be 1,000 ms. The number of RF cycles in the 1-μs pulse would depend on the radar RF frequency. If, for example, the radar operated at a frequency of 5 GHz, there would be 5,000 cycles of RF in the 1-μs pulsed waveform. Typical radar pulse power output is in the range of 100 kW to 5 MW.

Figure 5.1 showed the waveform and frequency spectrum for a pulsed CW signal. The spectra are $(\sin x)/x$-type voltage spectra or $[(\sin x)/x]^2$ power spectra, with the maximum main spectrum lobe width at the nulls equal to two times the reciprocal of the minimum RF pulse width. For example, if the RF pulse width is 1.0 μs, the width of the main spectral lobe at the nulls would be 2.0 ms. The sidelobes each have half that width. The first sidelobe is down in amplitude by about 13.5 dB from the main spectrum lobe. Higher-order sidelobes are progressively lower, falling off at a rate of about 20 dB per decade. The second sidelobe is down by 17.9 dB, the third side lobe is down by 20.8 dB, and the fourth side lobe is down by 23.0 dB.

Spectral lines are separated by the PRF. Thus, if the PRR is 1,000 Hz, there would be spectral lines every 1,000 Hz. The required bandwidth for this type of modulation is approximately the reciprocal of the pulse width.

For most radar and communication system applications, it is not possible or desirable to use a rectangular pulse with zero rise time and zero fall time. The more common approach is to provide a nearly trapezoidal pulse. Such a pulse has the advantage of reducing interference to other systems by reducing the energy that is transmitted in far-out sidelobes.

A knowledge of the exact spectrum for the trapezoidal pulse usually is not required. What is important is to know the envelope of the spectrum so that the maximum level of sidelobe signals can be determined as a function of frequency. This envelope can be determined using (5.4) and (5.5) [3].

$$F_1 = 1/\pi[\tau + (D_1 + D_2)/2] \tag{5.4}$$

where

F_1 = first frequency break point (megahertz)
D_1 = rise time of trapezoidal pulse (microseconds)
D_2 = fall time of trapezoidal pulse (microseconds)
τ = −3 dB pulse width (microseconds)

The equation for F_2 is as follows:

$$F_2 = (1/D_1 + 1/D_2)/2\pi \tag{5.5}$$

where

F_2 = second frequency break point (megahertz)
D_1 = rise time of trapezoidal pulse (microseconds)
D_2 = fall time of trapezoidal pulse (microseconds)

The predicted envelope of the frequency spectral has a 0-dB per frequency decade slope from the spectrum center to F_1. The envelope then falls off at a rate of 20 dB per decade (6 dB per octave) between F_1 and F_2. At frequencies greater than F_2, the envelope amplitude falls off at a rate of 40 dB per decade, or 12 dB per octave. The three examples in Figure 5.8 show the use of (5.4) and (5.5) in

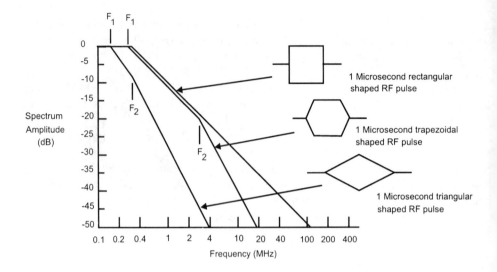

Figure 5.8 Examples of spectrum envelopes for RF pulses.

predicting the envelope of the frequency spectrum for pulse radar. The first example is the rectangular pulse with zero rise and fall times. For the case of a 1-μs pulse width, the value of F_1 is computed to be 0.318 MHz. The value of F_2 is infinity. Between those values, the slope of the envelope is 20 dB per decade. For this example, the spectrum is down only 30 dB at 10 MHz and 50 dB at 100 MHz on either side of the main spectral lobe.

In the second example in Figure 5.8, the same 1-μs half-power pulse width is assumed, but the rise and fall times are each 0.1 μs. For this case, F_1 is computed to be 0.289 MHz, and F_2 is 3.183 MHz. At frequencies greater than F_2, the falloff rate for the spectrum envelope is 40 dB per decade. The spectral envelope has dropped to about −65 dB at 40 MHz offset from the carrier or spectral center. In the third example in Figure 5.8, the same 1-μs pulse is assumed, but this time the rise and fall times are each 1 μs. We thus have a triangular-shaped pulse. For this case, the value of F_1 is calculated to be 0.159 MHz, and the value of F_2 is 0.318 MHz. We see from the plot that the spectrum envelope has dropped to −50 dB at 4.0 MHz from the spectral center.

5.9.2 High-Power Impulse Generators and Ultra-Wideband, High-Power Microwave Generators

Recent developments in high-power impulse generators and ultra-wideband, high-power microwave (HPM) generators have made possible some new types of pulsed CW radars with very short duration pulses and wide frequency spectra. In the case of impulse type generators, it is possible to generate pulses with peak powers of more than 10^{10}W with pulse widths of less than 1 nanosecond (10^{-9} sec). The spectrum for the generated pulse that feeds the antenna has its peak at 0 frequency and frequency lines that extend to many gigahertz. If we assume a 0.5-ns pulse width, the first null in the spectrum is at 2 GHz, the second null is at 4 GHz, the third null is at 6 GHz, and so on.

Very wideband antennas are used with video pulse systems. A TEM horn antenna is one of the better antennas for such systems. A typical lower frequency for such a system is 50 MHz. The result is that the antenna converts the pulse to a single-cycle RF pulse with some ringing. The result is a modified, but very wide, frequency spectrum. The HPM systems also generate very high powers (10^9W is typical). An RF pulse is generated at microwave frequencies, which may contain only a few RF cycles and may have pulse durations of only a few nanoseconds. Again very wide frequency spectra are generated.

It is expected that there may be many applications for very short duration, extremely high power pulse systems. A chief advantage for such systems, in the case of radar, is very high range resolution. These very high power, short pulse systems also may have important applications as impulse jammers. They also may be used for "zapping" electronic components via electromagnetic pulse (EMP).

5.9.3 Chirp Pulse Modulation

Another type of modulation often used when high range resolution is needed is known as chirp. The RF frequency is swept in a linear fashion during the time that the pulse is present.

As an example of a chirp pulse modulation, a 10-μs pulsed RF signal might be used with the frequency changed from 5.0 GHz to 5.5 GHz during the 10-μs period. That produces a wideband signal with a frequency spectral width in excess of 0.5 GHz. The received chirp signal may be processed by a pulse compression filter, yielding a very short duration output pulse. That type of pulse has the advantage of providing high range resolution with signal processing gain.

5.9.4 Phase Code Modulated Pulse Modulation

A third type of pulsed RF waveform that can be used for improved resolution is a phase code modulated pulse of RF. This type of waveform is shown in Figure 5.9. A biphase modulated system would use two phase states, 0- and 180-degree phase shift. Phase change intervals could be as short as a few nanoseconds or could be much longer, depending on the code used and the duration of the pulsed RF signal. When the reflected signal is received, it can be processed by a phase detector and matched filter to produce a very short duration output pulse.

The type of code illustrated in Figure 5.9(a) is a seven-unit Barker code, only one of many possible codes that can be used. It has the advantage that the matched

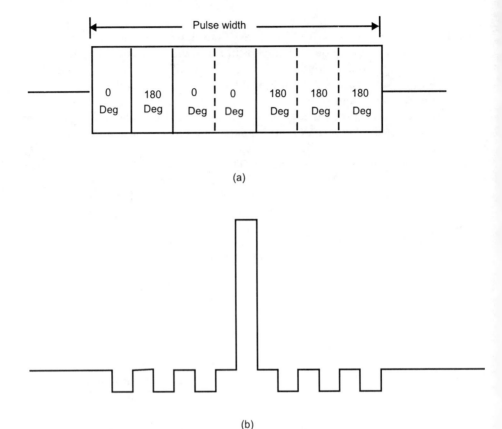

Figure 5.9 Example of phase code modulated pulse for radar: (a) details of modulated pulse and (b) matched filter response following detection.

filter detected output will have all sidelobes at unity level in one direction, and the main-lobe seven units high in the opposite direction. The detected output is illustrated in Figure 5.9(b). When received and processed by the matched filter, the main output pulse will be one-seventh the width of the input pulse.

Other longer codes usually are used for larger pulse compression ratios and larger integration improvement. The longest Barker code is a 13-unit code. The peak-sidelobe ratio in that case is $-20 \log N = -22.3$ dB. Other types of codes are much longer than Barker codes and can be used if larger compression ratios are desired.

5.9.5 Continuous-Wave Modulation

Radars also may use CW modulation, in which case the amplitude, frequency, and phase are held constant. This type of modulation does not permit range measurements, but it does permit angular measurements and measurements of radial velocity.

5.9.6 Frequency-Modulated CW Modulation

Another type of waveform used for radar is FMCW modulation. The amplitude of the wave is held constant and the frequency is swept from a low frequency to a high frequency within a band in a sawtooth fashion. FMCW modulation permits measurements of range as well as measurements of angle and radial velocity. Figure 5.10 illustrates concepts for FMCW radar. The range measurement is proportional to the frequency difference between the signal being transmitted and the received signal reflected from the target.

5.10 SINGLE-CHANNEL TRANSMITTER SYSTEM

Figure 5.11 shows a simplified system block diagram for a single-channel transmitter system. The input system is assumed to be a microphone for voice. The output of the input device is at baseband frequencies and is assumed to be bandpass or lowpass filtered to a frequency band of 300 to 3,300 Hz. The frequency band is

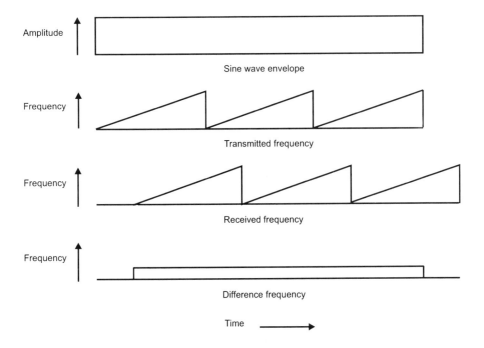

Figure 5.10 Example of FMCW modulation for radar.

132 | RF Systems, Components, and Circuits Handbook

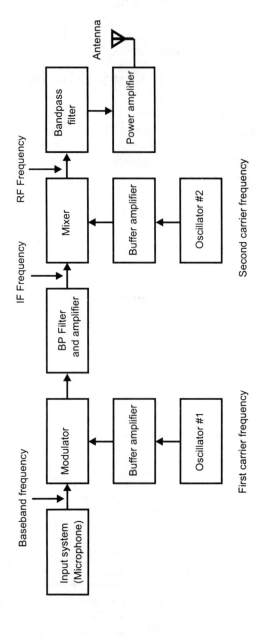

Figure 5.11 Single-channel transmitter system block diagram.

consistent with the bandwidth of a telephone channel. The baseband signal is fed as one input to a modulator. The second input is a local oscillator operating at an IF frequency, for example, 30 MHz. The carrier is modulated by the baseband signal. The modulation type may be amplitude, frequency, or phase modulation.

The output of the modulator is first amplified and filtered, then fed as one input to a mixer circuit that acts as a frequency up-converter. The second input to the mixer is the output of an RF oscillator, commonly referred to as the LO. A buffer amplifier is used between the oscillator and the mixer for reverse isolation and frequency pulling considerations.

The output of the mixer is an RF signal with the desired carrier frequency being the sum of the two oscillator frequencies. For example, if the frequency of the oscillator that feeds the modulator is 30 MHz and the frequency of the second oscillator that feeds the mixer is 120 MHz, the bandpass-filtered mixer output will be 150 MHz.

There also will be undesired outputs from the mixer, including the difference frequency, which for this case would be 120 MHz minus 30 MHz or 90 MHz, the two oscillator frequencies, harmonics of the oscillator frequencies, and spurs or intermodulation products. The spurs or intermodulation products can be the result of mixing of harmonics with fundamental signals and can include both sum and difference frequencies. The mixer type is chosen to eliminate, to the extent possible, the two oscillator signals by using balanced mixers. The bandpass filter also is designed to suppress harmonics, fundamental frequencies, difference frequencies, and spurs.

The modulated RF signal is then fed to a power amplifier chain, where the power level is increased to the desired transmitter output power level. The output power may be 100W or more. The output of the power amplifier feeds the transmit antenna.

5.11 FREQUENCY DIVISION MULTIPLEX TRANSMITTER SYSTEM

Some of the material presented in this section and in Sections 5.12 and 5.13 is quoted or adapted from [4].

In communication systems such as microwave relay systems, satellite relay systems, and undersea cable systems, the requirement is that many hundreds of telephone channels or other signals be transmitted on a single communication link. One way to accomplish that task is to use FDM. For such types of systems, many subchannels are used in parallel in the main allocated channel. In the past, analog modulation has been used for each of those subchannels. Modulation types are often SSB suppressed carrier AM or narrowband frequency modulation FM.

Figure 5.12 shows a simplified system block diagram for a FDM transmitter. The figure assumes that the inputs are telephone channels. Each channel is fed to a modulator, which has as its second input a CW signal from an LO. Each oscillator

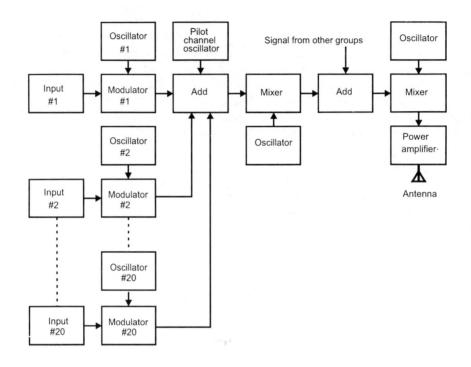

Figure 5.12 Simplified FDM system block diagram.

has a different frequency to provide the necessary separation of frequency channels. A typical group of modulators would have about 20 channels and 20 different crystal oscillators. Crystal oscillators are used rather than LC oscillators because of greatly improved frequency accuracy and stability. The oscillator frequencies typically may be separated by only about 10 kHz.

The outputs of each of the modulators in the group plus a pilot channel oscillator signal are added using an RF add circuit. The combined signal feeds a mixer circuit acting as a frequency up-converter.

The next step is to combine the outputs from the different groups of modulators. The signal from each group is different, since the mixers associated with each group use different oscillator frequencies. The combined signal is then frequency converted by another mixer to the final microwave frequency band used by the system. A power amplifier chain is used to increase the power level of the signal to the desired transmitter output level. The power amplifier connects to the transmit antenna.

5.12 SAMPLE CIRCUITS AND ANALOG-TO-DIGITAL CONVERTER CONCEPTS

Figure 5.13 shows an analog signal being sampled by a gate circuit. The output of the gate is connected to an A-to-D converter system. The sampling rate must be

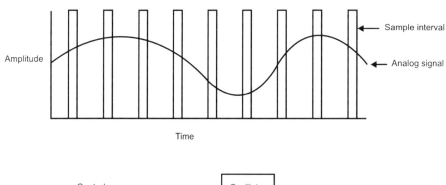

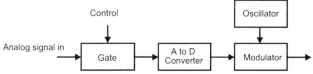

Figure 5.13 Sample and A-to-D converter concepts.

more than twice the highest frequency component in the analog signal and is termed the Nyquist sampling rate. Higher sampling rates are often used. The measured amplitude is expressed as a binary code with the number of bits required depending on the necessary resolution. For example, if 128 amplitude levels are used, the required number of bits per sample is 7, that is, $2^7 = 128$. The more bits used, the more accurate will be the representation of the analog signal. Using more bits, however, means more bandwidth is required and therefore lower signal-to-noise ratio.

The binary bits next are fed to a modulator. The modulator is frequently a phase modulator in which the carrier frequency is supplied by an IF signal oscillator, and the phase of the signal with respect to the oscillator signal is determined by the digital signal.

5.13 TIME DIVISION MULTIPLEX TRANSMITTER SYSTEM WITH PULSE CODE MODULATION

Figure 5.14 shows key elements and concepts for a TDM transmitter system with PCM. In communication systems such as microwave relay systems, the requirement is that many hundreds of telephone channels or other signals be transmitted on a single communication link. A second way to do that, in addition to FDM is to use TDM. This approach has been found to be superior to FDM. Advantages include reduced cost, improved reliability, and improved signal-to-noise ratio. For that reason, many FDM systems used for long-distance telephone service recently have been replaced by TDM systems.

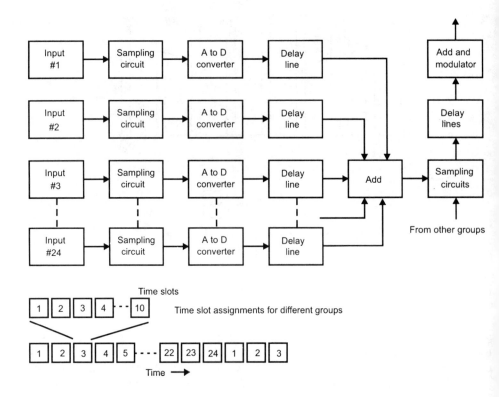

Figure 5.14 TDM system with pulse code modulation.

With TDM, a single wideband channel is used for the many hundreds of input channels by assigning time slots to each of the input channels of data. In essence, the system takes turns using the channel. This concept of time sharing the channel is illustrated at the bottom of Figure 5.14, which shows 24 major time slots and 10 time slots in each major time slot. Thus, there are a total of 240 time slots for digital data. The repetition rate for major time slots must be greater than twice the highest frequency component in the signals that are being sampled and A-to-D converted. Thus, if we are dealing with telephone signals with 3.3 kHz as the highest frequency component, the time-slot repetition rate must be greater than 6.6 kHz.

A simplified system block diagram for a TDM transmitter of this type is shown at the top of Figure 5.14. Inputs are fed to gate circuits to select the times for viewing the signals. The gates are followed by A-to-D converters and delay lines for spacing the digital data in time. The 24 delay line outputs are combined using an add circuit. A sampling circuit and set of delay lines follow that add circuit and are used to place digital data pulses from the different groups in their proper time slots in each major time slot. For simplicity, Figure 5.14 does not show the other elements of the system that follow the modulator. Those elements include a mixer and associated oscillator, a power amplifier chain, and the transmitter antenna.

5.14 TWO-STATE MODULATION TYPES FOR BINARY SIGNALS

A number of two-state modulation types can be used for binary signals. Three such systems are illustrated in Figure 5.15: on-off amplitude keying, frequency shift keying, and two-state phase-shift keying.

5.14.1 On-Off or Two-State Amplitude Keying

One possible two-state modulation type for binary signals is on-off or two-state amplitude keying, one of the earliest types of modulation used. It has been used in the past for such communication modes as telegraphy and teletype or telex. In modern systems, two-state and four-state amplitude modulation can be used in combination with pulse phase modulation to increase the number of bits per baud or bits per pulse that are sent.

5.14.2 Frequency Shift Keying

Frequency shift keying (FSK) is a type of frequency modulation in which only two distinct frequency states are used. The amplitude of the RF waveform is constant.

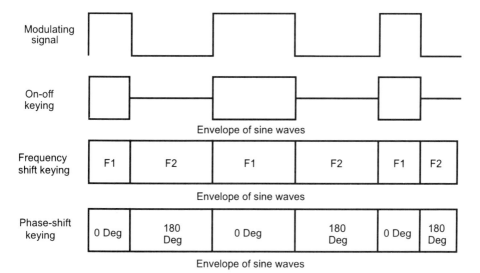

Figure 5.15 Two-state modulation types for binary signals.

In practice, the frequency shift is only a few kilohertz [1]. FSK is used for transmission of many types of digital data. One important application is for telegraphy and telex, two forms of communication that employ typewriter-like machines operating at a maximum speed of about 60 words per minute to send written messages from one point to another [1].

5.14.3 Binary Phase-Shift Keying

BPSK is another form of two-state modulation. Both the frequency and the amplitude remain constant, and only the phase of the RF waveform is changed. The two states are 180 degrees apart. The reference normally is a signal derived from an oscillator and is sometimes referred to as coherent-phase signal. A similar reference is used in both the transmitter and the receiver.

It is also possible to send BPSK as a differential phase-shift signal. The information sent is determined by the phase change between successive pulses. If there is no change in phase, the signal may be called a "zero." If there is a 180-degree change in phase, the signal may be called a "one." With differential phase-shift modulation, the phase reference at the receiver is the signal itself delayed by one pulse period. This delay may be achieved by either analog delay lines or digital delay lines.

The coherent method of BPSK is known to be about 1 dB better in terms of signal-to-noise ratio than that of the differential method of BPSK. The advantage of using a coherent reference approaches 3 dB for multilevel modulation. Accordingly, coherent detection is preferred in those applications in which small losses in signal-to-noise ratio are significant, as in the case of the downlink for satellite communication [3].

In practice, we seldom have a true noise-free coherent reference. Usually, the phase reference uses a phase-locked loop (PLL) oscillator that has some jitter with respect to the transmitter reference. The result is that only partially coherent reception actually can be claimed [3].

In some cases, the differential-phase system has advantages over the coherent system. One case is when there are multipath induced phase changes during the signal pulse period. In that case, having a fixed phase-reference is not as good as having a reference that has been effected by the same multipath phase changes as the signal being detected.

5.15 FOUR-STATE AND EIGHT-STATE PHASE-SHIFT KEYING

It is also possible to use a four-state modulation with the four states being 90 degrees apart. An example would be +0, +90, ±180, and −90 degrees. This type of modulation provides two bits of information for each pulse sent. The performance of this type of modulation is the same as that of BPSK for thermal noise, but it degrades more rapidly with CW interference and linear delay distortion [5]. Four-state phase modulation is illustrated in Figure 5.16.

Both coherent and differential QPSK can be produced with the 00 signal represented by a change of +0 degree from the phase reference, 01 can be represented by a change of +90 degrees, 10 can be represented by a change of −180 degrees, and 11 can be represented by a change of −90 degrees.

Another possible phase-shift keying system is eight-state phase modulation or octonary phase-shift keying (OPSK). In OPSK, phase-shift steps are separated by 45 degrees. Example states are +45, +90, +135, +157.5, ±180, −135, −90, −45, and 0 degrees. With OPSK, three bits of information are sent per pulse. This mode of modulation is used with a number of digital microwave radio systems operating in the 6- and 11-GHz bands.

5.16 SIXTEEN PHASE-STATE KEYING (16-PSK)

In 16 phase-state keying (16-PSK), 16 distinct phase states are used. Each phase step is separated by 22.5 degrees. With 16-PSK four bits of information are sent per pulse. No 16-PSK systems are in commercial use at this time because of the

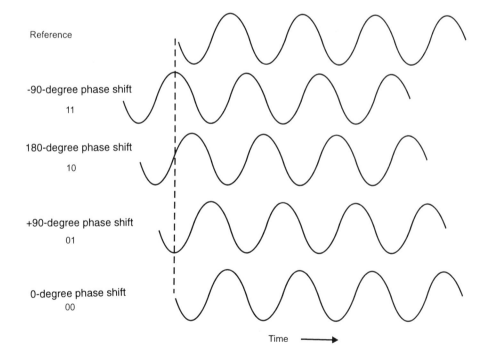

Figure 5.16 Four-state phase modulation for sending two bits per pulse.

140 | RF Systems, Components, and Circuits Handbook

superior performance of the 16 amplitude-phase keying (16-APK) modulation method.

5.17 SIXTEEN AMPLITUDE-PHASE KEYING

Figure 5.17 shows the 16 possible state combinations for a modulation system that uses four possible phase states and four possible amplitude states per pulse. This type of modulation permits four bits per pulse to be sent. This type of modulation has been found to be a reliable method of communication for high-capacity TDM systems.

At the top of Figure 5.17, we show a reference and two of the 16 possible phase and amplitude states. These are −45 degrees phase shift with level 1 amplitude and −135 degrees phase shift with level 2 amplitude. Each of the 16 possible

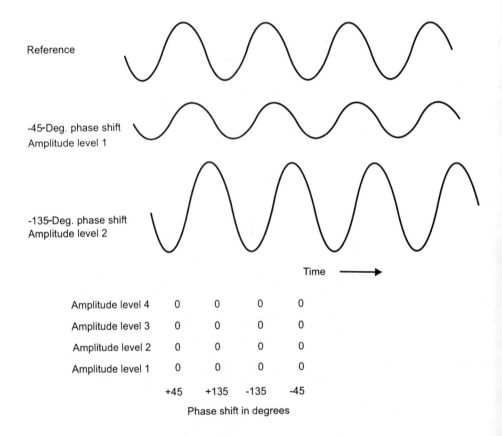

Figure 5.17 Phase angle and amplitude level combinations for 16-APK.

combinations of phase shift and amplitude is shown at the bottom of Figure 5.17 in a state table.

References

[1] Kennedy, George, *Electronic Communication Systems*, 3rd ed., New York: McGraw-Hill, 1985, Chaps. 3, 4, and 5.
[2] Kennedy, George, loc. cit., pp. 484–486.
[3] Skolnik, Merrill I., Editor, *Radar Handbook*, New York: McGraw-Hill, 1980, pp. 29–18 to 29–22, "Electromagnetic Compatibility" by John P. Murray.
[4] Kennedy, George, loc. cit., pp. 537–542.
[5] Keiser, Bernhard E., and Eugene Strange, Digital Telephony and Network Integration, 2nd ed., New York: Van Nostrand, Reinhold, 1995, Chap. 6.

Selected Bibliography

Krauss, Herbert L., Charles W. Bostian, and Frederick H. Raab, *Solid State Radio Engineering*, New York: John Wiley & Sons, 1980, Chap. 8.

Noll, A. Michael, *Introduction to Telephones and Telephone Systems*, 2nd ed., Boston: Artech House, 1991, Chap. 3.

Van Valkenburg, Mac E., ed., *Reference Data for Engineers*, 8th ed., SAMS, Carmel, IN: Prentice Hall Computer Publishing, 1993, Chap. 23, "Analog Communication" by Michael B. Pursly.

Ibid., Chapter 24, "Digital Communication" by Michael B. Pursly.

Ibid., Chapter 1, "Frequency Data" by Frederick Matos.

CHAPTER 6

RF Amplifiers, Oscillators, Frequency Multipliers, and Mixers

This chapter briefly discusses RF amplifiers, oscillators, frequency multipliers, and mixers. These key components are found in communication, radar, and other important RF systems. The discussions here are at the systems level. More detailed discussions at the components and circuits level are provided in Part 2.

6.1 AMPLIFIERS

Amplifiers are important building blocks for RF systems, for example, in the following applications:

- Front-end, low-noise RF amplifiers for receivers;
- IF amplifiers for receivers;
- Audio, video, and other LF amplifiers;
- Buffer amplifiers for oscillators;
- Gain stages to augment gain lost due to passive components such as filters, mixers, and multipliers;
- Transmitter RF amplifier chains;
- Transmitter output power amplifiers.

Each of these applications is illustrated in Figure 6.1, which shows simplified system block diagrams for communication transmitter and receiver systems. Some of the information presented in this section is adapted from material presented in [1].

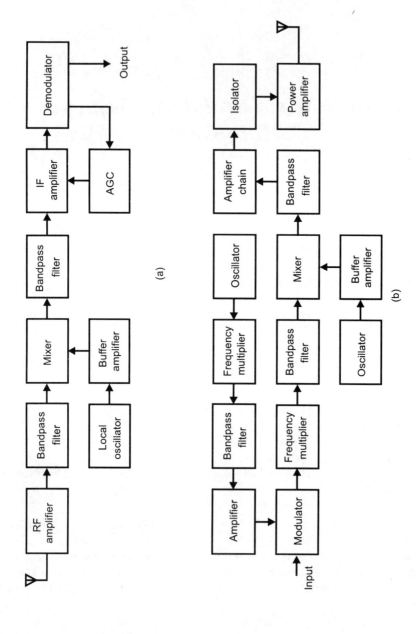

Figure 6.1 Simplified system block diagrams for receiver and transmitter systems: (a) receiver system and (b) transmitter system.

6.1.1 Front-End Low-Noise RF Amplifiers for Receivers

Front-end, low-noise RF amplifiers are an important part of many receiver systems. At MF and lower frequencies, there is no need for front-end, low-noise RF amplifiers for receivers. The external noise received by the antenna is so large that little is gained by having a low-noise amplifier precede the mixer.

Some, but not all, HF receivers also use low-noise, front-end amplifiers.

VHF and higher frequency receivers nearly always use low-noise, front-end amplifiers. The receiver noise from a mixer can be much greater than the noise that is received by the antenna. Having a low-noise amplifier that precedes the mixer thus can greatly improve the noise and sensitivity performance of the receiver system. It was pointed out earlier that it is the first stage of a receiver that largely determines the system noise figure if that stage has reasonably high gain.

Low-noise RF amplifiers are constructed with either bipolar junction transistors (BJTs) or field-effect transistors (FETs). In each case, class A amplifiers are used. The term class A refers to the fact that the transistor is so biased that it is in conduction at all parts of the RF cycle. In contrast, a class B amplifier has a conduction angle of 180 degrees; a class AB amplifier has a conduction angle slightly greater than 180 degrees; and a class C amplifier has a conduction angle of much less than 180 degrees. The conduction angles for the four classes of amplifiers are illustrated in Figure 6.2.

Class A amplifiers usually consist of single-transistor gain stages. Class B and class AB amplifiers usually are two-transistor or two-tube "push-pull" type power amplifiers. Push-pull amplifiers are made up of two active devices, with one device operating for one-half of the RF cycle and the other device operating for the other half of the cycle. They frequently use transformers to combine the outputs. Class C amplifiers may be either single-transistor or single-tube circuits or push-pull type power amplifiers. In most cases, class C amplifiers use resonant circuits to convert the short-duration pulses to sine waves. Parallel devices are sometimes used in power amplifiers to increase the output power capability of the amplifiers.

Class A amplifiers have a maximum possible efficiency of 50%, whereas class B amplifiers have a maximum efficiency of 78%. Class AB amplifiers have maximum efficiency slightly greater than class B amplifiers. The efficiency of a class C amplifier can be as high as 90% or more. Efficiency expressed as a percentage is 100 times the ratio of the RF output power to the DC input power.

Because class A amplifiers normally are used for small-signal amplification, efficiency is not very important. Class B and class AB amplifiers normally are used for large-signal amplification where efficiency is important. They normally are designed as linear amplifiers where the output is directly proportional to the input. They each may be wideband amplifiers. Class C amplifiers, on the other hand, are nonlinear amplifiers that may use tuned circuits on the output. Those tuned circuits are sometimes called *tank circuits*.

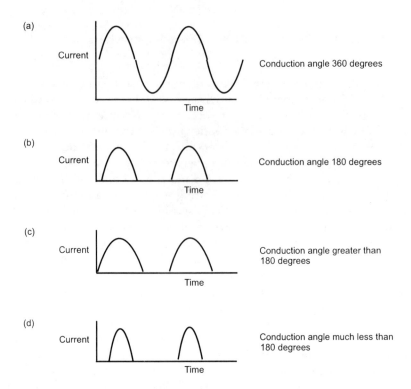

Figure 6.2 Conduction angles for different classes of amplifiers: (a) class A amplifier current; (b) class B amplifier current; (c) class AB amplifier current; and (d) class C amplifier current.

Silicon BJT RF amplifiers typically are used at the lower RF frequencies because of higher gain capability and more linear operation. The frequency is limited to 8–10 GHz, depending on the application. FET RF amplifiers made from gallium arsenide (GaAs) typically are used at microwave and higher frequencies because of their higher frequency capability and lower noise figure. The main reason for superior performance for these FETs at higher frequencies is the higher mobility of GaAs compared to silicon. FETs can use either silicon or GaAs in their construction, whereas BJTs use only silicon.

BJTs and FETs are discussed in detail in Chapter 18.

The BJT RF amplifier is a current-controlled amplifier with the collector current determined by the base current. The FET amplifier, on the other hand, is a voltage-controlled amplifier with the drain current determined by the gate voltage. Each of these devices is diagrammed in Figure 6.3.

In each case, it is necessary to bias the transistor to the desired operating current or voltage using dc bias networks. It is also necessary to provide both input

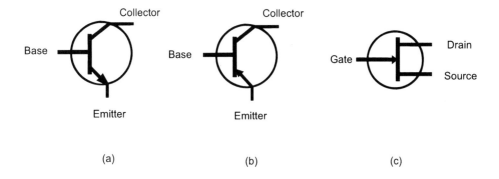

Figure 6.3 Diagrams for BJT and FET devices: (a) NPN BJT; (b) PNP BJT; and (c) FET.

and output impedance matching circuits. Some matching circuits are designed to maximize power output, and some are designed for flat gain over a large bandwidth. In the case of the front-end low-noise amplifier, the matching may be for minimum noise figure. The design of impedance matching circuits for transistors using S-parameters is discussed in Chapter 18. The input impedance matching requirement for maximum gain is that the impedance of the transmission line that feeds the amplifier be transformed to the complex conjugate of the device input impedance. The output impedance matching requirement is that the impedance of the transmission line or load fed by the amplifier be transformed to the complex conjugate of the device output impedance. In each of these cases, the assumption is that S_{12} be essentially zero. For example, if the transistor input impedance at a given frequency is $25 + j\,15\Omega$, the complex conjugate impedance would be $25 - j\,15\Omega$. With a 50Ω characteristic impedance for the transmission line that feeds the amplifier, the required transformation would be from $50 + j\,0\Omega$ to $25 - j\,15\Omega$.

Impedance matching is done at VHF and lower frequencies using lumped constant LC circuits. At UHF and microwave frequencies, very small LC circuits sometimes are used in monolithic and MIC amplifier chips and ceramic packages. It is also possible to use microstrip and coax impedance matching circuits at microwave frequencies.

Figure 6.4(a) shows an example design for a low-noise transistor RF amplifier. It uses a separate bias source to set the bias current and operating point for the transistor. It also uses lumped element input and output impedance matching networks.

Some of the following material is quoted or adapted from [2].

In the ideal case, the signal input for a small-signal RF amplifier is simply increased in amplitude at the output without changing the frequency components of the signal. In practice, there will be some distortion and harmonic generation due to the nonlinear characteristics of the transistor. Bandpass filters are used to minimize or eliminate (suppress) signals outside the desired signal band.

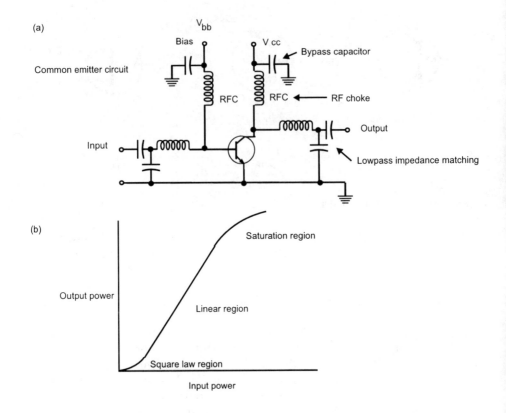

Figure 6.4 Example of a BJT RF amplifier: (a) RF transistor amplifier using impedance matching and (b) transfer function for a typical transistor.

Figure 6.4(b) shows the transfer function for a typical transistor. Because of its nonlinear nature, the transistor generates harmonics. It also can act as a mixer for multiple frequency signals, generating sum and difference frequencies. The mixing products form nonlinearities known as second- and third-order intermodulation distortion. Other forms of distortion generated by a transistor amplifier include cross-modulation distortion, and composite triple-beat distortion. As an example of the frequency components that are involved with a three-frequency signal at the input, the second-order distortion components include 3 dc components, 6 sum and difference components, and 3 second harmonic components. The third-order distortion components include 3 third harmonic components, 12 intermodulation components, 4 triple-beat components, 3 self-compression components, and 6 cross-compression or cross-expansion components.

In Figure 6.5, a typical transistor transfer curve shows the relationship between the fundamental, the second-order, and the third-order components of the signal.

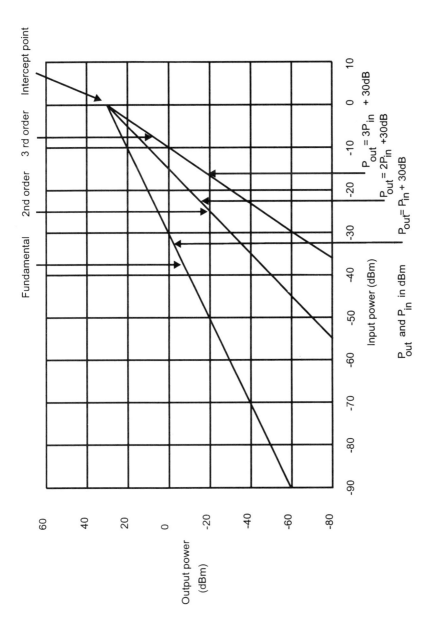

Figure 6.5 Fundamental, second-order, and third-order amplifier response curves for an RF amplifier. (*After:* [2].)

If the fundamental input power versus output power response of an amplifier is plotted on a log-log scale, it will have a 1-to-1 slope in the linear operating region. A plot of the second-order intermodulation products of the amplifier, plotted on the same scale, will have a slope of 2 to 1 (corresponding to a square relationship), and the third-order products will have a slope of 3 to 1 (corresponding to a cubic relationship).

The third-order spurious products are the most troublesome, because they may fall within the bandpass of even moderate bandwidth amplifiers.

The intercept point for the curves in Figure 6.5 generally is defined as the point where the extensions of the fundamental and the third-order responses intersect on the output power scale. The second-order response generally intersects at about the same point as well, unless the amplifier design suppresses even-order responses (e.g., push-pull stages).

The response curve shown in Figure 6.5 is for a device with a gain of 30 dB, a power output of +20 dBm at the 1-dB gain compression point, and a third-order intercept point of +30 dBm. As an example of the use of this plot, assume that the amplifier is driven to +15 dBm output. The second-order product will be suppressed 30 dB below the intercept point to 0 dBm. The third-order product will be suppressed 45 dB below the intercept point to −15 dBm.

For another example, assume that the amplifier is driven to +0 dBm output power. The second-order product will be suppressed 60 dB below the intercept point to −30 dBm. The third-order product will be suppressed 90 dB below the intercept point to −60 dBm.

One of the problems that we have with high-gain RF amplifiers is that they like to oscillate. Any student who has designed and built an RF amplifier knows about this problem.

A number of ways are used to prevent oscillations and provide the needed stability for RF amplifiers. Amplifiers are always designed so stages and overall systems have stability factors greater than 1.0, usually greater than 1.5 for unconditional stability. It is common practice to place amplifiers in metal containers or shields, so individual stages are shielded from each other. This type of shielding greatly reduces unwanted coupling between stages. Also, use of Eccosorb or other radar absorbing material (RAM) attached to covers (approximately 20–30 mils thick) usually reduces feedback and eliminates possible oscillations.

Another way to improve stability is to use neutralizing circuits. The goal is to provide a negative feedback signal that is equal in amplitude and opposite in phase to the positive feedback signal that is causing oscillations.

It is possible to buy packaged RF amplifiers from suppliers that specialize in amplifier design and production. Many such sources are available. These amplifiers may be in packages involving small 50Ω coaxial connectors, or they can be purchased as chips or flatpack devices with pins. Table 6.1 lists the characteristics for a low-noise GaAs MMIC amplifier manufactured by the ANZAC division of M/A-COM, which is owned by AMP.

Table 6.1
Characteristics for ANZAC Low-Noise GaAs MMIC Amplifier

Characteristics	Values		
Model Number	AM-280		
Guaranteed specifications (from −55°C to +85°C)			
Frequency range	1.1 to 1.7 GHz		
Minimum gain (at 25°C)	1.1–1.3 GHz	20 dB	
	1.3–1.5 GHz	18 dB	
	1.5–1.7 GHz	16 dB	
Gain variation with temperature	±3 dB max		
VSWR			
Input	1.1–1.5 GHz	2.0:1	
	1.5–1.7 GHz	2.3:1	
Output	1.1–1.7 GHz	2.0:1	
Maximum noise figure	At 25°C	At 85°C	At −154°C
1.1–1.5 GHz	1.7 dB	2.1 dB	1.5 dB
1.5–1.7 GHz	2.0 dB	2.5 dB	1.8 dB
Output power for 1 dB compression	10 dBm min		
Operating characteristics			
Impedance	50Ω nominal		
Intermodulation intercept point			
(for two-tone output power up to 0 dBm)			
Second order	+35 dBm typ		
Third order	+22 dBm typ		
Bias power			
V_{D1} = 2.0–5.0 Vdc at I_{D1} = 20–50 mA			
V_{D2} = 2.0–5.0 Vdc at I_{D2} = 20–50 mA			
(Set VG1 and VG2 in the range of 0 Vdc to −3 Vdc to achieve desired ID1 and ID2 bias setting.)			
Die size	0.058 × 0.048 × 0.010 in		
	(1.47 × 1.22 × 0.25 mm)		

6.1.2 IF Amplifiers

Most of the gain of a receiver is provided by the IF amplifiers. That is because it is easier to construct narrowband bandpass filter at IF frequencies than at RF frequencies, and amplifier stages have higher gain at the lower frequencies. These amplifiers follow the mixer and the LO, which converts the signal frequency from RF to IF.

At lower frequencies, it is possible to use single conversion and only one IF frequency. For example, at MF the AM receiver uses only one IF frequency at 455 kHz. At the higher frequencies, it is common to use two or more frequency conversions and two or more IF amplifier frequencies. For example, a communication system might use a first IF at 30 MHz and a second IF at 455 kHz. The higher frequency for the first IF makes it easier to construct filters to suppress the image frequency, and the low frequency for the second IF makes it easier to make the required precision narrow band filters for providing optimum noise and signal bandwidth.

The image frequency is an undesired frequency that, when mixed with the LO signal, yields a signal at the IF frequency, just as the desired signal frequency does. Image frequency and its suppression are illustrated in Figure 6.6.

Figure 6.6(a) shows a simplified frequency spectrum for the output of the receiver antenna, along with a spectral line corresponding to the LO signal (f_{LO}). Two signals provide a frequency difference equal to the IF frequency: a signal frequency lower in frequency than the LO signal and an image frequency higher in frequency than the LO signal.

Figure 6.6(b) shows a simplified superheterodyne receiver that includes a bandpass filter for suppression of the image frequency. The filter has its bandpass centered at the desired signal and is sufficiently narrow that it presents large attenuation to the image frequency.

In some cases, the required IF bandwidth is very small. In other cases, such as for radar, this bandwidth must be quite large. In the latter case, it is common

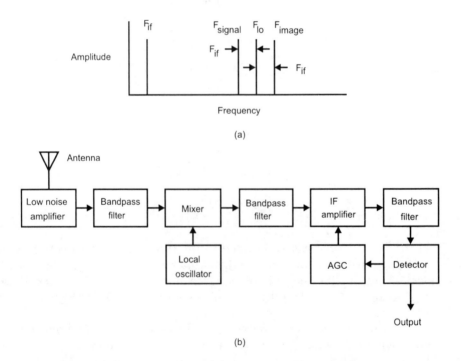

Figure 6.6 Superheterodyne receiver concepts: (a) frequency spectra showing desired signal frequency, LO frequency, and image frequency and (b) simplified superheterodyne receiver block diagram showing bandpass filter for suppression of image frequency.

RF Amplifiers, Oscillators, Frequency Multipliers, and Mixers | 153

to use a second IF frequency such as 70 MHz and a first IF frequency of 300 MHz or more. In DBS receivers, the first IF frequency is 950–2,150 MHz, depending on the satellite system.

The type of system discussed in the foregoing paragraphs that involves one or more mixers for down-conversion and IF amplifiers is called a *superheterodyne* receiver. This is the most frequently used type of receiver. There are a few applications, however, where superheterodyne receivers are not used. The receiver used in those cases is called a *homodyne* receiver. In a homodyne receiver, all the gain is provided at the RF frequency, and no IF amplifiers are used. An example of such a system is a very wide bandwidth receiver used in an ELINT system.

IF amplifiers usually include the capability for automatic gain control (AGC). A voltage is fed back from the demodulator following the last IF amplifier that is proportional to the amplitude of the detected signal. That voltage is applied to one or more amplifier stages to adjust the gain of those stages. An AGC loop is illustrated in the circuit in Figure 6.6(b).

The IF amplifiers used are sometimes BJT amplifiers involving a single transistor per stage, or they may be integrated circuits. In modern receivers, integrated circuit IF amplifiers are the most frequently used type.

Figure 6.7(a) shows a block diagram for a two-stage IF amplifier chain with the bandpass filtering and impedance matching provided by single-tuned magnetic transformers. Such a system frequently is used in simple AM receivers. Figure 6.5(b) shows a block diagram for an IF amplifier using discrete components and a three-electrode ceramic filter. This type of IF amplifier frequently is used in modern receiver systems rather than the older tuned transformer concept. Some televisions

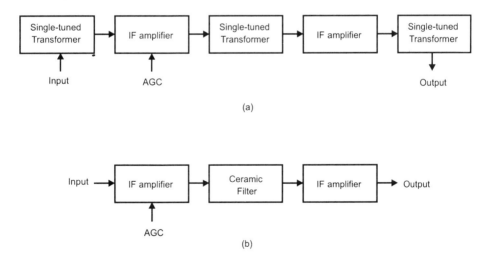

Figure 6.7 Example IF amplifiers: (a) two-stage IF amplifier with single-tuned magnetic transformers and (b) two-stage IF amplifier with a ceramic filter.

use SAW filters for better selectivity. Many other types of IF amplifier circuits could be shown, but these two are adequate to illustrate the main concepts of IF amplifiers.

6.1.3 Audio and Other LF Amplifiers

Many possible types of audio and other LF amplifiers are used in RF systems following demodulators or preceding modulators, including class A RC-coupled amplifiers, class A transformer-coupled amplifiers, emitter followers, and class AB push-pull amplifiers.

6.1.4 Transmitter RF Amplifier Chains

RF transmitters often use low-power modulators followed by amplifier chains that raise the power levels from at most a few milliwatts to the required drive level for the final power amplifier. In many cases, this cascade of amplifier may include a frequency up-converter mixer stage for shifting the frequency from IF frequency to the desired RF one prior to transmission. In some cases, this amplifier chain may include a frequency multiplier stage. Block diagrams for these types of amplifier chains are shown in Figure 6.8.

The total gain for the CW amplifier chain is the sum of the gains for the individual amplifiers expressed in decibels. The output frequency is the same as the input frequency.

In a transmitter RF amplifier chain with a frequency converter, the output frequency is the sum of the input frequency and the LO frequency for upper sideband upconversion. It may also be the difference frequency, depending on LO frequency. For example, the input frequency may be 300 MHz and the LO frequency 1,000 MHz. The output frequency would be 1,300 MHz for the sum frequency or 700 MHz for the difference frequency. Again, the gain of the amplifier chain will be the sum of the gains or losses for each of the cascaded stages, expressed in decibels.

In a transmitter RF amplifier chain with a frequency multiplier, the output frequency is N times the input frequency, where N is the multiplication factor. For example, the chain may use a times-3 frequency multiplier. If the input frequency is 300 MHz, the output frequency will be 900 MHz.

6.1.5 Transmitter RF Power Amplifiers for Communication Systems

The final stage for the communication transmitter is the power amplifier. Transmitter RF power amplifiers for communication systems include both transistorized power amplifiers and tube-type power amplifiers. Transistor amplifiers include both single-ended RF amplifiers and push-pull amplifiers. Transistor types used in these amplifiers include both BJTs and FETs.

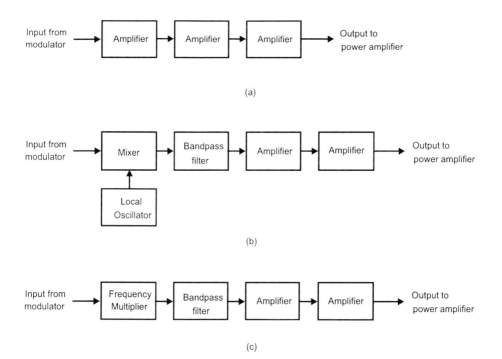

Figure 6.8 Example frequency amplifier chains for transmitters: (a) CW transmitter RF amplifier chain; (b) transmitter RF amplifier chain with frequency converter; and (c) transmitter RF amplifier chain with frequency multiplier.

Figure 6.9 shows a schematic diagram for a class AB push-pull RF power amplifier. This amplifier uses two NPN BJTs. The bases for the transistors are connected to the secondary of an input transformer. This secondary winding is center tapped and is biased by means of a voltage supply, a series resistor, and a

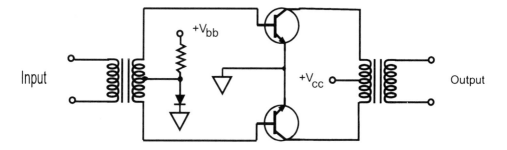

Figure 6.9 Class AB push-pull transistor amplifier circuit.

forward-conducting diode to the turn-on voltage for the transistor. That helps avoid distortion caused by the nonzero turn-on characteristic for BJTs. Having a conduction angle slightly greater than 180 degrees, as needed for class AB operation, is made possible by this type of bias.

The collectors for the two transistors are connected to the primary winding of the output transformer. This is a center-tapped winding, with the center tap connected to +V. With this type of amplifier, the two transistors take turns conducting for a half-cycle each. The result of the transformers is that a full-wave signal of low distortion is produced from the two half-wave signals.

Very high power transistor power amplifier systems may be made by combining the output power from many individual power amplifiers using RF power combiner circuits. With the use of a large number of transistors and combiners, it is now possible to provide output powers in the kilowatt range at frequencies up to about 1.0 GHz. Progressively smaller power levels can be provided at microwave frequencies.

Figure 6.10 shows a block diagram for a high-power transistor amplifier system consisting of 16 individual class AB power amplifiers. Groups of four amplifiers are combined using 4-to-1 RF power combiner circuits. The outputs of the four combiners are then combined using another four-way power combiner. The result is that the power output for the system is 16 times the power output of the single amplifiers less the losses in the power combiners.

Tube-type power amplifiers for communication transmitter systems are of two types. Grid-type vacuum tube amplifiers use triodes, tetrodes, or pentodes. At microwave frequencies, tubes used in communication transmitter systems include traveling wave tubes (TWTs) and klystrons. Vacuum-tube devices and circuits are discussed in Chapter 19. Some of the information here about tube-type power amplifiers is adapted from [3,4].

Both communication-type, wideband TWT amplifiers and power solid-state amplifiers are used for point-to-point and mobile communication systems for surface installations. Transistor amplifier chains are used for the lower frequency bands, while TWTs are used for microwave frequencies. With TWTs in the 5-18 GHz range, the amplifier gains typically are about 50 dB. Output powers are available for as low as 25W to as high as 900W. These are driven by transistor amplifier chains.

Satellite uplinks, which require fairly high transmitter powers, usually use either TWT power amplifiers or klystron power amplifiers. Satellite downlinks currently use TWT amplifiers rather than solid-state amplifiers. The main reason for that is the higher efficiency for the TWT amplifiers. TWT amplifier efficiencies typically are as high as 60%, whereas solid-state amplifiers for the same frequencies typically are only about 30% to 40% efficient. The reason that efficiency is so important for satellite downlink transmitters is that the dc power for the amplifiers must be provided by solar cells. A saving in dc power means a big saving in cost of the satellite.

RF Amplifiers, Oscillators, Frequency Multipliers, and Mixers | 157

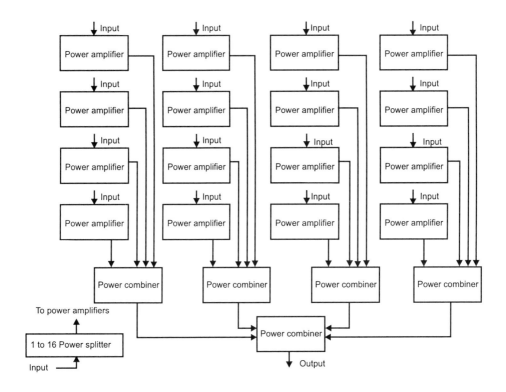

Figure 6.10 High-power transistor amplifier using 16 individual power amplifiers that are combined using RF power combiners.

Klystrons are used for UHF television transmitters. Output powers typically are in the range of 30–60 kW. Typical gains are 30–50 dB. TWTs also can be used for broadcast, including both surface installations and spaceborne installations.

High-powered grid-type triodes and tetrodes are used as power amplifiers for broadcast at MF frequencies and for other low frequency applications. They produce output powers in the range of 10–500 kW.

6.1.6 RF Power Amplifiers and Oscillators for Radars, Navigation, and Electronic Countermeasure Applications

One type of power amplifier used for surface radar installations is a TWT/crossed-field amplifier (CFA) chain. A CFA is a very efficient, high-power amplifier. Because a typical gain for this type of device is between 10 and 20 dB, it is necessary to drive this tube with a fairly high-powered amplifier of high gain such as a TWT. Thus, the term TWT/CFA refers to a CFA driven by a TWT amplifier.

As an example of power levels and gains, a typical CFA has an output power of 1 MW (60 dBw) and a gain of 16 dB. The required input power from the TWT thus would be 60 − 16 = 44 dBw, or 25 kW. A typical coupled-cavity TWT (CCTWT) could have an output power of 25 kW and a gain of 44 dB. The required input to the TWT then would be 1W, which can be supplied by a transistor amplifier chain. If an up-converter is the RF signal source that produces 1.0 dBm of output power, the required gain of the transistor amplifier chain would be 30 dB.

A second microwave tube used for surface installation search and surveillance radar is a high-power klystron. This type of high-powered microwave tube amplifier is available over the frequency range from UHF to K_a-band. Varian Associates makes narrowband, high-powered, pulsed klystrons with peak output powers in the range of 1–5.5 MW over the frequency range of 0.9–5.9 GHz. An example is an amplifier operating in the 2.7–3.0 GHz band with an output peak power of 1.5 MW (61.8 dBw), an average power output of 3 kW (34.7 dBw), and a typical gain of 50 dB. The required input or driving power from a transistor amplifier chain would be 61.8 − 50 = 11.6 dBw, or 14.4W.

The third microwave tube possibility for the surface installation search and surveillance radar is the high-power CCTWT. This type of high-power microwave-tube amplifier also is available over the frequency range from UHF to K_a-band. For example, a tube provides a peak output power of 240 kW (53.8 dBw) at UHF. Gains are in the range of 30 to 65 dB. For example, assume a gain of 50 dB. The required drive power from a transistor amplifier chain is thus 3.8 dBw or 2.4W.

The fourth microwave-tube possibility for the surface installation search and surveillance radar is a high-power twystron. A twystron has an input section that is like that of a klystron and an output section like that of a TWT. It has the advantage of having a larger bandwidth capability than a klystron but with similar power and efficiency characteristics. Twystrons are available over the same frequency range as the klystron (UHF to K_a-band). A typical output power level is about 1 MW. A typical gain is about 50 dB. Again, the required driving power is only about 10W, which may be provided by a transistor amplifier chain.

A fifth way to provide the needed power for a search and surveillance radar operating at L- or S-band is to use many solid-state, pulse-power amplifiers that feed individual elements of a phased-array antenna. If, for example, 100 amplifiers are used, with each providing 100W output, the total power level would be 10 kW if properly phased.

Search and surveillance radars are also used with airborne and spaceborne installations. The type of microwave amplifiers used are TWTs and klystrons. The power levels used for these applications typically are lower than the examples shown for the surface installations. TWTs may be of the wideband type using helix slow-wave structures. Such tubes are available with pulse power outputs of 1 to 2 kW over the frequency range from 2 to 18 GHz. Bandwidths may be of the order of 2:1 to 3:1 (6–18 GHz), so complete radar and ECM bands are covered.

A wide range of powers is available for pulsed klystron amplifiers used for airborne and spaceborne radars. These may be as small as 1 kW or as large as 100 kW. Such pulsed amplifiers are available over the frequency range of 0.4–36 GHz. Gains are in the range of 40–60 dB. Again, the drivers for such amplifiers can be transistor amplifier chains.

Surface-based fire-control radars use TWT/CFA amplifier chains and TWT amplifiers. Fire control radars for airborne installations also use pulsed TWTs. They also may use pulsed magnetrons. Magnetrons are small crossed-field microwave oscillators with good power-to-weight ratios. The required inputs for these devices are high-voltage dc pulses. Magnetrons are very efficient and rugged. Typical output power levels are in the range of 10 kW to 1 MW.

Lower power magnetrons also are the tubes used to supply the RF energy in most, if not all, microwave ovens.

Weather radars are used in aircraft for bad-weather avoidance. The systems may use either TWT amplifiers or magnetron oscillators. TWT amplifiers are driven by transistor amplifier chains.

Missile homing as used by airborne systems usually is done using small magnetron oscillators. These are the simplest and lowest cost type radar systems.

Speed measurement by the police usually is done using small solid-state microwave Gunn diode oscillators. These very simple transmitters produce a few tens of watts.

In the past, navigation beacons used only magnetrons. These are now being replaced by solid-state amplifiers.

Both surface-based and airborne ECM systems use wideband helix-type TWTs. CW power levels typically are on the order of 200W or less. Dual-mode systems may be used with higher pulse power capability provided in addition to CW operation. Again, these devices are driven by low-power solid-state amplifier chains.

6.2 OSCILLATORS AND FREQUENCY SYNTHESIZERS

This section discusses oscillators and frequency synthesizers. Some of the information presented in this section is adapted from information presented in [5].

6.2.1 Transistor Feedback Oscillators

Communication transmitters and receivers typically use transistor feedback oscillators as frequency generators. Basically they are transistor amplifiers with positive feedback so that a part of the output signal is fed back to the input as a signal with the proper phase and amplitude to permit stable oscillation. The frequency of oscillation is determined mainly by the resonant circuit that is used in the feedback circuit.

Two main types of transistor oscillators are used in communication systems at lower frequencies: LC oscillators and crystal oscillators. There are also dielectric resonator oscillators (DROs) and those that use ceramic resonators. Others use transmission-line elements to provide the necessary resonant or tank circuits.

6.2.1.1 LC Oscillators

LC oscillators are frequently used in communication systems to generate signals. Types of LC oscillators frequently used include the Colpitts oscillator, the Clapp oscillator, the tuned- input tuned-output JFET oscillator, and the differential-pair oscillator. These oscillators may be fixed-frequency oscillators that are mechanically tuned using variable capacitors, or they may be voltage-controlled LC oscillators (VCOs) that use varactor diodes to provide variable capacitance, occasionally in parallel with fixed capacitors.

LC oscillators can be used as LOs for mixers and frequency converters, as carrier frequency inputs to modulators, and as frequency modulators. Their main weakness is that they do not provide very good frequency accuracy or stability or good phase noise performance, necessary for digital communication systems.

6.2.1.2 Quartz Crystal Oscillators

Quartz crystal oscillators are used extensively in communication and radar system transmitters. They can operate either in a series resonant mode or in a parallel resonant mode, depending on the design of the oscillator. Such crystals are very high-Q devices that can provide very high accuracy for frequency control. (The Q of a crystal is the center frequency divided by the 3-dB or half-power bandwidth.) Accuracies are in the range of one part in 10^5 to one part in 10^7, in other words, 0.1 to 10.0 parts per million (ppm). The frequency range for these oscillators is typically 10^5 Hz (100 kHz) to 10^8 Hz (100 MHz).

Crystal oscillators can be tuned over a small frequency range using trimmer capacitors. These may be either mechanically tuned capacitors or voltage-tuned capacitors (varactors). Crystal oscillators are used for the same functions as LC oscillators.

Figure 6.11 shows a Colpitts oscillator with the crystal in series with the feedback path to the emitter. The crystal in this case is operating in the series resonant mode. Other frequently used crystal oscillators are the Miller oscillator and the Pierce oscillator (discussed in Chapter 16).

6.2.1.3 Yttrium Iron Garnet Resonator Oscillators

One type of UHF and microwave oscillator that is sometimes used is a yttrium iron garnet (YIG) resonator oscillator. YIG oscillators use highly polished YIG spheres

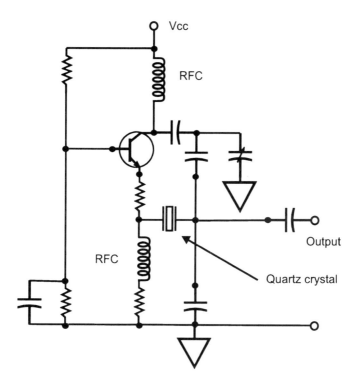

Figure 6.11 A Colpitts crystal oscillator.

as resonant devices with magnets to determine their resonant frequencies. Electromagnets can be used to tune the devices to the desired frequency of oscillation. The oscillators are used by transmitters as LOs for mixers or frequency converters. YIG devices and circuits are discussed in more detail in Chapter 16.

6.2.1.4 Dielectric Resonator Oscillator

DROs are microwave oscillators that use the unique properties of high-Q dielectric materials as a resonator to stabilize the frequency of free-running sources. Typical frequency stability is about 5 ppm per degree centigrade. Typical output power levels for oscillators of this kind are 0 to +20 dBm. DROs can operate from 1 to about 35 GHz. They have good phase noise properties.

6.2.2 Negative Resistance Two-Terminal Oscillators

A second class of oscillator sometimes used at UHF and microwave frequencies is a negative resistance oscillator. This class of oscillator includes tunnel diode oscillators, Gunn diode oscillators, IMPATT or avalanche diode oscillators, trapped plasma avalanche transit time (TRAPATT) diode oscillators, and limited space-charge accumulation (LSA) mode oscillators. The frequency of operation for these diode oscillators is determined by the use of cavity resonators. The Gunn diode and the IMPATT diode both are capable of a few watts output power. In general, the negative resistance-type oscillators have poor phase noise and poor frequency stability.

6.2.3 Frequency Synthesizers

Frequency synthesizers frequently are used with transmitters and receivers when there is a need for accurate multiple frequency selection. The two main types are direct frequency synthesizers and indirect frequency synthesizers. In direct frequency synthesizers, a number of crystal oscillators are used along with frequency multipliers, mixers, and switches. Frequency multipliers provide multiple frequencies from a single source that may be selected by an RF switch. Mixers provide sum and difference frequencies. By selection of the proper combination, it is possible to produce any one of hundreds of frequencies with the same accuracy as that of the crystal oscillators. Direct frequency synthesizers tend to be more complex, more costly, and have poorer noise characteristics than indirect frequency synthesizers.

Indirect frequency synthesizers frequently are used in all types of communication systems involving frequency selection. A system of this type is shown in Figure 6.12. The system in the figure uses a 100-kHz crystal oscillator as a reference frequency source for a PLL system. This oscillator is followed by a divide-by-4 frequency scaler, which provides a 25-kHz output signal. That 25-kHz signal is fed to a phase detector that compares its phase with that of the second input from a programmable frequency divider. The output of the phase detector is fed through a loop filter to a varactor-tuned LO. For the system in Figure 6.12, the oscillator frequency range is 98.8–118.6 MHz. The output of this oscillator initially is divided in frequency by 8 by a frequency prescaler. The output of the circuit is then fed to the programmable frequency divider, which has possible divider factors of 494 to 593. The output frequency is 25 kHz when the system is phase locked.

This circuit provides LO frequency output in the frequency range of 98.8–118.6 MHz tunable in steps of 25 kHz with an accuracy equal to that of the crystal oscillator. PLL circuits of this kind are available in integrated circuit form at low cost. They are, thus, finding application in many different types of low-cost receivers as well as in transmitters. They also are finding use in cellular devices.

PLLs and prescalers are available as integrated circuits from many companies. Prescalers are available that operate to 12 GHz (Hewlett-Packard Co.).

Frequency synthesizers can be made for UHF and microwave frequencies as well as for VHF and lower frequencies. Some microwave systems use YIG-tuned oscillators. Most use varactor-tuned oscillators, since the tuning rate is faster and

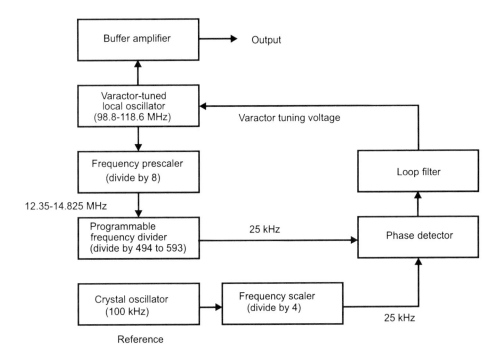

Figure 6.12 An indirect frequency synthesizer. (*After:* [6].)

the tuning circuitry is easier to construct and simpler. Also, YIG-tuned oscillators can suffer from tuning hysteresis.

It is common practice to use a buffer amplifier between the oscillator or frequency synthesizer that feeds the carrier frequency to the modulator circuit. Doing so provides isolation and a good impedance match between the VCO and the equipment to which it is connected. The use of a buffer amplifier also increases gain to provide sufficient power out for a mixer it may be driving. An alternative is to use an isolator (discussed in Chapter 11).

6.3 FREQUENCY MULTIPLIERS

The nonlinearity inherent in any semiconductor diode or transistor can be used to multiply frequency. The most popular diode frequency multipliers use either varactor diodes or step-recovery diodes. Frequency multipliers of this type are discussed in this section. Some of the information presented here is adapted from [5,7].

6.3.1 Varactor Diode Frequency Multipliers

Figure 6.13 shows a varactor-diode frequency tripler. Input and output impedance-matching circuits are used with this circuit. The input and output ports are coupled to the diode through series-tuned circuits, causing the input current and the output current and voltage to be essentially sinusoidal.

The varactor diode is a nonlinear voltage-variable capacitor. One or more so-called idler circuits are used with the circuit. These series-tuned circuits are placed in parallel with the diode and are resonant at harmonic frequencies other than the desired output frequency. Idler circuits are in practice empirically selected and adjusted to improve efficiency. They are resonant at other harmonics, presenting an impedance to reflect unwanted harmonic energy back to the input.

Hyperabrupt snap-off varactor diodes multiply by high factors with better efficiency than ordinary varactor diodes, so they are used wherever possible. GaAs varactors often are used at the higher frequencies. A varactor multiplier of this type can have an efficiency for a 60-GHz doubler of greater than 50%.

The maximum output power for the varactor diode multipliers ranges from more than 10W at 2 GHz to about 25 mW at 100 GHz. Tripler efficiencies range from 70% at 2 GHz to about 40% at 36 GHz. One of the current applications for multiplier chains is to provide a low-power signal to phase-lock a Gunn or IMPATT oscillator.

6.3.2 Step-Recovery Diode Frequency Multipliers

A step-recovery diode is a silicon or GaAs p-n junction diode with construction similar to that of a varactor diode. It stores charge when conducting in the forward direction. When reverse bias is applied, the diode very briefly discharges the stored energy in the form of a sharp pulse that is rich in harmonics. The duration of the pulse typically is only 100 to 1,000 picoseconds (ps) (1 ps = 10^{-12} sec), depending on diode design.

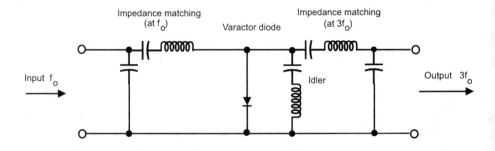

Figure 6.13 Varactor diode tripler circuit. (*After:* [5].)

Step-recovery diodes are frequently used in frequency multipliers. The circuit is essentially the same as that used for the varactor frequency multiplier, except that an inductor is used in series with the step-recovery diode. During the RF cycle, when the charge is completely drained from the diode, it switches from a conducting to a nonconducting state. This abrupt change in impedance then causes the inductor, which has stored energy, to generate a voltage impulse and hence a relatively flat frequency spectrum. Thus, the circuit is essentially an impulse generator. The desired harmonic frequency component is then extracted by the narrowband-tuned circuit and appropriate idler networks.

Figure 6.14 shows a typical multiplier chain using step-recovery diodes and varactors. The first stage is a transistor crystal oscillator followed by an amplifier with 35W output power at 160 MHz. That is followed by a step-recovery diode with a tuned circuit that is tuned to the tenth harmonic of 160 MHz, or 1.6 GHz. The output power of this stage is 3.5W, and the output frequency is 1.6 GHz. The next stage is a step-recovery diode with a tuned circuit that is tuned to the fifth harmonic of 1.6 GHz or 8.0 GHz. The output power of this stage is 0.7W, and the output frequency is 8 GHz. The last two stages of the circuit use varactor diodes. The first of those stages is a tripler with an output frequency of 24 GHz and an output power of 1,100 mW. The last stage is a doubler with an output frequency of 48 GHz and an output power of 275 mW.

Frequency multipliers at microwave frequencies use microstrip, stripline, coaxial line, or waveguide. Step-recovery diodes are not available for frequencies above about 20 GHz, whereas varactors can be used well above 100 GHz.

Step-recovery diodes are available for powers in excess of 50W at 300 MHz, 10W at 2 GHz, and 1W at 10 GHz. Multiplication ratios up to 12 are commonly available. Efficiency can be in excess of 80% for triplers at frequencies up to 1 GHz. The efficiency drops to about 15% for a 5-times multiplier with an output frequency of 12 GHz.

6.3.3 Transistor Multipliers

Transistor multipliers also are frequently used in RF systems. These multipliers can exhibit less loss than varactors or SRDs. In a class A multiplier using a BJT, the circuit is similar to that of a small-signal amplifier with the output circuit tuned to the desired harmonic of the input signal.

A class A doubler can be made using an FET with a square-law transfer characteristic. Again, the circuit can be designed as a small signal amplifier with the output circuit tuned to the second harmonic of the input signal. A typical efficiency for this type of frequency multiplier is about 16%.

Class C frequency multipliers are possible using transistors. The design principles and procedures generally are the same as for class C power amplifiers. The output circuit is tuned to the desired harmonic of the input signal.

166 | RF Systems, Components, and Circuits Handbook

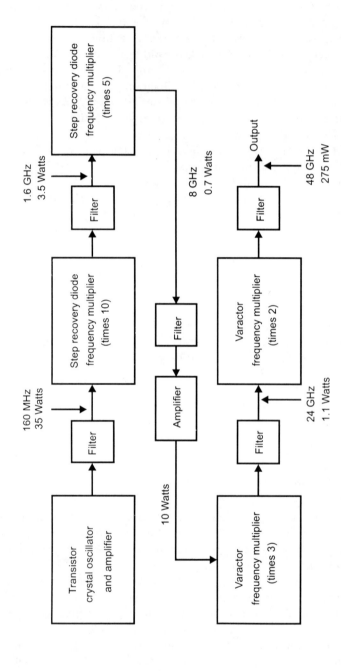

Figure 6.14 Frequency multiplier chain using step-recovery diodes and varactor diodes. (*After:* [7].)

6.4 MIXERS

Mixers are one of the most important of all RF components. Some of the information presented in this section is adapted from [6,8,9].

6.4.1 Diode Mixers

An important application for diodes is in mixer circuits. A number of different types of diode mixers are used in modern RF systems, including single-ended diode mixers, balanced diode mixers, double-balanced diode mixers, and triple-balanced diode mixers. Diode mixers can be used at both IF and microwave frequencies.

6.4.1.1 Single-Ended Diode Mixers

Figure 6.15(a) shows the circuit diagram for a single-ended diode mixer. A single diode is shown in series with the RF and LO inputs, a bias source, and a circuit

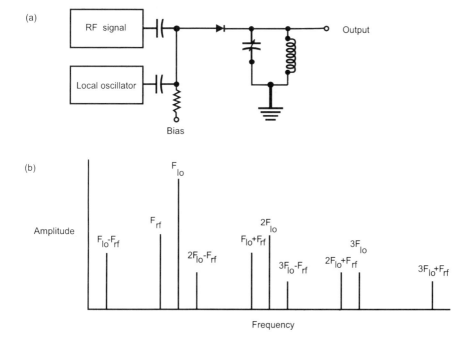

Figure 6.15 Single-diode mixer: (a) circuit diagram for a single-ended diode mixer and (b) partial frequency spectrum for a single-ended diode mixer.

tuned to the desired IF frequency. Such a circuit has a number of disadvantages compared with other types of mixers at the lower frequencies. It has a relatively high noise figure, a high conversion loss, high-order nonlinerarities, no isolation between the LO and the RF inputs, and large output current at the LO frequency. Its main application is at microwave frequencies, where other types of mixers may not be practical.

At microwave frequencies, the single-ended diode mixer may use a Schottky-barrier type diode or, in some cases, a point-contact diode. The diode may be mounted in a waveguide so that it provides a complete dc path for rectification, without causing serious reflections in the waveguide. It is located a quarter-guide wavelength from a short where there is a region of high electric field strength. The diode output is passed through a dielectric RF bypass capacitor. The IF output to the IF amplifier is via a coax cable. The LO signal is introduced using a sidearm with a tuning screw to adjust tuning.

Waveguide is only one media that can be used for the single-ended diode mixer. Such mixers also are made in stripline, coax, and microstrip.

Figure 6.15(b) shows a partial frequency spectrum for a single-ended diode mixer. Note that the unfiltered spectrum includes the RF input frequency, the LO input signal, and all harmonics of the LO signal. It also includes the sum and difference frequencies on either side of the LO frequency and its harmonics. Other frequency components are not shown in the diagram. The amplitudes shown for the different frequency components are not to scale but are representative of the frequencies involved.

6.4.1.2 Single-Balanced Diode Mixers

Figure 6.16(a) shows a single-balanced diode mixer. Four diodes are used to provide a switching or clamping circuit. The RF signal can be fed to the mixer through a series resistor, which in turn is connected to both the diode switch and the output filter and load resistor. The control of the switch is by means of a transformer that is fed by the LO having an output voltage larger than the RF signal. During one-half of the cycle, the output is effectively zero. During the other half of the LO signal cycle, the diodes are back-biased and the switch is an effective open circuit. The output produced by this mixer thus has a chopped waveform. If the RF frequency is higher than the LO frequency, the chopped signal output would be a series of RF pulses at the LO frequency repetition rate. Other versions of the single-balance mixer also are possible.

Figure 6.16(b) illustrates the frequency spectrum produced by the single-balanced diode mixer. The magnitude of the LO frequency and its harmonics are very small but are shown as frequency reference points. The components are largely eliminated by this type of mixer. The magnitude and the frequency of the RF signal and the sum and difference frequencies on either side of the LO frequency and

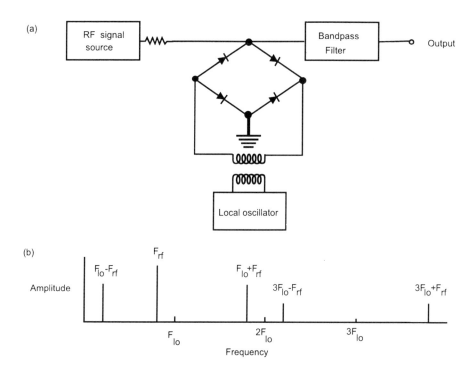

Figure 6.16 Single-balanced diode mixer: (a) single-balanced diode mixer circuit; (b) partial frequency spectrum for a single-balanced diode mixer. (*After:* [6].)

odd harmonics of this frequency are shown. Other frequencies also are generated but are not shown. A bandpass filter is used to select the desired frequency components from the mixer and to eliminate any possible image response.

The single-balanced mixer has the disadvantage that a component of the RF frequency appears at the output. That is not the case for the double-balanced mixer.

6.4.1.3 Double-Balanced Diode Mixer

Figure 6.17(a) illustrates the double-balanced diode mixer, which uses two center-tapped transformers or power dividers along with four diodes. The RF signal is introduced using one transformer, and the LO signal is introduced using the other. As in the case of the single-balanced diode mixer, the LO signal is assumed to be larger in amplitude than the RF signal and controls the on-off cycle of the diodes. Only two of the four diodes conduct at a time for each half of the LO cycle. The result is a chopped waveform with no dc component.

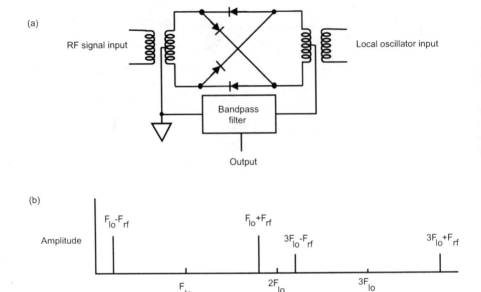

Figure 6.17 Double-balanced diode mixer: (a) double-balanced diode mixer circuit and (b) partial frequency spectrum for a double-balance diode mixer. (*After:* [6].)

Figure 6.17(b) illustrates the frequency spectrum produced by the double-balanced diode mixer. The output spectrum for the double-balanced diode mixer will contain only the frequencies $nF_{lo} \pm F_{rf}$, with n odd. Neither F_{lo} nor F_{rf} appears in the output (theoretically). That is an important advantage for this type of mixer. A bandpass filter is used to select the desired frequency components from the mixer. If ferrite toroidal-core transformers are used for lower frequency realizations, bandwidths of 400:1 can be achieved. Microwave mixers have smaller frequency ranges. For example, a low-frequency system might cover the 1–400 MHz range. For example, a microwave system might cover the 2–26 GHz range. These mixers typically have a conversion loss of about 7 dB and a SSB noise figure within 1 dB of the conversion loss. Frequency coverage is available from as low as 0.02 MHz to as high as 40 GHz. The isolation of the LO from the RF port is around 40 dB decreasing at higher frequencies. The two-tone third-order intermodulation products typically are down 40 to 50 dB from the desired IF components.

The following information is adapted from a technical note in the ANZAC RF and Microwave Signal Processing Components catalog [9].

Diode-type double-balanced mixers, as shown in Figure 6.17, belong to the general classification of resistive switching mixers, wherein an LO input signal is applied that is sufficiently large to cause strong conduction of the alternate diode pairs, thereby changing them from a low- to a high-resistance state during each half of the LO cycle. A virtual ground is, therefore, switched or commutated between the RF/IF transformer windings at a rate corresponding to the LO frequency. Since that switching causes a 180-degree phase reversal of the RF to IF port transmission during each half of the LO cycle, the mixing process is called biphase modulation.

For low-frequency operation, these devices typically use ferrite-core flux coupled transformers, which exhibit leakage inductance and stray capacitance. That limits upper frequency operation to approximately 4 GHz. For higher frequency operation, true transmission line realizations of the transformer functions will allow four-diode mixer operations to beyond 18 GHz. A system of this type is shown in Figure 6.18.

The low-frequency performance for the microwave mixer in Figure 6.18 is determined by the highpass nature of the RF and LO transmission line structure. A detailed discussion of transmission lines and terms is presented in Chapter 11.

Overlapping RF-IF or LO-IF frequency coverage is difficult to attain for the circuit in Figure 6.18 because the IF output encounters both the RF and LO structures in series for the IF signal path. To produce an overlapping IF range, a more complex eight-diode mixer was developed, as shown in Figure 6.19. Examination of this structure reveals that the LO is switching two diode pairs at a time, which are in series with the RF-IF signal path. By tracing out the RF-to-IF signal connections for each half of the LO input cycle, we see that biphase modulation is again being performed. The IF port can be seen to be an RF and LO null. The principle advantage of this design is its large RF-IF frequency range overlap. Its disadvantage is that it uses twice as many diodes and requires 3 dB more LO drive.

Figure 6.20 is a schematic diagram of a termination insensitive mixer (TIM), which consists of a transmission line hybrid network driving two sets of diodes. Isolation between each hybrid's opposite ports allows the LO to control independently the switching action of alternately conducting diode sets. The reverse bias applied to the off diodes is determined only by available LO input power and not

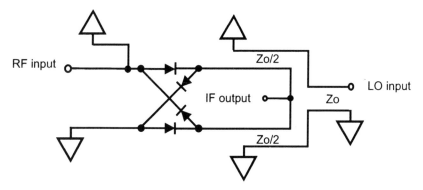

Figure 6.18 Typical four-diode microwave mixer schematic using transmission lines. (*After:* [9].)

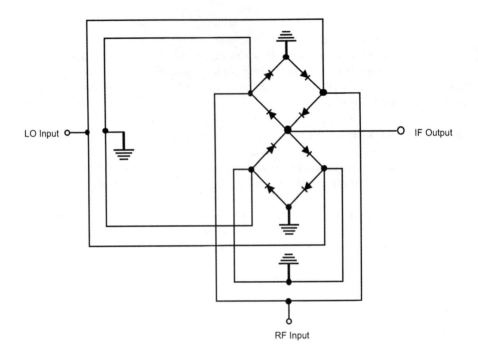

Figure 6.19 Typical eight-diode microwave mixer schematic. (*After:* [9].)

by the diode's forward potential, as in the conventional ring-type mixers. An internal resistor absorbs mixer-generated, even-order LO frequency terms and improves LO VSWR by terminating the hybrid port opposite its LO input. This circuit feature improves performance by closely approximating a square-wave LO drive.

Other types of microwave mixers use hybrid junctions. A balanced mixer using a 3-dB quadrature (90-degree) coupler is shown in Figure 6.21. Other balanced mixers use the 180-degree "magic T." If all conditions were perfect, there would be little difference between the two types of hybrids. In practice, it is virtually impossible to match the mixer diodes perfectly at all frequencies and under all LO drive conditions. Therefore, some degradation in the hybrid performance can be expected as a result of reflections from the diode mounts. In the case of the 90-degree hybrid mixer, input VSWR at either port generally will be low since the reflections from the mixer diodes will be shunted out the opposite port. The isolation between the signal port and the LO port will be strictly a function of the return loss of the diode mount. Despite the fact that the isolation is low, the AM noise cancellation, which is a function of the amplitude and phase balance of the hybrid itself and not the return loss of the diodes, generally is as high with a 90-degree type hybrid as with the magic-T type hybrid. Those factors, combined with the ability to operate over octave and multi-octave bandwidths with ease, make

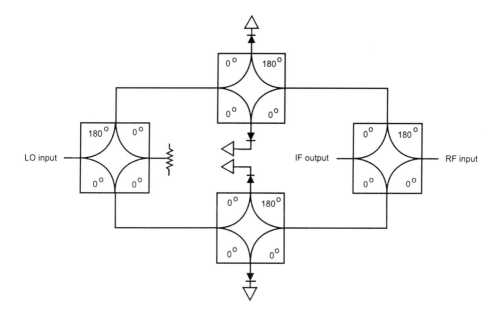

Figure 6.20 TIM schematic. (*After:* [9].)

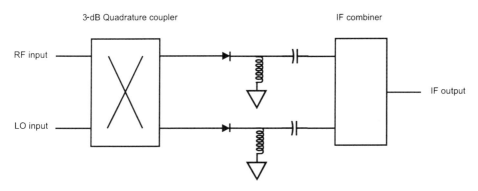

Figure 6.21 Balanced mixer using a 90-degree hybrid coupler and two diodes.

the 90-degree hybrid the most frequent choice for broadband microwave mixers [10].

The performance characteristics for a few diode mixers, as listed in the ANZAC catalog, are shown in Table 6.2 to illustrate mixer performance potential.

6.4.2 Transistor Mixers

It clearly is possible to make transistor mixers using either BJT or FET devices. These can be single-ended mixers, balance mixers, and double-balanced mixers.

Table 6.2
Some Performance Characteristics for Mixers

Characteristics	Standard RF Mixers		Termination-Insensitive Mixers	
Model number	MD-108	MD-150	MD-162	MD-164
Frequency range				
RF/LO (MHz)	5–500	700–2,000	1,000–7,000	500–9,000
IF (MHz)	DC-500	DC-300	10–2,000	10–2,000
Conversion loss (dB typical)	5.6	6.2	6.0	6.5
Isolation				
LO-RF (dB typical)	45	35	25	22
LO-IF (dB typical)	40	20	20	27
RF-IF (dB typical)	25	24	22	25
LO drive (dBm)	+7	+7	+13	+13

Figure 6.22 shows two examples of single-ended FET mixers. The circuit in Figure 6.22(a) adds the LO signal and the RF signal at the gate input to the mixer. The circuit in Figure 6.22(b) has the LO applied to the source terminal of the FET and the RF signal applied to the gate terminal of the FET.

One advantage for transistor mixers is that they can provide conversion gain rather than conversion loss. BJTs can have conversion gains on the order of 20 dB,

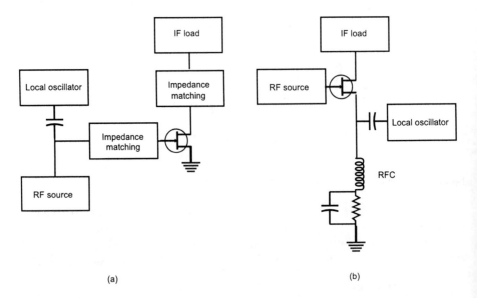

Figure 6.22 Example FET mixers: (a) single-gate-input-type mixer and (b) gate and source input type FET mixer.

while FETs can have conversion gains on the order of 10 dB. FETs produce less intermodulation and cross-modulation distortion; for that reason, they are preferred over BJTs for high-frequency mixers. Both JFETs and MOSFETs are used, with MOSFETs generally exhibiting higher power gain.

MESFETs are used at microwave frequencies for transistor mixers.

References

[1] Krauss, Herbert L., Charles W. Bostian, and Frederick H. Raab, *Solid State Radio Engineering*, New York: John Wiley & Sons, 1980, Chaps. 4, 12, and 13.
[2] Dye, Norm, and Helge Granberg, *Radio Frequency Transistors: Principles and Practical Applications*, Boston: Butterworth-Heinemann, 1993, Chapter 1.
[3] Sivan, L., *Microwave Tube Transmitters*, London: Chapman and Hall, 1994.
[4] Whitaker, Jerry C., *Power Vacuum Tubes Handbook*, New York: Van Nostrand Reinhold, 1994.
[5] Krauss, Bostian, and Raab, op. cit., Chap. 5.
[6] Krauss, Bostian, and Raab, op cit., Chaps. 6 and 7.
[7] Kennedy, George, *Electronic Communication Systems*, 3rd ed., New York: McGraw-Hill, 1985, Chap. 12.
[8] Kennedy, op. cit., pp. 341–343.
[9] *1989 ANZAC RF and Microwave Signal Processing Components Catalog*, Adams Russell Components Group, Burlington, MA, pp. 234–241.
[10] Howe, Harlan, Jr., *Stripline Circuit Design*, Norwood, MA: Artech House, 1974, pp. 268–269.

Selected Bibliography

Vendelin, G. D., *Design of Amplifiers and Oscillators by the S-Parameter Method*, New York: John Wiley & Sons, 1982.

Vendelin, George D., Anthony M. Pavio, and Ulrich L. Rohde, *Microwave Circuit Design Using Linear and Nonlinear Techniques*, New York: John Wiley & Sons, 1990.

Medley, Max W., Jr., *Microwave and RF Circuits: Analysis, Synthesis and Design*, Norwood, MA: Artech House, 1993.

Maas, Stephen A., *Microwave Mixers*, Norwood, MA: Artech House, 1986.

CHAPTER 7

Modulators and Demodulators

This chapter discusses two additional major devices used in communications, radar, and other RF systems: modulators and demodulators. Modulators are used in transmitter systems, demodulators in receivers. Demodulation devices are sometimes called detectors.

7.1 MODULATORS

This section discusses some of the more important types of modulator systems, including AM, FM, and phase modulators. Some of the information presented here is based on [1–3].

7.1.1 Modulators for Conventional Amplitude Modulation

7.1.1.1 Plate-Modulated Class C Amplifiers

Vacuum tube power amplifier stages are used predominantly for high-power AM modulators at VHF and lower frequencies. That is because of the need for transmitter power greater than that possible with transistor amplifiers. Tube types include triodes and tetrodes. A triode has a cathode for emitting electrons, a single control grid for controlling the flow of electrons, and a plate for collecting electrons. Thus, it has three electrodes. A tetrode has those same elements plus a second grid, called a screen grid, between the control grid and the plate. Tetrodes have the advantage that the screen grid greatly reduces coupling between the plate and the grid circuit, thereby reducing chances for oscillation and distortion and eliminating the need for a neutralizing feedback circuit. Power outputs for standard AM broadcast at MF typically are in the range of 10–100 kW. Thus, transmitter systems of this type require high-power audio frequency (AF) modulators, large dc power supplies, and large cooling systems. Because efficiency is important, class C operation normally is used for the output stage.

Figure 7.1 is a block diagram of a high-power AM transmitter system used for standard AM broadcast at MF. frequencies. The modulator is a plate-modulated class C RF power amplifier that uses a high-power triode or tetrode vacuum tube as the output stage. The output stage has a high-power audio transformer in series with the high-voltage supply. That permits the plate voltage to the amplifier to be modulated up and down in amplitude in accordance with the input signal.

The signal source for this transformer is an amplifier chain consisting of a first-stage AF processing and filtering circuit, an AF preamplifier, an AF class B power amplifier, and a high-power class B modulator amplifier. Up to one-third of the total output power for the system must be supplied by the high-power class B modulator amplifier.

In Figure 7.1, the carrier signal source starts with a crystal oscillator, which generates a CW signal at the desired RF frequency. That is followed by a class A RF buffer amplifier and a class C RF power amplifier. The RF carrier signal is fed to the control grid of the modulated class C power amplifier by means of a transformer with the appropriate signal level and bias so the amplifier operates in class C. A tuned output transformer is used to convert the pulse-type signal from the plate-modulated class C RF output amplifier to a modulated sine-wave output.

7.1.1.2 Grid-Modulated Class C Amplifiers

Another type of amplitude modulator sometimes used for high-power transmitters is a grid-modulated class C amplifier. In a grid-modulated triode class C amplifier, the RF input is provided by one transformer and the AF input by a second transformer connected in series with the first.

The grid of the triode is biased negatively, and a neutralizing feedback signal is provided to it from the output transformer. The output transformer is a single-tuned transformer with the primary tuned to the desired output frequency. The secondary typically is untuned.

The triode current is a series of current pulses corresponding in frequency to the RF input signal and in amplitude to the AF input. The tuned output transformer converts the signal to a sine-wave signal by means of the so-called fly-wheel action of the parallel resonant circuit.

7.1.1.3 Collector-Modulated Class C Transistor Amplifiers

Transistors also are used in AM class C power amplifiers. Figure 7.2 illustrates the case of a collector-modulated class C push-pull transistor amplifier. Such amplifiers can be used for AM for lower power applications. In the system shown in

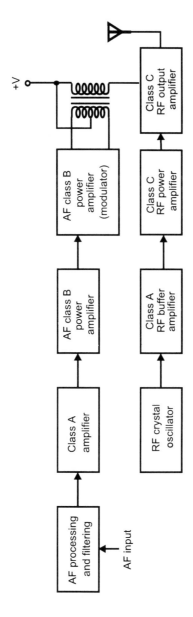

Figure 7.1 Block diagram of a AM transmitter.

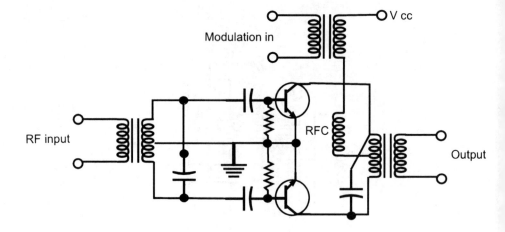

Figure 7.2 Collector-modulated class C transistor push-pull amplifier. (*After:* [2].)

Figure 7.2, the modulating signal is fed to the primary winding of an untuned AF transformer. The secondary winding of the transformer is in series with the V_{cc} power supply. It is connected to the center tap of the primary winding of the output transformer with the use of an RF choke. The voltage applied to the transistor collectors moves up and down in response to the modulation input. The RF signal is provided to the bases of the BJT devices with a tuned center-tapped RF transformer. The base of each transistor is biased to ground. With the bias set at 0V, the operation of the transistor amplifiers is large-angle class C.

7.1.1.4 Other Types of Conventional Amplitude Modulators

Other types of conventional amplitude modulators occasionally are used, including FET gate-biased AM systems in which the AF signal provides modulation of the gate bias for a class C FET power amplifier. This type of system typically uses an RF choke for dc power input and a capacitively coupled-tuned, parallel-resonant, output circuit. FETs also can be used in push-pull modulator configurations.

7.1.2 Modulators for Double-Sideband Modulation

Figure 7.3 is a simplified system block diagram for a DSB transmitter system. The first RF frequency is provided by a crystal oscillator. The low-power signal is fed to a balanced modulator through a buffer amplifier. A typical frequency for the signal might be in the range of 10–70 MHz.

The second signal to the balanced modulator is the input modulating signal, which is fed to the modulator through a buffer amplifier. The signal is assumed

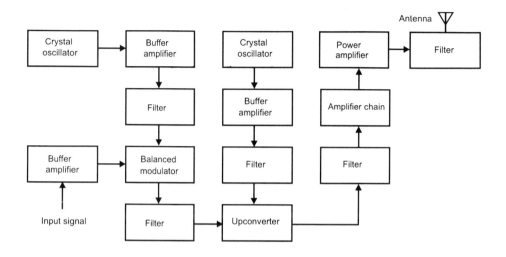

Figure 7.3 DSB transmitter system.

to be an audio, a video, or a digital signal. A balanced modulator is the same circuit as a balanced mixer, discussed in Chapter 6.

The output of the balanced modulator is a set of sidebands on either side of the carrier, with the carrier having very low amplitude. The desired set of sidebands is selected by an output filter. An option might be to inject a small amount of carrier signal through an add circuit following the balanced modulator to make it easier for the receiver to lock on to the carrier frequency. A typical transmitter output frequency might be in the VHF, UHF, or microwave frequency range. The output of the modulator is then passed through an up-converter, which converts the signal to the desired output frequency.

The next stage is an amplifier chain that increases the power level of the up-converted signal to the desired drive level for the power amplifier. Amplifiers for the chain are class A and class B amplifiers.

The power amplifier can be a solid-state amplifier, a triode or tetrode, or a microwave tube power amplifier such as a klystron or a traveling-wave tube. Output to the antenna is through a suitable lowpass or bandpass filter.

Figure 7.4 shows a balanced modulator and the associated frequency spectrum for a DSB system. For simplicity, a single-frequency sine wave is assumed for the input modulating signal. The passband of the modulator output filter is indicated on the spectrum diagram.

7.1.3 Vestigial-Sideband Modulators

Figure 7.5 shows a VSB modulator. The block diagram looks like that of the DSB modulator. The difference is that the modulator includes a balanced modulator

182 | RF Systems, Components, and Circuits Handbook

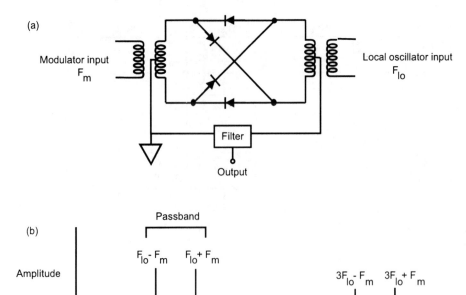

Figure 7.4 A balance modulator for a DSB system: (a) schematic diagram of balanced modulator and (b) frequency spectrum for DSB modulation.

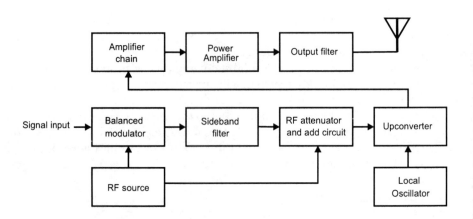

Figure 7.5 Block diagram of a VSB modulation system.

with sideband filter and reduced carrier injection. The sideband filter removes only part of one sideband. The remainder is sent along with the reduced carrier and the other full sideband. The up-converter, amplifiers, power amplifier, and output filter are the same as those described for DSB modulation.

7.1.4 Modulators for Single-Sideband Modulation

Figure 7.6 shows an SSB transmitter. The block diagram looks like that of the DSB transmitter. The difference is that the modulator includes a balanced modulator with either a complete sideband filter or a balanced modulator with SSB by a phase-shift method. The up-converter, amplifiers, power amplifier, and output filter are the same those described for DSB and VSB modulation.

Figure 7.7 shows a balanced modulator and associated frequency spectrum used for SSB modulation. The modulating signal is assumed to be a single-frequency sine wave. The passband for the output filter is indicated on the diagram for the upper sideband. The lower sideband also could be selected, if desired, rather than the upper sideband.

The sideband-suppression filter must have very sharp cutoff characteristics, and the IF must be quite low for most SSB applications. In a typical example, the filter's response must change from near zero attenuation to near full (30 dB or more) attenuation over a range of only 600 Hz. To obtain a filter response curve with skirts as steep as those suggested, the Q of the filter (reactance/resistance) must be very high. Possible filter types include LC filters, crystal filters, ceramic filters, mechanical filters, and SAW filters. Because of Q limitations, LC filters cannot be used for IF values greater than about 100 kHz. Mechanical filters have

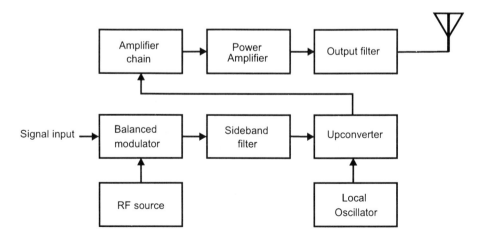

Figure 7.6 Block diagram of a transmitter using the filter method of SSB modulation.

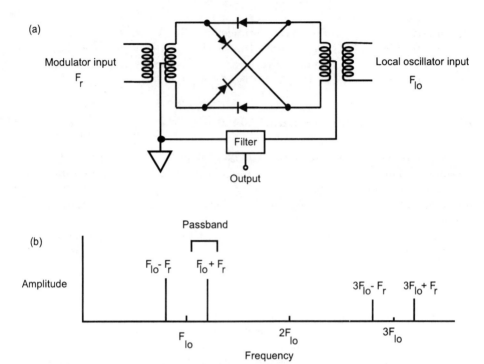

Figure 7.7 Example of a balance modulator: (a) balanced modulator circuit and (b) sample frequency spectrum for SSB suppressed carrier modulation.

been used at frequencies up to 500 kHz and crystal filters and ceramic filters up to about 30 MHz. SAW filters can be used up to 2 GHz.

The phase-shift method of producing SSB suppressed-carrier signal is shown in Figure 7.8. This method avoids filters and some of their attendant disadvantages. The audio input signal is applied to two all-pass networks with phase shifts that differ by 90 degrees over the frequency range of interest. The signals are then applied to two balanced modulators along with in-phase and quadrature (90-degree out of phase) signals of the desired RF frequency. The in-phase and quadrature signals can be obtained by digital frequency division of the output of a variable-frequency oscillator operating at four times the output frequency. Not shown in Figure 7.8 is a gating circuit that sets the initial state of the flip-flops and thus determines which sideband is produced. The outputs of the two balance modulators are summed and then amplified to the desired level.

The operation of an SSB modulator that uses the phase shift method is demonstrated as follows:

The equation of a wave with the carrier removed is

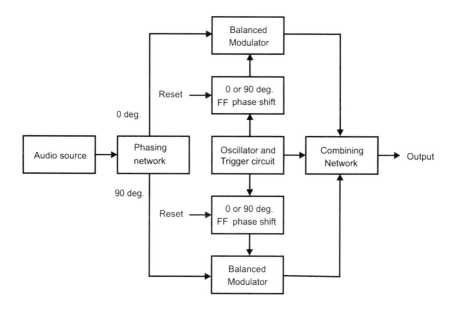

Figure 7.8 Phase-shift method of producing SSB suppressed-carrier modulation. (*After:* [1].)

$$e_1 = mE \sin \omega_m t \sin \omega_c t \quad (7.1)$$

This is the case for the output of the first modulator. When both modulating and carrier frequencies are shifted 90 degrees, as in the case of the second balanced modulator, the equation for a wave with the carrier removed is

$$e_2 = mE \cos \omega_m t \cos \omega_c t \quad (7.2)$$

Adding (7.1) and (7.2) gives

$$\begin{aligned} e_1 + e_2 &= mE \sin \omega_m t \sin \omega_c t + mE \cos \omega_m t \cos \omega_c t \\ &= mE/2[\cos(\omega_c - \omega_m)t - \cos(\omega_c + \omega_m)t] \\ &\quad + mE/2[\cos(\omega_c - \omega_m)t + \cos(\omega_c + \omega_m)t] \\ &= mE[\cos(\omega_c - \omega_m)t] \end{aligned} \quad (7.3)$$

Equation (7.3) corresponds to the equation of the lower sideband. If the polarity of one of the modulating voltages or one of the RF voltages is reversed, the other sideband would appear at the output terminals.

Possible variations of SSB are SSB with full carrier and SSB with reduced carrier. The carrier can be added after generation of the SSB signal.

7.1.5 Modulators for Frequency-Division Multiplex

Figure 7.9 is a simplified block diagram of an FDM modulator system, used to send many signals in parallel by using different frequencies for each channel. In this system, the signals are sent using SSB suppressed-carrier modulation with the filter method for generation of SSB. Each channel has its own amplifier, oscillator, balanced modulator, and sideband filter. Each generated SSB signal is added along with a pilot signal and passed through a group bandpass filter. Table 7.1 lists oscillator frequencies and SSB modulator output frequencies for the 12 voice channels and pilot channel.

The next step up from a group is the basic supergroup, which consists of five groups. By proper choice of up-converter oscillator inputs, these groups occupy frequency channels as shown in Table 7.2.

These five groups are added along with a pilot channel at 547.92 kHz.

Supergroups can be combined to form mastergroups, supermastergroups, and so on. The total set of sideband frequencies then can be shifted up in frequency to the desired transmit frequency.

7.1.6 Modulators for Standard Frequency Modulation

Frequency-modulated signals often are produced at low power levels and amplified by amplifier chains. The modulation can be accomplished either directly by variation

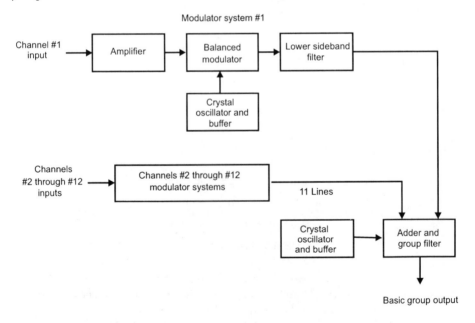

Figure 7.9 Basic 12-channel group translating equipment for FDM. (*After:* [4].)

Table 7.1
Frequencies for the 12 Voice Channels and Pilot Channel

Channel Number	Crystal Oscillator Output Frequency (kHz)	SSB Modulator Output Frequency Band (kHz)
1	108	104.6 to 107.7
2	104	100.6 to 103.7
3	100	96.6 to 99.7
4	96	92.6 to 95.7
5	92	88.6 to 91.7
6	88	84.6 to 87.7
7	84	80.6 to 83.7
8	80	76.6 to 79.7
9	76	72.6 to 75.7
10	72	68.6 to 71.7
11	68	64.6 to 67.7
12	64	60.6 to 63.7
Pilot channel	104.08	

Table 7.2
Frequency Channels of Groups

Group Number	Frequency Bands Used (kHz)
1	312–360
2	360–408
3	408–456
4	456–504
5	504–552

of the frequency of an oscillator by the input signal or indirectly by phase modulation and other methods. To achieve good linearity, most frequency modulators produce a smaller modulation index or frequency deviation than is desired in the transmitter output. Frequency multipliers multiply frequency deviation and modulation index, as well as frequency. Thus, they often are used in the transmitter chain.

Figure 7.10 is an example of an FM modulator system. In this system, an AF source provides the modulating signal to an FM source consisting of a varactor-tuned crystal oscillator. The FM source has as its output a center frequency f with frequency deviation df. The signal is fed to the frequency multiplier having a multiplication factor of k. The output of this stage has a center frequency of kf and a deviation of k df. That signal is then up-converted to the desired transmit

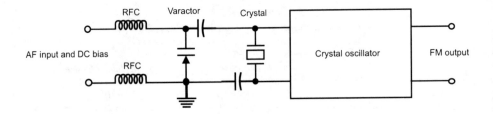

Figure 7.10 Varactor-tuned crystal oscillator FM modulator.

frequency and amplified with an amplifier chain and a power amplifier. The output then is filtered and fed to the antenna.

The varactor-tuned crystal oscillator in Figure 7.10 is a direct FM modulator, a system frequently used in portable and mobile transmitters. Frequency multiplication often is used with this system because of the small frequency deviation that is possible.

Another FM modulator is the phase-shift modulator. The phase-shift method of producing FM is shown in Figure 7.11.

The phase-shift modulator in Figure 7.11 is an indirect FM modulator. This system uses a controllable conductance in combination with a fixed reactance to vary the phase delay of signals. Since this modulator directly varies the phase of the signal rather than the frequency, it is necessary to integrate the input AF signal before the modulator. The PLL method for producing FM is shown in Figure 7.12. This is also an indirect FM modulator. Again, the system produces phase modulation

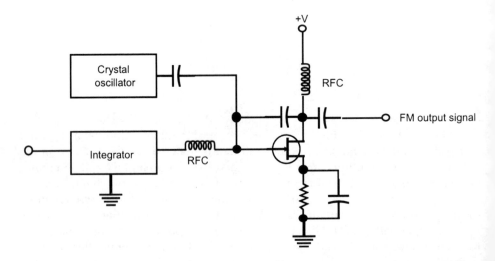

Figure 7.11 Phase-shift method of producing FM.

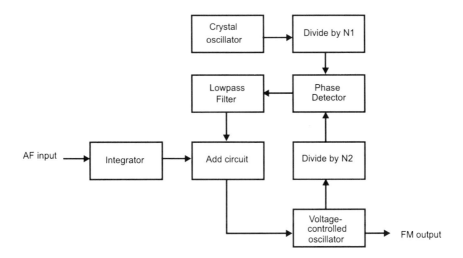

Figure 7.12 PLL method for producing standard FM.

rather than FM, and it is necessary to integrate the AF signal before modulation. The signal is introduced into the system as an error voltage. The system responds to the injected error voltage by adjusting the VCO to produce the specified phase shift between its output and the reference signal.

It is common practice with FM modulator and receiver systems to use preemphasis and deemphasis as a means of improving the link signal-to-noise ratio. Noise has a greater effect on the higher modulating frequencies than on the lower ones. If the higher frequencies are artificially boosted in amplitude before modulation, noise will have less effect. This process is known as preemphasis. To recover the audio signal in its original form, we must reduce the amplitude of the higher frequencies at the receiver to the same degree they were boosted. That process is known as deemphasis. The net result of using this combination is that the signal-to-noise ratio is improved for the higher frequencies.

7.1.7 Modulators for Frequency-Shift Keying

FSK modulators work the same as standard frequency modulators except that only two frequency states are used. This type of modulation frequently is used for digital data transmission.

7.1.8 Modulators for Phase-Shift Keying

Modulators for PSK were introduced in Chapter 5. Some of the information presented here is adapted from [3].

7.1.8.1 Binary Phase-Shift Keying

With BPSK, the signal phase is keyed between two phase states 180 degrees apart. Each pulse represents just one bit of information (1 or 0). An example of a modulator for such a system is shown in Figure 7.13.

In the system in Figure 7.13, a balanced modulator is used with one input from an LO at an assumed frequency of 70 MHz and the other input from a two-level amplitude modulator. The amplitude modulator is controlled by a logic circuit. The output of the balanced modulator normally is fed to an amplifier.

7.1.8.2 Quaternary Phase-Shift Keying

Four-phase states are used with QPSK with phase states separated by 90 degrees. Each pulse represents two bits of information (00, 01, 10, or 11). An example of such a modulator is shown in Figure 7.14. In this system, an LO at an assumed frequency of 70 MHz feeds a 90-degree phase splitter. The outputs of the phase splitter, in-phase and quadrature signals, are fed to two balanced modulators, which are also fed by two two-level amplitude modulators. The amplitude modulators are controlled by a logic circuit. The outputs of the balanced modulators are fed to an in-phase power combiner circuit. The four possible vectors of +45, +135, −45, and −135 degrees are shown in Figure 7.14.

7.1.8.3 $\pi/4$-DQPSK Modulation

An important variation of QPSK modulation is $\pi/4$-DQPSK modulation. This is the Telecommunication Industries Association (TIA) standard for digital cellular telephones. The D in DQPSK indicates that the four-state phase modulation is a differential phase modulation in which the phase reference is derived from the previous RF pulse. Information is sent by changes in phase from pulse to pulse rather than in absolute phase. The $\pi/4$ term means that the minimum phase-shift with respect to the reference is 45 degrees. Possible differential phase states thus are +45, +135, −135, and −45 degrees.

7.1.8.4 Eight-State Phase-Shift Keying

Eight phase states are used with 8-PSK with the phase states separated by 45 degrees. Each pulse represents three bits of information. An example of a modulator for such a system is shown in Figure 7.15. In this system, an LO at an assumed frequency of 70 MHz feeds a quadrature phase power divider. The outputs of the phase splitter are fed to two balanced modulators, which also are fed by two four-level amplitude modulators. The amplitude modulators are controlled by a logic circuit.

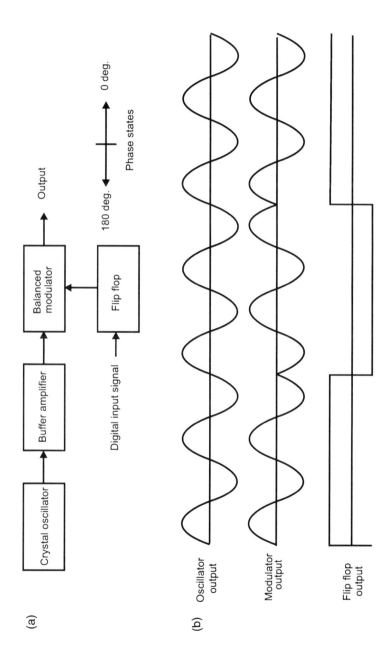

Figure 7.13 BPSK modulator: (a) circuit diagram and (b) sample waveforms.

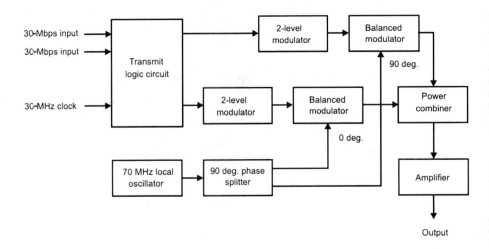

Figure 7.14 QPSK modulator. (*After:* [3].)

The outputs of the balanced modulators are fed to a power combiner circuit. The output of that circuit is fed to an amplifier. The eight possible vector angles are +22.5, +67.5, +112.5, +157.5, −22.5, −67.5, −112.5, and −157.5 degrees.

7.1.8.5 16-Quadrature Amplitude Modulation

It is possible to use 16-PSK, providing four bits per pulse capability, but this type of system currently is not used because another 16-state keying system is better. The improved system is 16-QAM. This system uses four phase states and four possible amplitudes. This 4 × 4 combination results in 16 possible states. Figure 7.16 shows a block diagram for a 16-QAM system and lists 16 possible amplitude and phase combinations. Recently, 16-QAM modulation has been used extensively in microwave relay systems.

7.1.9 Modulators for Pulse Code Modulation Time-Division Multiplex Modulation

A 24-channel PCM TDM modulator system is shown in Figure 7.17. The 24-channel groups have a sampling rate of 8,000 samples per second, 8 bits (256 sampling levels) per sample, and a pulse width of about 0.625 ms. The sampling interval is 125 ms, and the period required for each pulse group is 5 ms. Each 125-ms frame is used to provide 24 adjacent channel time slots, with the twenty-fifth slot used

Modulators and Demodulators | 193

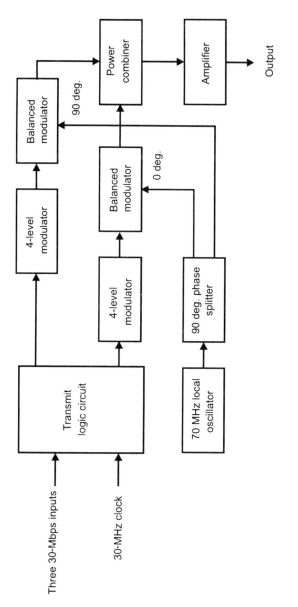

Figure 7.15 8-PSK modulator. (*After:* [3].)

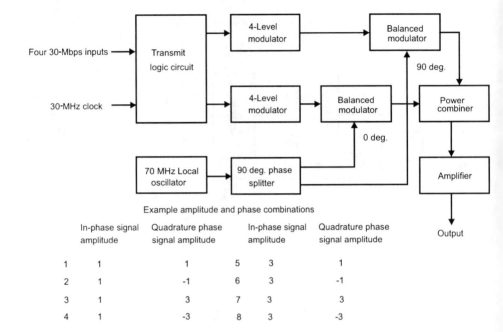

Figure 7.16 Block diagram of a 16-QAM modulation system. (*After:* [3].)

for synchronization. Sampling is done simultaneously on each channel. Sampling circuits are followed by delay lines, each with a different delay. Delays are 0, 5 μs, 10 μs, 15 μs, and so on, to the twenty-fourth channel, which is delayed 115 μs. The outputs of the delays are added to provide a single bit stream. The process is repeated 8,000 times per second. The bit stream is used to frequency- or phase-modulate the carrier of the transmitter.

The second multiplex level, illustrated in Figure 7.18, provides 96 channels. The bit rate is 6.312 Mbps. Some of the bits are used for synchronization and others for housekeeping functions. The method of producing secondary multiplex levels consists essentially in dividing by 4 the pulse widths in the primary level signal and using the slots thus vacated to combine four primary streams, using delay lines and an adder to convert the four bit steams to one. The bit streams are then used to frequency- or phase-modulate the carrier of the transmitter.

7.1.10 Time-Division Multiple Access

Multiple access means that a multiplicity of participating signal sources can use the modulation system simultaneously by occupying time slots in a time frame. The

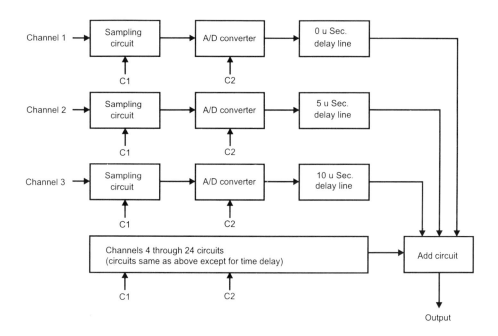

Figure 7.17 24 Channel PCM TDM modulator system.

TIA standard for North America digital cellular telephones uses TDMA. Details about this important system are presented in Chapter 9.

7.2 DEMODULATORS OR DETECTORS

This section discusses amplitude, frequency, and phase demodulators and detectors. These circuits follow the IF amplifiers and are used to recover the baseband signals. Some of the information presented in this section is adapted from [5,6].

7.2.1 Amplitude Modulation Detectors

An AM detector is a circuit that converts an amplitude-modulated RF or IF signal to an audio, video, or pulse signal of the same form that was originally used to modulate the transmitted signal. One important type of AM detector is the envelope detector circuit. A simple form of that type of AM detector is shown in Figure 7.19(a). The circuit involves a diode rectifier followed by a lowpass filter. The rectifier converts the full-wave signal to a half-wave signal of either positive or negative polarity, depending on the direction of the diode. The lowpass filter is simply a resistor and capacitor in parallel following the rectifier. That allows for a

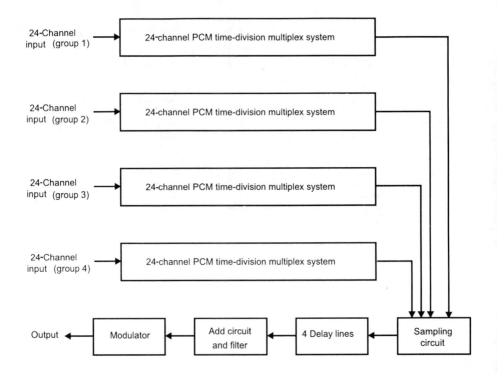

Figure 7.18 96-channel PCM TDM modulator system.

fast rise time and a slow fall time, as desired. The combination makes the detector a peak-type detector, and the circuit output follows the envelope of the modulated signal. A form of distortion known as diagonal clipping results if the fall time is made too slow.

A slightly more complex envelope detector that is often used with AM receivers is shown in Figure 7.19(b). That circuit includes capability for providing volume control for the output signal and an AGC output to control the gain of the receiver.

7.2.2 Product Detectors

A product detector (Figure 7.20) is simply a balanced modulator or balanced mixer circuit, as discussed in Chapter 6. A product detector can be used to demodulate SSB signals, DSB signals, and standard AM signals. The mixer may be either a diode mixer or a transistor mixer. It has one input from the IF amplifier output and the second input from a reference crystal or other oscillator. The frequency of the reference oscillator is the same as the carrier frequency. The difference frequency for the mixer is the desired modulation frequency, which can be obtained using a simple lowpass filter.

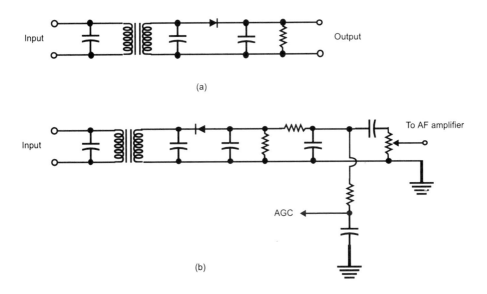

Figure 7.19 AM detectors or demodulators: (a) diode envelope detector and (b) practical AM receiver detector.

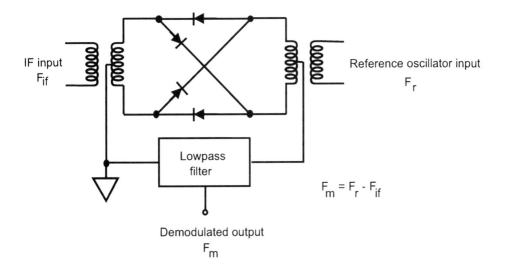

Figure 7.20 Product detector.

In the case of SSB detection, the requirements on frequency and phase of the reintroduced carrier at the phase detector are not severe, since only speech is transmitted. Small errors in frequency and phase of the recovered audio signal go unnoticed. The carrier oscillator frequency usually can be tuned manually for best reception.

Detection of DSB signals requires the reintroduced carrier to be in exact frequency and phase with the carrier that would have been there if the signal were standard AM to avoid distortion. If the frequency is incorrect, the upper and lower sidebands will produce different beat frequencies with the carrier. If the phase of the carrier is not correct, the amplitude of the audio output is reduced. If a pilot reduced-amplitude carrier is transmitted with the DSB wave, it can be used to synchronized the VCO in a PLL.

Detection of AM waves with a product detector provides performance superior to that of the simple envelope detector because it is a coherent process. Although the carrier component is present in the AM wave, the product detector requires a separate carrier input of the same frequency and phase. That is obtained from the VCO output of a narrowband PLL that is locked to the carrier component of the AM wave.

A coherent I and Q detector, shown in Figure 7.21, is a circuit often used in radar receivers. The circuit does not use a PLL but rather a fixed frequency oscillator as the signal source for two balanced mixers. One of the two mixers uses a 90-degree phase-shift circuit so the two signals to the balanced mixers are in phase quadrature. The balanced mixers are followed by lowpass filters.

One reason for having I and Q outputs is so the phase angle of the signal can be detected as well as the amplitude. When there is a Doppler frequency shift, the amplitude of the detected signals will change at the Doppler frequency rate.

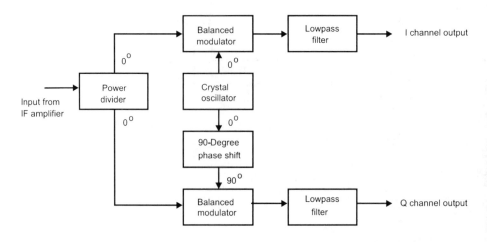

Figure 7.21 I and Q product detectors.

By signal processing, the Doppler frequency can be measured using the two detected signals.

7.2.3 Frequency Modulation Detector Concepts

A number of FM demodulators or detectors use diodes and transformers. One important type is the Foster-Seeley discriminator, illustrated in Figure 7.22. The main shortcoming of the Foster-Seeley discriminator is that any AM on the incoming signal will be demodulated. A good limiter therefore must precede the discriminator for satisfactory operation. That requirement has eliminated the Foster-Seeley discriminator from almost all mass production entertainment receiver circuits in favor of the ratio detector.

The Foster-Seeley discriminator uses a transformer system that includes a tuned primary circuit, a tuned center-tapped secondary circuit, and a coupling capacitor between one side of the primary winding and the center tap of the secondary winding. The two ends of the secondary winding are connected to half-wave rectifiers and lowpass filters. The output is between the two filters.

The operation of the Foster-Seeley discriminator can be understood using the vector diagrams in Figure 7.23. In those diagrams, the two voltages V2 and –V2 are the voltages across the two parts of the secondary winding. The voltage V1 is the voltage at the primary winding. The vector sum of V1 and V2 is Va. Likewise,

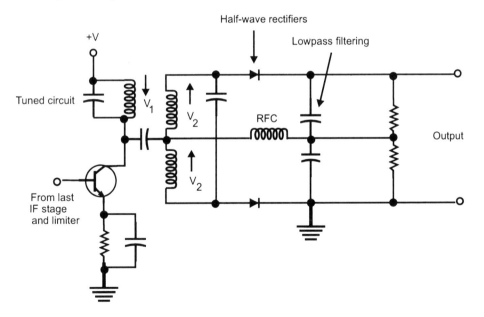

Figure 7.22 Foster-Seeley discriminator. (*After:* [5].)

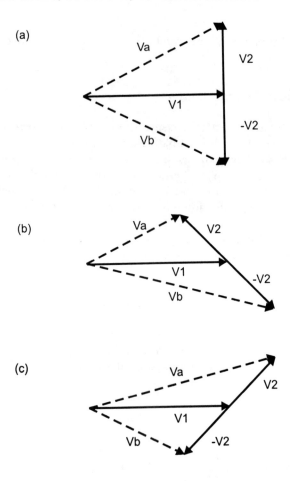

Figure 7.23 Vector diagrams for Foster-Seeley frequency discriminator: (a) at resonance frequency; (b) below resonance frequency; and (c) above resonance frequency. (See Figure 7.22 for V_1 and V_2.)

the vector sum of V1 and −V2 is Vb. At resonance, Va and Vb are equal, as shown in Figure 7.23(a). Below resonance, the vectors are as shown in Figure 7.23(b). Va in that case is less than Vb, as a result of the change in phase shift with respect to V1 for V2 and −V2. Above resonance, the vectors are as shown in Figure 7.23(c). Va in that case is greater than Vb. Again, that is a result of the change in phase shift with respect to V1 for V2 and −V2.

Figure 7.24 shows the circuit diagram for one type of ratio detector. This circuit is not very sensitive to AM on the incoming signal and does not require a limiter.

The discriminator in Figure 7.24 uses the same phase-shift and vector addition concepts as the Foster-Seeley discriminator. The main differences are the shift in

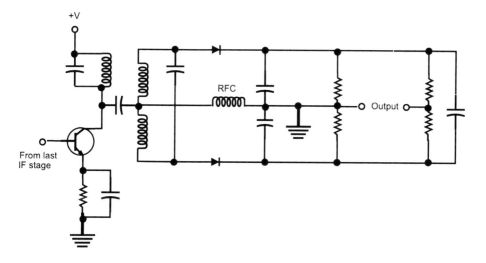

Figure 7.24 Basic ratio detector circuit. (*After:* [5].)

position for the ground, the addition of a resistor and capacitor network across the two rectifier outputs, and the location of the output terminals.

A number of other possible ratio detectors are also used as FM demodulator systems. They are similar to the system in Figure 7.24, with only small changes in implementation.

Another important type of FM detector is the PLL detector, illustrated in Figure 7.25. As the frequency changes, the error signal needed to track the frequency changes. Thus, that error signal is a measure of the signal frequency. It is used as the output signal for the circuit. Performance is at least as good as that of the ratio detector.

Another FM detector sometimes used is a zero-crossing detector. In one version, the IF output signal is first hard-limited and converted into triggers at the zero crossings. The zero crossings are used to trigger a monostable multivibrator

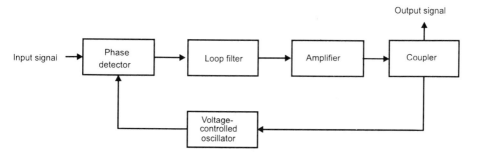

Figure 7.25 PLL FM detector.

with a fixed pulse on time. The two outputs from the multivibrator are then fed to lowpass filters. The outputs of the filters are fed to a differential amplifier that indicates the difference between the two signals.

7.2.4 Phase Detectors

A double-balance diode mixer normally is used for the detection of BPSK. One input for the mixer is the phase-modulated carrier frequency from the IF amplifier. The other input is the reference carrier frequency from a crystal oscillator or other oscillator system. The output of the mixer is a detected signal. Synchronization systems can be used to obtain the correct frequency and phase angle for the phase reference. A system of this kind is shown in Figure 7.26.

Figure 7.27 shows the case of a coherent demodulator for QPSK. This demodulation provides two bits per baud (pulse).

In the circuit in Figure 7.27, the input from the IF amplifier is fed through a bandpass filter to a three-way power splitter. One output from the power splitter is used for carrier recovery. The carrier is then fed to a coupler whose outputs are in phase quadrature. The other two outputs from the input power splitter are fed to two balanced mixers. The two reference signals for those mixers are from the quadrature phase coupler.

The two mixers are followed by lowpass filters, which provide signals to two-way power dividers. One output from each divider is fed to a symbol-timing-recovery circuit. The other output is fed to detectors and timing circuits, which determine the one or zero states for the signals and provide the necessary time delay so that, when added, the parallel signal is converted to a serial digital signal.

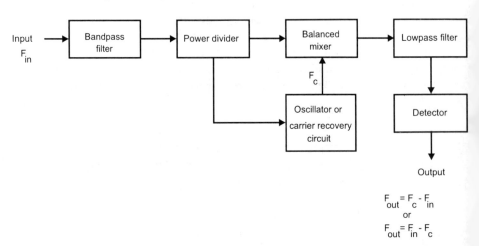

Figure 7.26 Phase detector for BPSK.

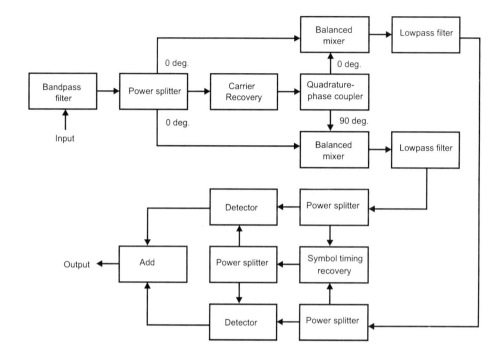

Figure 7.27 Block diagram of a QPSK demodulator.

References

[1] Krauss, Herbert L., Charles W. Bostian, and Frederick H. Rabb, *Solid State Radio Engineering*, New York: John Wiley & Sons, 1980, Chap. 8.
[2] Kennedy, George, *Electronic Communication Systems*, 3rd ed., New York: McGraw-Hill, 1985, Chaps. 3–5.
[3] Keiser, Bernhard E., and Eugene Strange, *Digital Telephony and Network Integration*, 2nd ed., New York: Van Nostrand Reinhold, 1995, Chap. 6.
[4] Kennedy, George, *Electronic Communication Systems*, 3rd ed., New York: McGraw-Hill, 1985, pp. 537–542.
[5] Krauss, Bostian, and Rabb, op. cit., Chaps. 9 and 10. 5. Kennedy, op. cit., Chaps. 3, 4, 6.
[6] Ibid., Ch. 6.

CHAPTER 8

Older Communication Systems

This chapter presents several examples of some of the older types of communication systems, include HF ground-to-air communication systems, VHF and UHF ground-to-air communication systems, VHF and UHF mobile communication systems, FM broadcast systems, microwave relay systems, and satellite relay systems. Chapter 9 discusses newer types of communication and navigation systems, including cellular radio systems, fiber optics communication systems, and global positioning satellite receiver systems, while Chapter 10 presents examples of radar systems. The components and subsystems used in each of these types of systems are mainly those discussed in Chapters 6 and 7. An important objective of this chapter is to show applications for the previously described components and subsystems.

8.1 HF COMMUNICATION SYSTEM USING SINGLE-SIDEBAND MODULATION

Figure 8.1 illustrates an HF communication link between a ground station and an aircraft using ionospheric reflection propagation. This is an important means of communication for aircraft flying over an ocean. The types of communication systems involved are discussed in the following paragraphs.

Figure 8.2 is a block diagram of an HF communication system transmitter and receiver system operating in the 2–30 MHz frequency range. A combination transmitter and receiver system of this kind is called a transceiver. The system in Figure 8.2 is assumed to operate with SSB modulation.

The input to the transmitter is from an input system that includes a microphone and a buffer amplifier. The input system is followed by a SSB, suppressed carrier modulator. The second input to the modulator is a CW signal from a crystal oscillator operating at a frequency of 1.5 MHz. The SSB modulator is followed by

206 | RF Systems, Components, and Circuits Handbook

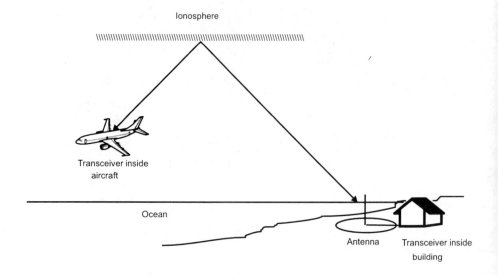

Figure 8.1 Illustration of an HF communication link between an aircraft and a ground station.

a mixer up-converter. The LO for the mixer is a frequency synthesizer, which gives the transmitter the ability to operate on any one of many assigned frequencies with the precision of a crystal reference.

The mixer converts the modulated signal to the desired transmitting frequency. The mixer is followed by an amplifier chain consisting of a number of amplifiers in series, which bring the power level of the modulated signal to the level required to drive the output power amplifier. The power amplifier is assumed to be a push-pull class B power amplifier that uses either BJT or FET transistors. The output power to the antenna is assumed to be in the range of 100–1,000W.

Possible antenna systems for an HF communication system (discussed in Chapter 13) usually involve an impedance matching circuit and possibly a balanced to unbalanced (balun) transformer circuit. The same antennas can be used by both the transmitter and the receiver system by using a duplexer to select either the transmit or receive mode. The transmit and receive antennas normally use a spark-gap-type lightning protection circuit to protect the transmitter and receiver circuits from lightning strikes.

The lower portion of Figure 8.2 is a block diagram of an HF receiver. The receiver is connected to the antenna system by way of the duplexer. The first stage of the receiver is a low-noise RF amplifier. The output of the RF amplifier is connected to a tunable bandpass filter and mixer circuit. The filter is used to limit the band of frequencies seen by the receiver to only frequencies near the desired signal. It is designed specifically to reject the image frequency, which, when mixed with the LO signal, will be converted to the IF amplifier frequency.

Older Communication Systems | 207

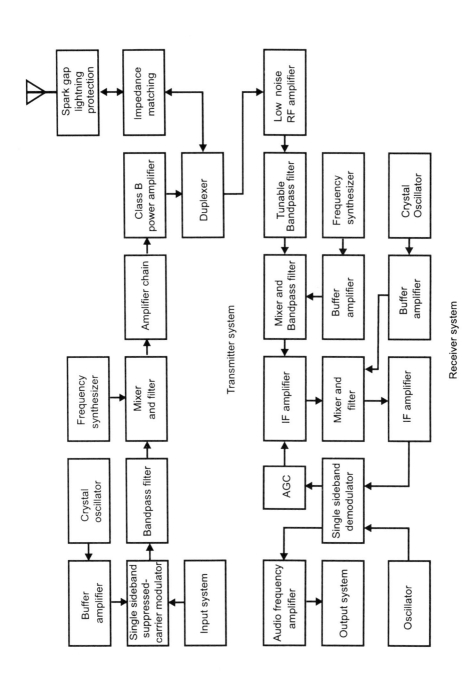

Figure 8.2 HF communication system using SSB modulation.

The system in Figure 8.2 uses dual conversion with two IF amplifier frequencies. Double conversion can provide better image rejection and better narrowband filtering capability than single conversion, at a cost of more complexity. The mixers are assumed to be diode-type double-balanced mixers. The first LO is a frequency synthesizer, which gives the system the ability to operate at any one of a number of assigned frequencies with the accuracy of a crystal oscillator.

A filter at the output of the mixer is used to select the lower sideband, which is at the IF amplifier frequency of 1.5 MHz. The IF amplifier is assumed to consist of two or more transistor class A amplifiers with bandpass filters. An alternative system uses integrated circuit IF amplifiers. Typical filters are single-tuned transformers, ceramic filters, SAW filters, or mechanical filters. The IF amplifier has AGC, with control signals provided by the output amplitude detector. The output of the first IF amplifier is fed to a second mixer and filter combination. The LO for this second mixer is a fixed-frequency crystal oscillator having a frequency of 1.4 MHz. The difference frequency output for the second mixer and filter combination thus is 100 kHz. That signal is fed to the second IF amplifier.

The output of the second IF amplifier is connected to the SSB demodulator. The second input to the demodulator is from a tunable 100-kHz oscillator (the same frequency as the IF amplifiers). That allows the oscillator to be tuned to essentially the same frequency as the carrier frequency of the signal. The demodulator acts as a mixer down-converter to shift the signal down to baseband, and that is the recovered signal.

The output of the SSB demodulator is amplified by an audio frequency amplifier, the output of which is then fed to the output system. The output system may consist of a speaker and headphones.

An amplitude detector also is used at the output of the IF amplifier chain. The detector provides an AGC signal that is fed back to the IF amplifier to control the gain of the system.

8.2 VHF OR UHF GROUND-TO-AIR COMMUNICATION SYSTEM USING EITHER AMPLITUDE MODULATION OR NARROWBAND FM

Figure 8.3 illustrates a VHF or UHF ground-to-air communication system. VHF systems normally use AM, while UHF systems normally use narrowband frequency modulation. Systems of this kind are used by all aircraft flying over land or near land. One or more transceivers are located in the ground stations and in the aircraft.

Figure 8.4 is an example of a VHF or UHF communication transceiver. The system shown uses narrowband FM modulation. A system of this type can be used for both ground-to-air communication and base station-to-ground mobile communications.

The frequency band used for ground mobile units is the 150–156.8 MHz range. The frequency band for FM ground-to-air communication is 225–400 MHz. The system depicted in Figure 8.4 is used for voice-only communication.

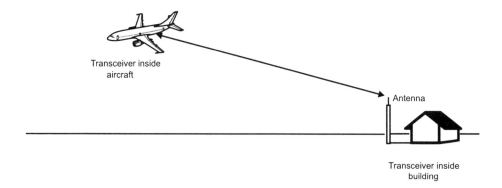

Figure 8.3 Ground-to-air communication system.

The input to the system normally is from a microphone. The main signal conditioning used is filtering and preemphasis. That is followed by a frequency modulator and a frequency multiplier. The output of the frequency multiplier is fed to a bandpass filter. The output of that filter is fed to a balanced mixer and filter. The second input to the mixer is from a frequency synthesizer. The balanced mixer feeds an amplifier chain, which in turn drives a class B transistor power amplifier. The output of the power amplifier is fed to a duplexer. In the transmit mode, the duplexer connects the power amplifier to the antenna system. This system includes impedance matching, lightning protection, and a monopole, dipole, or similar antenna.

In the receive mode, the antenna is connected by the duplexer to a low-noise RF amplifier. The output of this amplifier is fed to a tunable bandpass filter, which feeds a mixer and filter circuit. The second input to the mixer is from a frequency synthesizer and buffer amplifier. The output from the mixer is at the first IF frequency, assumed to be 10.7 MHz. The output is fed to an IF amplifier chain, which includes the necessary bandpass filtering. The amplifier chain has AGC, so the detected output will be essentially constant regardless of input signal strength.

The output of the first IF amplifier is fed to a second mixer and filter. The LO for that mixer is from a crystal oscillator and buffer amplifier. The output of the mixer is fed to the second IF amplifier, which is assumed to have a center frequency of 455 kHz.

The output of the second IF amplifier is fed to a frequency demodulator, which is followed by a filter and an AF amplifier. The output of the amplifier is fed to a signal processing circuit, which provides for deemphasis. The output of that circuit is fed to the output systems, including speakers and possibly headphones.

Many other two-way communication systems use narrowband FM, including ship-to-shore communication systems, ship-to-ship communication systems, and ship-to-air communication systems. The frequency of operation for these systems includes both VHF and UHF.

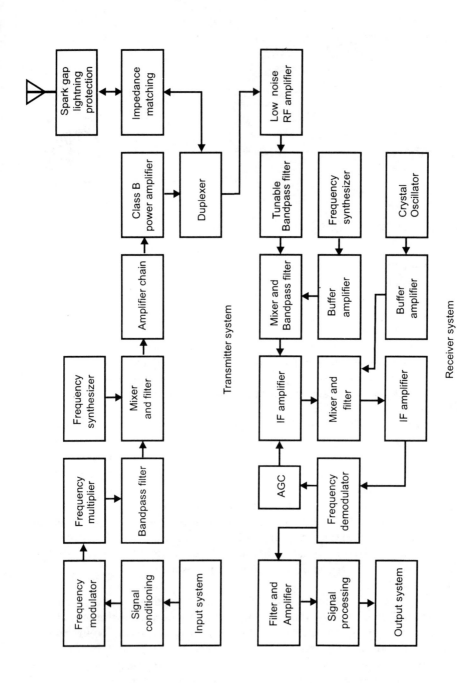

Figure 8.4 Example of a VHF or UHF FM mobile communication system.

8.3 FREQUENCY MODULATION BROADCAST SYSTEMS

Some of the information presented in this section is adapted from [1]. Figure 8.5 illustrates an FM broadcast system, which operates in the 88–108 MHz range. The transmitter generally uses an elevated site for the transmit antenna. That is done to extend the range of the system. The frequencies used are essentially line-of-sight frequencies. Even with an elevated site, signal blockage by buildings and trees is a possibility. In that case, the communication link will experience diffraction propagation loss as well as free-space propagation loss.

A block diagram of the transmitter system is shown in Figure 8.6(a). The system has the capability for stereo operation, so it is necessary to include a signal conditioning circuit with the transmitter for FM multiplex generation.

The input to the transmitter system is from the input system, which includes record players, tape players, compact disc players, switching circuits, controls, and microphones. Microphones for left and right positions with respect to the source are used. In addition, a subsidiary communication authorization (SCA) generator also may be used. The SCA system is used for transmitting background music for stores, restaurants, and so on. The input system signal is fed to a signal conditioning circuit, and the output of this system is fed to a frequency modulator and filter.

The frequency modulator is assumed to be a direct frequency modulator using a varactor-modulated crystal oscillator. That is followed by a frequency multiplier and filter that feeds a balanced mixer. The frequency multiplier is used to increase the modulation index and the frequency of the signal.

The balanced mixer has as its second input a CW signal from a crystal oscillator and buffer amplifier. The frequency output of the balanced mixer is selected by a bandpass filter to be the desired sideband. The output signal is amplified by an amplifier chain and then fed to a 10-kW class B tetrode-tube power amplifier. The output of that amplifier is fed through a coaxial cable to an impedance-matching circuit that includes the capability of feeding an array of six or more wideband turnstile antennas. The turnstile antenna consists of a pair of dipoles placed in the same plane at right angles to each other and fed with equal amplitudes and 90 degrees phase difference. This antenna array provides a pancake-type pattern that is omnidirectional in azimuth and provides a small elevation beam angle. The polarization is horizontal. The antenna covers the full 88–108 MHz frequency band for FM broadcasting.

An FM transmitter system uses frequency preemphasis. The higher frequencies are boosted at the transmitter and correspondingly reduced at the receiver. That is done because noise has a greater effect on the higher modulating frequencies than on the lower ones.

The FM receiver system shown in Figure 8.6(b) typically uses a dipole receiving antenna. Other possible types include Yagi antennas and whip antennas. The output of the antenna is connected to a low-noise RF amplifier, which is used to provide a low-noise figure for the receiver system. The output of the RF amplifier is fed to

212 | RF Systems, Components, and Circuits Handbook

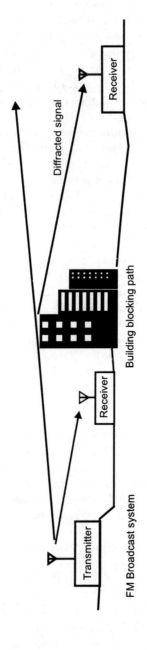

Figure 8.5 Illustration of an FM broadcasting system.

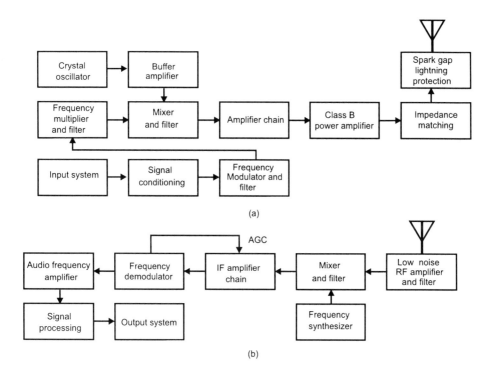

Figure 8.6 Example of an FM broadcast system: (a) transmitter system and (b) receiver system.

a tunable bandpass filter and a down-converter mixer. The mixer can be fed by a frequency synthesizer that acts as the LO for the mixer. The output of the mixer is connected to a bandpass filter. The output of that filter is fed to an IF amplifier chain. The IF amplifier chain includes bandpass filters to band-limit the received signal. That is followed by a frequency and amplitude demodulator system. The amplitude demodulator is used for generating the AGC signal. The frequency demodulator is used to demodulate the FM signal. Frequency deemphasis is used with this circuit.

The output of the FM demodulator is fed through an AF amplifier to a stereo FM multiplex demodulator. The AF amplifier amplifies not only audio frequencies but also frequencies up to about 75 kHz. The outputs from the signal processing circuits are fed to the output systems, such as speakers or headphones.

Signals from the input system are fed to the stereo FM multiplex generator shown in Figure 8.7. The left and right channel inputs are fed to a matrix circuit that provides sum and difference signals. The sum signals are in the frequency range of 50 Hz to 15 kHz. Those are fed to an adder circuit. The difference signals are fed to a balanced modulator that generates DSB modulation with no carrier. The second input to the balance modulator is a CW signal with a frequency of

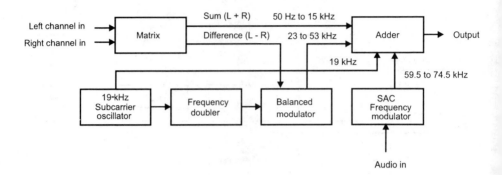

Figure 8.7 Stereo FM multiplex generator with SCA. (*After:* [1].)

38 kHz. The CW signal is generated by a 19-kHz oscillator followed by a doubler. The resulting DSB signal has a spectrum from 38 − 15 kHz to 38 + 15 kHz, or 23 to 53 kHz. This signal is also fed to the adder. The third signal to the adder is the output of the SCA frequency modulator, which has a subcarrier frequency of 67 kHz and a maximum deviation of ±7.5 kHz. The resulting spectrum is in the range of 59.5–74.5 kHz. The fourth signal to the adder is the 19-kHz subcarrier oscillator signal that is used as a pilot signal. The sum of the four signals is then fed to the frequency modulator.

Figure 8.8 is a block diagram of a stereo FM multiplex demodulator circuit. The outputs of the stereo FM multiplex demodulator system are fed to the output systems, which are speaker systems, headphones, and others.

8.4 MICROWAVE RELAY SYSTEMS

An important type of microwave communication system that is used extensively is ground-to-ground microwave relay. Such systems are used to send telephone, radio, television, and other types of signals over long distances. Some of the information presented here regarding microwave relay systems was adapted from [2,3].

A microwave relay system is illustrated in Figure 8.9.

Microwave relay systems currently operate in C-band and the K_u-band. The 4-GHz band uses frequencies from 3.7 to 4.2 GHz. The 6-GHz band uses frequencies from 5.925 to 6.425 GHz. The width of each of these bands is 500 MHz. Each band is subdivided into a number of channels, with channels in the 4-GHz band being 20 MHz wide, while channels in the 6-GHz band are 30 MHz wide.

The K_u-band systems use the 11-GHz band and the 18-GHz band. The 11-GHz band provides operation from 10.7 to 11.7 GHz. The 18-GHz band provides operation from 17.7 to 19.7 GHz.

The C-band systems have the advantage that they are not as affected by weather. The K_u-band systems, on the other hand, can experience significant attenuation

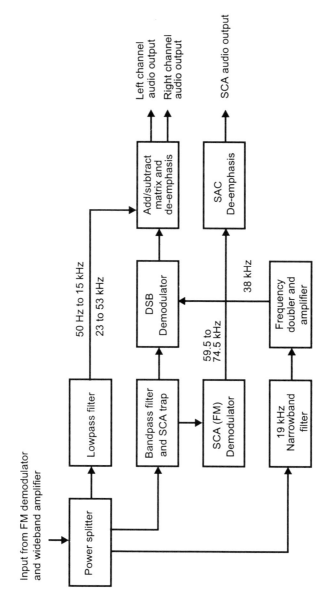

Figure 8.8 Stereo FM multiplex demodulator system. *(After:* [1].)

Figure 8.9 Illustration of a microwave relay system.

due to rain or snow. Typical repeater spacings for C-band systems are in the range of 20 to 30 miles. K_u-band systems normally have spacings of about 15 miles.

Figure 8.10 is a signal-flow diagram for a 6-GHz microwave relay repeater system. The diagram shows three relay stations, each with two antennas. One antenna of the pair points in the A direction, while the other antenna points in the B direction. Each antenna has both transmission and reception capabilities. Two frequencies are assumed for each relay station: 6.0 GHz and 6.4 GHz. Station 1 receives from the B direction at 6.4 GHz and transmits in the A direction at 6.0 GHz. It also receives from the A direction at 6.4 GHz and transmits in the B direction at 6.0 GHz.

Station 2 receives from the B direction at 6.0 GHz and transmits in the A direction at 6.4 GHz. It receives from the A direction at 6.0 GHz and transmits in the B direction at 6.4 GHz. The process of switching transmit and receive frequencies every station continues along the line of stations, as shown for relay station 3.

Figure 8.11 is a block diagram of a microwave relay repeater system. The microwave relay repeater receives a modulated microwave signal from one repeater and transmits the signal to the next repeater. Two-way operation is provided. The frequency difference for the two directions of communication is typically a few hundred megahertz at the 4- to 6-GHz frequency band. Two antennas are used, one facing in the A direction and the second facing in the B direction. The antenna types typically used are either parabolic dish antennas or hoghorn antennas. A hoghorn antenna is a combination of a parabolic reflector and a horn antenna with the reflector focus at the horn center directly below the reflector.

The antennas are connected by means of waveguide to circulators, which are used at the junction of receivers and transmitters. Thus, in the ideal case, the

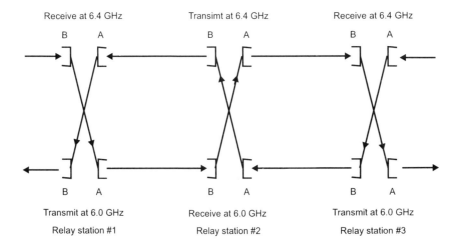

Figure 8.10 Signal-flow diagram of microwave relay systems.

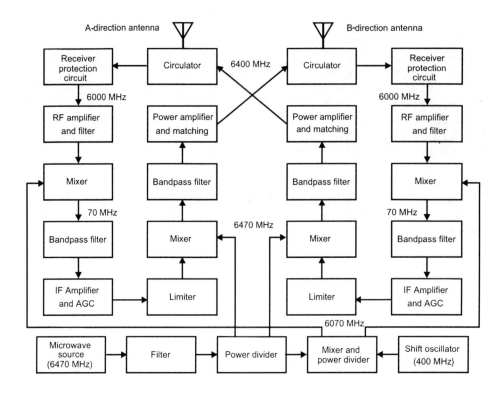

Figure 8.11 Block diagram of a microwave relay system. (*After:* [3].)

transmitter is connected to the antenna but not to the receiver, and the receiver is connected to the antenna but not to the transmitter. (Circulator systems are discussed in Chapters 11 and 12.) In the practical case, there is small coupling between the two systems but not enough to cause the receiver to be degraded in its operation by the transmit signal.

In the system shown in Figure 8.11, the transmit frequency is at 6,400 MHz and the receive frequency is at 6,000 MHz. The first block following the circulator is a receiver protection circuit. A protection circuit might include a sparkgap and zener diodes, which would conduct if the input voltage should exceed the breakdown or zener voltages.

The output of the protection circuit connects to a low-noise RF amplifier, which is followed by an image rejection bandpass filter. That is followed by the receiver mixer. For the example here, the LO frequency is 6,070 MHz and the input signal is 6,000 MHz. The difference frequency is 70 MHz.

The output of the receiver mixer is selected by a 70-MHz bandpass filter and fed to the 70-MHz IF amplifier chain. That is where most of the gain for the receiver takes place. An AGC circuit is used with the system to control the amplitude of the output signals. An amplitude limiter follows the IF amplifier to prevent spurious amplitude modulation.

The next stage following the limiter is the transmitter mixer. In the system shown, the LO frequency for this stage is 6,470 MHz. The difference frequency for the mixer is 6,400 MHz, which is selected by a bandpass filter that follows the mixer. The output of the filter is fed to the power amplifier. Examples of power amplifiers include FET power amplifiers and TWT amplifiers.

The output of the A channel power amplifier is fed to the circulator for the B-direction antenna. There, the signal is fed to the B-direction antenna. The operation of the two channels corresponding to the A and B directions is similar.

The source for LO signals is shown at the bottom of Figure 8.11. Two oscillators are used, one at the microwave frequency and one at the shift frequency. A typical modern system likely would use a VHF transistor crystal oscillator with a varactor frequency multiplier for the microwave oscillator. Multiplication factors are of the order of 20 to 40, and the power output is about 200 mW. The shift oscillator also may be of this type but with smaller multiplication factors required. In the system shown, the output of the microwave source is at 6,470 MHz. It is fed to a three-way power divider. Two of the outputs are fed to the transmitter mixers, and one is fed to a mixer that has as its second input the shift oscillator output. The shift oscillator frequency is 400 MHz, and the difference frequency in that case is 6,070 MHz. The difference frequency is selected by a bandpass filter and fed to a two-way power divider. The output of the divider is to the two receiver mixers.

Until recently, the type of modulation used by microwave relay system typically was analog FM, and the multiplex mode used was frequency multiplex. More recently, the telephone companies have switched over to time division multiplex systems, with phase modulation or QAM modulation used in the time slots. There clearly are reliability and performance advantages for switching to that type of modulation.

For an example of the type of performance achieved by modern microwave relay systems, consider the case of the AT&T DR6-30-135 system operating in the 6-GHz band and utilizing 64-QAM. This system was introduced in 1984, and 64-QAM modulation provides 4 bits per pulse. The total capacity for the system is 14,112 two-way digital voice circuits.

8.5 SATELLITE RELAY COMMUNICATION SYSTEMS

Satellite relay communication is one of our most important forms of communication. Its main use is for telephone and television relay. Some of the information presented here regarding satellite relay was adapted from [2,4].

Figure 8.12 illustrates a satellite relay communication system. A single ground station transmits to a satellite relay, which receives the signal and then retransmits it to a number of ground stations. Each station uses high-gain parabolic antennas.

220 | RF Systems, Components, and Circuits Handbook

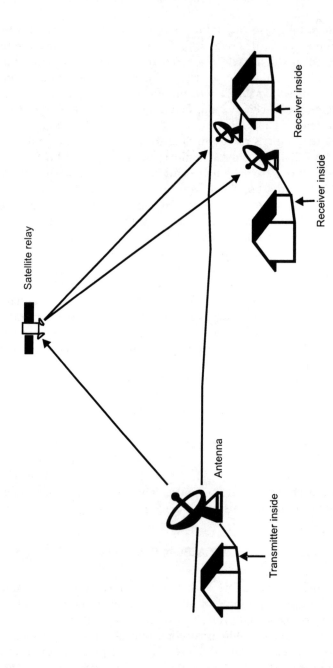

Figure 8.12 A satellite relay communication system.

Figure 8.13 is a block diagram of one important type of early satellite relay system, the so-called transparent relay. In this system, the relay simply amplifies the signal and shifts the frequency.

In Figure 8.13, the signal from the ground station is received by the satellite using a high-gain directional antenna that points at the Earth. The signal received by the antenna is fed to a low-noise RF amplifier. The output of the amplifier is then fed to a balanced mixer and filter, where the signal frequency is changed to the transmitter frequency. A frequency shift of 1-2 GHz is typical. The signal is then amplified by an amplifier chain to the desired level for driving the TWT transmitter power amplifier. The output signal is bandpass filtered and then fed to the transmitting antenna. In some systems, the same antenna is used for both transmit and receive by use of a suitable duplexer or circulator.

It was recognized early that multiple-access ability was one of the most valuable features of satellite relay systems. Enhancement of that capability required the addition to the basic functions of the transparent relay of such functions as beam switching, transponder switching, and signal processing. Examples of signal processing additions included demodulation, remodulation, buffering, and storage.

The most important improvement in the evolution of satellite relay systems has been the availability of more weight and size for the satellites as a result of more powerful launchers, resulting in more power and larger antennas. That, in turn, has resulted in a great reduction in the required size of ground-station antennas and a reduction in transmitting power. The introduction of very small antenna terminals (VSATs) allowed the development of numerous independent communications networks linking industries, banks, and other institutions. Currently most of

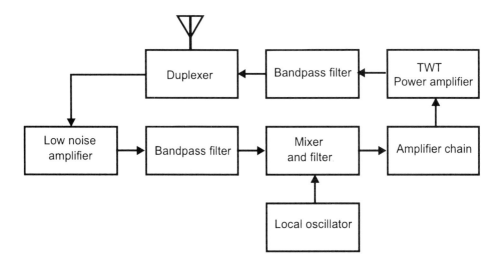

Figure 8.13 Single conversion type satellite relay (transparent relay).

the peripheral terminals operate with and through central stations, or hubs, on Earth. It is envisioned that the functions of the hub will be available onboard more sophisticated satellites.

The use of satellites has facilitated worldwide television distribution for special events in real time and, in most cases, with delays as required by the different time zones. In the United States, the nationwide distribution of television programs to cable network has become an industry of its own. For rural areas not reached by conventional television broadcast stations and not served by cable television systems, the higher levels of power provided by satellites have made possible direct reception of television from satellites by receive-only Earth terminals equipped with parabolic dish antennas with diameters of 2–5m.

The recent introduction of satellite relay digital television is an important step in reducing the required size of earth-station antennas for satellite television. That permits antennas with diameters of less than 1m.

Another development is direct television broadcast from satellites operating in the K_u band (10.7–12.75 GHz) with transmitter power levels between 100W and 250W. These broadcasts can be received with antennas as small as 0.5m in diameter.

Satellite relay communication is very desirable for mobile stations. An early satellite relay mobile system was the MARISAT system, established in 1967, followed by the INMARSAT global system. Current systems are able to provide ship-to-shore and air-to-ground communication capability on a global scale. Satellite systems capable of serving Earth vehicles such as trucks, trains, and automobiles currently are being implemented. These systems will be able to provide not only communications but also such functions as determination of location, determination of status of critical shipments, and so on.

The majority of communications satellites are of the geostationary type. The orbital height is $35.863 \cdot 10^6$m (22,278.5 mi). The orbital plane coincides with that of the Earth's equator. The transmission delay for a single hop ranges from 0.238 to 0.275 sec. That has been found to be acceptable for telephone communications. Echo cancelers are used to keep echos under control.

There are cases in which inclined orbits are used to provide high-latitude coverage. Highly elliptical orbits are typically used in such systems, with the apogee over the northern hemisphere. For that case, the satellite moves much slower near the apogee than it does near the perigee. The orbit is inclined 63.5 degrees to permit zero rotation of the line of apsides. Under those conditions, a satellite with a 12-hr period spends about 8 hr near geosynchronous altitude. Thus, three such satellites can provide continuous service at altitude. The former Union of Soviet Socialist Republics uses a system of this kind known as MOLNIYA.

Polar orbits are also used for noncommunications-type missions, such as Earth observation, weather, and surveillance.

The optimum frequency region for satellite communication systems can be shown to be between 0.8 and 8 GHz. Most of the early satellite relay systems have operated in that region. The earliest frequency bands used for telephone relay were the same bands used for ground-based microwave relay. The 4-GHz band (3.7 to 4.2 GHz) is used for the downlink, and the 6-GHz band (5.925 to 6.425 GHz) is used for the uplink.

Expansion of satellite communications systems has resulted in the use of the 10–15 GHz band as well. Some of those systems operate at 10.95 to 11.2 GHz plus 11.45 to 11.7 GHz for the downlink and 14.0 to 14.5 GHz for the up-link.

In the near future, it is anticipated that there also will be much use of the 17- and 30-GHz bands for satellite relay.

Earlier types of transponders typically used bandwidths of 36 MHz. The number of transponders per satellite typically was limited to 12 for single-polarization satellites and 20 to 24 for dual-polarization satellites. Current systems use much larger numbers of transponders. For example, the INTELSAT VI uses 46 transponders. The introduction of the 11- and 14-GHz bands and high-speed digital transmission has brought about the use of wider bandwidth systems (80 to 200 MHz) in addition to the 36–40 MHz systems.

FM originally was used for radio signals transmitted to both Comstar I and Comstar IV. In 1982, the equipment at the Earth stations was changed to use SSB-AM of the uplink wave. SSB-AM is considerably more efficient in its use of spectrum than is FM. The number of voice channels handled by each transponder channel was increased to 7,800. With 12 transponders used, that increased the capacity of the satellites to 93,600 two-way voice circuits.

There is a trend now for converting satellite relay systems to TDM in much the same way that there was a switch over to this mode for ground-based microwave relays. This will not increase the number of voice circuits per satellite, but it will add to the reliability and the performance of the circuits.

The capacity of modern communication satellites is impressive. For example, Comstar IV has 24 transponder channels, each capable of carrying 670 Mbps. The total capacity of the satellite thus is 1.4 Gbps, which corresponds to a capability to transmit 220,000,000 pages of text per hour.

Future satellites are expected to use onboard signal processing functions in addition to amplification. Digital transmission and digital techniques are expected to be used extensively. Digital transmission offers the following major advantages:

- Guaranteed error control;
- Reduced sensitivity to channel nonlinearity;
- Efficient tradeoff of power and bandwidth;
- Flexibility with regard to multiplexing diverse signals;
- Easy combination of the functions of transmission switching and routing;
- Signal regeneration capability;
- Implementation with rugged hardware.

8.6 SATELLITE RELAY EARTH STATIONS

There are three categories of Earth stations for satellite relay communications: stations that transmit and receive, stations that receive only, and stations that transmit only. The first type of station is used in two-way communication systems. The second type is used for receiving television broadcast from satellites. The third type is used in data collection systems. Satellite communication Earth-station transmit power ranges from a few watts to 10 kW or more. Solid-state systems are used at low power levels, and klystrons are typically used for high-power levels. Low-noise receivers are used in all cases.

Figure 8.14 is a simplified system block diagram for a two-way satellite communication system earth station.

The antenna in a system like that depicted in Figure 8.14 typically is a parabolic dish antenna. The antenna size varies considerably. Some high-capacity systems use antenna diameters as large as 32m. Most antennas are somewhat smaller.

The antenna is followed by a duplexer. There are a number of possible types of duplexers, including circulators. The receive signal from the duplexer is first amplified by a low-noise amplifier, which is followed by down-converters and IF amplifiers. Dual conversion normally is used in the receiver system, with the first IF amplifier having a typical center frequency of about 300 MHz, and the second IF amplifier having a typical center frequency of about 70 MHz.

The IF amplifier is followed by a demodulator circuit. The type of demodulator depends on the type of modulation used. Examples are FDM/FM (single-carrier case), FDM/FM/frequency division multiple access (FDMA) (multiple-carrier case), and TDMA.

The demodulator is followed by a demultiplex system (deMUX), which is followed by a terrestrial interface unit with connections to and from the serving office.

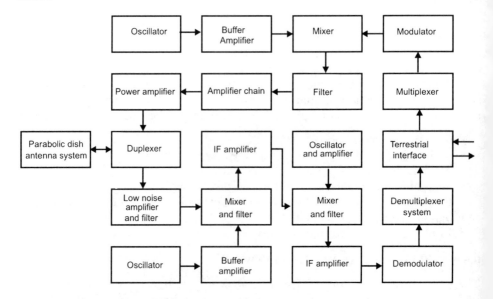

Figure 8.14 A block diagram of a two-way Earth station.

In the transmit mode, signals are fed to the terrestrial interface unit from the serving office. The signals are routed to a multiplexer system. The output of this system is fed to a modulator. Again, the type of modulator used depends on the desired modulation. Again, example modulations are FDM/FM (single-carrier case), FDM/FM/FDMA (multiple-carrier case), and TDMA. The output of the demodulator is fed to the up-converter and amplifier system, which is followed by the power amplifier. A typical system uses a multiple-cavity klystron. The output of the power amplifier is fed to the duplexer, which routes the signal to the antenna.

For reasons of simplicity, a number of important system blocks are not shown in Figure 8.14, including pointing controls for the antenna, high voltage power supplies for the power amplifier, cooling systems, monitoring systems, control systems, among others.

References

[1] Kennedy, George, *Electronic Communication Systems*, 3rd ed., New York: McGraw-Hill, 1985, Chapters 5 and 6.
[2] Noll, A. Michael, *Introduction to Telephones and Telephone Systems*, Boston: Artech House, 1991, Chap. 3.
[3] Kennedy, George, *Electronic Communication Systems*, 3rd ed., New York: McGraw-Hill, 1985, pp. 548–552.
[4] Van Valkenburg, Mac E., ed., *Reference Data for Engineers*, 8th ed., SAMS, Carmel, IN: Prentice Hall Computer Publishing, 1993, Chap. 27, "Satellite and Space Communications" by Pier Bargellini and Geoffrey Hyde.

CHAPTER 9

Current and Future Commercial Communication Systems

This chapter discusses some of the newer types of business and personal communication systems, cellular telephone systems, global position satellite receivers, and fiber optic communication systems. Business and personal communication systems covered include cordless telephones, pagers, citizens band AM radios, and VHF and UHF FM business and personal two-way radios. Some of the key features of these systems are the use of superheterodyne receivers, frequency synthesizers, and extensive use of digital integrated circuits for data processing, logic circuits, coding, decoding, control of displays, and data storage. Modulation types include both analog and digital systems. Wherever possible, digital integrated circuits have replaced traditional RF and analog circuits.

9.1 BUSINESS AND PERSONAL COMMUNICATION SYSTEMS

Much of the information presented in this section is derived from [1].

9.1.1 49-MHz Cordless Telephones

Figure 9.1 shows a cordless telephone system. The system includes a base unit connected to a two-wire telephone line and a handset that the user can carry several hundred feet from the base unit. The base unit is designed to hold the cordless telephone handset in either a flat or an upright position when it is not in use.

Figure 9.2 is a simplified system block diagram of the cordless telephone system shown in Figure 9.1. The base unit includes an antenna, a low-powered (10 mW) FM transmitter, and an FM receiver. Full duplex operation is used so the

228 | RF Systems, Components, and Circuits Handbook

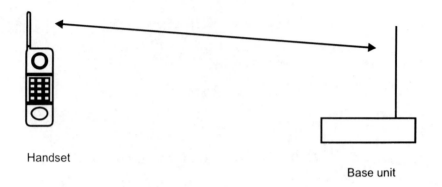

Figure 9.1 A 49-MHz cordless telephone system.

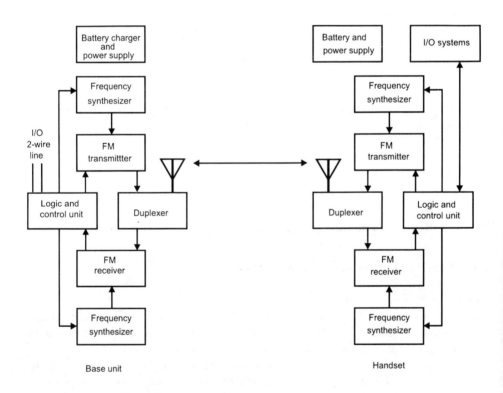

Figure 9.2 Simplified system block diagram of a cordless telephone system.

transmitter and the receivers are on slightly different frequencies in the 49-MHz band. A duplex system uses different frequencies for transmitting and receiving, so it is possible to transmit and receive at the same time. A simplex system uses the same frequency for transmit and receive, so it cannot transmit and receive simultaneously.

The base unit for the cordless telephone also includes a power supply, a battery recharger for the handset, and a logic and control unit. The handset includes a rechargeable battery, a low-power (10 mW) FM transmitter, an FM receiver, a dialing unit, a speaker, a microphone, and a logic and control unit.

A modern system such as the RadioShack model ET-532 cordless telephone has 25 channels for operation. It has a built-in 25-channel selector that automatically seeks the best channel to deliver maximum performance with minimum interference. Even if other cordless telephone systems are operating near enough to cause interference, the selector and the use of multiple channels avoid that problem.

This model uses a liquid crystal display (LCD) with low-battery indicator and memory index that shows channel and phone numbers up to 12 digits. A warning indicator lets the user know when the handset is not in range of the base unit. A base-to-handset page helps the user locate a misplaced handset.

A 65,000-combination digital security coding scrambler helps to prevent other nearby cordless phones from accessing the user's line. The unit's memory stores up to 30 numbers, and up and down scroll features allow the user to review numbers stored in memory.

The keypad of the ET-532 is large and lighted, so it is easy to see, even in the dark. Flash buttons are provided for such services as call waiting and three-way calling. The system has compact, flexible whip antennas that are designed to resist breakage. It also provides one-touch redial capability and hearing aid compatibility.

9.1.2 900-MHz Cordless Telephones

Considered to be the top-of-the-line cordless telephone systems, 900-MHz cordless telephones are more expensive than 49-MHz systems, but they have some improved features. Currently there are three main types of 900-MHz cordless telephones: analog systems, which provide a continuous signal that varies in intensity, like traditional radio broadcast systems; digital-type systems, which send the signal as a series of computer codes to increase security and sound quality; and digital spread spectrum (DSS) systems, which offer the longest range capability with the highest degree of call security.

One example of 900-MHz digital cordless telephone is the Sony model SPP-ID910. This unit provides superior range and sound quality, 20-number speed-dial memory, and digital security. The unit's automatic channel-hopping feature finds another, clear channel if the one being used encounters any interference. The SPP-ID910 also accommodates Caller ID service (the handset LCD screen displays the identity of the caller and the calling number while the phone is ringing, even if call waiting is in effect). Other features include base-to-handset paging and an illuminated keypad.

An example of a 900-MHz DSS cordless telephone is the RadioShack model ET-909. This unit uses 100 channels, and its range is so great that the user can take the handset and work in the yard or visit neighbors. A warning light indicates when the handset is too far from the base unit. The ET-909 has a 10-number memory, base-to-handset paging, redial capability, hold capability, and automatic search capability for clear channel operation. Its most important features are extended-range capability and improved security provided by spread spectrum operation.

9.1.3 Pager Systems

Over 10 million people in the United States subscribe to paging services. Most of the earlier pager systems were fairly bulky and were worn on the user's belt. Newer pagers are so small they can be carried in a shirt pocket or on the wrist, like a watch. Older pagers simply beeped, and the user had to telephone the paging office. These days, pagers not only beep, but some also vibrate, for noiseless paging (the user selects which mode to use). Newer pagers present information in the form of LCDs, including the telephone number of the paging party and, in some systems, short messages. Other pagers can provide up to 8 sec of voice message.

In some pager systems, the radio signal sent to the pager is carried on a subcarrier of a local FM radio station. The transmit frequency in that case is in the 88-108 MHz band. In other systems, the radio signal is sent from a base station with operation at a frequency near 900 MHz. For example, in Utah AT&T Wireless Services uses 35 base stations to cover the main population centers and most of the major highways in the state. Their operating frequency is 931.2125 MHz. Because of the availability of elevated transmitter sites, coverage of about a 40-mi radius is possible for each station.

Long-distance paging is also available. For example, AT&T Wireless Services uses communication satellites to carry paging requests across the continent to metropolitan areas, where local paging signals are then sent by radio.

There are many manufacturers of paging equipment, and the designs and costs of each type of system differ; generally, the smaller the device, the higher the cost. Also, the more costly units provide the greatest information capability.

A pager package includes an antenna, a receiver, a digital logic circuit, an LCD circuit, a sound or speaker circuit, a vibrator unit, and a battery. All that equipment fits into a package only a few cubic inches in volume (a wristwatch-like system takes up less than 1 in^3)—a remarkable degree of miniaturization for an RF system.

An example of a numeric-only pager is one made by RadioShack that costs $59.99. It is a dependable, one-button-operation pager with silent vibrator alert. Its flexible memory holds eleven 10-number messages or eight 20-number messages. An elapsed-time clock shows how long it has been since each page came in. It reminds the user for up to 12 hours that a new page has been received. Memory retention keeps the pages in memory when power is off. It has low-battery alert and uses one AA battery. The pager, which has a large slanted display with backlit display, is only about 2 in^3 in size.

A very compact alphanumeric pager made by RadioShack can display messages and phone numbers. It has 16 message slots, 120 digits per message, and simple 3-button operation. It includes an alarm clock, calendar, and message stamp. Its automatic backlighting makes messages easy to read, and it protects or deletes messages as desired. A variety or alert tones are possible. The cost of this unit is $119.99.

9.1.4 Citizens Band Radios

Citizens band (CB) radio makes it possible to transmit and receive over distances of up to about 10 mi. CB systems operate in the 26.96–27.23 MHz band and use either standard amplitude modulation or SSB modulation. This band is called citizens band because no FCC license or special skills are required for operation.

An example of a base station or vehicle-mounted CB transceiver made by RadioShack is a 40-channel AM mobile unit with digital signal processing. The advanced signal processing virtually eliminates some types of noise and interference. It provides RF gain, squelch, and tone controls. By squelch is meant the ability to turn the sound on only when the desired signal is above a predetermined level in the receiver. The size of this unit is $1\text{-}3/4 \times 5\text{-}1/2 \times 8$ in.

The new RadioShack CB radio model TRC-485 provides both AM and SSB capability. This unit is great for talking to other SSB stations when extra range is important. It features dual watch, so the user can monitor one channel while listening to another. It features automatic modulation control, AGC, and adjustable RF gain. The LCD shows the exact frequency being used as well as the channel number.

All of the newer CB radio systems use dual ceramic filters in the IF amplifiers for superior selectivity and to prevent adjacent channel interference. (Filters and other types of components are discussed in detail in Part 2.)

CB walkie-talkies are also available. The size of a typical unit is only $7\text{-}1/4 \times 3 \times 1\text{-}1/2$ inches. It can have high/low power capability with a maximum input power of 5W. It has most of the features of the mobile units, including use of 40 channels, dual ceramic filters, automatic noise limiter (ANL), squelch, and LED.

There are also 49-MHz FM walkie-talkies, which require no license for operation. Walkie-talkies are low-power, short-range systems with a typical range up to 1/4 mi. These systems use dual conversion superheterodyne receivers and provide up to five channels. The units are small enough to fit into a pocket or purse. A typical five-channel unit has a size of $6\text{-}1/8 \times 2\text{-}5/8 \times 1\text{-}1/16$ in.

9.1.5 VHF and UHF FM Business and Personal Two-Way Radio

Handheld VHF-FM and UHF-FM business two-way radios are available for ranges of up to several miles. These units require an FCC license to operate. The license is easy to get, and virtually any business, school, hospital, clinic, church, or organization can qualify. No test is required.

An example of a VHF system is one made by Motorola that provides 1W of output power. A typical frequency set at the factory is 154.60 MHz, but other assigned operating frequencies are easily set. A 2W UHF-FM unit has a frequency of 464.550 MHz. The size of each unit is 6-7/8 × 2-9/16 × 1-3/8 in. Both the VHF and the UHF versions use PLL frequency synthesizers, rechargeable Ni-Cd batteries, and flexible quarterwave antennas.

9.2 CELLULAR TELEPHONE SYSTEMS

Much of the following section is based on [2–13].

Cellular telephone service is a newer type of communication system that has had remarkable growth and acceptance by the public. It was designed initially for communication between base stations and ground-based vehicles. In the early 1980s, the number of subscribers was forecast to be 4 to 5 million in the United States by the mid 1990s. That estimate has proved far too low; by 1994, the actual number had exceeded 20 million subscribers in the United States and nearly 50 million subscribers worldwide. The great increase in subscribers is the result of the development of small, hand-held portable cellular telephones. No longer is the mobile telephone confined to vehicles. Portable cellular phones have become the equivalent of super cordless telephones. At the time of this writing, there are three main types of cellular telephones: systems that use FDMA; systems that use TDMA; and systems that use code division multiple access (CDMA) (spread spectrum modulation). FDMA systems, sometimes called the first-generation systems, are analog systems that use FM. TDMA systems, the second-generation systems, were developed to increase the number of voice channels that can be provided in the limited bandwidth allocated for cellular telephones. The basic North American digital system increases by a factor of 3 over the basic analog system the number of channels that can be used. The CDMA system uses spread spectrum modulation. It provides even greater improvement in channel capacity over the basic analog system, with a theoretical improvement of up to a factor of 20.

Table 9.1 lists the key characteristics of five main types of cellular systems now in operation or being tested. (Several other systems are also available.)

9.2.1 The Concept of Spatial Frequency Reuse

A key problem faced by the developers of cellular telephones was the problem of insufficient available bandwidth for the number of users. A partial solution to this

Table 9.1
Characteristics of Five Main Types of Cellular Telephone Systems

Characteristics	AMPS North America	IS-54 North America	IS-95 North America	GSM Europe	PDC Japan
Frequency RX in MHz	869–894	869–894	869–894	935–960	940–956 1,477–1,501
TX in MHz	824–849	824–849	824–849	890–915	810–826 1,429–1,453
Access method	FDMA	TDMA/FDM	CDMA/FDM	TDMA/FDM	TDMA/FDM
Duplex method	FDD	FDD	FDD	FDD	FDD
Channel spacing (kHz)	30	30	1,250	200	25
Number of channels per user	832	832	10	124	1,600
Users per channel	1	3	118	8	3
Bit rate (Kbps)		48.6	1,228.8	270.83	42
Modulation	FM	Π/4 DQPSK	BPSK/ OQPSK	GMSK (0.3 Gaussian)	Π/4 DQPSK

Source: [2].

problem was spatial frequency reuse. The concept of spatial frequency reuse can be illustrated as follows. Assume an analog FM system with 70 channels and a square service area 100 mi on a side. The number of telephone calls that can be handled in parallel by this system is equal to the number of channels (70). Now assume that the 100-mi × 100-mi service area is divided into 10-mi × 10-mi cells and the transmitter power is reduced. The 100-mi × 100-mi service area would then contain 100 cells. Then assume there is one base station in each cell and let each base station use 10 of the 70 channels. Adjacent cells would use different channels to avoid interference.

The spacing of cells of the same type and the reduced transmitter power provide the needed attenuation to prevent interference between cells using the same frequencies. The same 70 channels can now handle 1,000 telephone calls in parallel. If the cell size is reduced to 2-mi by 2-mi cells and the transmitter power is further reduced, there would be 2,500 cells, each with base stations that use 10 channels. The result would be the ability to service 25,000 calls. As the cells become even smaller, the power levels used by the base stations and the mobile or portable units must be decreased to avoid interference with other systems operating on the same frequencies.

These concepts have been implemented in both analog and digital cellular telephone systems. The cells usually are hexagonal in shape, to better approximate circles of constant radius. Clusters of cells are used. This arrangement of cells and clusters is illustrated in Figure 9.3.

In Figure 9.3, the base station is located at the center of the cell site. Each cell site is identified by a letter that indicates the cell cluster and by a number that indicates the cell site within the cluster. For example, cell cite C4 is located in cell 4 in cell cluster C.

A second approach to base-station location is to have each cell site located at the edge of a cell where three cells meet. A single antenna mast can be used with three directional antennas. Each antenna covers a 120-degree sector, and each covers a cell. This approach reduces the number of cell sites needed by a factor of 3.

Seven cells are used for each cluster in Figure 9.3. Other cluster sizes sometimes used are 4, 12, and 14 cells per cluster. In each case, there is a large separation of cell sites of the same type. That results in large attenuation of signals of the same frequency from an adjacent cluster. A requirement is that the interference from other cells be at least 18 dB below the desired signals.

Not all cells in a service area are the same size. Cells in heavily populated areas are made smaller so more users can be accommodated. As the population density decreases, the cell size can be made larger. In rural areas, the cell size can be quite large (up to 50 km in diameter).

9.2.2 Propagation Characteristics of Cellular Telephone Systems

The frequency bands used for the first-generation U.S. analog cellular radio system is 824–849 MHz for transmission from the mobile units to the base stations and 869–894 MHz for transmission from the base stations to mobile units.

The FCC provides for two groups of frequencies. The A group is used by nonwireline common carriers, and the B group is used by wireline common carriers. The channel width for both analog and digital cellular systems is 30 kHz. Thus, a total of 832 transmit channels and 832 receive channels are provided in the 824–894 MHz band. Transmit and receive channels for a given mobile unit are separated by 45 MHz. Twenty-one channels in the A group and 21 channels in the B group are used as shared data channels or setup channels. Usually there will be one setup channel per cell site.

Cellular radio recently was given new frequency allocations at 1.5 GHz. The 1.7–2.2 GHz segment was designated specifically for development of wireless personal communications service (PCS) systems under the definitions of the Future Public Land Mobile Telecommunications System (FPLMTS) [2]. PCS systems include portable computers, cordless telephones, pagers, and cellular phones.

Current and Future Commercial Communication Systems | 235

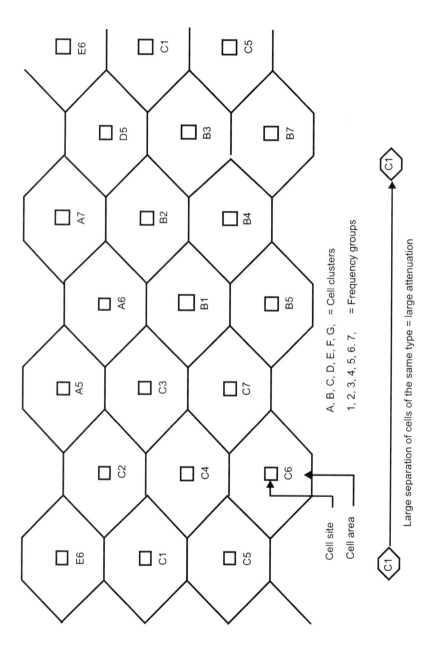

Figure 9.3 Cells and cell clusters for reusing channels. (*After:* [12].)

At the frequencies used by cellular telephones and the signal paths involved (ground wave and multipath), the attenuation tends to increase as the fourth power of range. In contrast, free-space propagation attenuation (involving no interaction with the ground) increases as the second power of range. Thus, the power received at the 4-mi range from the cellular transmitter would be a factor of 4^4, or 256 (24 dB), below the power received at the 1-mi- range. Similarly, the power received at the 10-mi range would be a factor of 10^4, or 10,000 (40 dB), less than the power received at the 1-mi range. The rapid rate of attenuation increasing with distance is one of the key features that make spatial frequency reuse possible [3].

An important propagation problem for operation in the cellular environment is the problem of multipath. Usually, a number of propagation paths are present between the transmitter and the receiver. The strongest path will be the direct path. Other signals received are reflected signals both from the ground and from other reflecting objects. The resulting signal is the vector addition of all those signals. At some ranges, the reflected signals may add in phase, and the resulting signal will be greater than the direct signal. At other ranges, the reflected signals may add 180 degrees out of phase from the direct signal, and the resulting signal will be less than the direct signal. With a moving mobile unit, the signal will fade in and out as a function of time.

9.2.3 Advance Mobile Phone Service

Development of the Advance Mobile Phone Service (AMPS) cellular telephone system was started in 1983. It is still in use in both the United States and Canada as the primary cellular telephone system in less heavily populated areas. In addition, it serves as the alternative system in heavily populated areas for users without digital instruments.

Table 9.2 lists the parameters for the AMPS system.

Figure 9.4 is a block diagram of a cellular analog-type radiotelephone (R/T). This system includes a stub or short whip antenna that is used for both transmit and receive. The antenna is connected to a duplexer, which permits the transmitter to be connected to the antenna when transmitting, and the receiver to be connected to the antenna when receiving. The system can transmit and receive at the same time.

When the system is receiving, the input signal first is fed to an LNA, followed by a mixer that downshifts the signal to the IF frequency. The second input to the mixer is from a frequency synthesizer.

The output of the mixer feeds the IF amplifier circuit, where the signal is amplified and bandpass filtered. The output of the IF amplifier chain connects to a frequency demodulator circuit. The demodulated signal is sent to a logic and control unit. The signal output is amplified and fed to a speaker circuit.

For transmit operation, the voice signal is received from a microphone. The signal is amplified and then fed to a frequency modulator. The output of the

Table 9.2
Parameters of the AMPS System

Category	Parameters
System-Related Parameters	
Number of channels	832 (2 groups of 416 channels; each group includes 21 signaling channels)
Cell radius	2–20 km
Mobile receiver frequency range	869–894 MHz
Mobile transmitter frequency range	824–849 MHz
Channel spacing	30 kHz
Mode of transmission	Full duplex
Voice transmission	Frequency modulation with peak frequency deviation of ±12 kHz
Data transmission	Signaling-channel data are frequency modulated; Manchester-encoded binary at 10 Kbps; peak frequency deviation is ±8 kHz; burst-type digital messages transmitted at 10 kbps on voice channel
Number of cells	50 (typically, on a fully developed system)
Maximum base station ERP	100W per channel
Frequency separation	45 MHz between transmit and receive channels
Communication System Parameters	
Speech quality	Toll quality (similar to that of wireline telephone)
Speech processing	2:1 syllabic compander
Grade of service	2% blocking probability
Mobile Unit Parameters	
Tx RF output	3W (nominal)
TX RF power control	10 steps of 4-dB attenuation each; minimum power:−34 dBW
Rx spurious response	−60 dB 15 kHz away from center of channel
Number of synthesizer channels	832 (tunable to any channel)
Rx noise figure	6 dB measured at the antenna port
Rx sensitivity	−116 dBm

(*After:* [4].)

modulator feeds a mixer up-converter, which is followed by an amplifier chain and a power amplifier. The output of the power amplifier is fed to the duplexer, where the transmit signal is directed to the antenna.

Other elements include a touch-tone dialing unit and a battery power supply.

AMPS cellular R/Ts differ in many ways from basic two-wire conventional telephone sets that are connected to a local exchange. Some of the key differences are as follows:

- An R/T requires a portable source of power in the form of a rechargeable battery in order to function. This battery usually is recharged overnight when the system is not in use.

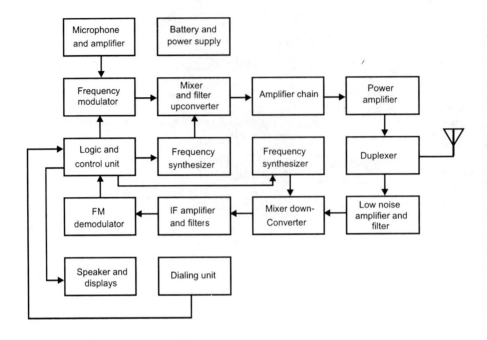

Figure 9.4 Block diagram of a cellular analog R/T.

- The local exchange used by conventional telephones has been replaced by a cell-site base station. The base station is connected to a mobile telephone network, which may be a part of the fixed network used for conventional telephones.
- Both the R/T and the base station need radio antennas, which must be suitable for the radio frequencies licensed for use by the R/T in operation.
- Two radio channels must be allocated to each mobile station or mobile phone in order to have duplex operation. Both a forward- and a return-path radio channel are required. The forward channel refers to the path from the base station to the mobile station; the reverse channel is the path from the mobile station to the base station. Both the base station and the mobile station require radio transmitters. The weakest path is the return path, because the mobile unit has a low-gain antenna and limited radio power, which are needed to conserve battery power and extend operational time between charging. The power level for the mobile transmitter is controlled by the base station.
- The mobile phone carries it own telephone number in internal memory. That allows subscribers to roam over the network, provided they can set up communication control with the base station where they are operating. A mobile phone also contains a radio receiver, a transmitter, and tuning frequency synthesizer circuits, which take and give instructions to the local memory and control module.

Current and Future Commercial Communication Systems | 239

- The ringer is controlled through the circuits mentioned in the preceding paragraph. The R/T has no cradle.
- The base station and the mobile station keep in touch by various "handshaking" protocols. A call is initiated by depressing a specific send button on the keyboard.

Figure 9.5 is a drawing of a cellular telephone. An example of such a system is the RadioShack model CT-500. It is just 5-5/8 in high and weighs only 8.6 oz. Its slim battery has 60 min of talk time and 10 hr of standby time. It has a selectable traditional electronic ringer and a vibrator alert when the user wishes to be discreet. Also provided are 50 memories with 9 programmable, one-touch dialing settings for emergency or frequently called numbers. It uses a seven-character LED with alphanumeric memory so the user can store both name and number for easier recall. It has timers for tracking talk time cumulatively or by individual call. Meters include a signal-strength meter and a battery-voltage meter. Other features include tone dialing, high-performance retractable antenna, and an internal ac charger adapter.

Figure 9.6 is a simplified system block diagram of the total analog cellular telephone system. The figure shows only a few of many cell sites (base stations) and a few of many mobile subscriber units (mobile units or mobiles). Mobile units can be vehicle-mounted or handheld. Two-way radio links are provided between the cell sites and the mobile units. Separate links are provided for voice and data (commands and control). Two-way voice trunks and data links also are provided between the mobile telecommunication switching office (MTSO) and the cell sites. The MTSO is also connected to direct distance dialing networks, which connect

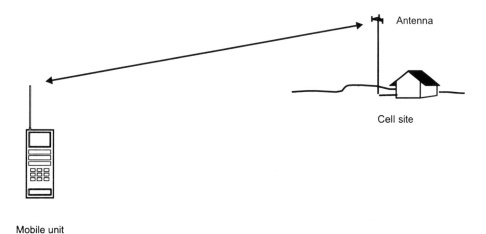

Figure 9.5 A cellular telephone.

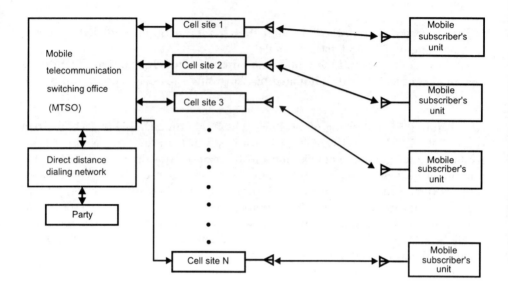

Figure 9.6 Simplified system block diagram of total cellular telephone system. (*After:* [13].)

to the party. The trunks may be microwave relay links, wirelines, or fiber optic links.

The partitioning of control functions among AMPS control elements is as follows. The MTSO provides standard local switching, radio channel management, maintenance of remote units, cell-site and mobile control, message administration, mobile location, and handoff synchronization. Each cell site provides cell-site radio control, location-data collection, mobile control, message relaying and reformating, and switchhook and fade supervision.

Figure 9.7 is a simplified system block diagram of a cell site with one mobile unit. Each cell site has one or more antennas, a combiner and duplexers that allow both the transmitter and the receiver systems to be connected to the antenna, a set of transceivers for each forward and reverse voice channel, a transceiver for the data channel, logic and control circuits, and an interface to the landline voice and data channels to and from the MTSO. The cell-site transceivers consist of an FM receiver with an associated frequency synthesizer, an FM transmitter with an associated frequency synthesizer, and other circuits.

The remote mobile unit consists of an antenna, a duplexer, an FM receiver and associated frequency synthesizer, a logic and control unit, an FM transmitter with associated frequency synthesizer, a battery and power supply, and an input/output (I/O) system. The I/O system includes a dialing system, a microphone, a speaker, and display systems.

The procedures followed in calling to and from a mobile unit are described in the following paragraphs. Much of the following sequence is quoted or adapted from [5].

Current and Future Commercial Communication Systems | 241

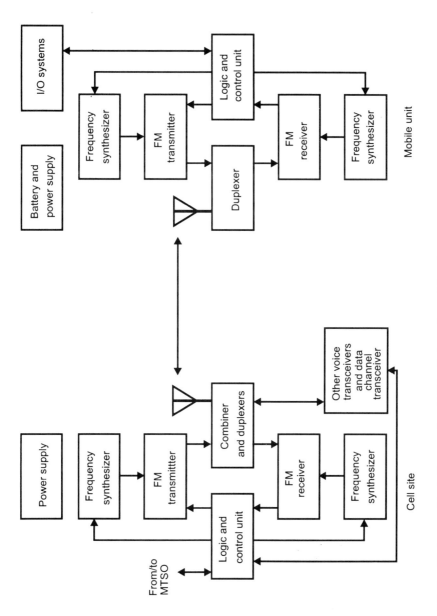

Figure 9.7 Simplified system block diagram of an AMPS cell site and a mobile unit.

9.2.3.1 Call to a Mobile Unit

1. Paging

When a mobile unit is being paged, a call first is sent from the party's central office. It is then routed by the wire line network to the home MTSO of the mobile unit. The MTSO receives the digits, converts them to the mobile unit's identification number, and instructs the cell sites to page the mobile unit over the forward setup channels. The paging signal thus is broadcast over the entire service area so many cell sites are involved.

2. Cell-Site Selection

The base station at each cell site has a transceiver for each voice channel assigned to it. The transceivers normally all share the same antennas. The base station or cell site also contains scanning receivers for measuring the signal strengths of mobiles that enter the service area of the cell. Certain channels are used as setup channels rather than voice channels. When a mobile unit is turned on, it looks for the strongest setup channel and monitors that channel. When the mobile unit detects that it is being called (paged), it quickly samples the signal strength of all the system's setup channels so it can respond through the cell that provides the strongest signal at its current location.

3. Page Reply

The mobile unit then responds to the cell site it selected over the reverse setup channel. The selected cell site then reports the page reply to the MTSO over its dedicated landline data link.

4. Channel Designation

The MTSO next selects an idle voice channel and an associated landline trunk in the cell site that handled the page reply. It then informs the cell site of its channel choice over the appropriate data link. The serving cell site then informs the mobile unit of its channel designation over the forward setup channel. The mobile unit then tunes to its assigned channel and transponds the supervisory audio tone (SAT) over the voice channel. Upon recognizing the SAT, the cell site places the associated landline trunk in an off-hook (in-use) state, which the MTSO interprets as successful voice-channel communication.

5. Alerting

Upon command from the MTSO, the serving cell site sends a data message over the voice channel to ring the mobile telephone. A signaling tone from the mobile unit causes the cell site to confirm successful alerting to the MTSO. The MTSO, in turn, provides audible ringing to the calling party.

6. Talking

When the mobile user answers, the cell site recognizes removal of signaling tone by the mobile and places the landline trunk in an off-hook state. That is detected at the MTSO, which removes the audible ringing circuit and establishes the talking connection so that conversation can begin.

9.2.3.2 Call from a Mobile Unit

1. Preorigination

The user enters the digits to be dialed into the mobile unit's memory.

2. Cell-Site Selection

After the mobile unit is placed in an off-hook state, cell-site selection occurs. The process involves scanning the setup channels used for access in the mobile serving area and selecting the strongest one. The selected channel usually is from the closest cell site.

3. Origination

The stored digits, along with the mobile unit's identification, are sent over the reverse setup channel selected by the mobile unit. The cell site associated with the setup channel receives the information and relays it to the MTSO over its landline data link.

4. Channel Designation

The MTSO designates a voice channel and establishes through the cell site voice communication with the mobile. The MTSO also establishes routing and charging information at this time by analyzing the dialed digits.

5. Call Completion

The MTSO completes the call through the wireline, fiber optic, or satellite relay network using standard techniques.

6. Talking

When the MTSO establishes a talking connection, communication takes place when the called party answers.

9.2.3.3 Handoff

1. New-Channel Preparation

Because the mobile unit is moving, it may be necessary to change cell sites during conversations or calls. That is done by handoff. Location information gathered by the cell site serving the mobile unit, as well as by surrounding cell sites, is sent to the MTSO over the various cell-site landline data links. The data are analyzed by the MTSO. If the MTSO determines that a handoff to another cell site is desirable, it selects an idle voice channel and an associated landline trunk at the receiving cell site. It also informs the new cell site to enable its radio. The cell site then sends the SAT.

2. Mobile Retune Command

A message is sent to the current serving cell site informing it of the new channel and new SAT for the mobile. The serving cell site sends this information to the mobile over the voice channel.

3. Channel/Path Reconfiguration

The mobile unit sends a brief burst of signaling tone and turns off its transmitter. It then retunes to its new channel and transponds the SAT it finds there. The old cell site, having recognized the burst of signaling tone, places an on-hook signal on the trunk to the MTSO. The MTSO reconfigures its switching network, connecting the other party with the appropriate landline trunk to the new serving cell site. The new serving cell site, on recognizing the transponded SAT on the new channel, places an off-hook signal on the associated landline trunk. The MTSO interprets the two signals as a successful handoff.

9.2.3.4 Mobile-Initiated Disconnect

1. Release

The mobile unit sends a signaling tone (ST) and turns off its transmitter. The ST is received by the cell site, which places an on-hook signal on the appropriate landline trunk.

2. Idle

In response to the on-hook signal, the MTSO idles all switching office resources associated with the call and sends any necessary disconnect signals through the wireline network.

3. Transmitter Shutdown

As the final action in the call, the MTSO commands the serving cell site over its landline data link to shut down the cell-site radio transmitter associated with the call. All equipment used on the call now can be used on other calls.

9.2.3.5 System-Initiated Disconnect

1. Idle

In response to the disconnect signal received from the wireline network, the MTSO idles all switching office resources associated with the call.

2. Ordered Release

The MTSO sends a release-order data link message to the serving cell site. The cell site sends that command to the mobile unit over the voice channel. The mobile unit confirms receipt of the message by invoking the same release sequence as with a mobile-initiated disconnect.

3. Transmitter Shutdown

When the MTSO recognizes successful release by the mobile unit, it commands the serving cell site to shut down the radio transmitter.

9.2.4 Narrowband Advanced Mobile Phone Service

As a result of the rapid growth of cellular service in many areas, channels have become scarce. Not all communities, however, require the considerable increases in spectrum efficiency afforded by digital systems. As a result, there has been some interest in a development by Motorola known as Narrowband Advanced Mobile Phone Service (NAMPS). NAMPS uses the digital control methods of AMPS, but the voice bandwidth of channels is decreased by a factor of 3. The required bandwidth is thus changed from 30 kHz to 10 kHz, and the number of available voice channels is increased by a factor of 3. The reduction in channel bandwidth is achieved by using smaller FM deviation than is used in AMPS. The smaller deviation means less FM improvement, but the smaller bandwidth means less noise. The performance from the user's point of view is essentially the same as that achieved with AMPS [6].

9.2.5 North American Digital Cellular Telephone Systems

Digital cellular telephone systems have been developed to greatly increase the number of subscribers that can be served with the limited bandwidth available for cellular telephone use. The three main types of digital telephone systems in use today are the American digital cellular (ADC) system, the Japanese digital cellular (JDC) system, and the European digital cellular system known as the Global System for Mobile Communications (GSM). These systems all use TDMA modulation. While the three systems are similar, there are a few important differences. Some of the information presented here is adapted from [7].

9.2.5.1 IS-54

The IS-54 digital cellular telephone system uses the same frequency bands as are used for AMPS, the same cell structure, and the same overall control system design. One difference is the type of modulation used for voice channels. The AMPS system uses analog FM for voice, while the IS-54 digital system uses $\pi/4$ DQPSK modulation. It uses a digital traffic channel structure consisting of six time slots, each with a duration of 6.667 ms. Those six time slots occupy 40 ms and are called a frame. Twenty-five frames per second are thus sent, each containing 1,944 bits. The total bit rate is thus 48,600 bps.

Each full-rate voice channel uses two equally spaced time slots, such as 1 and 4, 2 and 5, or 3 and 6. Thus, three times as many voice channels are provided by the digital cellular system as are by the analog system.

With no timing advance, the offset at the subscriber's set between the reverse and the forward frame timing is 207 symbol periods, where a symbol is 2 bits. Thus, time slot 1 of frame N in the forward direction occurs 207 symbol periods after time slot 1 of frame N in the reverse direction. Table 9.3 lists the IS-54 slot formats.

Table 9.3
IS-54 Slot Formats

Characteristics	Values
For mobile unit to cell site:	
Guard time	6 bits
Ramp time	6 bits
Data (user information or FACCH)	16 bits
Synchronization and training	28 bits
Data (user information or FACCH)	122 bits
SACCH	12 bits
CDVCC	12 bits
Data (user information or FACCH)	122 bits
For cell site to mobile unit:	
Synchronization and training	28 bits
SACCH	12 bits
Data (user information or FACCH)	130 bits
CDVCC	12 bits
Data (user information or FACCH)	130 bits
Reserved	12 bits

The ramp time is the time needed for the transmitter to reach full output from a quiescent condition. User information refers to speech bits or user data bits. CDVCC consists of an 8-bit digital verification color code plus four error-correcting bits. FACCH is a signaling channel for the transmission of control and supervision messages between the cell site and the subscriber set. It is sent in place of user information whenever needed, and is used where rapid action is required, for example, in handoff. SACCH is a signaling channel that is sent continuously for the transmission of control and supervision messages. It is used for functions that can be performed over a longer period of time, for example, information about channel-signal quality.

The subscriber set derives the timing for its transmit symbol, TDMA, frame, and slot clocks, from a common source that tracks the cell-site symbol rate.

Figure 9.8 is a simplified system block diagram for a digital-type cellular telephone mobile transceiver.

The voice input to the transmitter is from a microphone. Analog speech is digitized and passed though a vector sum-excited linear predictive (VSELP) coder. The coder reduces the data rate substantially at the expense of a higher demand on the processing speed. Coded data passes through a channel coder and a modulator block, where the following functions are performed:

- Error correction and detection bits are added.
- Control channel data are added.
- Training sequence data are added.

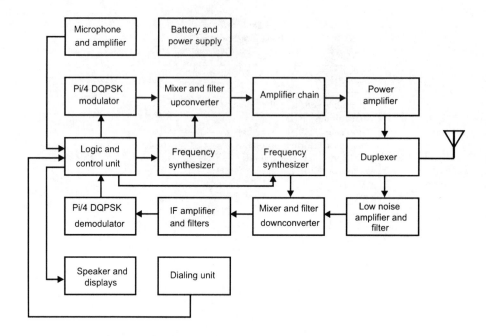

Figure 9.8 Simplified system block diagram for a digital-type cellular telephone mobile transceiver.

- Guard bits are added.
- A channel coder interleaves the data over two time slots for better interference rejection.
- Channel-coded output data are provided as the input to a $\pi/4$ DQPSK modulator.
- The RF-modulated data are transmitted as fixed-size packets.

The modulated data are filtered and amplified by the RF power amplifier. The output of the amplifier is fed to a duplexer, which routes the signal to the antenna.

In the receive mode, the signal from the cell site is received by the antenna and fed through the duplexer to a tuned RF section. It is then down-converted to an IF frequency, where it is amplified and filtered again. The IF signal is then passed through a DQPSK demodulator, and the data are recovered. A channel decoder strips off the coding information before the burst data pass through a VSELP speech decoder for the original speech recovery. Finally, the speech is amplified and fed to the speaker. The overall system block diagram for the IS-54 digital cellular telephone system cell site and a remote mobile unit is shown in Figure 9.9.

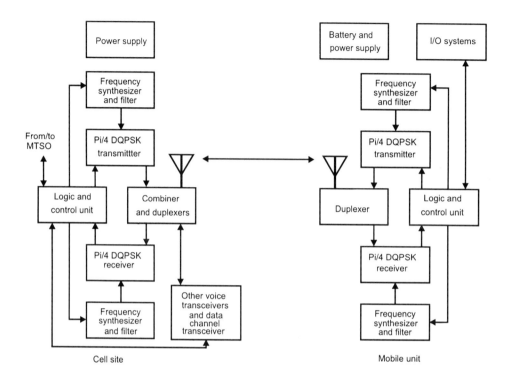

Figure 9.9 Cell site and mobile unit system block diagram.

The procedures followed in calling to and from an IS-54 digital-type mobile unit follow. Much of the information presented here is quoted or adapted from [8].

Mobile Initialization Sequence

1. Mobile unit power on.
2. Scan primary set of dedicated control channels for either system A or system B and find the one having the highest power level.
3. Update digital overhead information (SID, paging channel number, etc.).
4. Scan paging channels, find the strongest one, and lock on to it.
5. Determine access channel numbers, and so on.
6. Start monitoring paging channel continuously.

Paging

1. Mobile unit idle.
2. Mobile unit receives paging message.

3. Mobile unit scans reverse access channel and finds the strongest one. It then responds to the cell site and waits for the cell-site response over the forward access channel.
4. Cell receives the message, makes landline connection through MTSO, and sends voice channel assignment over forward access channel.
5. Mobile unit receives voice channel assignment and sends acknowledgment over SACC.
6. Mobile unit turns off its transmitter, adjusts power level and digital mode of operation, sets CODEC rate, and so on, then sends CDVCC to the cell site.
7. After receiving CDVCC, cell sends ringing message to mobile unit.
8. Mobile phone rings and mobile unit goes off-hook.
9. Conversation starts.

Call Origination by Mobile Unit

1. Mobile unit idle.
2. Mobile unit stores the desired party's telephone number in its memory and presses "send."
3. Mobile unit scans the reverse access channels and finds the strongest one. Retrieving the access attempt parameters from the forward access channel allows the unit to make sure the reverse access channel can be used.
4. The mobile unit transmits the desired telephone number along with its own identity over the reverse channel and waits for the cell site to respond.
5. Cell site receives the message, makes landline connection through the MTSO, and sends voice channel assignment over the access channel.
6. Mobile unit sends an acknowledgment, adjusts its power level, tunes to an assigned RF channel, selects digital mode for its transceiver, sets the speech CODEC rate, and so on, then sends CDVCC to the cell site.
7. After the cell site receives the CDVCC, it sends a ringing message to the mobile.
8. The mobile phone rings, and the mobile unit goes off-hook.
9. Conversation starts.

Mobile-Assisted Handoff

1. Mobile unit is in the conservation mode.
2. Mobile unit detects low signal strength and informs the cell site.
3. The serving cell alerts the MTSO. The MTSO commands adjacent cells to locate the mobile unit. Mobile unit is asked to make measurements of signal quality on a specific channel during the off-time of the frame when not receiving data. Thus, the MTSO knows which is the best cell for the mobile unit.

4. MTSO commands old cell site for handoff, and old cell site transmits handoff message to the mobile unit over FACCH.
5. Mobile unit adjusts system parameters and turns to the new voice channel.
6. Conversation continues. The change is made so fast that users usually are not aware it is taking place.

9.2.5.2 Extended Time-Division Multiple Access

Extended TDMA (E-TDMA) is a development of Hughes Network Systems, Inc., Germantown, Maryland. It uses half-rate speech coding and the TIA IS-54 TDMA format discussed in Section 9.2.5.1. Each of the six slots of an IS-54 frame is used as a separate channel rather than using three pairs of slots for messages. Thus, without other features it achieves a spectrum efficiency six times that of the analog system rather than three times. In addition, the E-TDMA system uses digital speech interpolation (DSI) and silence-interval detection on the speech channel. With that combination, it is possible to provide more than a tenfold improvement in capacity over the analog system.

E-TDMA works by treating each slot of an IS-54 frame as being available for speech, forward or reverse control, packet data, or special applications. For example, a system might use 12 frequency channels that would support only 12 voice channels of speech in the analog system and 36 channels of voice in the IS-54 TDMA system. In the E-TDMA system, nine of the slots are control slots, with the remaining 63 slots being used for speech slots. With DSI, the number of required voice channels can be reduced by about a factor of 2. As an example, the system might combine six channels into three channels. The net result in the example here would be to have the 63 slots being used for speech slots support 126 voice channels. This is a factor of 10.5 better utilization of bandwidth than that provided with the AMPS.

Some of the information here is adapted from [9].

9.2.6 PDC Japan Digital Cellular System

Japan has established a digital cellular standard based on the United States TIA IE-54 system but adapted to the 25-kHz channels in use in Japan. Its characteristics were summarized in Table 9.1. The frequencies used by the Japanese system are different from those used by the North American system, being 940–956 MHz for receive and 810-826 MHz for transmit in one band and 1,477–1,501 MHz for receive and 1,429–1,453 MHz for transmit in a second band. The total number of channels is 1,600. Some of the information here is adapted from [10].

9.2.7 GSM Europe System Digital Cellular System

The GSM was designed to be used in all the countries of Europe. It differs in some important details from the North American digital systems. It uses a home-location register and a visited-location register. GSM characteristics are summarized in Table 9.4.

MSK is a form of frequency-shift keying in which the peak frequency deviation equals ±0.25 times the bit rate and coherent detection is used.

The information presented here is based on [11].

9.2.8 TIA IS-95 North America Digital Cellular System

The IS-95 system uses CDMA modulation. It has the potential for improving the efficiency of spectrum utilization over analog of the AMPS type by a factor of up to about 20. That is clearly a major improvement over the analog type of system and also over the IS-54 digital cellular system. This system, which was developed by QUALCOMM, Inc., is often called a third-generation cellular telephone system.

The QUALCOMM CDMA standard uses a 1.25-MHz bandwidth per channel. That amounts to 10% of the frequency spectrum allocated to each carrier in any given serving area. The 45-MHz separation between uplink and downlink directions is the same as for other cellular systems. Direct sequence phase code modulation is used, with a pilot carrier consisting of a 32,768-bit beacon code being sent from each cell site. Time shifts in increments of about 200 ms are used to distinguish the cells from one another. The pilot carriers provide the synchronization signal for the pseudorandom noise code streams. The modulation is DQPSK.

Table 9.4
GSM Characteristics

Characteristics	Values
Carrier separation	200 kHz
Users per carrier	
Full-rate voice	8
Half-rate voice	16
Transmission rate in channel	271 Kbps
Transmission rate per user	
Full rate	33.88 Kbps
Half rate	16.94 Kbps
Modulation	Gaussian minimum shift keying (GMSK)
Channel equalization	Adaptive, to 10-ms delay
Channel coding	Convolutional
Frequency hopping	Mandatory capability for mobile units
	Optional capability for cell sites
Frequency hopping rate	217 hops/s

The QUALCOMM CDMA cellular system uses a variable-rate speech coder (QSELP). Forward error correction is incorporated into the transmission in each direction.

Studies have shown that the interference contribution from neighboring cells is about one-half that of sources in a user's own cell. That fact makes it possible to use the same channels in every cell rather than using a seven-cell group with a different channel in each of the seven cells. That by itself leads to a sevenfold spectrum efficiency improvement over the analog system.

The number of pseudorandom codes used determines the number of voice channels that can be used in the 1.25-MHz channel bandwidth. The number of codes per channel for the IS-95 system is 118, so there can be 118 channels per user. That means the total number of users per cell site can be as high as 1,180, which would be about 2.8 users per 30-kHz bandwidth. With the 7-to-1 advantage of being able to use all channels in every cell, that leads to an advantage of 19.6 times as many voice channels for the CDMA system as for the AMPS-type system. A less optimistic advantage of about 17 to 1 results from using 100 codes per 1.25-MHz channel.

Cell-site hardware includes a single output amplifier per 1.25-MHz channel. No power combiner is needed as in the other digital or analog cellular systems where multiple transceivers are used. A typical ERP of 45W is used for pilot, setup, and 36 users.

The cell site sends power-control commands to each mobile unit on both an open-loop and a closed-loop basis. With an open loop, the mobile unit measures the signal strength received by the cell. If it is small, the mobile unit increases its power so that the sum of the receive and transmitted power expressed in dBm is a constant. With closed-loop control, the cell site measures the power received from the mobile unit and gives commands to adjust the power level of the mobile unit. An 80-dB dynamic range of power adjustment is provided in the mobile unit. With the dynamic range, the cell site attempts to maintain all received signals from the mobile units at the same power level. An 800-bps stream is used to achieve this control with a 2.5-ms response time. Such quick response is important because of the fluctuating nature of the signal caused by movement of the mobile unit and changing multipath.

Handoff is initiated by the mobile set. The new cell switches to the code the mobile unit is using, while the mobile unit still is receiving signals from the old cell. This type of handoff is called a soft handoff.

Registration can be accomplished on an automatic basis. Upon turn-on or entry into a new cell, a mobile set signals the cell site. The procedure prevents excessive paging load on the system.

The transmit power required from the mobile unit can be appreciably lower than in analog service, thus allowing fewer, larger cells in areas with low traffic density areas. The transmitter has a typical level of 10 mW under urban conditions, with 300 mW seldom being exceeded. Some of the information presented here is adapted from [11].

9.3 GPS RECEIVERS

This section discusses briefly global positioning satellite (GPS) receivers. GPS systems are recognized as being very important and as having increased importance in the future. Some of the information presented here is from [14,15].

GPS is an all-weather, worldwide, accurate, three-dimensional navigation system. It consists of a constellation of satellites that transmit signals, a network of ground stations, and the user receivers. The position of a GPS receiver is determined by using simultaneous measurement of the distances to four satellites.

The satellite constellation consists of 21 satellites and 3 in orbit spares. The satellites are deployed in a 10,900–nautical mile circular orbit with a 12-hour period. Four satellites are located in each of six planes inclined at 55 degrees to the plane of the Earth's equator. Each satellite continuously broadcasts pseudorandom codes on two frequencies: L1 at 1,575.42 MHz and L2 at 1,227.6 MHz. L1 is modulated with two types of code, the coarse/acquisition (C/A) code and the precision (P) code. L2 carries only the P code. The network of ground stations consists of monitoring stations at widely spaced, precisely known locations. Transmissions from the satellites are received and data forwarded to the master station. The master station prepares signal-coding corrections as needed and changes orders for the satellite control facility that uploads data to the satellites.

The final components for the system are the user's receivers. A generic GPS receiver consists of an antenna, one or more receivers, a microprocessor, memory, command and display units, and an appropriate power supply. A system of this type is shown in Figure 9.10. The antenna is usually an omnidirectional antenna with near-hemispheric coverage. Examples are monopulse, dipoles, microstrip patch, and spiral helix. Circular polarization is desirable.

System architectures include single-channel sequential tracking, single-channel multiplexed tracking, or single-channel-per-satellite tracking. With a single-channel sequential receiver, one of the required four satellites is tracked continuously for a number of seconds before the next satellite is acquired and tracked. Repeated sequencing through the four satellites yields enough information to provide the desired fix. With single-channel multiplexing, the sequencing rate is so high that individual bits from the signals of the four satellites are viewed in an almost simultaneous fashion. This approach requires less time to deliver a first fix but suffers signal loss due to self-jamming introduced by the multiplexing process. With multiple-channel receivers, at least four channels are required, one dedicated to each satellite. In practice, at least five channels are used, the fifth channel to acquire the next satellite needed for a continuous update. With a larger number of channels, more satellites can be tracked. Special precision and backup capability are the principal benefits of receivers with more than five channels.

Current and Future Commercial Communication Systems | 255

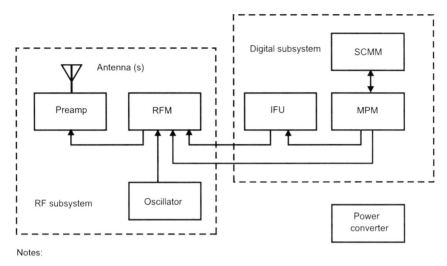

Notes:

RFM = RF module (dual conversion from L-band to 10.28 MHz 1 st IF)

IFU = IF module (IF signal processing, code correlation, and generation of I/Q signal along with carrier/frequency track)

MPM = Microprocessor module (microprocessor for signal processing, control, and display functions)

SCMM = Semiconductor memory module

Figure 9.10 Simplified system block diagram of a GPS receiver.

To prevent an enemy from using the GPS system against the United States, the ground control station can deliberately introduce satellite timing and position errors into satellite transmissions. That strategy, called selective availability (SA), reduces the accuracy of civilian and unauthorized users to 100m 90% of the time. That is enough for navigation but inadequate for weapon delivery. A further security measure can be achieved by encryption of the precision-code, yielding the P(Y) mode. The effects of SA can be virtually eliminated in a local area by use of differential GPS (DGPS), a technique requiring two receivers, one of which is at a known location. DGPS is widely used for precise surveying work even when SA is invoked.

GPS has many current and future uses. One current use is for aircraft navigation. Another is for providing improved accuracy in surveying. An exciting future application (already in use in Japan and being worked on by Ford and General Motors in the United States) is the provision of position location for automobiles driving down the streets of a city or down a highway. The antenna for such a system may be embedded in the trunk or the roof. A computer would be used to display local maps and details of position. Some other future users for GPS may include trucks, moving vans, trains, ambulances, fire and police departments, ships, crop-spraying vehicles, and military ground units.

256 | RF Systems, Components, and Circuits Handbook

There are many manufacturers for GPS receivers, and each manufacturer may produce several different models. Most of the systems produced provide five or six channels. Most are designed to track the same number of satellites as channels. Many track only the L1 frequency (1,575.42 MHz). Typical accuracies are about 15m for a single measurement. Accuracies for differential measurements typically are in the range of 1m to 3m. A typical time for first fix is about 2 min. A typical antenna is a microstrip patch antenna. A typical size for the receivers is 8" × 8" × 4", and a typical weight is about 5 lb (some units are larger, some smaller). Most units are designed for operation on land, on ships, or in aircraft. Applications typically include navigation, tracking, and surveying.

9.4 FIBER OPTIC COMMUNICATION SYSTEMS

Some of the information presented here is adapted from [16]. Fiber optic communication systems have become very important in recent years. Such systems may be used wherever coaxial cable or other transmission lines are used and may replace other means of communication, such as microwave relay systems. The main advantages for fiber optic cables include lower weight, lower cost, and larger bandwidth capability.

The attenuation characteristics of silicon fibers are shown in Figure 9.11. It shows that the attenuation for silicon fibers depends largely on the wavelength of the signal. The lowest attenuations are obtained at near-infrared wavelengths.

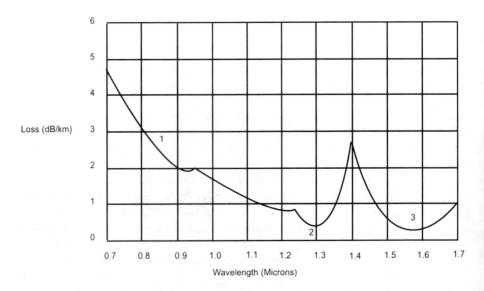

Figure 9.11 Attenuation as a function of wavelength for silicon fibers.

Infrared-emitting diodes are of major interest and are used as transmitters for fiber optic systems. The lowest attenuation is just above and just below 1.4 μm. The attenuation peak at 1.4 μm is a hydroxyl (OH) absorption band.

Attenuation for silicon fibers increases appreciably above 1.6 μm. Fibers based on fluoride and chalcogenide may exhibit ultra-low loss in the 2–5 μm range, possibly allowing repeaterless transoceanic fiber systems. If the optical energy has only one path through the fiber, there will be no pulse broadening. If, on the other hand, the dimensions of the fiber are large enough, there can be multipath, which results in broadening of the pulse. Pulse broadening limits the data rate that may be sent in an optic fiber.

Several types of optical fibers can be used. Plastic fibers are cheap but can be used only for short distances. They have diameters on the order of 1 μm or more. Plastic fibers are well suited for use with light-emitting diodes (LEDs). They are useful in rugged environments but are not found in telecommunications applications. Plastic-clad silica fibers have diameters less than about 600 μm. Like plastic fibers, plastic-clad silica fibers can have very large numbers of propagation paths and thus are used only for relatively short distances. They are inexpensive and may use inexpensive emitters such as LEDs.

Glass fibers have diameters less than 200 μm and can be classed as multimode, graded index, and single-mode fibers. The multimode, step-index fiber is most useful for distances up to several kilometers in systems that use low-cost LED emitters. The graded-index fiber finds common use in systems several kilometers in length to repeater systems, such as cable television, as well as some long-haul systems. The single-mode, step-index fiber has the smallest dispersion or multipath and is very useful for long-haul telecommunication applications. It requires use of laser diode emitters or light sources. This type of fiber has a diameter of only 5–10 μm. Because of its low dispersion, it can allow pulse rates of 10 Gbps and higher to be sent over distances well beyond 100 km. Single-mode optical fibers are being used in submarine cable systems installed across the Atlantic and Pacific oceans by using repeaters. The most commonly used sources in optical-fiber communication are the LED and the diode laser. Each of these systems is discussed in detail in Part 2. Diode lasers are capable of 5–100 mW on a CW basis, whereas LEDs typically have power ratings of 1–10 mW. Diode lasers can be modulated at frequencies of 1–25 GHz, whereas LEDs have maximum modulation rates less than about 200 MHz.

Intensity modulation of a laser can be used to transmit information digitally. This mode has been used most often in the past. It is also possible to use polarization modulation for a CW source.

The output of a fiber optic cable is received and detected by a photodetector followed by an amplifier. Examples of noncoherent optical detectors are photomultipliers, photoconductors, photodiodes, and photo transistors. Photodiodes are reversely biased p-n diodes or p-i-n diodes. They convert optical power directly into electric current. Avalanche photodiodes provide an additional internal gain on the order of 10 to 100. In all cases, the minimum detectable signal (MDS) is determined by the level of internally generated noise for the detectors.

Repeaters are used when the required distance is greater than that which can be covered by a single link. Such repeaters operate by detecting optical pulses, regenerating them, and then modulating a new light source. Such a system is shown in Figure 9.12.

Fiber optic communication systems have had a remarkable growth and public acceptance in only a few short years. Fiber telecommunication lines now connect most major metropolitan centers with one another throughout North America, Europe, parts of the Mideast, Japan, and Australia. Extensions to other metropolitan centers elsewhere in the world are either under construction or being planned. The replacement of major portions of copper wire subscriber-loop plants with fiber is under active consideration throughout the world for the removal of existing bandwidth limitations. Such replacement is largely a matter of economics.

Current and Future Commercial Communication Systems | 259

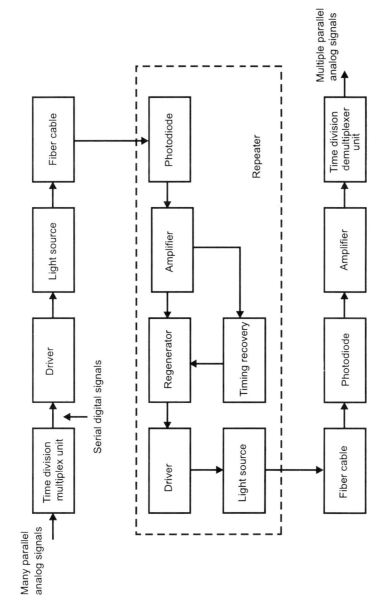

Figure 9.12 Use of a repeater in a fiber optic link.

References

[1] Radio Shack, 1997 Catalog, pp. 6–59.
[2] Williams, Mark, and Daniel Ong, "PCS and RF Components," Applied Microwave & Wireless, Spring 1995.
[3] Macario, Raymond C. V., Cellular Radio Principles and Design, New York: McGraw-Hill, 1993, pp. 47–53.
[4] Mehrotra, Asha, Cellular Radio Analog and Digital Systems, Boston: Artech House, 1994, pp. 146–157.
[5] Keiser, Bernhard E., and Eugene Strange, Digital Telephony and Network Integration, 2nd ed., New York: Van Nostrand, Reinhold, 1995, pp. 631–634.
[6] Keiser and Strange, op. cit., p. 635.
[7] Ibid., pp. 235–239.
[8] Mehrotra, op. cit., pp. 308–323.
[9] Keiser and Strange, op. cit., pp. 239–244.
[10] Ibid., p. 257.
[11] Ibid., p. 258.
[12] Noll, A. Michael, *Introduction to Telephones and Telephone Systems*, Second Edition, Norwood, MA: Artech House, 1991, Chap. 6.
[13] Van Valkenburg, Op. Cit., Chap. 46, "Cellular Telecommunication Systems" by William C. Y. Lee.
[14] Herskovitz, Don, "Global Positioning System Receivers," Microwave Journal, Sept. 1994.
[15] Van Valkenburg, Mac E., ed., Reference Data for Engineers, 8th ed., SAMS, Carmel, IN: Prentice Hall Computer Publishing, 1993, p. 37-6.
[16] Keiser and Strange, op. cit., pp. 333–369.

CHAPTER 10

Radar Systems

10.1 RADAR CROSS-SECTION OF TARGETS

The RCS, σ, of a target is a measure of the effective cross-section area of the target if the assumption is made that the radar power to the target is scattered equally in all directions (through 4π steradians). The RCS depends not only on the size of a target but also on the shape of the target, the materials involved, the aspect angle at which the target is viewed, and the frequency of the radar signal. Some targets have very small radar cross-sections, even though they have large physical areas, because the signal is not scattered back toward the radar. On the other hand, some very small targets have very large RCSs because most of the energy that reaches the target is scattered back in the direction from which it came. A corner reflector is an example of that type of target.

The only target whose RCS is independent of viewing aspect angle is a conducting sphere. For that reason, metallic spheres often are used to calibrate radar systems. Figure 10.1 shows the ratio of the RCS of a sphere divided by the physical cross-section as a function of circumference and wavelength. As shown in the figure, there are three main regions of RCS. The first region is the Rayleigh region, where the RCS is much smaller than the physical cross-section of the sphere and falls off as the fourth power of the circumference-wavelength ratio. The second region is the resonance region, where the amplitude of the RCS oscillates up and down with increasing circumference over wavelength. The third region is the optical region, where the ratio of RCS to the cross-sectional area is nearly 1.

For an example of the use of Figure 10.1, assume operation at a wavelength of 1.0m and a sphere diameter of 0.1m. The circumference is 0.314m, and the circumference-to-wavelength ratio is 0.314. Operation is thus in the Rayleigh region. From Figure 10.1, the ratio of the RCS to the physical cross-section area is about 0.1. The cross-section area is 7.85×10^{-3} m^2; thus, the RCS is 7.85×10^{-4} m^2. That

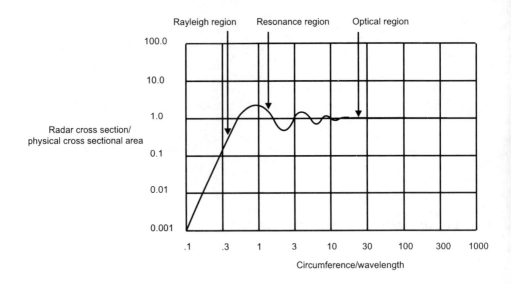

Figure 10.1 RCS for a sphere. (*After:* [1].)

is a very small RCS. On the other hand, if we assume operation at a wavelength of 0.03m and that same sphere, the circumference-to-wavelength ratio is 10.47. Operation is in the optical region, and the ratio of the RCS to the physical cross-section area is about 1.0. The RCS of the sphere thus is about 7.85×10^{-3} m^2. We see a factor-of-10 difference in the RCS for the two frequencies. When a sphere is used to calibrate a radar, the size of the sphere usually is chosen so that operation will be in the optical region.

The general ideas regarding radar scattering in the three different regions apply to objects of other shapes as well as to spheres. Hence, all target RCSs are wavelength sensitive. All targets except spheres have RCS values that change with aspect angle. For example, an aircraft typically has a much larger RCS when viewed from a 90-degree aspect angle than when viewed nose on. Its RCS is characterized by many peaks and nulls as the azimuth aspect angle changes. A typical RCS pattern for an aircraft is shown in Figure 10.2.

The lobbing structure of the RCS signature or pattern changes greatly with frequency. At lower frequencies, there are fewer peaks and nulls than there are at the higher frequencies. The RCS can be viewed as the vector sum of many individual scatters located at different distances and therefore having different phase angles.

It is possible to reduce greatly the RCS of military aircraft by designing the aircraft so no surfaces provide specular reflections, such as is done with the Stealth bomber. It is also possible to reduce the RCS of targets by using special RAM coatings. RAMs are frequency and thickness dependent. The thickness should be nearly one-quarter wavelength at the frequency of interest. Table 10.1 lists some

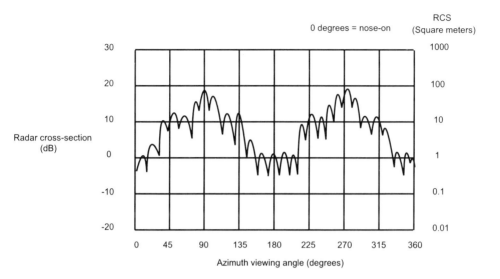

Figure 10.2 RCS for an aircraft as a function of azimuth viewing angle with 0-degree elevation angle.

RCSs for targets at microwave frequencies. The values given are average or typical values. Peak values could be 5–10 dB greater [1].

The RCSs for most targets tend to increase with increasing frequency. They are greatest when large flat surfaces are viewed normal to the surfaces. Note that the RCS of a pickup truck is typically larger than that of even a large aircraft. That is because the pickup truck has flat surfaces that can be normal to the viewing angle, whereas the aircraft has curved or slanted surfaces with respect to the viewing angle. A large naval ship has a very large RCS for the same reason. A human has an RCS of about 1 m^2, while a bird has an RCS of about 0.01 m^2. A group of birds would have a peak RCS many times that of a single bird. In a similar way, a large swarm of insects would have a significant RCS, even though that of a single insect is very small.

It is typical to have large RCS fluctuations as a moving target is viewed over a period of time. That is due to the fact that at one period of time the target may present an RCS near a peak in the scatter pattern, while a short time later it may present a null in the scatter pattern.

10.2 RADAR CLUTTER

Another important type of RCS is the RCS of the earth and objects on the Earth as viewed by the radar. When that signal is not desired, it is referred to as clutter. Some of the information in this section is adapted from [1–3].

Table 10.1
RCSs of Some Targets at Microwave Frequencies (3–10 GHz)

Type of Target	RCS (m^2)
Aircraft	
Conventional unmanned winged missile	0.5
Small single-engine aircraft	1.0
Small fighter aircraft	2.0
Four-passenger jet aircraft	2.0
Large fighter aircraft	6.0
Medium bomber or medium jet airliner	20.0
Larger bomber or large jet airliner	40.0
Jumbo jet airliner	100.0
Ships	
Small open boat	0.2
Small pleasure boat	2.0
Cabin cruiser	10.0
Large naval ship	10,000.0
Land Vehicles	
Bicycle	2.0
Automobile	100.0
Pickup truck	200.0
Nonmetallic Targets	
Insect	0.0001
Bird	0.01
Human	1.0

(*After:* [1].)

Figure 10.3 shows the geometry of main beam radar clutter. The clutter patch from which the RCS is determined is shown in the plan view. The size of the clutter patch depends on the antenna azimuth beam angle, the elevation or grazing angle, and the pulse width.

RCS for clutter is computed from an estimate of per-unit surface-area RCS times the projected antenna beam and range resolution footprint area, as given by (10.1):

$$A = 2R \tan(\theta_b/2) \times 150\text{m} \times \tau/\cos\theta_{el} \tag{10.1}$$

where

R = range (meters)

θ_b = antenna azimuth beam angle

τ = radar pulse length (microseconds)

θ_{el} = antenna beam elevation grazing angle

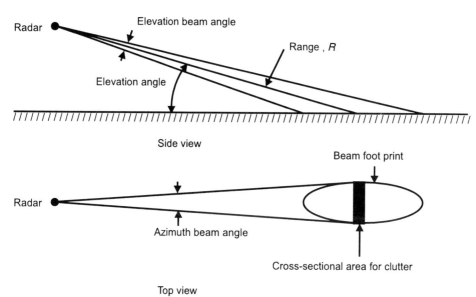

Figure 10.3 Geometry for radar clutter.

For an example of the clutter RCS for land, assume the following:

Per-unit RCS = −30 dB = 0.001
Range = 100 km
Antenna azimuth beam angle = 2 degrees
Radar pulse length = 1 μs
Grazing angle = 10 degrees

Substituting those value into (10.1), the area of the clutter patch is

$$A = 2R \tan(\theta_b/2) \times 150 \text{ meters} \times \tau/\cos\theta_{el} \quad (10.2)$$
$$A = 2 \times 100{,}000 \times 0.017 \times 150 \times 1/0.985 = 517{,}766 \text{ m}^2$$

The RCS is

$$\text{RCS} = \sigma = 0.001 \times 517{,}766 = 517.8 \text{ m}^2$$

There are a number of ways to help reduce the clutter in radar systems. One way is to use highly directional antennas with low sidelobes. Another way is to use short-duration pulses. Each of these methods helps reduce the area or the volume of the clutter and, therefore, the RCS of the clutter.

Figure 10.4 shows typical land clutter reflectivity or RCS per unit area expressed in decibels. The value depends on the grazing angle and the type of surface. For microwave frequencies, it tends to be independent of frequency. Types of surfaces shown are flatland, farmland, wooded hills, and mountains.

For an example of the use of Figure 10.4, assume a grazing angle of 3.0 degrees and wooded hills. For that condition, the indicated RCS per unit area is −20 dBsm. Thus, if the clutter patch is 100,000 m^2, the RCS is 0.01 times that area, or 1,000 m^2. Figure 10.5 shows typical sea clutter reflectivity or RCS per unit area as a function of grazing angle and frequency. The values shown are for medium sea conditions. The values would increase as the wind speeds increase and the sea state becomes rougher. There is very little backscatter for a calm sea.

For an example of the use of Figure 10.5, again assume a grazing angle of 3.0 degrees and operation at X-band. We can see from the figure that the per-unit RCS is about −30 dBsm. If the clutter patch is 100,000 m^2, the RCS is 0.001 times that area, or 100 m^2.

Another type of radar target is weather (clouds, fog, rain, or snow). Weather-detecting radars are used both on the ground and on aircraft to detect weather features. Rain or snow can be viewed as clutter when it interferes with the detection of other desired targets. Radar signals are backscattered from range resolution cells or volumes having lengths defined by the pulse width (150 m/μs), and cross-sectional areas are defined by the range and the azimuth and elevation beam angles.

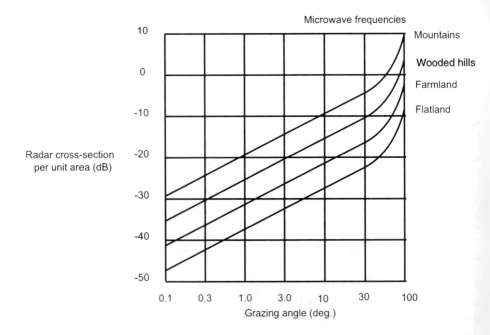

Figure 10.4 Land-clutter reflectivity versus grazing angle. (*After:* [2].)

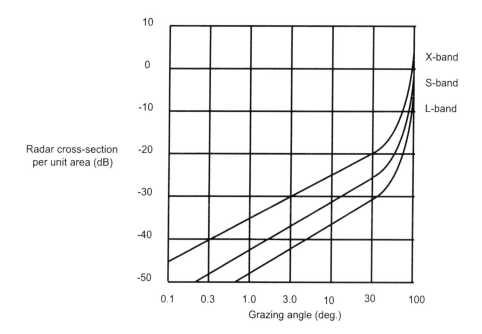

Figure 10.5 Sea-clutter reflectivity versus grazing angle. (*After:* [2].)

Thus, if for a range resolution cell at a range of 100 km, an elevation beam angle of 2 degrees, an azimuth beam angle of 4 degrees, and a pulse width of 2 μs, the volume of the clutter cell would be 150 m/μs × 2 μs × 2 tan 1 degree × 2 tan 2 degrees = 300 × 3,492 × 6,993 = 7.3 × 10^9 m^3. If the RCS per cubic meter is 10^{-8} m^2 per cubic meter of clutter cell, the clutter RCS would be 73 m^2.

Figure 10.6 shows the volume reflectivity of rain as a function of precipitation rate and wavelength. We see that the RCS increases rapidly with increasing frequency. Thus, if we have a precipitation rate of 10 mm/h, the per-unit RCS is about 10^{-12} m^2 per cubic meter with a wavelength of 100 cm (frequency = 300 MHz). At a wavelength of 2 cm (frequency = 15 GHz) and the same precipitation rate, the per-unit RCS is about 10^{-5} m^2 per cubic meter. That is a factor of 10^7 difference for a factor of 50 increase in frequency.

Snow has an RCS similar to that of rain for very low precipitation rates but can have as much as a factor of 10 or more increase in RCS for high precipitation rates (>30 mm/h).

10.3 RADAR MEASUREMENTS

10.3.1 Range Measurements

The main types of measurements made by radars include the range or distance to the target, the pointing angle to the target, the radial velocity of the target with

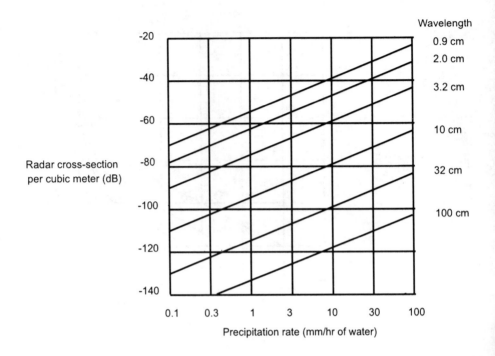

Figure 10.6 Volume reflectivity of rain. (*After:* [2].)

respect to the radar, and the RCS of the target. Figure 10.7 shows concepts for range measurements using a pulse radar. (Block diagrams for radars of this kind are presented in Section 10.5.)

The radar transmits a train of short-duration RF pulses using a directional transmitting antenna. The pulse width must be less than the expected round-trip time delay.

The pulse width determines the range resolution of the radar. If small range resolution is needed, the pulse width must be small. The range resolution is the minimum distance between two targets that can be seen by the radar as separate targets.

A small part of the power transmitted is reflected back to the radar receiving antenna. The received signal is fed to the radar receiver, where it is amplified, down-converted in frequency, and detected. The two trains of pulses are shown in Figure 10.7. The radar measures the delay time between the two sets of pulses. The range can then be computed using (10.2).

$$R = Tc/2 \qquad (10.3)$$

where

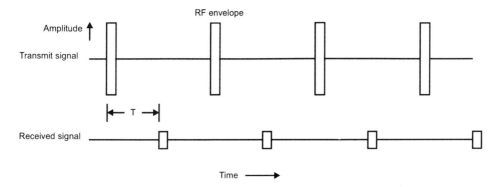

Figure 10.7 Concepts for range measurements using pulse-type radar.

R = range (meters)
T = delay time (seconds)
c = speed of light = 3×10^8 m/s

From (10.2), we see that the radar range is equal to 150 m/μs of delay. For example, if the measured delay is 200 μs, the radar range to target is $200 \times 150 = 30,000$m = 30 km.

Figure 10.8 shows concepts for range measurements using FMCW-type radar. The radar transmits a CW signal with the frequency changing as a function of time. A triangular frequency pattern is shown in Figure 10.8. A directional transmitting antenna transmits the signal toward the target. A small part of the power transmitted

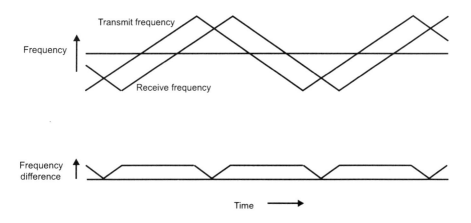

Figure 10.8 Concepts for range measurements using FMCW radar.

is reflected back to the radar receiving antenna. The two frequency patterns are shown in Figure 10.8.

The received signal is fed to the radar receiver, where it is mixed with the transmitter signal, and the difference frequency is amplified and detected. The difference in frequency between the transmit signal and the receive signal is a measure of the time delay between the two signals. The range then can be computed using (10.2).

For example, assume that the transmitted signal changes linearly from 3.1 GHz to 3.2 GHz in a period of 0.001 sec. The rate of change of frequency is thus $10^8/0.001 = 10^{11}$ Hz/s. If the measured frequency difference is 50 MHz, the corresponding delay is $5 \times 10^7/10^{11} = 5 \times 10^{-4}$ sec = 500 μs. Substituting values into (10.2), the corresponding range is

$$R = TC/2$$
$$R = 500 \times 150 = 75 \text{ km}$$

10.3.2 Velocity Measurement Using CW Radar

Figure 10.9 shows the concepts for target radial velocity measurements using a CW radar. The radar transmits a CW signal with constant frequency and amplitude. A small part of the power transmitted is reflected back from the target to the radar receiving antenna. The signal has a frequency equal to the transmitted frequency plus or minus the Doppler frequency, depending on whether the target is moving toward the radar (plus) or away from the radar (minus).

The received signal is fed to the radar receiver mixer, where it is mixed with part of the transmitter signal that is reflected back from the antenna. A circulator or diplexer typically is used to direct the transmit signal to the antenna and the receive signal to the receiver mixer.

The difference frequency output from the mixer is the Doppler frequency. The signal is selected by a bandpass filter and then is amplified by a beat-frequency amplifier and fed to an indicator. The two-way radar Doppler frequency is given by

$$f_d = 2v_r f_0/c \qquad (10.4)$$

where

v_r = radial velocity with respect to the radar

The radial velocity of the target with respect to the radar can be computed as follows:

$$v_r = f_d c/(2f_0) \qquad (10.5)$$

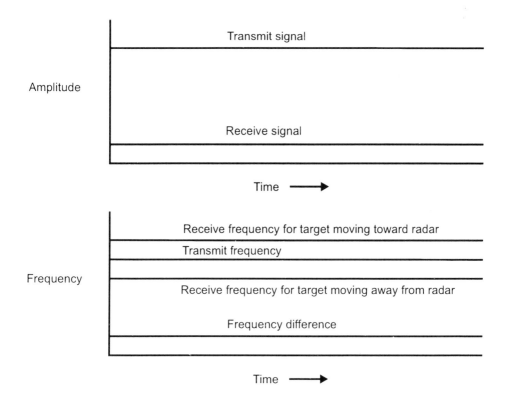

Figure 10.9 Concepts for target radial velocity measurements using a CW radar.

where

v_r = target radial velocity with respect to the radar
f_d = measured Doppler frequency
c = speed of light = 3×10^8 m/s
f_0 = transmitted frequency

For an example of the use of (10.4), assume that the transmit frequency is 9.2 GHz and the measured Doppler frequency is 1,000 Hz. Then, from (10.4)

$$v_r = 1{,}000 \times 3 \times 10^8 / (2 \times 9.2 \times 10^9)$$
$$v_r = 16.3 \text{ m/s} = 58.7 \text{ km/h}$$
$$= 36.5 \text{ mi/h}$$

10.3.3 Velocity Measurements Using FMCW Radar

Figure 10.10 shows the concepts for target radial velocity measurements using an FMCW radar. The radar transmits a CW signal with triangular frequency modulation as a function of time, as shown in the figure. A directional transmitting antenna transmits the signal toward the target. A small part of the power transmitted is reflected back to the radar receiving antenna. The received signal is fed to the radar receiver, where it is mixed with the transmitter signal, and the difference frequency is amplified and detected. The transmitted signal frequency and the received signal frequency are shown for the condition where the received signal is shifted in frequency by the Doppler effect. The difference frequency for each half of the cycle is shown. The difference between the two difference frequencies is equal to twice the Doppler frequency. The measured Doppler frequency can be used to determine the radial velocity of the target with respect to the radar using (10.3).

10.3.4 Velocity Measurements Using Pulse Type Radar

Figure 10.11 shows the concepts for velocity measurements using a pulse-type radar. The radar transmits a train of short-duration RF pulses using a directional transmitting antenna. A small part of the power transmitted is reflected back to the radar receiving antenna. The received signal is fed to the radar receiver, where it is amplified, converted in frequency, and detected using a coherent phase-sensitive detector. The two trains of pulses are shown in Figure 10.11.

With a phase-sensitive detector, the amplitude of the detected pulses will be constant only if there is no radial velocity for the target. If there is a radial velocity,

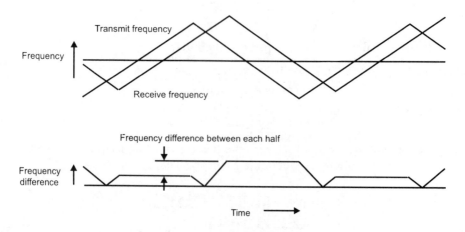

Figure 10.10 Concepts for velocity measurements using FMCW type radar.

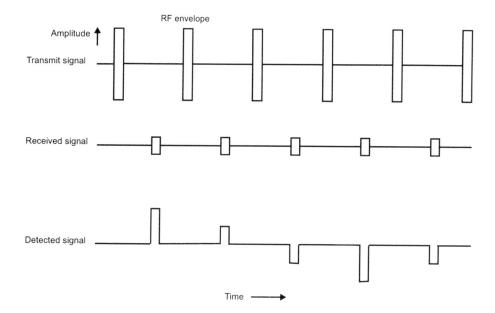

Figure 10.11 Concepts for velocity measurements using pulse type radar.

there will be a Doppler frequency shift, and the amplitude of the received pulse train will oscillate in amplitude at the Doppler frequency rate, as shown in Figure 10.11. The rate of oscillation can be detected using appropriate signal processing circuits. The measured Doppler frequency then can be used with (10.3) to determine the target velocity.

10.3.5 Angle Measurements for Radars

Angle measurements for radars depend on having highly directional antennas and means for reading out the pointing angles for the antennas. Very high mechanical precision is possible for such radars as the old but reliable FPS-16 tracking radar, as well as for many of the newer tracking radars.

Receiver antenna beams often are split, as shown in Figure 10.12. A process known as sequential lobbing, shown in Figure 10.12(a), can be used in tracking radars to permit pointing-error detection and pointing correction.

Another way to determine pointing errors is to use conical scan tracking, shown in Figure 10.12(b). As the pencil beam is scanned in a circle with a squint angle, the return signals have amplitudes that change as a function of the scan angle. The information then can be used to correct the pointing for minimum-error signal.

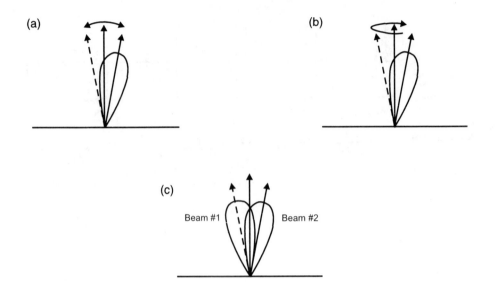

Figure 10.12 Concepts used for high-accuracy angle tracking and measurements: (a) single-beam sequential lobbing; (b) single-beam conical scan; and (c) two-beam monopulse tracking.

Monopulse tracking, shown in Figure 10.12(c), is frequently used for high-accuracy angle measurement and tracking. Monopulse radar uses split beams similar to that used for sequential lobbing. The difference is that measurements can be made on a single pulse rather than on two or more sequential pulses. That leads to greater accuracy since errors due to fading are avoided.

Monopulse radars may use a single beam for transmit and two or more beams for receive. A four-beam system frequently is used for tracking in both azimuth and elevation. Again, the idea is to detect pointing errors by the difference in amplitude of the signals received in the several antennas. The detected errors then can be used to correct the pointing of the tracking radar antennas. Mechanical readouts of pointing angles are provided.

The following sections discuss briefly a number of complete radar systems that use these measurement methods.

10.4 CONTINUOUS-WAVE RADAR SYSTEMS

10.4.1 Continuous-Wave Radars for Target Velocity Measurement

Some of the information in this section is adapted from [4]. Moving targets are detected in a background of earth scatter of the radar signal on the basis of Doppler frequency filtering. The radar return signal from the Earth has a Doppler frequency different from that from the moving target. The Doppler frequency for the Earth is not usually zero, as one might expect, but some low frequency corresponding to the motion of waves on a sea or movement of trees and brush due to wind.

With this system, there is no capability for measuring the range to the target but only the velocity with respect to the radar. One important example of this type of radar is the radar used by the police to measure the speed of automobiles. These small solid-state microwave radars can be either vehicle-mounted or hand-held. The receive radar frequency is the transmitter frequency plus or minus the Doppler frequency. If the target is moving toward the radar, the receive frequency is increased above that of the transmitted frequency. If it is moving away from the radar, the received frequency is decreased, that is, a negative Doppler shift.

Figure 10.13 shows the case of a simple CW radar with zero IF frequency. This system uses a single antenna. The signal is transmitted with frequency f_0. The scattered signal from the target is received at a frequency $f_0 + f_d$, where f_d is the Doppler frequency. The two signal frequencies mix in the mixer, resulting in a beat frequency equal to the Doppler frequency, f_d. The signal is amplified and fed to the indicator circuit. The indicator may be a counter and display, or it may be a more complex system.

Figure 10.14 is the schematic diagram for an example of high-performance CW radar that uses an IF amplifier. The microwave frequency is generated by a solid-state Gunn oscillator. The output of the oscillator is fed to the transmit antenna. A part of the signal also is fed to a mixer. The second input to the mixer is from an LO that is at the IF amplifier frequency, assumed to be at 70 MHz. The sum frequency is selected by a sideband filter and fed to the receiver mixer. The second input to the receiver mixer is from the receiver antenna. The signal frequency input from the antenna is the transmit frequency plus or minus the Doppler frequency. The difference output frequency from the receiver mixer is selected by a bandpass filter and fed to the IF amplifier. That frequency is the IF amplifier frequency plus or minus the Doppler frequency.

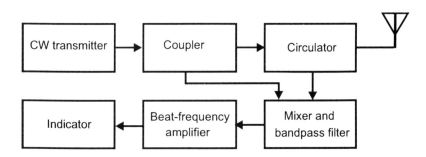

Figure 10.13 Simple block diagram of CW radar.

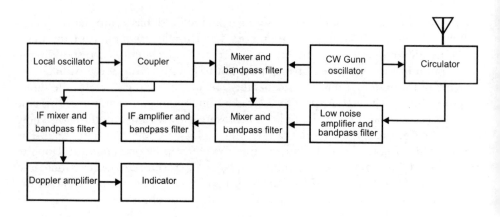

Figure 10.14 CW radar with IF amplifier-type receiver.

After amplification and filtering by the IF amplifier chain, the signal is fed to a detector that consists of a balanced mixer. The second input to the mixer is a signal from the LO. The difference frequency from the mixer is the Doppler frequency. The signal is filtered and fed to a Doppler amplifier. The output of the amplifier is to a Doppler frequency indicator.

A possible variation of the circuit in Figure 10.14 is to use a low-noise amplifier and a bandpass filter between the receiver mixer and the receive antenna. A second variation is to use separate antennas for transmit and receive functions. That variation usually provides greater transmit-to-receive isolation than that obtainable through the use of a circulator.

10.4.2 Frequency-Modulated Continuous-Wave Radars

Another important type of radar is FMCW radar. Figure 10.15 shows the case of an FMCW radar altimeter. The generated waveform may sweep from a low frequency in a microwave band to a higher frequency in the band and back to the low frequency in either a linear or a sinusoidal fashion. The frequency band from 4.2 to 4.4 GHz is reserved for radio or radar altimeters. That allows for a maximum frequency deviation or frequency sweep of 200 MHz. In the past, radars also operated at UHF. With the FMCW radar altimeter, the range or altitude is measured by the difference in the transmitted frequency and the delayed return frequency from the Earth.

Equation (10.5) provides the beat frequency or frequency difference for the FMCW radar altimeter.

$$f_r = 4R f_m \, \Delta f / c \qquad (10.6)$$

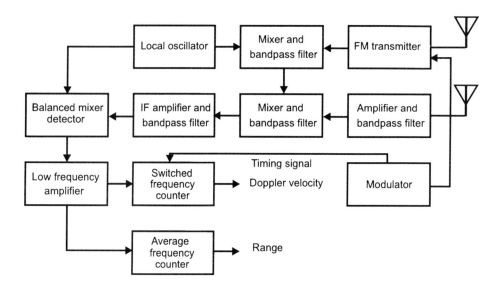

Figure 10.15 FMCW radar altimeter.

where

f_r = beat frequency (hertz)
R = range (meters)
f_m = modulation rate (hertz)
f = frequency excursion (hertz)

For example, assume $R = 100$m, $f_m = 120$ Hz, $\Delta f = 70$ MHz and $c = 3 \times 10^8$ m/s. Using those values,

$$f_r = 4 \times 100 \times 120 \times 70 \times 10^6 / (3 \times 10^8) = 11{,}200 \text{ Hz}$$

The modulator provides the triangular sweep pattern and timing for the FM transmitter. This unit also provides a timing signal for the switched frequency counter.

A solid-state FM transmitter is used with a typical output power level of 2–10W. The main output from this transmitter is to a transmit antenna, which may be a surface-mounted antenna such as a cavity-backed slot antenna. A part of the transmitter power also is coupled to a mixer. The second input to the mixer is from an LO that is at the IF amplifier frequency. The output of the mixer is then fed to a sideband filter that selects the upper sideband. That signal then is fed to the receiver mixer, which includes an RF bandpass filter.

The receiver mixer is connected to the receiver antenna, which is also a surface-mounted antenna such as a slot antenna. The output of the mixer is fed to the IF amplifier chain. The amplifier chain includes bandpass filters and amplifier stages.

The output of the IF amplifier is fed to a balanced mixer. The second input of the mixer is from the LO. This detector recovers the beat-frequency signal, which is then amplified by a low-frequency amplifier and fed to two counters: the switched frequency counter, which measures the Doppler velocity, and the average frequency counter, which measures the range.

If Doppler is present, the beat frequency for the two modulation parts will be different. That provides the basis for measuring vertical velocity as well as vertical range. By proper signal processing, it is possible to measure altitude and vertical velocity accurately. The accuracy for the altimeters depends on the signal-to-noise ratio and the so-called quantization error for the counters used with the systems. The quantization error in turn depends on the frequency deviation used. The fixed or quantization error is independent of range and carrier frequency. This error, in meters, is approximately 75 divided by the frequency excursion, in megahertz. For example, if the frequency excursion is 70 MHz, the fixed or quantization error is 1.07m. With high signal-to-noise ratio, the total error will be nearly the same as the fixed error.

Fairly large antenna beam angles are used with radar altimeters, so a large spot on the Earth is illuminated when the aircraft is at high altitude. The spot becomes smaller as the altitude is decreased. This large beam angle is the result of using only a small antenna.

FMCW radars also can be used for target acquisition and tracking. These radars may use a linear swept triangular waveform. The range to the target is determined by the time difference in the delayed return frequency and the transmit frequency. The range resolution for the radar depends on the bandwidth of filters used to determine the range. A large bank of filters may be used to provide many range cells for detecting multiple targets. By proper signal processing, it is possible to reject the clutter signal return and to measure the range, angle, velocity, and RCS of the target to high accuracy.

10.5 MOVING-TARGET INDICATOR AND PULSE DOPPLER RADARS

Some of the information in this section is adapted from [4]. MTI radar takes advantage of the velocity-measuring capability of pulse radar to filter out unwanted radar signals that come from clutter. In an ideal MTI radar, only those signals having significant Doppler shift are received. In a practical MTI radar, there will be large clutter rejection, but filtering will not be perfect.

Pulse Doppler radars (PDRs) are similar in function to MTI radars. The main difference is in the PRF used by the two types of radars. MTI usually refers to a radar in which the PRF is chosen low enough to avoid multiple-time-around echoes. The problem here is that there can be ambiguous Doppler frequency measurements. PDR, on the other hand, has a high PRF, which may result in ambiguities in range. In other words, we do not know to which of several possible ranges the return signal corresponds. The main advantage of the PDR is that, because of the high PRF, there are no blind speeds or ambiguous velocity measurements.

Figure 10.16 is the block diagram for an MTI radar with a power amplifier transmitter. The power amplifier may be an amplifier chain with a modulator stage, several solid-state amplifiers in series, and a high-power output amplifier. Possible power amplifier types include CCTWT amplifiers, pulse klystron amplifiers, or twystron amplifiers.

The power amplifier input CW signal is modulated by a pulse modulator, which receives inputs from the timing circuit of the radar. The CW RF input to the amplifier system is from an up-converter that consists of a mixer and a bandpass filter. One input to the mixer is a CW RF signal from a stable local oscillator (a "stalo"). The second input to the mixer is a CW IF signal from a coherent oscillator (a "coho"). The sum of the two frequencies is selected by the bandpass filter.

The pulse-modulated output signal from the power amplifier is fed through the duplexer to a high-gain directional antenna. The antenna then radiates the signal in the direction of the target.

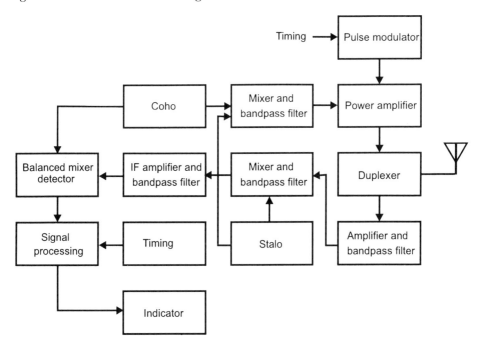

Figure 10.16 MTI radar with power amplifier transmitter. (*After:* [4].)

The return signal from the target is received by the same antenna as used for transmit. It is routed by the duplexer to a down-conversion mixer, which converts the received signal frequency to the difference between the stalo frequency and the received frequency. That difference is the coho frequency plus or minus the Doppler frequency, with the sign depending on the direction of travel of the target with respect to the radar. The signal is fed to the IF amplifier, where it is amplified to the desired level for detection and processing.

The output of the IF amplifier is fed to a phase detector, which has as its second input a CW coherent reference signal from the coho. The output of the phase detector is a train of bipolar video pulses, which are modulated in amplitude at the Doppler frequency.

The detected Doppler signal is then processed by a signal processing circuit that largely rejects clutter signals because they do not have the necessary Doppler frequency shift. Early radar systems used delay cancellation circuits for that function. Digital processing and computer systems are used in many modern systems. The output of the signal processing system goes to the indicator systems, which may involve visual displays and data processing systems. Figure 10.17 is the block diagram of an MTI radar with a power oscillator transmitter. In the figure, a high-power magnetron oscillator is used as the transmitter and is controlled by a trigger genera-

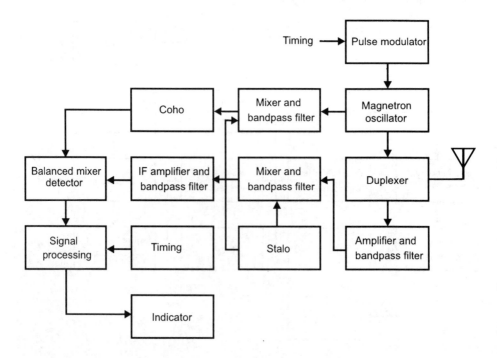

Figure 10.17 MTI radar with power oscillator transmitter. (*After:* [4].)

tor followed by a pulse modulator. A sample of the magnetron oscillator output signal is fed to a mixer, where it is mixed with the signal from a stalo. The difference frequency is selected and is used as a locking reference signal for a coho.

The operation for the balance of the system is the same as that discussed for the system shown in Section 10.6. The output of the magnetron oscillator is fed through a duplexer to a high-gain directional antenna. There it is transmitted to the desired target. The signal from the target is scattered back to the antenna.

The received signal next is passed through the duplexer to a down-converter mixer. The second input to the mixer is a CW RF signal from the stalo. The output signal is at the IF frequency plus or minus the Doppler frequency. The signal is amplified to the desired level by an IF amplifier. It then is fed to a phase detector that has as its second input a CW signal from the coho. The output of the phase detector is a train of bipolar video pulses that are modulated in amplitude at the Doppler frequency.

The detected Doppler signal is processed by a signal processing circuit that largely rejects clutter signals because they do not have the necessary Doppler frequency shift. Earlier radar systems used delay cancellation circuits for that function. Digital processing and computer systems are used in many modern systems. The output of this system goes to the indicator systems.

10.6 SIGNAL PROCESSING FOR MOVING-TARGET INDICATOR RADARS

Some of the information in this section is adapted from [5,6].

10.6.1 Delay-Line Cancelers

The simple older-generation MTI delay-line canceler shown in Figure 10.18 is an example of a time-domain filter. The delay line has a delay equal to 1 divided by the pulse repetition frequency. For example, if the pulse repetition rate is 500 pulses/sec, the delay must be equal to 2,000 μs (2 ms). Delay times of this magnitude cannot be achieved with standard electromagnetic transmission lines, but they can be achieved with acoustic signals in fused-quartz delay lines. Delay lines of this type were used in the early days of radar. They have a number of disadvantages compared to more modern digital delay lines, including larger size

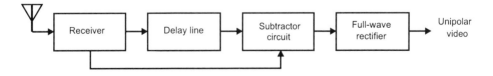

Figure 10.18 MTI receiver with delay-line canceler.

and weight and high attenuation, which must be overcome by added gain. Loss in acoustic delay lines for radar can be 40–60 dB.

Analog acoustic delay lines were supplanted in the early 1970s by storage devices based on digital computer technology. The use of digital delay lines requires that the output of the MTI receiver phase detector be quantized into a sequence of digital words. The compactness and convenience of digital processing allow the implementation of more complex delay-line cancelers with filter characteristics not practical with analog methods. Quantization also provides greater accuracy and ability to process inferior signal-to-noise ratios.

Figure 10.19 shows the frequency response of a single and a double delay-line canceler. The delay-line canceler acts as a filter that rejects the dc component of clutter. Because of its periodic nature, the filter also rejects signals with frequencies in the vicinity of the pulse repetition frequency and its harmonics. The target velocities that result in zero MTI response are called blind speeds. In practice, long-range MTI radars that operate in the region of L- or S-band or higher and are primarily designed for the detection of aircraft must usually operate with ambiguous Doppler and blind speeds if they are to operate with unambiguous range. To have unambiguous range, we must have a period between pulses that is greater than the maximum delay expected for the targets of interest.

The use of a number of delay-line cancelers in series offers improvement in the amount of cancellation possible. Thus, it is common to use double or triple cancellation. The use of more than one PRF offers additional flexibility in the design of MTI Doppler filters. It not only reduces the effect of the blind speeds,

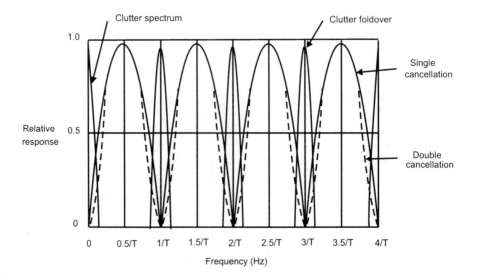

Figure 10.19 Frequency response of delay-line cancelers for single and double cancelers. (*After:* [6].)

it also allows a sharper low-frequency cutoff in the frequency response than might be obtained with a cascade of single-delay-line cancelers. This type of improvement is illustrated in Figure 10.20. Two delay-line cancelers are needed corresponding to the two PRFs, and they must be switched as the PRFs are switched.

10.6.2 Digital Signal Processing

A digital MTI processor is shown in Figure 10.21. Its operation is as follows. The output of the IF amplifier is fed to two phase detectors. The phase reference for the detectors is from a coho, with one of the reference signals shifted by 90 degrees from the other. One of the channels is the in-phase, or I, channel, while the other is the quadrature, or Q, channel. The outputs of the phase detectors are fed to sample and hold circuits followed by A/D converters. The outputs of the A/D converters then are fed to digital storage or digital delay line circuits as well as to subtraction circuits. The subtraction circuits compare the delayed signals with the undelayed signals in much the same way as is done with a delay-line canceler.

The output of each subtraction circuit is fed to a magnitude addition circuit. The method of combining signals is to add the absolute value of the I channel to the absolute value of the Q channel ($|I| + |Q|$). The addition of the Q channel removes the problem of reduced sensitivity due to blind phases.

The output of the magnitude addition circuit is fed to a D/A converter. The output of that circuit is fed to a data display unit.

Digital signal processing has some significant advantages over analog delay lines. As with most digital technology, it is possible to achieve greater stability, repeatability, and precision with digital processing than with analog delay-line

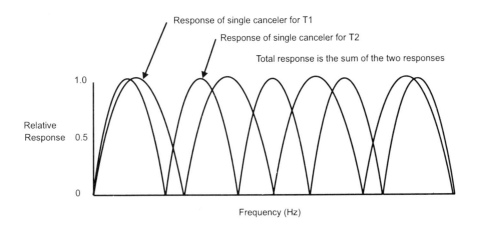

Figure 10.20 Frequency response of delay-line cancelers for $T_1/T_2 = 4/5$. (*After:* [6].)

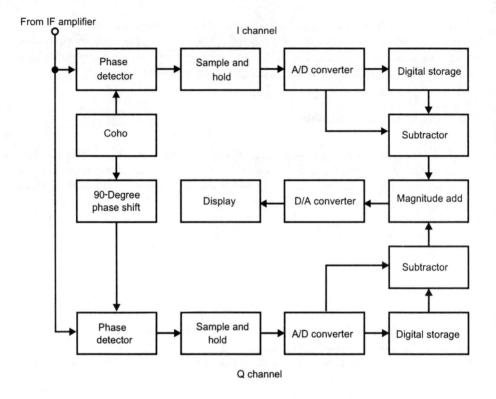

Figure 10.21 Simple digital MTI signal processor. (*After:* [6].)

cancelers. Thus, reliability is better. Other advantages include no special temperature control requirements, smaller size and weight, and larger dynamic range.

Digital filter banks are sometimes used with modern radars. These filters are based on the use of the fast Fourier transform (FFT). Since each filter occupies approximately $1/N$ the bandwidth of a delay-line canceler, its signal-to-noise ratio will be greater than that of a delay-line canceler by $5 \log_{10} N$. The division of the frequency band into N independent parts by the N filters also allows a measure of the Doppler frequency to be made. If moving clutter appears at other than zero frequency, that may be rejected.

10.6.3 MTI From a Moving Platform

When the radar itself is in motion, as when mounted on a ship, an aircraft, or a police car, the detection of a moving target in the presence of clutter is more difficult than if the radar is stationary. The Doppler frequency of the clutter is no longer near zero hertz. It is determined by the speed of the radar platform. The frequency spectrum of the clutter is also widened by the platform motion.

An MTI radar on a moving platform is called AMTI (the A originally stood for "airborne"). There are two basic methods for providing the Doppler frequency compensation that is needed. In one method, the frequency of the coho is changed to compensate for the shift in the clutter Doppler frequency. That can be accomplished by mixing the output of the coho with a signal from a tunable oscillator having a frequency approximately equal to the clutter Doppler frequency. The other method is to introduce a phase shift in one branch of the delay-line canceler.

10.7 TRACKING RADARS

Some of the information in this section is adapted from [7,8].

10.7.1 Monopulse Tracking Radars

Figure 10.22 is a block diagram for a two-coordinate (azimuth and elevation) amplitude-comparison monopulse tracking radar (MTR). The radar uses a parabolic dish antenna with a cluster of four feed horns. The four horns provide four partially overlapping antenna beams. Four hybrid junctions are used with the four horns to provide sum and difference signals. Hybrid junction 1 provides the sum and difference signals for the left-side pair of horns, and hybrid junction 2 provides the sum and difference signals for the right-side pair of horns. Hybrid junction 3 has as its input the two sum outputs from hybrid junctions 1 and 2. Its output is the sum for all four horns. The difference output for hybrid junction 3 is the azimuth difference signal. Hybrid junction 4 has as its inputs the two difference outputs from hybrid junctions 1 and 2. The sum output from that hybrid junction is the elevation difference signal.

When the radar is transmitting, the transmitter signal is fed through the ATR unit and the TR unit, which direct the signal to the sum channel. Each antenna horn is fed with equal amplitude and equal phase. The resulting transmit antenna beam thus is centered on the axis of the antenna dish. When the radar is receiving, the transmitter is disconnected by the ATR and the TR units, and the sum channel signal is fed to the radar receiver. In the simplified block diagram, in Figure 10.22, the receiver consists of a mixer followed by an IF amplifier, a video amplifier, and a range tracker. In practice, the mixer would be preceded by a low-noise RF amplifier for improved sensitivity. The mixer has as its second input a CW signal from the LO. An AGC circuit is used to control the gain of the IF amplifiers.

The elevation difference channel is fed to a mixer, followed by an IF amplifier and a phase-sensitive detector. The phase-sensitive detector has as its reference input the output of the sum channel IF amplifier. The output of the detector is the elevation pointing angle error.

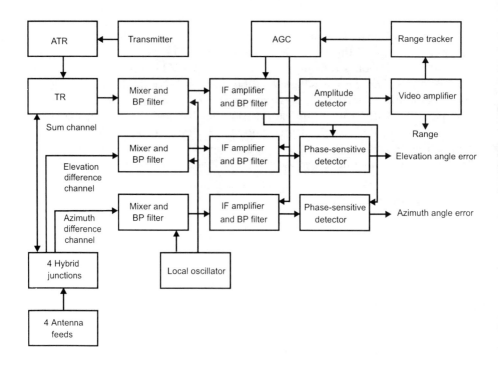

Figure 10.22 Block diagram of two-coordinate (azimuth and elevation) amplitude-comparison MTR. (*After:* [8].)

The azimuth difference channel is processed in a similar way as the elevation difference channel signal. It is fed to a mixer followed by an IF amplifier and a phase-sensitive detector. The output is the azimuth angle error.

The main advantage for this type of angle-tracking system is that an angle-pointing error is generated for each pulse. The system thus is much more accurate than a system that uses sequential lobbing or conical scan.

There are other monopulse tracking systems that use more than four antenna feed horns for improved accuracy. There are also phase-comparison monopulse systems that use two separate antennas. This type of system is sometimes called an interferometer radar. Angle error in this case is measured as the phase difference between the two received signals.

10.7.2 Tracking in Range Using Sequential Gating

One method for automatically tracking in range is based on the use of the split range gate. Figure 10.23(a) shows a detected echo pulse. Two range gates are used as shown in Figure 10.23(b). One is the early gate, the other the late gate. The

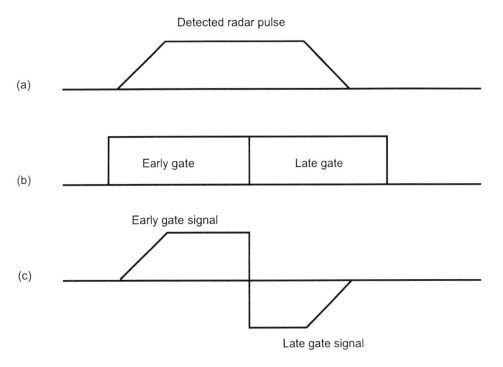

Figure 10.23 Split range gate tracking: (a) detected radar pulse; (b) early and late range gates; and (c) outputs from gates.

gates are automatically positioned in time so the received-signal energy is equal in the two gates. If the gates are too late in time, the early gate will receive energy for a larger period of time than will the late gate. The integration of the two outputs results in a positive error signal, a condition illustrated in Figure 10.23(c). If the gates are too early in time, the early gate will see less energy than the late gate and a negative error signal will be produced.

The use of range gates for automatic tracking provides for high-accuracy range measurement. Reported accuracies are within 0.8 times the length of the target. Range gating also improves the signal-to-noise ratio since it eliminates the noise from other range intervals. The optimum gate width is equal to the pulse length. In ECM systems, range gate pull-off (RGPO) can be used to defeat these systems. RGPO jammers use TWT memory loops with delay lines to transmit pulses with increasing time delay.

References

[1] Skolnik, Merrill I., *Introduction to Radar Systems*, 2nd ed., New York: McGraw-Hill, 1980, pp. 33–52.
[2] Ibid., pp. 33–51.

[3] Barton, David K., *Modern Radar System Analysis*, Boston: Artech House, 1988, pp. 123–147.
[4] Skolnik, op. cit., pp. 101–106.
[5] Barton, op. cit., 232–274.
[6] Skolnik, op. cit., pp. 106–148.
[7] Barton, op. cit., pp. 377–421.
[8] Skolnik, op. cit., pp. 152–186.

PART II
RF Components and Circuits

CHAPTER 11

Transmission Lines and Transmission Line Devices

11.1 TWO-WIRE TRANSMISSION LINES

Some of the information presented in this section is adapted from [1].

Many types of transmission lines can be used for RF applications. One type that is frequently used is the two-wire line. Figure 11.1 shows a two-wire transmission line.

Unshielded two-wire lines are often used when line pickup or radiation from the line is not too important. They have the capability for high-power handling and usually are constructed to provide either 300Ω or 600Ω characteristic impedance. The common two-wire television cable used for connecting a television antenna and the television receiver is an example of a low-power 300Ω two-wire line with a solid dielectric spacer between the lines. Higher power systems usually use an air dielectric with low-permittivity insulating spacers between the lines. The spacers are used to maintain the desired geometry of the line.

Two-wire lines are balanced lines. That means the fields around the conductors are symmetrical and there is no specific ground conductor. The coaxial line, on the other hand, is an unbalanced line.

The characteristic impedance for a transmission line is that impedance, which when used as the termination or load for the transmission line, produces no reflections but absorbs the total input power from the line. All other loads produce some reflection and standing waves on the line. The characteristic impedance for a two-wire transmission line is given approximately by (11.1). The equation applies where conductors are immersed in dielectric (the dielectric exists around the conductors and not just between them).

$$Z_0 = 276 \log_{10}(2D/d)/(\epsilon_r)^{1/2} \tag{11.1}$$

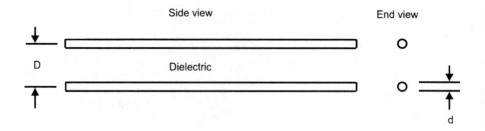

Figure 11.1 Two-wire transmission line.

where

ϵ_r = relative permittivity of dielectric medium
d = diameter of conductors
D = spacing between centers of conductors

For an example of the use of (11.1), assume the use of air as the dielectric spacer for the lines (ϵ_r = 1.0), a conductor diameter, d, of 1/8 in, and a spacing, D, of 1 in. Substituting those values into (11.1), we have

$$Z_0 = 276 \log_{10}(2/0.125) = 332.3\,\Omega$$

11.2 COAXIAL TRANSMISSION LINES

The coaxial line, or coax, is the most frequently used transmission line for conducting signals from point to point in RF systems. Figure 11.2 illustrates a coaxial transmission line.

The advantages of the coax transmission line include high power-handling capability and good shielding. Good shielding is important to prevent pickup of unwanted signals and to prevent radiation of signals that might interfere with other systems. Coaxial lines may be either flexible or rigid. The lines may use a solid dielectric, foam dielectric, or air dielectric, with spacers between inner and outer conductors. Most coax cables have either 50Ω or 75Ω characteristic impedances, although other values are available. The characteristic impedance for a coax line is given by (11.2).

$$Z_0 = [138/(\epsilon_r)^{1/2}] \log_{10}(D/d) \tag{11.2}$$

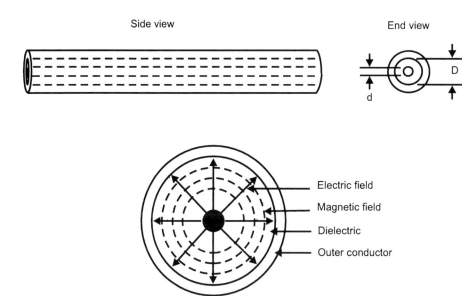

Figure 11.2 Coaxial cable transmission line.

where

ϵ_r = relative dielectric constant of dielectric medium surrounding center conductor

d = diameter of inner conductor

D = diameter of dielectric or inner diameter of outer conductor

For an example of the use of (11.2), assume the use of polyethylene as the dielectric spacer ($\epsilon_r = 2.3$), an inner conductor diameter of 0.106 in, and a dielectric outer diameter of 0.370 in. (These values happen to be those for the type RG-217/U 50Ω cable.) Substituting those values into (11.2), we have

$$Z_0 = (138/1.52) \log_{10}(0.370/0.106) = 49.3 \, \Omega$$

Table 11.1 lists the approximate characteristics for several types of coax lines. The values in the table are based on information presented in [1].

Some of the coaxial lines listed in Table 11.1 use polyethylene as the dielectric material. The velocity factor for those cables is 0.66. By that is meant that the speed of the wave in the cable is 0.66 times the speed of light.

Large-diameter rigid coax cables are used for high-power applications where low attenuation and high power-handling capability are desired. Each of the four

Table 11.1
Approximate Characteristics of Coax Cables

JAN or Other Types	Overall Diameter (in)	Impedance (Ω)	Attenuation per 100 ft (dB)	Average Power Rating (kW)
50Ω single braid				
RG-58C/U	0.195	50	5.6 at 100 MHz	0.20 at 100 MHz
			20 at 1,000 MHz	0.03 at 1,000 MHz
RG-213/U	0.405	50	2.3 at 100 MHz	0.7 at 100 MHz
			8.5 at 1,000 MHz	0.2 at 1,000 MHz
RG-218/U	0.870	50	1.0 at 100 MHz	2.0 at 100 MHz
			4.0 at 1,000 MHz	0.5 at 1,000 MHz
50Ω double braid				
RG-55B/U	0.206	50	4.5 at 100 MHz	0.2 at 100 MHz
			16 at 1,000 MHz	0.03 at 1,000 MHz
RG-214	0.425	50	2.0 at 100 MHz	0.7 at 100 MHz
			8.0 at 1,000 MHz	0.2 at 1,000 MHz
50Ω rigid				
Rigid	7/8	50	0.4 at 100 MHz	4.0 at 100 MHz
			1.4 at 1,000 MHz	1.3 at 1,000 MHz
Rigid	1-5/8	50	0.2 at 100 MHz	15.0 at 100 MHz
			0.7 at 1,000 MHz	3.0 at 1,000 MHz
Rigid	3-1/8	50	0.1 at 100 MHz	50.0 at 100 MHz
			0.4 at 1,000 MHz	10.0 at 1,000 MHz
Rigid	6-1/8	50	0.05 at 100 MHz	140.0 at 100 MHz
			0.15 at 1,000 MHz	30.0 at 1,000 MHz
75Ω single braid				
11A/U	0.412	75	2.1 at 100 MHz	0.7 at 100 MHz
			8.3 at 1,000 MHz	0.2 at 1,000 MHz
34B/U	0.630	75		1.3 at 100 MHz
				0.36 at 1,000 MHz
59B/U	0.242	75		0.3 at 100 MHz
				0.08 at 1,000 MHz
75Ω double braid				
6A/U	0.332	75	3 at 100 MHz	0.4 at 100 MHz
			12 at 1,000 MHz	0.12 at 1,000 MHz
216A/U	0.425	75	2.2 at 100 MHz	
			8.5 at 1,000 MHz	

Note: JAN = Joint Army-Navy.
Source: [1].

examples of rigid coaxial cables in Table 11.1 uses air as the dielectric material with teflon pins to support the center conductor. The velocity factor for those four cables is 0.81. Important large-size cables not shown in Table 11.1 are semirigid cables of the same diameter as the rigid cables listed. These cables have attenuations similar to those of the rigid cables at the lower frequencies but slightly higher attenuations at the higher frequencies. Examples of semirigid cables are Heliax™ cables, Styroflex™ cables, Spiroline™ cables, and Alumispline™ cables.

Very small diameter semirigid cables are used extensively in RF systems. They usually are made with solid copper outer and inner conductors and a teflon spacer ($\epsilon_r = 2.01$). These cables typically have characteristic impedance of either 50Ω or 75Ω. Small semirigid cables are used for the high microwave frequencies as well as for the lower frequencies.

Coax transmission lines have both an upper frequency limit and an upper power-handling limit, depending on the size of the coax. The larger the diameter of the coax cable, the lower the upper frequency limit but the higher the power limit. The upper frequency limit is the highest frequency that will support the TEM mode of operation, but it is just below the lowest frequency for which higher order modes (as supported in waveguide) are possible.

The maximum power limitation for cables depends on the size of the cables, the materials used, and the ambient temperature. The conduction losses of the cables result in heat, which must be dissipated. Operation of a polyethylene dielectric cable at a center conductor temperature in excess of 80°C is likely to cause permanent damage to the cable. A number of high-temperature cables, such as RG-211, RG-228, RG-225, and RG-227, can withstand inner conductor temperatures as high as 250°C.

The cables listed in Table 11.1 have upper frequency limits less than 6 GHz; the largest cable has a maximum, higher order mode-free frequency under 1 GHz. Microwave coax cables do exist for frequencies up to about 10 GHz. However, those cables have large per-unit length attenuations and low power ratings due to their smaller geometries. At microwave and higher frequencies, it is necessary to use waveguides for high-power and low-loss operation. Waveguides and waveguide-related circuits are discussed in Chapter 12.

11.3 COAXIAL CABLE CONNECTORS

A number of different types of connectors are used with coax cables. The most commonly used connectors are BNC connectors, TNC connectors, type N connectors, UHF connectors, and subminiature A (SMA) series connectors.

BNC connectors have a bayonet lock-coupling mechanism to provide a fast connect/disconnect coaxial termination. The BNC series range covers cable entry (flexible), printed circuit board (PCB), bulkhead, panel, and adapter versions, and most are available in 50Ω and 75Ω characteristic impedances. The most popular cable connectors are those with a solder center contact and clamp outer contact or a crimp center contact and crimp outer contact. A typical straight cable plug has a diameter of 0.56 in (14.3 mm) and a length of 1.08 in (27.5 mm). The maximum working voltage dc at sea level for this connector is 500V.

TNC connectors are threaded-coupling versions of the BNC connectors. The increased rigidity of the threaded coupling gives the TNC connector a more consistent performance than the BNC connector under adverse operating conditions. The TNC series covers cable connectors, bulkhead and panel styles, adapters, and receptacles. Most are available in 50Ω and 75Ω characteristic impedances. A typical cable plug connector has a diameter of 0.57 in and a length of 1.25 in.

Type N connectors also have a threaded coupling mechanism to provide a rigid coaxial termination. The series range covers cable connectors (flexible and semirigid), bulkhead, panel, and adapter versions; most are available in 50Ω and 75Ω characteristic impedance. The most popular cable entry versions are those with a solder center contact and clamp outer contact or a crimp center contact and crimp outer contact. A typical cable plug has a diameter of 0.81 in (20.6 mm) and a length of 1.44 in (36.5 mm). The maximum working voltage dc at sea level is 1,000V. The upper frequency limit is in the range of 10–12 GHz.

UHF connectors have threaded mating coupling with interlocking serrations to prevent accidental uncoupling in even the harshest environments. The upper frequency limit for UHF connectors is 500 MHz. A typical cable plug has a diameter of 0.72 in (18.2 mm) and a length of 1.56 in (39.7 mm). The maximum working voltage is 500V peak.

SMA connectors are semiprecision, high-frequency subminiature connectors. The inner diameter of the outer conductor is only 3 mm. They are characterized by a 1/4-36 mating thread and a butt-mating outer contact. They can accommodate semirigid and flexible cables with outer diameters from 0.047 in to 0.250 in, depending on the design. They provide repeatable electrical performance from 0 to 26 GHz and are widely used in microwave systems and subsystems where low attenuation and low voltage standing-wave ratio (VSWR) are required. This type of connector has a voltage rating of 355 to 500 Vrms, depending on the cable. A typical cable plug has an inside diameter of 0.14 in (3.6 mm) and a length of 0.43 in (10.9 mm).

Large-diameter rigid and semirigid cables use special connectors other than the types discussed here.

11.4 MICROSTRIP TRANSMISSION LINES

Figure 11.3 is a drawing of a microstrip transmission line. Such transmission lines are frequently used in microwave circuits.

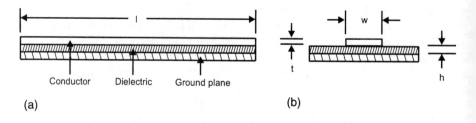

Figure 11.3 Microstrip transmission lines: (a) side view and (b) end view.

Microstrip consists of a thin, flat conductor above a ground plane with a solid dielectric spacer between the two lines. This type of line often is used in the construction of passive microwave circuits such as filters, couplers, power dividers, and power combiners. It also is used in the construction of active circuits such as amplifiers, mixers, switches, and phase-shifters. The characteristic impedance of these lines depends on the dimensions used. Often many different impedances are used on a single board (dielectric and ground plane). Application is limited to UHF, microwave, and higher frequencies where the wavelength is small.

Figure 11.4 shows the characteristic impedance of microstrip line as a function of the relative dielectric constant (ϵ_r) and the w-to-h ratio. The curves in the figure were calculated from quasi-TEM formulas presented in [1]. They assume that the thickness of the strip is negligible (t = 0).

For an example of the use of Figure 11.4, assume the use of glass-reinforced teflon (woven) as the dielectric spacer, sometimes called teflon fiberglass. It is one of the most commonly used materials for operation from UHF to Ku-band. It has a relative dielectric constant of about 2.55 and a dissipation factor or dielectric loss tangent of around 0.002. The thickness (h) for the dielectric material is assumed to be 0.10 in, and the width for the top conductor (w) is assumed to be 0.20 in. Thus, the w-to-h ratio is 2. The thickness of the top conductor is assumed to be 0.

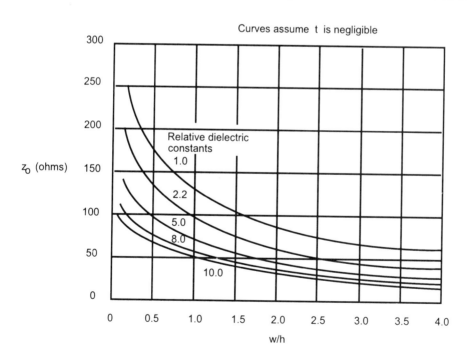

Figure 11.4 Characteristic impedance of microstrip line. (*After:* [1].)

Using these values, we see from Figure 11.4 that the characteristic impedance for the line is about 60Ω.

Perhaps one of the best all-around dielectric materials is woven quartz-reinforced teflon. It is similar to teflon fiberglass except that its electrical properties are greatly enhanced by the substitution of low-loss quartz for glass in the reinforcing cloth. Its dielectric constant is 2.47. Coupled with a loss tangent of 0.0006, that makes the material far superior electrically to other materials. It is, however, about six times more costly than teflon fiberglass.

With dielectrics, there is always a loss. The total current density is the sum of a conduction current and a displacement current density in time-phase quadrature. The dielectric-loss tangent is the ratio of the conduction current to the displacement current. For small loss tangents, the power factor is approximately equal to the loss tangent.

Line loss is the result of both dielectric losses and conductor losses. Dielectric losses increase proportionally to frequency, while conductor losses increase as the square root of frequency.

Power-handling capability is small for microstrip due to the small physical size of the line. A typical 7/32-in (0.22-in) top line using a 1/16-in (0.0625-in) thick teflon-impregnated fiberglass dielectric has a 50°C rise in temperature above 20°C, ambient for 300W CW power at 3 GHz. With the same line and pulse conditions, corona effects appear at the edge of the strip conductor for pulse power of roughly 10 kW at 9 GHz.

11.5 STRIPLINE TRANSMISSION LINES

Figure 11.5 is a drawing of a stripline transmission line.

Stripline transmission line, or stripline, differs from microstrip in that a second ground plane is placed above the conductor strip so that the center conductor is equally spaced from the pair of parallel ground planes. This type of line has a better shielding characteristic than microstrip. It can be used for many of the same types of applications as microstrip lines.

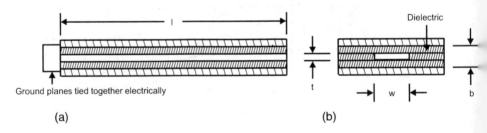

Figure 11.5 Stripline transmission line: (a) side view and (b) end view.

Figure 11.6 shows a plot of the square root of the relative dielectric constant of the dielectric times the characteristic impedance for stripline transmission lines as a function of w/b and t/b, where w is the width of the inner conductor, t is the thickness of this center conductor, and b is the spacing between outer conductors.

For an example of the use of Figure 11.6, assume a w-to-b ratio of 0.2 and a t-to-b ratio of 0.05. From the plot, the square root of the relative dielectric constant times Z_0 is seen to be 130. Assuming the use of a dielectric having a relative dielectric constant of 2.56, the characteristic impedance of the line is

$$Z_0 = 130/(2.56)^{1/2} = 130/1.6 = 81.25\Omega$$

The average power capability of stripline is primarily a function of the permissible temperature rise of the center conductor and surrounding laminate. It is therefore related to the dielectric used, its thermal conductivity, the electrical loss, the cross-section for the line, any case or supporting material, the maximum allowable temperature, and the ambient temperature. Reference [2] shows that for woven teflon fiberglass as the dielectric, a 0.125-in ground plane spacing, a 50Ω line, an ambient temperature of 25°C, and an operating frequency of 2.5 GHz, the maximum average power that can be used is 560W. That is based on a calculated temperature rise per watt CW of 0.42°C and a maximum operating temperature of 265°C.

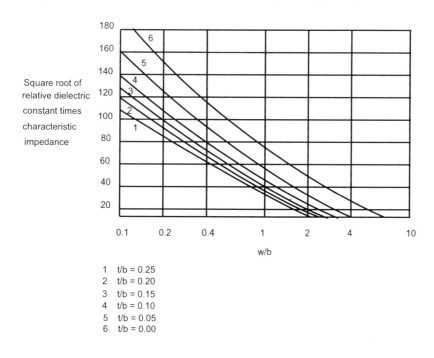

1 $t/b = 0.25$
2 $t/b = 0.20$
3 $t/b = 0.15$
4 $t/b = 0.10$
5 $t/b = 0.05$
6 $t/b = 0.00$

Figure 11.6 Characteristic impedance for stripline. (*After:* [1].)

300 | RF Systems, Components, and Circuits Handbook

The only dielectric material that would provide higher power capability than woven teflon fiberglass for the same example is microfiber teflon fiberglass. For this case, the change of temperature per watt CW is 0.38°C and the maximum operating temperature is 260°C. Therefore, the maximum average power capability is 620W [2].

Figure 11.7 shows the attenuation per inch of line length versus frequency and ground-plane spacing for a 50Ω stripline transmission line made of woven teflon fiberglass. For an example of the use of Figure 11.7, assume operation at a frequency of 10.5 GHz and a ground-plane spacing of 0.125 in. From the plot, notice that the attenuation or insertion loss is about 0.2 dB/in.

11.6 CHARACTERISTICS OF TRANSMISSION LINES

Some of the material in this section is quoted or adapted from [3].

11.6.1 Wave Velocity on Transmission Lines

The velocity of TEM waves on transmission lines, commonly called the phase velocity, is less than the free-space velocity of light. This velocity is given by (11.3).

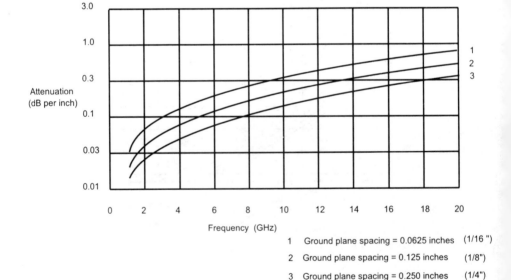

1 Ground plane spacing = 0.0625 inches (1/16 ")
2 Ground plane spacing = 0.125 inches   (1/8")
3 Ground plane spacing = 0.250 inches   (1/4")

Figure 11.7 Loss versus frequency for a 50Ω stripline using woven teflon fiberglass as the dielectric material. (*After:* [2].)

$$v = c/(\epsilon_r)^{1/2} \qquad (11.3)$$

where

v = wave or phase velocity
$c = 3 \cdot 10^8$ m/s
ϵ_r = relative dielectric constant or relative permittivity

For example, if the dielectric for a coax cable is polyethylene, the relative dielectric constant is 2.3. The velocity of the wave on the line is 0.66 times the speed of light, or $1.98 \cdot 10^8$ meters per second. Thus, the velocity factor is 0.66.

11.6.2 Reflection Coefficients

The voltage reflection coefficient for a load or termination is given by (11.4).

$$\rho = E_2/E_1 = [(Z_L/Z_0) - 1]/[(Z_L/Z_0) + 1] \qquad (11.4)$$

with

ρ = voltage reflection coefficient
E_2 = voltage of reflected wave
E_1 = voltage of incident wave
Z_L = load impedance
Z_0 = characteristic impedance of the system

The reflection coefficient has both magnitude and phase, because Z_L may be a combination of a resistance and a reactance, that is, $Z_L = R_L \pm jX_L$, and so is a vector quantity.

Another relationship of interest is

$$Z_L/Z_0 = (1 + \rho)/(1 - \rho) \qquad (11.5)$$

where the terms are as defined for (11.4). This equation can be used to find Z_L if the voltage reflection coefficient is determined by measurement.

11.6.3 Standing-Wave Ratio

Reflections back down the line from an unmatched load give rise to reflection coefficients and standing waves on the line. At points on the line, the direct and reflected waves will be in phase. At those points, the two signals add, producing a maximum standing wave. At other points on the line, 90 degrees electrically away from the point of maximum standing wave, the direct and reflected waves will be 180 degrees out of phase. At those points, the two signals subtract, producing a minimum. Equations for the standing-wave ratio (SWR) are as follows:

$$S = |E_{max}/E_{min}| = |I_{max}/I_{min}| \tag{11.6}$$

$$S = (|E_1| + |E_2|)/(|E_1| - |E_2|) = (1 + |E_2|/|E_1|)/(1 - |E_2|/|E_1|) \tag{11.7}$$

$$S = (1 + |\rho|)/(1 - |\rho|) \tag{11.8}$$

where all terms are as previously defined.

The magnitude of the reflection coefficient can be found from the SWR, as follows:

$$|\rho| = (S - 1)/(S + 1) \tag{11.9}$$

Figures 11.8 and 11.9 illustrate voltage standing-wave pattern on a lossless line for different load conditions.

Figure 11.8(a) shows the case of an open-circuit load. The direct and reflected voltages are equal and in phase at the load. That results in a voltage maximum at the load equal to twice the direct wave voltage. The current is zero at the load. One-quarter of a wavelength toward the generator, the direct and reflected voltage waves are 180 degrees out of phase, so they add to zero for the case of a lossless line, thus producing a null. The currents are in phase at that point, so there is a current maximum at that point. One-half wavelength from the load, the voltages are in phase and the currents are out of phase, so there exists another voltage maximum and current minimum or null. The standing-wave pattern repeats every half-wavelength up the line. The voltage reflection coefficient case is 1.0, and the SWR is infinity.

If the line has loss, the magnitude of the SWR decreases as we move up the line away from the load. That is because the direct wave increases, while the reflected wave decreases in magnitude.

Figure 11.8(b) shows the standing-wave pattern when the load is resistive and two times the magnitude of the transmission line characteristic impedance. The voltage reflection coefficient in that case is 1/3, and the SWR is 2. Again, voltage maximums occur at the load and at multiples of one-half wavelength up the line. Voltage nulls have one-half the voltage of the peaks. Current nulls are at the load and at multiples of one-half wavelength up the line.

Figure 11.8(c) shows that there are no voltage standing waves when the load is equal to the characteristic impedance of the line or the operating system. That

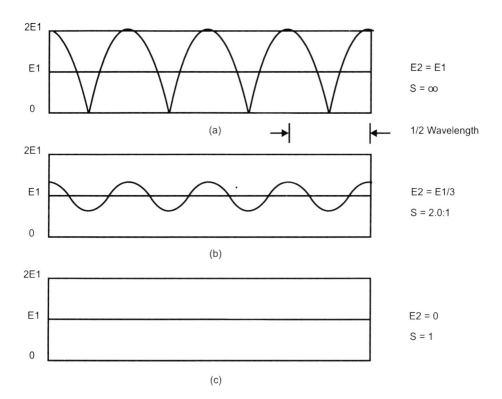

Figure 11.8 Standing waves for three resistive loads: (a) open-circuit load; (b) load is resistor with $R = Z_0$; and (c) load impedance = Z_0.

is the case a designer wants to achieve through impedance matching. For this case, there is no reflected voltage and no reflected power. Thus, the VSWR is 1.0.

Figure 11.9(a) shows the case of a load that is one-half the characteristic impedance of the line. The voltage reflection coefficient in that case is again 1/3, and the VSWR is again 2. The voltage minimum occurs at the load, and the first voltage maximum occurs one-quarter wavelength from the load. Voltage nulls have one-half the voltage of the peaks. Patterns are repeated every half wavelength up the line.

Figure 11.9(b) shows a load that is a short circuit or 0Ω impedance. For this case, the voltage is 0 at the load, and the current is a maximum. The voltage reflection coefficient is 1, and the SWR is infinity.

If the load is an inductive reactance, the voltage null is located at a distance from the load as though the inductive reactance were a short-circuited length of transmission line having a length less than a quarter-wavelength. If the load is a capacitive reactance, the voltage null is located a distance from the load as though

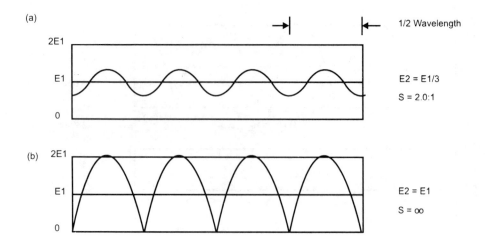

Figure 11.9 Standing waves for two additional resistive loads: (a) load is resistor with $R = Z_0/2$ and (b) short-circuit load.

the capacitive reactance were a length of line with an open-circuit for a load with a line length less than a quarter-wavelength.

11.7 THE SMITH CHART

11.7.1 Impedance and Admittance Coordinates

Figure 11.10 shows a simplified Smith chart [4]. The chart here differs in a number of ways from the usual Smith charts used in the solution of transmission line problems and in the design of impedance-matching circuits and other applications. It has the virtue that it is sufficiently small that it can be used in a handbook of this kind, and yet the numbers and letters are large enough to be easily read. Commercially available Smith charts are larger in size and have more lines and much better resolution and accuracy.

    A Smith chart can be used as either an impedance chart or an admittance chart. In either case, the values are normalized to an impedance of $1 + j0\Omega$. At the center of the chart, the resistance or the conductance has a normalized value of 1.0. Toward the top of the chart, along the vertical center line, the resistance or conductance circles have normalized values less than 1. Conversely, toward the bottom of the chart, along the vertical center line, the circles have normalized values greater than 1. The vertical center line corresponds to normalized reactance or susceptance of 0. The curved lines to the right of the vertical center line correspond to normalized inductive reactance lines or capacitive susceptance lines. The

Transmission Lines and Transmission Line Devices | 305

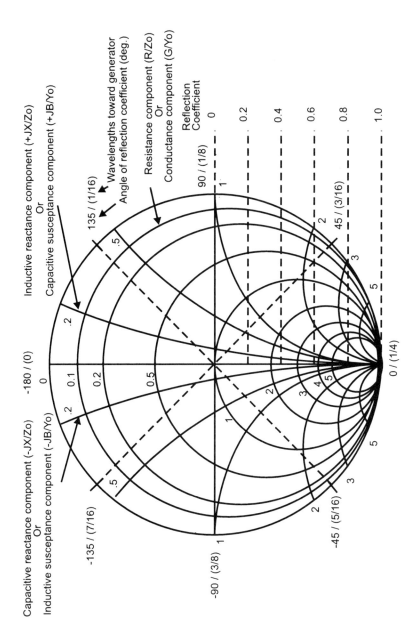

Figure 11.10 Simplified Smith chart.

curved lines to the left of the vertical center line correspond to normalized capacitive reactance lines or inductive susceptance lines. Thus, any point on the chart can be read as either a normalized impedance involving resistance and reactance or a normalized admittance involving conductance and susceptance.

11.7.2 Voltage Standing-Wave Ratio Circles

We can construct a voltage standing-wave circle on a Smith chart by drawing a circle with its center at the center of the chart and its radius such that the circle passes through the impedance or admittance of interest. The point on the center line where the circle crosses indicates the VSWR. If it crosses at 2.0, the VSWR is 2.0.

As a signal travels down a transmission line, its plot follows the standing-wave circle. The distance traveled can be determined with the scale identified as wavelength toward generator and located on the outer edge of the chart. Straight construction lines can be drawn from the center of the chart to the outside scales, with one passing through the load and the other passing through a point on the SWR circle of interest. The difference between wavelength readings indicates the wavelengths traveled.

11.7.3 Reflection Coefficients

We can determine the voltage reflection coefficient's angle for a given load impedance by first drawing a straight construction line from the center of the chart through the load to the scale on the outside of the chart. The angle is read from the angle of reflection coefficient scale. The magnitude of the coefficient is found by measuring the radial distance from the center of the chart to the impedance point of interest. A reflection coefficient scale is shown on the right of the chart.

The power reflection coefficient is the square of the voltage reflection coefficient. The return loss is a measure of how much of the transmitted power is returned to the transmitter due to reflection at the load. For example. if the voltage reflection coefficient is 0.316, the power reflection coefficient is 0.1, and the return loss is 10 dB.

11.7.4 Examples Using the Smith Chart

Figure 11.11 shows a normalized Smith chart with a 50Ω generator and a 50Ω coaxial cable connected to seven load impedances:

1. Open circuit;
2. $100 + j0\Omega$;

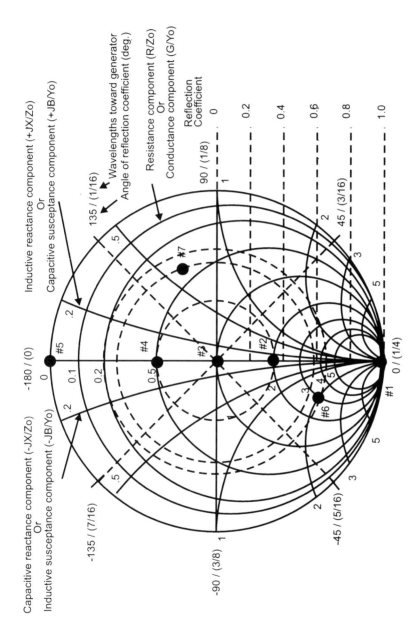

Figure 11.11 Example load impedances and standing-wave circles plotted on a Smith chart.

3. $50 + j0\Omega$;
4. $25 + j0\Omega$;
5. Short circuit;
6. $150 - j100\Omega$;
7. $20 + j30\Omega$.

The first five impedances are the same ones used for the plots in Figures 11.8 and 11.9.

Impedance 1 (open circuit) is located on the outer circle. The SWR is infinity and the voltage reflection coefficient is 1.0 ∠ 0 degree. The power reflection coefficient is 1.0, the return loss is 0 dB, the power transmission coefficient is zero, and the reflection loss is infinity.

Impedance 2 ($100 + j0\Omega$) is equal to $2.0 + j0.0$ when normalized to 50Ω and is located at $2.0 + j0.0$ on the Smith chart. The SWR is 2.0, and the voltage reflection coefficient is 0.333 ∠ 0 degree. The power reflection coefficient is 0.11, the return loss is 9.55 dB, the power transmission coefficient is 0.88, and the reflection loss is 0.52 dB. Thus, with a SWR of 2.0, only about 10% of the transmitted power is reflected back to the generator, and nearly 90% is transmitted. It is often a goal for impedance matching to have the resulting SWR 2.0 or less.

Impedance 3 ($50 + j0$) is equal to $1.0 + j0.0$ when normalized to 50Ω and is located at $1.0 + j0.0$, the center, on the Smith chart. The SWR is 1.0, and the voltage reflection coefficient is 0.0. The power reflection coefficient is 0.0, the return loss is infinity, the power transmission coefficient is 1.0, and the reflection loss is 0 dB. This is the ideal case for impedance matching where there is no power reflected to the generator and all the power is transmitted to the load.

Impedance 4 ($25 + j0$) is equal to $0.5 + j0.0$ when normalized to 50Ω and is located at $0.5 + j0.0$ on the Smith chart. The SWR is 2.0, and the voltage reflection coefficient is 0.333 ∠ 180 degrees. The power reflection coefficient is 0.11, the return loss is 9.55 dB, the power transmission coefficient is 0.88, and the reflection loss is 0.52 dB. We see that these values are the same as for impedance 2, except for the 180-degree phase shift for the reflection coefficient.

Impedance 5 (short circuit) also is located on the outer circle. The VSWR is infinity, and the voltage reflection coefficient is 1.0 ∠ 180 degrees. The power reflection coefficient is 1.0, the return loss is 0 dB, the power transmission coefficient is 0, and the reflection loss is infinity. These values are the same as for impedance 1, except for the 180-degree phase shift for the reflection coefficient.

Impedance 6 ($150 - j100$) is equal to $3.0 - j2.0$ when normalized to 50Ω and is located at $3.0 - j2$ on the Smith chart. The SWR is 4.4, and the voltage reflection coefficient is 0.625 ∠ −18.5 degrees. The power reflection coefficient is 0.4, the return loss is 4 dB, the power transmission coefficient is 0.6, and the reflection loss is 2.1 dB.

Impedance 7 ($20 + j30$) is equal to $0.4 + j0.6$ when normalized to 50Ω and is located at $0.4 + j0.6$ on the Smith chart. The VSWR is 3.4, and the voltage reflection coefficient is 0.55 ∠ 112 degrees. The power reflection coefficient is 0.3, the return loss is 5.2 dB, the power transmission coefficient is 0.7, and the reflection loss is 1.6 dB.

Figure 11.12 shows a 50Ω system with a 50Ω coax cable connected to four admittances. The normalized admittance for the chart is 1/50 = 0.02 siemens (S).

The five admittances shown in Figure 11.12 are as follows:

1. Open circuit;
2. Short circuit;
3. $1/(50 + j0)$ S = $0.02 + j0$ S;
4. $1/(50 - j50)$ S = $0.01 + j0.01$ S;
5. $1/(10 - j25)$ S = $0.014 + j0.34$ S.

These are normalized to 0.02S as follows:

1. $0.0 + j0.0$;
2. Infinity $+ j0.0$;
3. $1.0 + j0.0$;
4. $0.5 + j0.5$;
5. $0.7 + j1.7$.

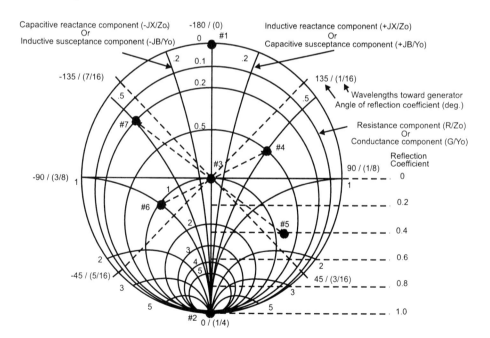

Figure 11.12 Example load admittances plotted on a Smith chart.

It is common practice when performing designs using the Smith chart to convert between an impedance chart and an admittance chart. That is done easily by constructing a straight line from the impedance point through the center of the chart to a point equal distance from the center of the chart as the impedance point. This is illustrated in Figure 11.12 for the case of admittances 4 and 5. First, impedance point 6 is located for $50 - j50\Omega$ normalized to $1 - j1$. That is converted to the admittance at point 4 by transferring through the center of the chart with a straight line to the corresponding point on the SWR circle. In a similar way, impedance point 7 is located for $10 - j25\Omega$ normalized to $0.2 - j0.5$. That is converted to the admittance at point 5 by transferring through the center of the chart with a straight line to the corresponding point on the SWR circle. The SWR values for the five admittances are as follows:

1. Infinity;
2. Infinity;
3. 1.0;
4. 2.62;
5. 4.2.

It is important that the impedance or the admittance at a point along a transmission line can be determined by locating that point on the SWR circle at the desired distance in wavelengths from the load, as illustrated in Figure 11.13.

In Figure 11.13, the load is assumed to have a normalized impedance of $Z_L = 1.0 - j1.9$ based on 50Ω, indicated by point 1 on the Smith chart. The SWR is about 5.0. If we move 0.0125 wavelength along the transmission line toward the generator, the normalized impedance is about $0.20 - j0.32\Omega$. This is found by moving on the 5.0 SWR circle from point 1 to point 2.

11.8 IMPEDANCE MATCHING USING THE SMITH CHART

11.8.1 Impedance Matching With a Quarter-Wave Transformer

The required characteristic impedance for a quarter-wave section of line to transform a load impedance to Z_0 of the main line is given by (11.10).

$$Z_T = (Z_0 Z_L)^{1/2} \tag{11.10}$$

where

Z_T = characteristic impedance of quarter-wave line
Z_0 = characteristic impedance of main transmission line
Z_L = impedance of the load

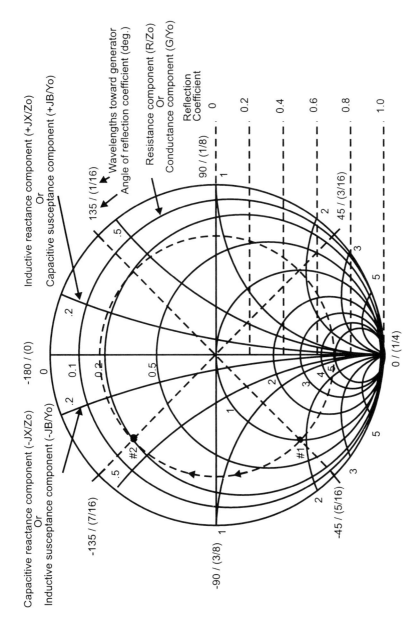

Figure 11.13 Impedance at a point along a transmission line, given the distance from a known impedance.

Equation (11.10) works only for purely resistive loads. For an example of the use of (11.10), assume that the load impedance is 39.8Ω and the main line characteristic impedance is 75Ω. Substituting values into (11.10), the characteristic impedance of the quarter-wave transformer line is

$$Z_T = (75 \cdot 39.8)^{1/2} = 54.6 \Omega$$

Figure 11.14 shows an example of impedance matching with a quarter-wavelength transformer. In the figure, the input line has a characteristic impedance of 50Ω. The load has an impedance of 100 − j50Ω. That impedance normalized to 50Ω is 2.0 − j1.0. By using a 50Ω line that is 0.213 wavelength long, the load impedance is transformed to a normalized impedance of 0.38 − j0.0. The quarter-wave transformer can then be used to convert the normalized impedance to 1.0 + j0.0. The characteristic impedance for the quarter-wave transformer is $(50 \cdot 19)^{1/2} = 30.8 \Omega$.

Figure 11.15 illustrates the use of a Smith chart for determining the distance for the load that the quarter-wave transformer should be placed. The normalized impedance of the load is 2.0 − j1.0Ω. That impedance is entered on the chart as point A. An SWR circle is then drawn and point B marked on the chart where the impedance first becomes real as we travel from the load toward the generator. The normalized resistance at that point is 0.38. The real resistance is 50 · 0.38, or 19.0Ω. The distance from the load traveled to reach that point is 0.213 wavelength.

Using (11.11), we compute the characteristic impedance for the quarter-wave transformer as follows:

$$Z_0' = (Z_0 Z_L)^{1/2} \qquad (11.11)$$
$$Z_0' = (50 \cdot 19.0)^{1/2} = 30.8 \Omega$$

The quarter-wave transformer provides a near-perfect match of impedance at only one frequency. It has a useful bandwidth that depends on the SWR that we can tolerate. An octave bandwidth is provided for the case of a VSWR of 1.5.

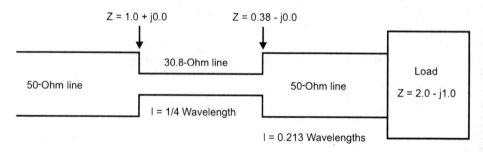

Figure 11.14 Example of impedance matching with a quarter-wavelength transformer.

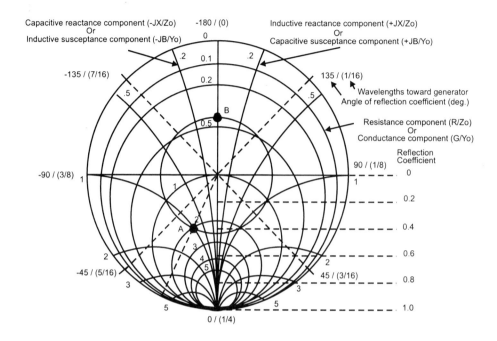

Figure 11.15 Example of the use of a Smith chart for impedance matching using a quarter-wavelength transformer.

11.8.2 Impedance Matching With a Short-Circuited Stub

Figure 11.16 shows impedance matching using a single short-circuited stub. This method of impedance matching frequently is used with transmission lines at microwave frequencies.

The stub, which is a short section of line, is connected to the main line at a calculated distance from the load. It has the same characteristic impedance as the main line and a length that is calculated. The method of selecting the location and the length of the stub is illustrated in Figure 11.17.

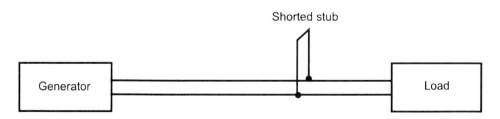

Figure 11.16 Short-circuited stub impedance matching.

314 | RF Systems, Components, and Circuits Handbook

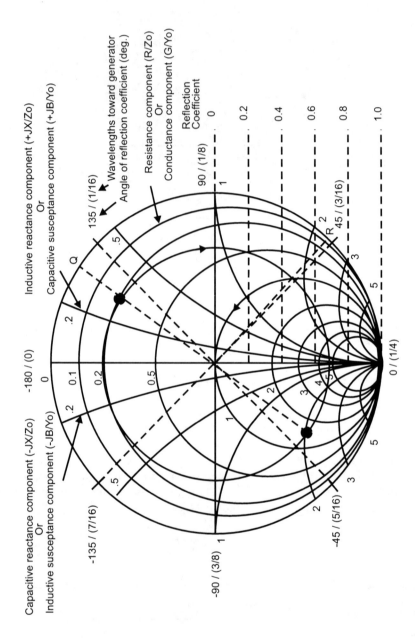

Figure 11.17 Example of the use of a Smith chart for impedance matching using a short-circuited stub.

For this example, assume a load of $75 - j100\Omega$ and a Z_0 of 50Ω. Thus, the normalized impedance is $1.5 - j2.0$. A VSWR circle drawn through that point shows that the VSWR without matching is about 4.7. The admittance for the load is found by transferring one-quarter wavelength around the VSWR circle to point 2. The normalized admittance is read as $0.2 + j0.3$. Next, move from that point on the VSWR circle to the point where the real part of the admittance is 1.0. The admittance at that point is $1.0 + j1.7$. The distance from point Q to point R is found from the chart to be about 0.13 wavelength. Therefore, the stub should be placed 0.13 wavelength from the load.

The susceptance of the stub should be $-j1.7$ at the frequency of interest to cancel out the $+j1.7$ of the admittance at point R. The length of the stub is found by moving from the position of the short when the chart is used as an admittance chart to the point on the outside of the chart where the susceptance is $-j1.7$. That distance is found to be about 0.08 wavelength. Adding $-j1.7$ to $1.0 + j1.7$ brings us along the 1.0 conductance circle to the $Y = 1.0 + j0.0$. That point is also the $Z = 1.0 + j0.0$ point.

Transmission line circuits normally are perfectly matched at only one frequency. At any other frequency, the match is not perfect. Usually, however, there is a substantial frequency range over which the match is good enough. Often, that frequency range is where the VSWR is 1.5 or less.

Figure 11.18(a) shows the case of a double-stub impedance-matching network. A double-stub matching system has the advantage that it can be placed almost any place on the main line that is convenient rather than at a critical point, as in the case of the single-stub system. The spacing of the stubs typically is one-eighth wavelength at the center frequency. The lengths of the shorted stubs A and B usually are selected by trial and error to reduce the VSWR to a low value. The type of stubs used typically involve adjustable shorting plungers. This type of system can achieve a perfect impedance match only if the conductive component of the impedance at the stub nearest the load and looking toward the load is less than $2/Z_0$. When this requirement is not satisfied, an impedance match still can be made by increasing the distance from the load by a quarter-wavelength. That is because of the impedance-transforming action of a quarter-wave line.

Figure 11.18(b) shows a double slug or sleeve tuner. The tuner uses either dielectric slugs or metallic sleeves, which can be moved along the line to adjust the phase of the reflections caused by the slugs or the sleeves. They can be adjusted in separation distance to adjust the magnitude of the reflections.

11.9 COAXIAL TERMINATIONS

There are three main types of terminations for coax lines: short-circuit terminations, Z_0 terminations, and open-circuit terminations. Short-circuit terminations are used in impedance matching stubs and in some measurement systems for system

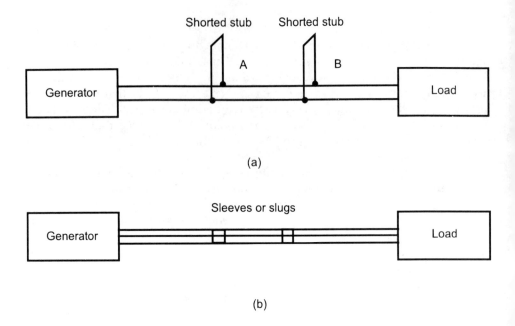

Figure 11.18 Double reflector matching: (a) double stub impedance matching and (b) double slug or double sleeve impedance matching.

calibration. Z_0 terminations also are used in measurement systems and in devices such as directional couplers and stripline isolators. High-power versions are used as transmitter dummy loads. Open-circuit line terminations are also used in impedance matching but not as often as shorted-stub terminations because of possible radiation problems. They are, however, used in coaxial measurement systems for calibration of vector impedance measurement systems, or network analyzers.

11.10 COAXIAL DIRECTIONAL COUPLERS

Figure 11.19 shows two coaxial-type direction couplers. Figure 11.19(a) shows a loop-type directional coupler, which is a wideband coupler that can work over several octaves. The coupling from the main line to the output line is through a combination of capacitive coupling and magnetic or inductive coupling. By proper design, the two coupling mechanisms can be made equal such that for a forward wave on the main line the two equal amplitude and in-phase waves on the output line add, and the two equal amplitude and 180-degree out-of-phase signals on the terminated line add to zero. Typical coupling usually is in the range of 10–30 dB and is selected by loop design.

Figure 11.19(b) shows a two-hole probe directional coupler. In this type of coupler, the operating bandwidth is more limited. The probes are placed a quarter-

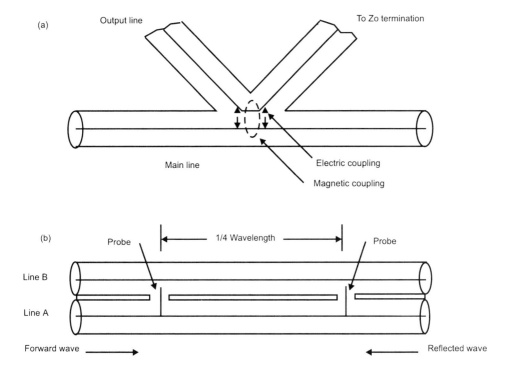

Figure 11.19 Coaxial-type directional couplers: (a) loop-type directional coupler and (b) two-hole probe-type directional coupler.

wave apart at the center frequency. A forward wave on line A is measured at the right side of line B where the two equal amplitude signals coupled to line B by the probes add in phase. These signals are 180 degrees out of phase at the left side of B and add to zero. On the other hand, the reflected signal is measured at the left side of line B using the same addition and subtraction method. Coupling is usually in the range of 10-30 dB and is selected by probe design.

11.11 BALUNS

Baluns are circuits or devices used to convert from a balance transmission line to an unbalanced transmission line or vice versa. A two-wire line is an example of a balanced line, while a coax is an example of an unbalanced line. Figure 11.20 shows two examples of baluns.

Figure 11.20(a) shows a very wideband balun using a modified length of coax. In this case, the transmission line on the left is a regular coaxial line, and the transmission line on the right is a two-wire line. The center portion of the line is

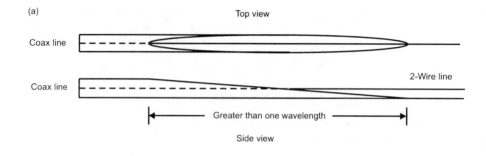

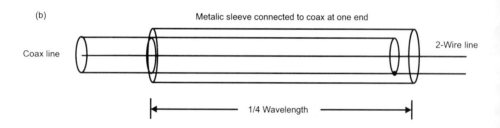

Figure 11.20 Two examples of baluns: (a) wideband balun and impedance transformer using a modified coax and (b) quarter-wave sleeve balun.

the balun. It has a length of one wavelength or greater at the lowest frequency of interest. The coax line is cut so the outer conductor gradually changes shape from a tube that surrounds the dielectric and the inner conductor to a half-sleeve and finally to a stripline to which a wire of the same diameter as the coax center conductor is connected. A balun of this type is sometimes used with log periodic or other wideband antennas having as much as a 10-to-1 frequency range.

Figure 11.20(b) shows a quarter-wave sleeve balun. In this case, a conducting sleeve is placed about the coaxial line spaced from the coax outer conductor over all but the left end where it is connected to the coax outer conductor. The short circuit of the sleeve and the outer conductor transforms to an open-circuit at the right end due to the quarter-wave transformer action. The inner conductor becomes one line of the two-wire line and the outer conductor is connected to a second line of the two-wire line. Again, the function of this circuit is to transform from an unbalanced line to a balanced line, where balanced and unbalanced refers to the electric field configurations.

11.12 TWO-WIRE TRANSMISSION LINE IMPEDANCE TRANSFORMER

A wideband, two-wire line impedance transformer can be constructed by slowly changing the spacing between the two wires. This transformer has a length of at least one-half wavelength. It can be used over many octaves of frequency, such as a 10:1 frequency range. The length of the circuit is determined by the lowest frequency or the largest wavelength to be used. The characteristic impedance of the lines is determined by the size of the wires and the wire spacing. It also is possible to make a wideband impedance transformer using microstrip or stripline using this same approach if the transition section is equal to or greater than a half-wavelength. It also is possible, but much more difficult, to make a coaxial wideband impedance transformer of this same type. Some of the information presented here is adapted from [3].

11.13 STRIPLINE AND MICROSTRIP CIRCUITS

Stripline and microstrip are used for many types of microwave circuits. This section discusses some of the more important of these circuits. Much of the information here regarding stripline circuits has been adapted from [2].

11.13.1 Shunt Stub DC Returns

Figure 11.21 shows a shunt stub as a DC return. The stub is connected to a ground at one end and to the main conductor at the other end. It has a length of one-quarter wavelength at the center frequency. With that length, the stub transforms from a short to an open-circuit at the main conductor attachment point for the microwave frequency, but it is a short circuit between the main line and the ground for DC. Shunt DC return stubs normally are formed using high-impedance lines.

11.13.2 Branch Line 90-Degree Hybrid Couplers

Figure 11.22 shows a two-arm branch line 90-degree hybrid coupler. The circuit consists of a main line, which is coupled to a secondary line by two

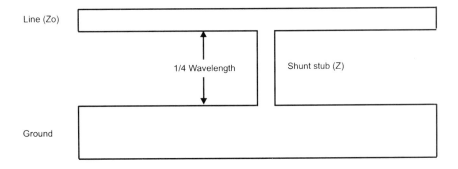

Figure 11.21 Shunt stub as a DC return. (*After:* [5].)

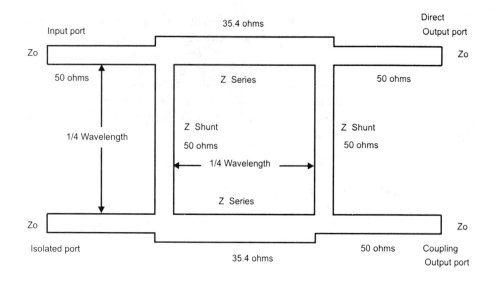

Figure 11.22 Branch line 90-degree hybrid coupler. (*After:* [6].)

quarter-wavelength lines spaced one-quarter wavelength apart, thus creating a square approximately one wavelength in circumference. The coupling factor is determined by the ratio of the impedance of the shunt and series arms, which also must be adjusted to maintain a proper impedance match over the band. Coupling to the secondary line typically is in the range of 3–26 dB. The phase difference between the two lines when used as a power divider is 90 degrees at the center frequency. The two-arm branch line 90-degree hybrid usually is used as a directional coupler, able to provide a measurement of both forward and reflected signals.

The circuit in Figure 11.22 shows a design in which $Z_0 = 50\Omega$ and the coupling factor is −3 dB. The direct output is also at −3 dB.

11.13.3 Stripline or Microstrip Rat Race Hybrid Coupler

Figure 11.23 shows the case of a 1.5-wavelength "rat-race magic T." It consists of a ring 1.5 wavelengths at the center frequency in circumference having four arms separated by 60 degrees of angular rotation. When used as a zero-degree phase-shift power divider, the input arm is arm 1. There are two output arms (arms 2 and 4) spaced one quarter-wavelength away, and a fourth terminated arm (arm 3) spaced a quarter-wavelength away from arm 4 and three-quarters of a wavelength away from arm 2. At center frequency, the output split from the common input arm 1 to the two output arms 2 and 4 is equal, and the phase relationship between them is zero degrees.

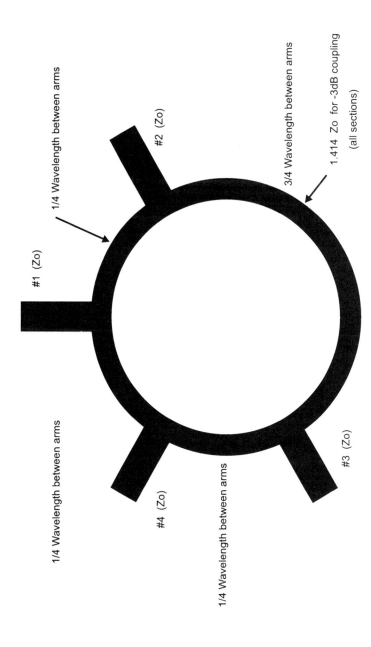

Figure 11.23 A 1.5-wavelength rat-race hybrid coupler. (*After*: [6].)

When used as a 180-degree phase-shift power divider, the input arm is arm 2, and arm 4 is terminated. The power split to the two output arms 1 and 3 is also equal; however, the phase shift between them is 180 degrees. In general, the ratrace coupler has more bandwidth than the branch-line coupler, but the choice between couplers usually is made based on the phase difference needed for a particular application. The branch-line coupler provides a 90-degree phase shift between outputs, whereas the rat-race coupler provides either a zero-degree phase shift between outputs or, if desired, a 180-degree phase shift between outputs. For a 50Ω input line, the impedance of the one-quarter and the three-quarter wavelength sections is the square root of 2 times 50Ω for equal power division.

11.13.4 Split Inline Hybrid Dividers and Combiners

Figure 11.24 shows two types of split inline hybrid dividers or combiners. These circuits can be used either to split the input power between two loads or to combine two input signals into a single output.

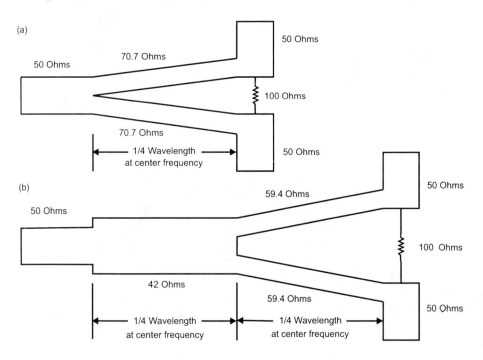

Figure 11.24 Two types of split inline hybrid dividers or combiners: (a) uncompensated inline hybrid divider and (b) compensated inline hybrid divider. (*After:* [6].)

Figure 11.24(a) is an uncompensated divider or combiner. The arm length is one-quarter wavelength at the center frequency. For a 50Ω input line, the impedance of the arms is 70.7Ω. A 100Ω resistor is placed between the output arms, as shown, for increased isolation between output ports.

Figure 11.24(b) shows a compensated inline divider or combiner. The difference between this combiner and the uncompensated system is the use of an additional quarter-wave section of line between the input line and the split. Different characteristic impedances are used for the arms, as shown. The VSWR characteristics for this divider or combiner are somewhat better than those of the uncompensated divider or combiner for the case of wideband operation. We see that a perfect VSWR is realized at only one frequency; however, a fairly wide bandwidth is available if we are willing to have a 1.7 or less VSWR. This is usually the case.

The systems shown are equal impedance systems for the two arms. It is also possible to make systems for coupling to lines of different impedance and to use more complex divider or combiner circuits that have multiple-octave coverage. Systems also are available that provide 1-to-4 and 1-to-8 division or combining ratios. These systems simply combine a number of the basic 1-to-2 ratio units.

11.13.5 Quarter-Wave Coupled-Line Directional Couplers

Figure 11.25 shows two stripline quarter-wave coupled-line directional couplers. The system shown in Figure 11.25(a) is an edge-coupled system; the system shown in Figure 11.25(b) has one line above the other separated by a dielectric spacer, normally considered broadside coupling in stripline. A fraction of the power on the main transmission line is coupled to the secondary line. This power varies as a function of the physical dimensions of the coupler and the direction of the propagation of the primary power. Port 1 is the input and port 4 is the output of the main line. Port 2 is the coupled output, while port 3 normally is terminated in the characteristic impedance and is isolated from the input port by 20 dB or so over an octave bandwidth. The output power from port 2 is proportional to the power traveling from port 1 to port 4. The usable bandwidth of this coupler is about one octave for coupling, which varies by ±10% from its nominal value.

11.13.6 90-Degree Coupled-Line Hybrid Coupler

The broadside-coupled directional coupler in Figure 11.25(b) can be used as a 3-dB, 90-degree hybrid coupler. This 90-degree hybrid is perhaps the most useful of all hybrids and is used in many microwave circuits. The symbol for this hybrid is shown in Figure 11.26(a). A typical frequency response for this hybrid coupler is shown in Figure 11.26(b).

The single-section quarter-wave directional coupler has a 90-degree or quadrature phase relationship between the outputs independent of frequency for a perfect

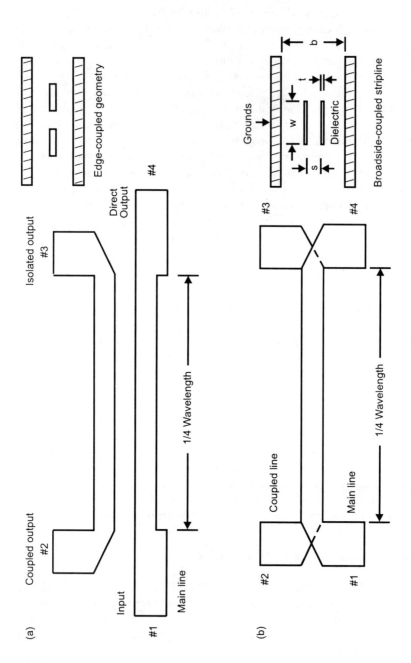

Figure 11.25 Stripline quarter-wave directional couplers: (a) edge-coupled directional coupler and (b) broadside-coupled directional coupler. (*After:* [7].)

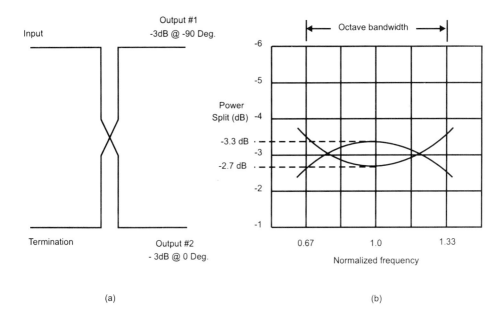

Figure 11.26 A 90-degree hybrid coupler: (a) symbol for 90-degree hybrid coupler and (b) typical frequency response for 90-degree hybrid coupler. (*After:* [7].)

device. In reality, this phase relationship is 90 degrees, ±2–3 degrees, over an octave, which is useful for many applications. It generally is constructed as a completely overlapped coupler with a crossover. Input power splits equally between output 1 and output 2. The fourth arm is terminated in Z_0. As in the case of the looser values of coupling, it frequently is desirable to couple at the center frequency in such a way as to provide a plus-or-minus tolerance around the nominal design frequency. Thus, an octave bandwidth −3dB coupler normally is designed for −2.7 dB midband coupling. The output at the through port will be −3.3 dB with no other losses present. At extremes of 2:1 band, coupled output will be −3.3 dB and direct output will be −2.7 dB, providing 3 dB, ±0.3 dB, over the octave bandwidth.

11.13.7 Stripline Lowpass Filters

Figure 11.27 shows a stripline lowpass filter. Also shown is the equivalent lumped element circuit. In the stripline circuit, inductors are narrow lines with small capacitance to ground and high characteristic impedances, while capacitors are fat lines with large capacitance to ground and low characteristic impedances. The two end lines are equivalent to LC series resonant circuits. All lengths are short compared

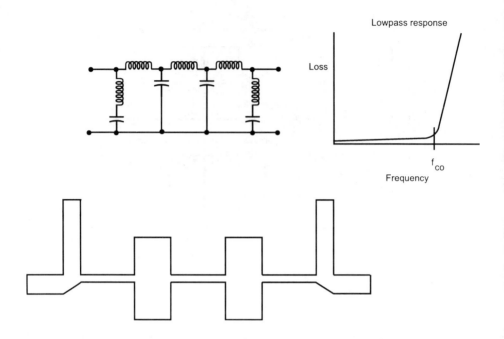

Figure 11.27 Example of a stripline lowpass filter. (*After:* [8].)

to a wavelength. They usually are less than 0.1 wavelength, to resemble lumped elements.

A lowpass filter is designed to provide low loss to an RF signal for frequencies below a selected design frequency, f_{co}. Above that cutoff frequency, the loss or attenuation increases rapidly with an increase in frequency, reaching a high value at a selected second frequency. Above the second frequency, the attenuation remains high for increasing frequency. Designs can be Chebychev or Butterworth. The rate of attenuation depends on the number of sections in the design.

11.13.8 Stripline Highpass Filters

Figure 11.28 shows a stripline highpass filter. The shunt inductors are narrow lines with small capacitances to ground. The lengths are critical and are determined by analysis. Overlap capacitors are used on the main line between the shunt inductors. All elements are short compared to a wavelength and usually are less than 0.1 wavelength.

A highpass filter is designed to provide high loss or high attenuation to an RF signal for frequencies below a selected design frequency. Above that frequency, the loss or attenuation decreases rapidly with increase in frequency, reaching a low value at a selected second frequency. Above the second frequency, the attenuation

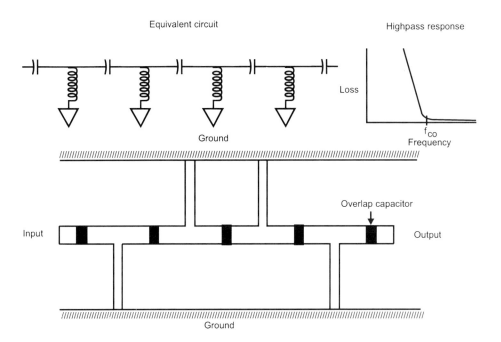

Figure 11.28 Example of a stripline high-pass filter. (*After:* [8].)

remains low for increasing frequency. Again, attenuation versus frequency depends on the number of elements in the filter and whether a maximally flat (Butterworth) or equal-ripple filter prototype is used.

11.13.9 Stripline Bandpass Filters

Figure 11.29(a) shows a side-coupled half-wave resonator bandpass filter. This circuit can be readily printed and has the advantage of providing dc isolation. The type of system shown is a folded configuration. There are a number of other possibilities for implementing systems of this type involving multiple resonators. Bandwidths from 3% to approximately 20% or more are possible.

Figure 11.29(b) shows a short-circuited quarter-wave stub bandpass filter. This direct coupled filter consists of a series of short-circuited quarter-wavelength stubs separated by quarter-wavelength sections of line. This type of bandpass filter is used mainly for the wide passband filters (30% or greater), whereas the circuit of Figure 11.29(a) is used mainly for medium and narrow bandwidth filters.

A stripline bandpass filter is designed to provide high loss or high attenuation to an RF signal for frequencies below a selected design frequency, f_1. Above that frequency, the loss or attenuation decreases rapidly with increase in frequency,

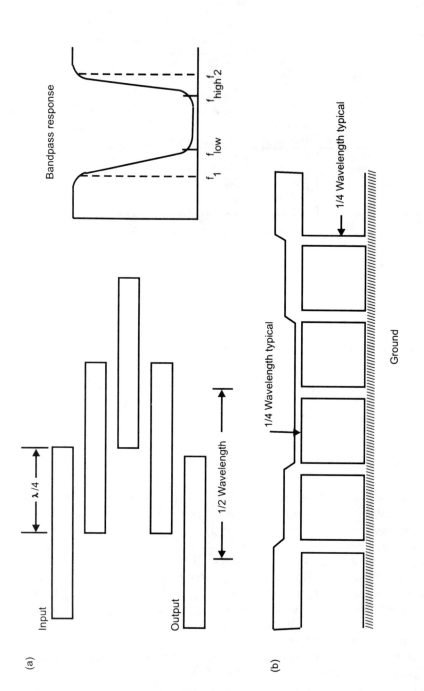

Figure 11.29 Stripline bandpass filters: (a) side-coupled half-wave resonator bandpass filter and (b) short-circuited quarter-wave stub bandpass filter. (*After:* [8].)

reaching a low value at a selected second frequency, f_{low}. Above the second frequency, the attenuation remains low until a third frequency, f_{high}, is reached. Above the third frequency, the attenuation increases rapidly, reaching a high value of attenuation at a fourth frequency, f_2. Above the fourth frequency, the attenuation remains high with increasing frequency. The bandwidth of stripline bandpass filter usually is defined for Butterworth or maximally flat filters as the width between half power points on the response curve. For Chebychev designs, it is the equal-ripple bandwidth, usually less than the 3-dB bandwidth.

11.14 FERRITE CIRCULATORS AND ISOLATORS

Figure 11.30 shows one version of a miniature Y-junction three-port ferrite circulator. This circuit consists of a stripline substrate, a circular ferrite disk, a single bias magnet in the case of microstrip or two magnets in the case of stripline, and a shielding and support can with top and bottom covers. With suitable magnetic field strength, a phase shift is applied to any signal fed in to the circulators in Figure 11.30. If the three striplines are arranged 120 degrees apart, as shown, with proper dimension, the input signal splits into two counterrotating signals that, through the magnetic field application, add in phase at one port and out of phase

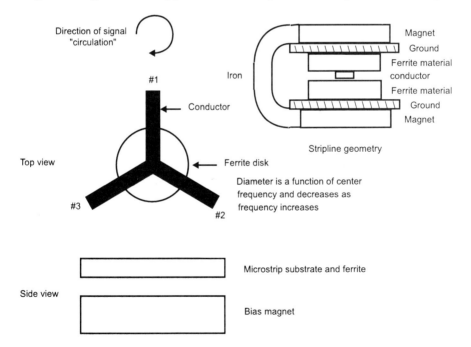

Figure 11.30 Diagram of a Y-type ferrite circulator.

at the third port. For input at port 1 with clockwise circulation, signals add at port 2, where all the signal emerges, and out of phase at port 3, which becomes isolated. Similarly, with port 2 as the input, all signal emerges from port 3. With port 3 as an input, all signal emerges from port 1. Thus, in a duplexer operation, the transmitter is connected to port 1, the antenna to port 2, and the receiver to port 3, as shown in Figure 11.31.

An isolator is a device that allows signals to pass in only one direction. Such an isolator can be obtained from the basic Y circulator by terminating one arm of the circulator in Z_0. Improved isolation is possible by connecting two or more isolators in series.

A four-port circulator is obtained by joining two Y-junction circulators, as shown in Figure 11.32. This type of system can be used as a switch for transceiver systems.

Figure 11.32(a) is a schematic diagram for a four-port ferrite switch for the condition in which the transmitter is connected to the antenna for transmission. Any reflected power from the antenna is routed to a dummy load rather than to the receiver.

Figure 11.32(b) shows the condition in which the direction of the magnetic field on the right-hand circulator is switched, and the combination routes the received signal from the antenna to the receiver.

M/A-COM, Inc., Burlington, Massachusetts, makes a small ferrite circulator unit that is similar in function to the unit shown in Figure 11.30. The M/A-COM unit is constructed on a ferrite substrate known as a Ferrodisc™. It is a true microstrip device using a single ferrite element with a circuit pattern on one face and the ground plane on the other. A two-element permanent magnet structure is included

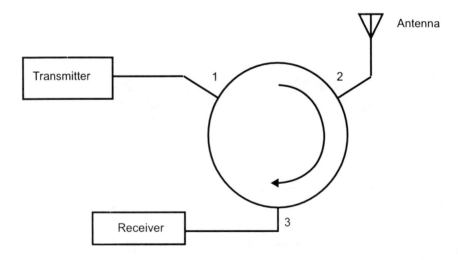

Figure 11.31 Duplexer arrangement using a circulator.

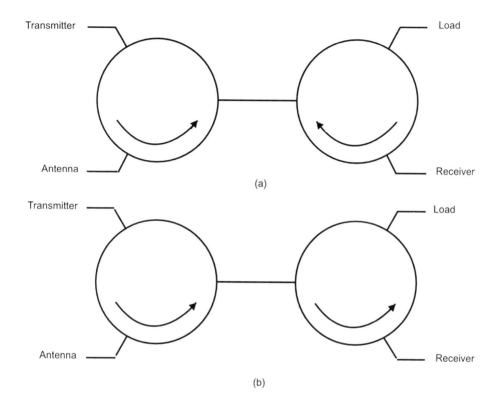

Figure 11.32 Schematic diagrams for ferrite switches: (a) transmit state and (b) receive state.

on the ferrite substrate. When one port is suitably terminated, typically in 50Ω, the device can be used as an isolator. Units are sold by M/A-COM covering the frequency range of 1.7–17.5 GHz. Typical insertion loss is 0.5 dB, and typical isolation is 18 dB. VSWRs of 1.3:1 are typical. Ferrite isolators are available from other sources that will work at frequencies as low as 100 MHz and above 17.5 GHz. Narrower bandwidth units typically exhibit lower insertion loss, higher isolation, and lower VSWR.

Some of the material in Section 11.14 includes information from [9].

11.15 COAXIAL ELECTROMECHANICAL SWITCHES

Coax electromechanical switches frequently are used in RF systems. The most common type is the single-pole, double-throw (SPDT) coaxial switch. An SPDT switch is used to switch a common coaxial input to either of two coaxial outputs or either of two coax inputs to a common output. An SPDT switch is shown in Figure 11.33.

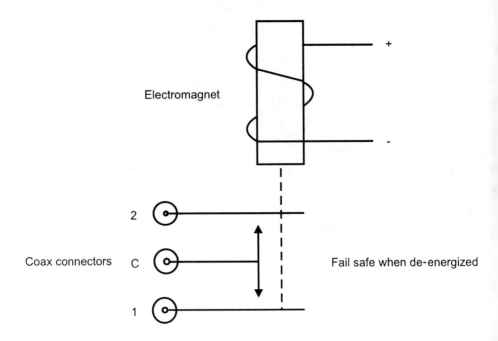

Figure 11.33 Coax electromechanical SPDT switch.

A typical standard SPDT switch (M/A-COM part no. 7524-6132-00) is useful over the frequency range of dc to 12.4 GHz or beyond. It has a CW power rating of 100W below 4 GHz, 80W in the 4–8 GHz range, and 75W in the 8–12.4 GHz range. Typical insertion loss is 0.3 dB below 4 GHz, 0.4 dB in the 4–8 GHz range, and 0.5 dB in the 8–12.4 GHz range. Nominal isolation between ports is 70 dB below 4 GHz, 65 dB in the 4–8 GHz range, and 60 dB in the 8–12.4 GHz range. Switching time is less than 30 ms. Actuating voltage is 20-30V dc and actuating current is 311 mA at 28 volts dc. The life of this type of switch is greater than 1,000,000 cycles.

A typical miniature SPDT electromechanical switch using SMA connectors (M/A-COM part no. 7530-6412-00) is available for operation from dc to 26 GHz. The power ratings are in the range of 60–20W, depending on frequency; insertion loss is in the range of 0.15–0.6 dB, depending on frequency; and isolation between ports is in the range of 80–60 dB, depending on frequency.

Another type of electromechanical switch sometimes used in RF systems is a transfer switch. A switch of this kind is shown in Figure 11.34. It has four ports and provides two separate RF paths that switch simultaneously when actuated. The paths between terminals 1 and 2 and between terminals 3 and 4 normally are closed, while paths between terminals 1 and 4 and between terminals 2 and 3 normally are open. When actuated, the reverse condition exists.

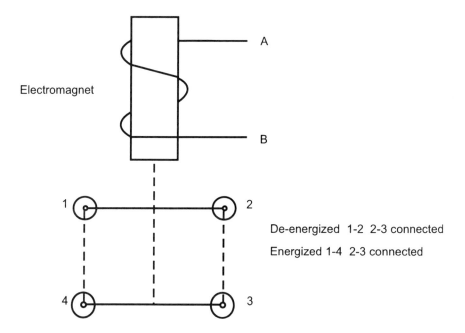

Figure 11.34 Electromechanical transfer switch.

A typical standard four-port transfer electromechanical switch (M/A-COM part no. 7525-6320-00) uses a type-N connector, and has a switching time of 50 ms max. Actuating voltage is 20–30V dc, and actuating current is about 373 mA at 28 volts dc. Operating frequencies are in the range of dc to 12.4 GHz. The CW power rating is in the range of 100–75W, the insertion loss is in the range of 0.3–0.5 dB, and the isolation between ports is in the range of 70–60 dB, depending on frequency. Miniature four-port transfer switches using SMA connectors are available for operation from dc to 18 GHz at lower power ratings.

It is also possible to obtain miniature and standard-size single-throw multiposition coax switches. These switches have a common input with multiple outputs or many inputs with a common output. The switches are used for selecting or combining signals. A typical miniature SPMT electromechanical switch provides SP6T operation from dc to 26.5 GHz in a very small package (1.75 × 1.75 in). The life of the switch is greater than one million cycles. Power ratings, insertion loss ratings, and isolation ratings are similar to the miniature SPDT switch.

11.16 PIN DIODE SWITCHES

A number of possible types of solid-state switches are used with transmission lines. Perhaps the most important of these are the PIN diode switches. Both series and shunt diode switches are used, as are series shunt combinations. When biased in the forward state, the equivalent circuit for the diode is a low-value resistor in series with package inductance. It acts like a short circuit to RF. When reverse biased, the equivalent circuit is a high-value resistor in series with the junction capacitance and that in parallel with the package capacitance. It acts almost as an open circuit to RF.

One important feature of PIN diode switches is that switching times are typically 1 μs or less. This short switching time is important for a number of radar applications.

An important application for PIN diode switches with radar is for phase-shifters used with phased array antennas. A system of this kind is shown in Figure 11.35.

An example phase-shifter system uses a 3-dB hybrid coupler plus two coax lines terminated with short circuits and two PIN diode switches. The 3-dB hybrid coupler has the property that a signal input at port 1 is divided equally in power between ports 2 and 3. No energy appears at port 4. The two PIN diode switches act to either pass or reflect the signals at ports 2 and 3. When the impedance of the diodes is such as to pass the signals, the signals are reflected by the short circuits located farther down the transmission lines. The signals at ports 2 and 3, after reflections from either the diode switches or the short circuits, combine at port 4. None of the reflected energy appears at port 1. The difference in path length with the diode switches open and closed is d. The two-way path, $2d$, is chosen to correspond to the desired increment of phase shift. An N-bit phase shifter can be obtained by cascading N such hybrids.

11.17 SPARKGAP SWITCHES FOR LIGHTNING PROTECTION

Coaxial gas-filled tube, sparkgap switches often are used for lightning protection on coaxial lines connected between antennas and transmitters or receivers. Nor-

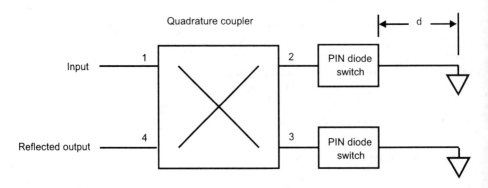

Figure 11.35 PIN diode phase-shifter.

mally, they are open-circuits and pass the RF signals that enter the units. If lightning strikes the antenna and a high voltage is produced on the transmission line, the gas tubes fire and produce an effective short circuit across the line. The outside conductor of the switch is connected to the earth by a good grounding rod or other means. That permits the lightning current to pass to the earth. The sparkgap switches thus act as protective means to the transmitter and receiver systems connected to the transmission lines.

References

[1] Van Valkenburg, Mac E., ed., *Reference Data for Engineers*, 8th ed., SAMS, Carmel, IN: Prentice Hall Computer Publishing, 1993, pp. 29–21 to 29–39, "Transmission Lines" by Tatsuo Itoh.
[2] Howe, Harlan, Jr., *Stripline Circuit Design*, Norwood, MA: Artech House, 1974, pp. 14–31.
[3] Kennedy, George, *Electronic Communication Systems*, 3rd ed., New York: McGraw-Hill, 1985, pp. 169–201.
[4] Smith, Philip H., "A Transmission Line Calculator," *Electronics*, Vol. 12, Jan. 1939, p. 2931.
[5] Howe, pp. 53–54.
[6] Ibid., Chap. 3.
[7] Ibid., Chap. 5.
[8] Ibid., Chap. 6.
[9] Kennedy, pp. 334–340 and pp. 345–346.

Selected Bibliography

Gupta, K. C., *Microstrip Lines and Slotlines*, 2nd ed., Norwood, MA: Artech House.
Hoffman, R. K., *Microwave Integrated Circuits*, Norwood, MA: Artech House. Kennedy, op. cit., pp, 334–340.

CHAPTER 12

Waveguides and Waveguide-Related Components

This chapter discusses waveguides and related components, including hybrid junctions, impedance-matching components, resistive loads and attenuators, directional couplers, ferrite isolators, ferrite circulators, ferrite switches, detectors, mixers, gastube switches, duplexers, and cavity resonators. Much of the information presented in this chapter is adapted from [1–4].

12.1 INTRODUCTION TO WAVEGUIDES

A waveguide is a hollow conducting tube used to transmit electromagnetic waves. Waveguides are an alternative to transmission lines at UHF and higher frequencies, providing much lower losses and much higher power-handling capability than transmission lines. They are not used at lower frequencies because of the very large size that would be required.

Any configuration of electric and magnetic fields that exists inside a waveguide must be a solution of Maxwell's equations. In addition, those fields must satisfy the boundary conditions imposed by the walls of the guide. To the extent that the walls are perfect conductors, there can be no tangential component of electric field at the walls. Many different field configurations can be found that meet those requirements. Each such configuration is termed a mode.

A critical examination of the various possible field configurations or modes that can exist in a waveguide reveals that they all belong to one of two fundamental types. In one type, the electric field is everywhere transverse to the axis of the guide and has no component anywhere in the direction of the guide axis. The associated magnetic field does, however, have a component in the direction of the axis. Modes of this type are termed transverse electric (TE) modes. In the other type of distribution, the situation with respect to the fields is reversed. The magnetic field in that case is everywhere transverse to the guide axis, while at some places the electric field has components in the axial direction. Modes of this type are termed transverse magnetic (TM) modes. The different modes of each class usually are designated by double subscripts, such as TE_{10}.

The behavior of a waveguide is similar in many respects to the behavior of a transmission line. Thus, waves traveling along a waveguide have a phase velocity and are attenuated. When a wave reaches the end of a waveguide, it is reflected back unless the load impedance is the same as the characteristic impedance of the waveguide. An irregularity in a waveguide produces reflections, just an irregularity in any other transmission line does. Again, reflected waves can be eliminated by the use of an impedance-matching network, exactly as with other forms of transmission lines. When both incident and reflected waves are present simultaneously in a waveguide, the result is a standing-wave pattern that can be characterized by an SWR.

In some respects, waveguides are different from other forms of transmission lines in their behavior. One difference is that a particular mode will propagate down a waveguide with low attenuation only if the wavelength of the wave is less than some critical value determined by the dimensions and the geometry of the guide. If the wavelength is greater than the critical cutoff value, the waves in the waveguide die out rapidly in amplitude even when the walls of the guide are of material that has high conductivity. Different modes have different values of cutoff wavelength. The particular mode for which the cutoff wavelength is greatest (lowest cutoff frequency) is termed the dominant mode.

12.2 RECTANGULAR WAVEGUIDES

The most frequently used type of waveguide has a rectangular cross-section, as illustrated in Figure 12.1(a). The width of the guide (indicated by a) is usually twice the height of the guide (indicated by b). In such a guide, the preferred mode of operation is the dominant or lowest order mode. Field configurations for the dominant mode are illustrated in Figure 12.1(b). This mode is the TE_{10} mode. Here, the electric field is transverse to the guide axis and extends between the top and bottom walls. The intensity of the electric field is maximum at the center of the guide and drops off sinusoidally to zero intensity at the edges. The magnetic field is in the form of loops that lie in planes at right angles to the electric field, that is, planes parallel to the top and the bottom of the guide. The magnetic field is the same in all these planes, irrespective of the position of the plane along the vertical axis. This field configuration travels along the waveguide axis. As it travels down the guide, the amplitude is reduced as a result of energy losses in the walls of the guide. The wave drops back in phase with distance, just as it does in analogous transmission lines.

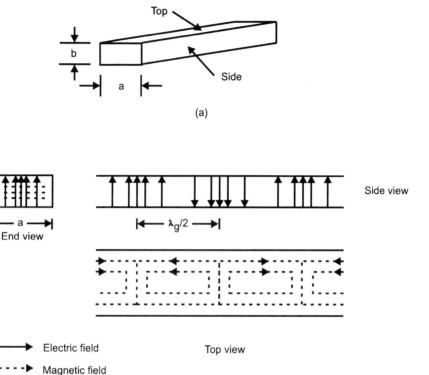

Figure 12.1 Rectangular waveguide and field configurations for dominant mode: (a) rectangular waveguide and (b) field configurations for the dominant or TE_{10} mode. (*After:* [4].)

In the term TE_{10}, the subscript 1 means that the field distribution in the direction of the long side of the waveguide contains one-half cycle of variation. The subscript 0 indicates that there is no variation in either the electric or magnetic field strength in the direction of the short side of the guide. For the dominant mode in the rectangular waveguide, the cutoff wavelength, λ_c, is exactly twice the width, a, of the guide, that is, $\lambda_c = 2.0a$. For example, to transmit 300 MHz where the wavelength is 1m, the guide width must be greater than 50 cm. To transmit 3 GHz where the wavelength is 10 cm, the guide width must be greater than 5 cm. If the frequency is 100 MHz, the wavelength is 3m, and the width of the waveguide must be greater than 1.5m. Clearly, it is not practical to use a waveguide for VHF and lower frequencies, but it is practical to use a waveguide for UHF and higher frequencies.

As pointed out earlier, each mode that can exist in a waveguide has its own cutoff wavelength. The useful frequency range for a waveguide is somewhat less than the frequency range between the dominant mode cutoff frequency and the next higher mode cutoff frequency.

In Figure 12.1, the length, $\lambda_g/2$, indicates one-half guide wavelength. The axial length, λ_g, corresponds to one cycle of variation of the field configuration in the axial direction. It is related to the free-space wavelength, λ, and the cutoff wavelength, λ_c, according to the following equation:

$$\text{Guide wavelength} = \lambda_g = \lambda/[1 - (\lambda/\lambda_c)^2]^{1/2} \quad (12.1)$$

where

λ = wavelength = c/f

c = speed of light

f = frequency

The phase velocity, v_p, is the distance the wave travels in 1 sec. It is related to the velocity of light, c, by the following equation:

$$v_p/c = \lambda_g/\lambda \quad (12.2)$$

It is seen that the velocity of propagation always exceeds the velocity of light. As the frequency is lowered so that it approaches the cutoff value, the phase velocity increases and becomes infinite at cutoff.

The phase velocity v_p is an apparent velocity deduced from the rate of phase change with position along the axis. The actual velocity with which a pulse of energy travels is the group velocity, v_{gr}, and is related to v_p and c by the following equation:

$$v_p v_{gr} = c^2 \quad (12.3)$$

Thus, the group velocity is less than the velocity of light to the extent that the phase velocity is greater. Table 12.1 lists selected rectangular waveguides, their useful frequency range, their dimensions, and their designations. For an example of the use of Table 12.1, assume operation in X-band at a frequency of 10 GHz. The required waveguide would be number 6, which has a frequency range of 8.2–12.4 GHz. Outside dimensions for this waveguide are 25.4 × 12.7 mm, and the wall thickness is 1.3 mm. The inside dimensions can be determined by subtracting twice the wall thickness. The RETMA type number is WR90, and the JAN type number is RG-52/U.

Figure 12.2 is a physical picture of wave propagation in a rectangular waveguide. We can consider that the fields inside the waveguide are the result of a pair of electromagnetic waves that travel back and forth between the sides of the guide following a zigzag path, as illustrated in Figure 12.2(a). Each time such a wave

Table 12.1
Characteristics For Example Rectangular Waveguides

No.	Useful Frequency Range (GHz)	Outside Dimensions (mm)	Wall Thickness (mm)	RETMA Type No.	JAN Type No.
1	1.12–1.70	169 × 86.6	2.0	WR650	RG-69/U
2	1.70–2.60	113 × 58.7	2.0	WR430	RG-104/U
3	2.60–3.95	76.2 × 38.1	2.0	WR284	RG-48/U
4	3.95–5.85	50.8 × 25.4	1.6	WR187	RG-49/U
5	5.85–8.20	38.1 × 19.1	1.6	WR137	RG-50/U
6	8.20–12.40	25.4 × 12.7	1.3	WR90	RG-52/U
7	12.40–18.00	17.8 × 9.9	1.0	WR62	RG-91/U
8	18.00–26.50	12.7 × 6.4	1.0	WR42	RG-53/U
9	26.50–40.00	9.1 × 5.6	1.0	WR28	RG-96/U
10	40.00–60.00	6.8 × 4.4	1.0	WR19	—
11	60.00–90.00	5.1 × 3.6	1.0	WR12	RG-99/U
12	90.00–140.0	4.0 diam.	2.0 × 1.0	WR8	RG-138/U
13	140.0–220.0	4.0 diam.	1.3 × 0.64	WR5	RG-135/U
14	220.0–325.0	4.0 diam.	0.86 × 0.43	WR3	RG-139/U

Note: RETMA = Radio Electronic Television Manufacturers' Association; JAN = Joint Army-Navy.
Source: [2].

strikes the conducting side wall, it is reflected with reversal of the electric field and with an angle of reflection equal to the angle of incidence.

The guide wavelength for the situation is the distance along the axis between the points in the guide where the positive crests coincide. In the top example in Figure 12.2(a), the free-space wavelength is much less than the cutoff wavelength, and the phase velocity in the waveguide is only slightly greater than the speed of light. The guide wavelength also is only slightly greater than the free-space wavelength. These relationships are shown in Figure 12.2(b). In the bottom example in Figure 12.2(a), the free-space wavelength is near the cutoff wavelength, resulting in the guide wavelength and the phase velocity being much greater than the free-space wavelength and the speed of light. These relationships also are shown in Figure 12.2(b). When the free-space wavelength is equal to or greater than the cutoff wavelength, the wave does not propagate down the guide but simply moves back and forth between the sides of the guide.

The fields inside a waveguide induce currents that flow on the inner surfaces of the walls and that can be considered to be associated with the magnetic flux adjacent to the wall. The direction in which the current flows at any point in the wall is at right angles to the direction of the adjacent magnetic flux. In the sides of the guide, the current everywhere flows vertically for the case of the TE_{10} mode, since the magnetic flux in contact with the side walls lies in planes parallel to the

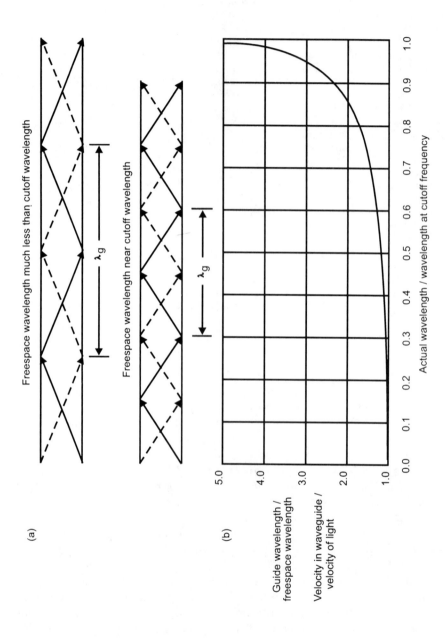

Figure 12.2 Guide wavelengths and velocities in waveguides: (a) paths followed by waves in waveguides and (b) plot of ratios of guide wavelength to free-space wavelength and velocity in waveguide to the speed of light. (*After*: [4].)

top and bottom sides of the guide. In the top and bottom of the guide are a transverse component of current proportional to the axial component of the magnetic field and an axially flowing current component proportional at any point to the transverse magnetic field.

The current in the guide walls penetrates in accordance with the laws of skin effect (discussed in Chapter 14). Accordingly, the depth of penetration is inversely proportional to the square root of the frequency. At the very high frequencies at which waveguides are used, penetration is very small, and the walls provide practically perfect shielding.

A hole, joint, or slot in the waveguide wall introduces the possibility that energy will leak from the guide to the outside. When that happens, the fields inside the guide are affected, thereby introducing an irregularity with resulting reflection. The coupling thus introduced by a hole in the guide wall may be either to the electric or the magnetic field inside the guide. Electric coupling occurs when electrostatic flux lines that normally would terminate on the guide wall are able to pass through the hole to the outside or to another piece of the waveguide. Magnetic coupling results when the hole or the slot interferes with the current flowing in the guide wall. With either type of coupling, both electric and magnetic fields are present outside the main waveguide. Thus, electric flux leaking through the hole induces current on the outer surface of the guide that produces a magnetic field. Again, when magnetic flux leaks through a hole, the associated interference with the flow of current in the wall produces a voltage across the hole that gives rise to an electric field that extends beyond the guide primary.

The nature and the magnitude of the coupling in any particular case depend on the size, shape, and orientation of the coupling hole and on the thickness of the guide wall. The factors involved can be understood by considering the effects produced by long narrow slots oriented in various ways, as illustrated in Figure 12.3.

Slot 1, which is transverse to the magnetic field inside the guide, produces a minimum of interference with currents in the guide wall and introduces little or no magnetic coupling. It does, however, permit electric coupling if the slot width is great enough in proportion the wall thickness to permit a reasonable number of electric flux lines to pass through the slot. The electric coupling will be negligible if the slot is in the nature of a joint representing two surfaces fitted together or is very narrow. Similarly, the long, narrow slot 4 produces little magnetic coupling because it is transverse to the magnetic flux and therefore interferes only negligibly with the flow of current in the guide wall. It does not produce electric coupling because there is no electric field terminating on the side wall for the TE_{10} mode. Such a slot, therefore, has negligible effect, even if it is quite long.

In contrast, slot 5, while causing no electric coupling, introduces a substantial amount of magnetic coupling to outside space because its long dimension is parallel to the magnetic field in the guide. Hence, this slot is oriented in a manner so as to permit easy escape of magnetic flux lines and to interfere to a maximum extent

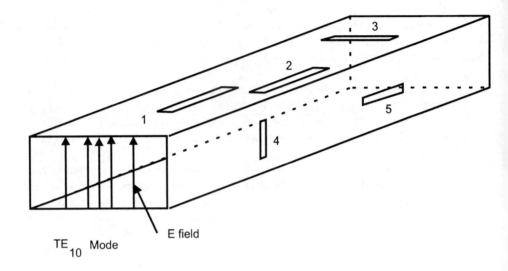

Figure 12.3 Waveguide with slots in walls and TE_{10} mode. (*After:* [4].)

with the wall currents. This coupling is fully effective even if the slot is quite narrow; it is necessary only that the slot interrupt the flow of current in the wall.

Slots 2 and 3 in Figure 12.3 also give rise to magnetic coupling because they interfere with the flow of current in the guide wall. In the case of slot 2, the amount of coupling will be greater the farther the slot is to the side of the center line of the guide. Slots 2 and 3 will also simultaneously introduce electric coupling to the extent that the slot is wide enough in relation to the wall thickness to permit the passage of electric flux. In the case of slot 2, the electric coupling becomes less the farther the slot is from the center line because the intensity of the electric field terminating on the top and bottom sides of the guide becomes less as the side walls are approached.

Table 12.2 lists average attenuations for some rectangular waveguides operating in the dominant TE_{10} mode. The propagation of energy down a waveguide is accompanied by a certain amount of attenuation as a result of the energy dissipated by current induced in the walls of the guide. The magnitude of the current at any point is determined by the intensity of the magnetic field adjacent to the wall at that point. The resistivity that the induced currents encounter is determined by the skin depth of the wall and is therefore proportional to the square root of the frequency and to the square root of the resistivity of the material of which the wall is composed.

Skin depth and loss in conductors are discussed in Chapter 14.

Table 12.2 shows that in the frequency range of 1.12–1.70 GHz, the average attenuation is only 0.0052 dB/m. In contrast, in the 220.0–325.0 GHz frequency range, the average attenuation is 8.80 dB/m. Thus, we see the very strong effects

Table 12.2
Sample Average Attenuations and CW Power Ratings for Waveguides

JAN Type	RETMA Type	Useful Frequency (GHz)	Average Attenuation (dB/m)	CW Power Rating (KW)
RG-69/U	WR650	1.12–1.70	0.0052	14,600
RG-104/U	WR430	1.70–2.60	0.0097	6,400
RG-48/U	WR284	2.60–3.95	0.019	2,700
RG-49/U	WR187	3.95–5.85	0.036	1,700
RG-50/U	WR137	5.85–8.20	0.058	635
RG-52/U	WR90	8.20–12.40	0.110	245
RG-91/U	WR62	12.40–18.00	0.176	140
RG-53/U	WR42	18.00–26.50	0.37	51
RG-96/U	WR28	26.50–40.00	0.58	27
	WR19	40.00–60.00	0.95	13
RG-99/U	WR12	60.00–90.00	1.50	5.1
RG-138/U	WR8	90.00–140.00	2.60	2.2
RG-135/U	WR5	140.0–220.0	5.20	0.9
RG-139/U	WR3	220.0–325.0	8.80	0.4

Notes: RETMA = Radio Electronic Television Manufacturers' Association; JAN = Joint Army-Navy. The values shown are for copper waveguides except for the last five, which are for silver waveguides.
Source: [2].

that frequency and skin depth have on the attenuation in a waveguide. We also see the very large range of power ratings for waveguides as a function of frequency. An equation for maximum power in a waveguide is as follows:

$$P_{max} \text{ for } TE_{10} = 3.6 \cdot a \cdot b \cdot (\lambda/\lambda_g) \text{ MW}$$

with a and b in inches. This holds for an air-filled waveguide at atmospheric pressure with the breakdown strength of air being 29 kV/cm.

It should be pointed out that the attenuation in the waveguide is not constant over the useful operating band. There is a particular frequency for which the attenuation is a minimum. On either side of that minimum, the attenuation increases, the result of two opposing tendencies. As the frequency is lowered, the skin depth becomes greater, causing the effective resistivity of the walls to decrease. At the same time, as the frequency approaches the cutoff value for the mode in question, the group velocity decreases. That causes the magnetic fields adjacent to the walls to become rapidly stronger for a given rate of energy flow down the guide. The operating range of the waveguide is thus chosen to be well above waveguide cutoff for the dominant or lowest order frequency mode and well below the next higher mode. Then either the average attenuation over this band is measured, or the theoretical average is predicted. From Table 12.2, we see that the useful bandwidth of a rectangular waveguide is much less than an octave. The ratios of the highest frequency to the lowest frequency in the band for the 14 waveguides shown in Table 12.2 are about 1.5.

A variation of the rectangular waveguide is a flexible waveguide. Flexible waveguides usually have about five times the attenuation of rigid waveguides due to the corrugations and so are used only for special cases where flexibility is needed.

12.3 HIGHER ORDER MODES IN RECTANGULAR WAVEGUIDES

The dominant mode is only one of many field configurations that can exist in a waveguide. Configurations can be of either TE or TM types. Figure 12.4 shows field configurations in the transverse plane for the first four higher modes in a rectangular waveguide with sides $a = 2b$. Figure 12.4(a) shows the TE_{20} mode. The 2 in the

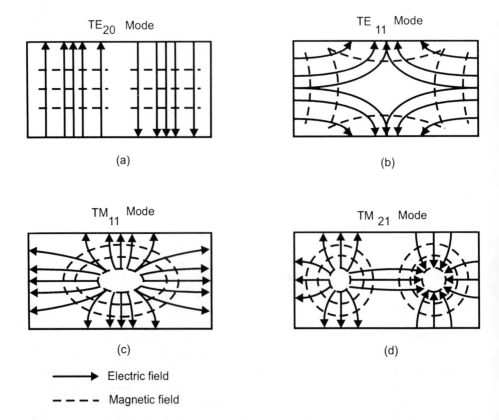

Figure 12.4 Field configurations in the transverse plane for the first four higher order modes in a rectangular waveguide: (a) TE_{20} mode; (b) TE_{11} mode; (c) TM_{11} mode; and (d) TM_{21} mode. (*After:* [4].)

subscript indicates that there are two half-cycles of the electric field configuration in the transverse plane in the direction of the long side of the rectangle. The 0 in the subscript indicates that there are zero half-cycle variations of the electric field in the direction of the short side of the rectangle. The cutoff wavelength for this mode is equal to a, where a is the dimension of the long side. Recall that for the dominant or TE_{10} mode the cutoff wavelength is $2a$. The TE_{11} mode is the first higher order mode that will be encountered.

Figure 12.4(b) shows the TE11 mode. The first 1 in the subscript indicates that there is only one half-cycle of the electric field configuration in the transverse plane in the direction of the long side of the rectangle. The second 1 in the subscript indicates that there is one half-cycle of the electric field in the direction of the short side of the rectangle. The cutoff wavelength for this mode is approximately equal to 0.89a, where a is the long side dimension. Figure 12.4(c) shows the TM_{11} mode. Here we see one half-period variation for the electric field in each of the directions. The cutoff wavelength for this case is the same as for the TE_{11} mode. Figure 12.4(d) shows the TM_{21} mode. Here we see two half-period variations for the electric field in the direction of the long side and one half-period variation for the electric field in the direction of the short side. The cutoff wavelength for this case is 0.71a. The cutoff wavelength in the general case is given by the following relation:

$$\lambda_c = 2a/[(m^2) + (na/b)^2]^{1/2} \quad (12.4)$$

Here a is the long dimension, b the short dimension for the rectangular guide, and m and n are the first and second subscripts describing the mode, respectively.

In the case of a square guide where $a = b$, the cutoff wavelengths are 2a for the TE_{10} and TE_{01} modes, 1.4a for the TE_{11} and TM_{11} modes, and a for the TE_{20} mode.

Any actual configuration of electric and magnetic fields existing in a waveguide can be regarded as the sum of a series of modes that are superimposed on one another. Modes in waveguides are thus analogous to the harmonics of a periodic wave, since a periodic wave of arbitrary shape always can be considered to be represented by the sum of a series of properly chosen harmonic components.

12.4 LAUNCHING THE TE10 MODE USING A COAXIAL LINE INPUT

Figure 12.5 illustrates the launching of the TE_{10} mode using a coaxial line as the input with the center conductor extending into the waveguide. In the figure, the center conductor extends all the way to the top. A variation is to have the center line extension extend only part way into the inside of the guide. That type of arrangement is known as a voltage probe. A voltage probe acts as a small antenna to launch the wave.

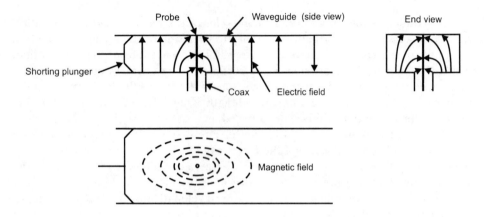

Figure 12.5 Launching of TE_{10} mode using a coaxial line and a voltage probe. (*After:* [4].)

Current in the center conductor generates a magnetic field in the guide that lies in a plane parallel to the top and bottom sides of the guide. At the same time, electric field lines are produced as shown. The TE_{10} mode is the largest single component in the field configuration. The difference between the field configuration of this mode and the actual field present is accounted for by the presence of higher order modes of smaller amplitude. Impedance matching is used to provide maximum power transfer from the coaxial line to the waveguide. (Characteristic impedance and impedance matching for a waveguide are discussed later.)

Figure 12.6 shows coupling from a coaxial line to a waveguide using a small loop. In the figure, the loop couples magnetic field into the waveguide.

One method of suppressing unwanted waveguide modes is to use metal vanes located on the short sides of the guide. Because of the field configuration, the

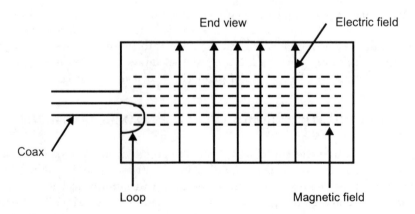

Figure 12.6 Coupling from a coaxial line to a waveguide using a small loop.

vanes do not affect the TE_{10} mode, but they do interfere with both the electric and the magnetic fields of any TM or TE_{0n} modes that might be present. It is also important to choose waveguide sizes such that higher order modes see waveguide cutoff. Modes that are beyond cutoff and so cannot propagate are sometimes termed evanescent modes.

12.5 CHARACTERISTIC WAVE IMPEDANCE FOR WAVEGUIDES

In a TEM mode transmission line, one can define a characteristic impedance that is determined by the geometry of the line and that holds for all frequencies. In a similar manner, the waveguide has a characteristic impedance. However, that impedance is a function of frequency. For the TE_{m0} modes, the characteristic wave impedance for the rectangular waveguide is

$$Z_0 = 377/[1 - (\lambda/\lambda_c)^2]^{1/2} \qquad (12.5)$$

where

λ = free-space wavelength

λ_c = cutoff wavelength for the guide

For example, assume the cutoff wavelength for a waveguide is 10 cm and the free-space wavelength of the signal is 8 cm. Substituting values into (12.5), the characteristic wave impedance for the TE_{m0} modes would be

$$Z_0 = 377/[1 - (8/10)^2]^{1/2} = 628.3 \Omega$$

For the case of the TM_{mn} modes, the characteristic wave impedance for the rectangular waveguide is

$$Z_0 = 377[1 - (\lambda/\lambda_c)^2]^{1/2} \qquad (12.6)$$

where the terms are the same as defined for (12.5). For example, assume the cutoff wavelength for a waveguide is 10 cm and the free-space wavelength of the signal is 8 cm. Substituting values into (12.6), the characteristic wave impedance for the TM_{mn} modes would be

$$Z_0 = 377[1 - (8/10)^2]^{1/2} = 226.2 \Omega$$

As in the case of TEM transmission lines, if the load impedance matches the characteristic impedance of the waveguide, there is no reflection at the load.

12.6 OTHER TYPES OF WAVEGUIDES

12.6.1 Ridged Waveguides

Figure 12.7 shows the case of ridged waveguides. Rectangular waveguides are sometimes made with single ridges located inside the guide on either the top or the bottom walls, as shown in Figure 12.7(a), or they are made with double ridges located inside the guide on both the top and the bottom of the guide walls, as shown in Figure 12.7(b). The ridges are located midway between the two sides. The principal effects of such ridges are to lower the value of the cutoff frequency and to increase the useful bandwidth of the waveguide. By those means, it is possible to achieve nearly an octave bandwidth capability. It should be noted, however, that ridged waveguides generally have more attenuation per unit length than rectangular waveguides without ridges.

12.6.2 Circular Waveguides

Circular waveguides have the advantage that they can be used in application where rotation about the axis of the guide is required, as in the case of a rotating antenna with a waveguide feed. They have the disadvantage that there is only a very narrow

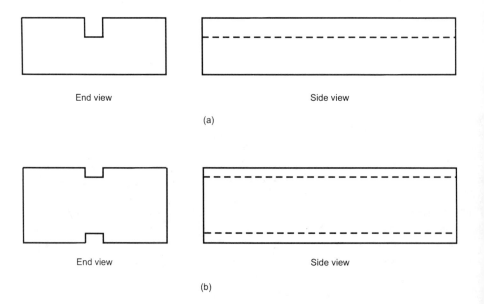

Figure 12.7 Ridged waveguide: (a) ridged waveguide with single ridge and (b) ridged waveguide with double ridge. (*After:* [2].)

range between the cutoff wavelength of the dominant mode and the cutoff wavelength of the next higher mode. Thus, the frequency range over which single-mode operation is assured is relatively limited. Also, because of the circular symmetry, the circular waveguide posses no characteristic that positively prevents the plane of polarization of the wave from rotating about the guide axis as the wave travels. As a result, circular waveguides are used only where it is necessary to have a rotating joint in a waveguide system.

Field configurations for the two most important circular modes are illustrated in Figure 12.8: the TM_{01} mode in Figure 12.8(a) and the TE_{11} mode in Figure 12.8(b). The TE_{11} mode is the dominant mode, and the TM_{01} mode is the first higher order mode.

Table 12.3 lists the cutoff wavelengths in terms of the guide radius, r, for the first five circular waveguide modes.

The guide wavelength in a circular guide is greater than the wavelength in free space, just as in the rectangular guide. The velocity of phase propagation is λ_g/λ times the velocity of light in all cases. A wave traveling down a circular guide is attenuated as a result of power dissipated in the walls by the induced wall currents, exactly as in the case of a rectangular guide.

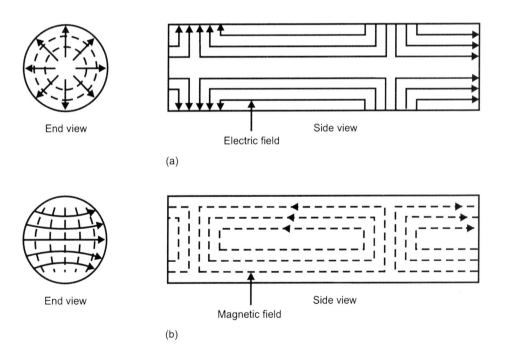

Figure 12.8 Circular waveguide: (a) circular waveguide with TM_{01} mode and (b) circular waveguide with TE_{11} mode. (*After:* [2].)

Table 12.3
Cutoff Wavelengths in
Circular Waveguides

Mode	Cutoff Wavelength
TE_{11}	3.42 r
TM_{01}	2.61 r
TE_{21}	2.06 r
TE_{01}	1.64 r
TM_{11}	1.64 r

12.7 WAVEGUIDE HARDWARE

12.7.1 Waveguide Flanges

A typical waveguide has a flange at either or both ends. At the lower frequencies, the flanges are brazed or soldered onto the waveguide. At higher frequencies, a much flatter, butted, plain flange is used. When two waveguides are connected, the two flanges are bolted together. Care must be taken to ensure that near-perfect mechanical alignment is achieved, so there are no undesirable reflections of the signal at those junctions. Waveguides with smaller dimensions sometimes are provided with threaded flanges, which are somewhat easier to align.

A second type of waveguide flange combination uses a plain flange on one side and a choke flange on the other end. With the choke flange, a short circuit is reflected to the junction of the waveguides using a half guide wavelength slot. Thus, an electrical short is placed at a surface where a mechanical short circuit would be difficult to achieve. Unlike the plain flange, the choke flange is frequency sensitive. A typical design for such a choke flange might provide a 10% bandwidth.

12.7.2 Rotary Joints

Rotating joints are often used in radar. An example would be where a radar is connected to a horn antenna feeding a parabolic reflector that must rotate for tracking. A rotating joint involving a circular waveguide is the most common type. The rotating part of the waveguide is circular and carries the TM_{01} mode, whereas the rectangular waveguide leading in and out of the joint carries the TE_{10} mode. The circular waveguide has a diameter that ensures that modes higher than the TM_{01} cannot propagate. The dominant TE_{11} in the circular guide is suppressed by a ring filter, which tends to short-circuit the electric field for that mode, while not affecting the electric field of the TM_{01} mode, which is everywhere perpendicular to the ring. A choke gap is left around the circular guide joint to reduce any mismatch that may occur. Impedance matching of some type is often provided at each circular-rectangular waveguide junction to compensate for reflections.

12.7.3 Tapered Transition Sections of Waveguides

A tapered section of waveguide is used when it is necessary to join waveguides having different dimensions or different cross-sectional shapes. Some reflections take place from a tapered section, but they can be reduced to low levels if the tapered section is made gradual, so the section has a length of two or more wavelengths at the lowest frequency of interest.

A tapered section of waveguide transforms the characteristic impedance of the waveguide. The impedance is directly proportional to the short dimension of the rectangular waveguide. The taper transition section thus can be used as an impedance-matching section for connecting to resistive loads.

12.7.4 Flexible Waveguides

Flexible waveguides are sometimes used when it is necessary to have a waveguide section capable of movement, which may include bending, twisting, stretching, and operation with vibration. The flexible waveguide must not cause undue attenuation or reflections and must be able to operate continuously for extended periods of time. Several different types of flexible waveguide are used, including copper or aluminum tubes having an elliptical cross-section, small transverse corrugations, and transitions to rectangular waveguides at the two ends. These tubes transform the TE_{11} mode in the flexible waveguide into the TE_{10} mode at either end. Such a waveguide is of continuous construction; thus, joints and separate bends are not required. It may have a polyethylene or rubber outer cover for environmental protection; it also bends easily but cannot be twisted readily. Power-handling ability and SWR are fairly similar to those of rectangular waveguides of the same size, but attenuation in decibels per meter is about five times greater than in equivalent rectangular waveguides.

12.7.5 Waveguide Accessories

Manufacturers' catalogs show a large number of accessories that can be used in waveguide systems: 90-degree bends of several types, circular to rectangular waveguide transitions, 90-degree twist sections, H-plane T-junctions, and E-plane T-junctions. T-junctions (particularly the E-plane variety) often are used for impedance matching in a manner identical to the short-circuited TEM-transmission line stub. The vertical arm is provided with a sliding piston or plunger to produce a short circuit at any desired point. Other accessories include terminations, filters, couplers, power dividers, hangers, feed-throughs, pressure windows, gas inlets, straight sections, and flex-twist sections.

12.8 WAVEGUIDE HYBRID JUNCTIONS

Figure 12.9 shows a waveguide hybrid T-junction, also known as a *magic T*. A hybrid T-junction has some very interesting and highly useful properties. Its basic property is that arms 3 and 4 both are connected to arms 1 and 2 but are isolated from each other, provided that each arm is terminated in a correct impedance. If a signal is applied to arm 3, it divides equally between arms 1 and 2 but with opposite phase for the two arms. Ideally, none of the signal enters arm 4. An input signal at arm 4 likewise divides equally between arms 1 and 2, but with the same phase for each arm. Again, ideally there is no signal coupled to arm 3. The operation for this coupler is the same as that of the stripline or microstrip rat-race coupler.

If two signals enter arms 1 and 2, they add in arm 4. On the other hand, the output in arm 3 for the two signals is the difference between the signals.

An example application of the magic T is as a front end for a microwave receiver. In that case, arm 3 is connected to the antenna, arm 4 is connected to the LO, arm 1 is terminated in a matched impedance, and arm 2 is connected to a mixer.

A second type of hybrid junction looks very different from the magic T and yet has very similar properties. This waveguide system, known as a hybrid ring or rat race and shown in Figure 12.10, consists of a rectangular waveguide bent in the E plane to form a complete loop whose median circumference is 1.5 guide wavelengths. It has four arms connected as shown with arms 1 and 4 separated by

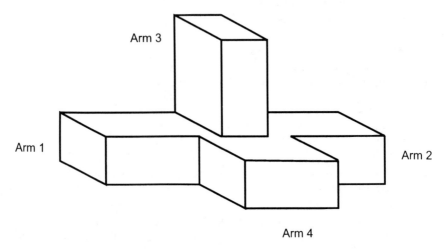

Figure 12.9 Waveguide hybrid T-junction (magic T). (*After:* [2].)

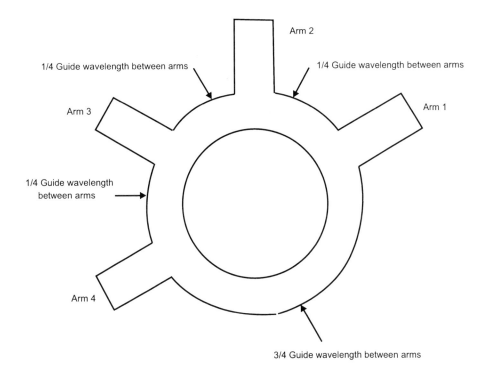

Figure 12.10 Waveguide hybrid ring or rat race. (*After:* [2].)

three-quarters guide wavelengths and the other arm spacings one-quarter guide wavelengths. If a signal is applied to arm 1, it divides evenly, with half traveling clockwise and the other half traveling counterclockwise. The two signals reaching arm 4 have traveled the same distance and add in phase. A part of the signal thus travels out through arm 4. A signal starting from arm 1 and reaching arm 2 has traveled a distance of $\lambda_g/4$ if traveling counterclockwise and 1.25 λ_g if traveling clockwise. Recall that phase repeats every wavelength. Thus, the two signals are in phase at arm 2 and add at that point. A part of the signal thus travels out arm 2. A signal starting at arm 1 and reaching arm 3 has traveled a half guide wavelength in the counterclockwise direction and a full guide wavelength in the clockwise direction. These two signals are 180 degrees out of phase and cancel. In a similar way, it can be shown that arm 3 is connected to arms 2 and 4 but not to arm 1.

If one signal enters arm 2 and a second signal enters arm 4, they add at arm 3, since they each have traveled the same distance. On the other hand, on reaching arm 1, the signals subtract since they have traveled distances separated by one-half guide wavelength.

Thus, the behavior of the rat-race coupler is very similar to that of the magic T, although for different reasons. The two types of systems can be used interchangeably, with the magic T having the advantages of smaller bulk, ease of manufacture, and better isolation over waveguide band. The hybrid ring has dimensions that are frequency dependent, hence performance also is frequency dependent. That is not so in the case of the magic T.

12.9 WAVEGUIDE IMPEDANCE MATCHING

Impedance matching can be accomplished in waveguides by various types of obstacles placed in the waveguide. By those means, the equivalent of parallel capacitors, parallel inductors, or parallel resonant LC circuits can be added. They can be used in the same way that shorted or open-circuited stubs are used in TEM-mode transmission lines to provide impedance matching.

Three types of obstacles are illustrated in Figure 12.11. Figure 12.11(a) shows two types of waveguide irises or apertures with the openings parallel to the long side of the guide. This type of obstacle is used for the equivalent of parallel capacitance. Figure 12.11(b) shows two types of waveguide irises or apertures with the

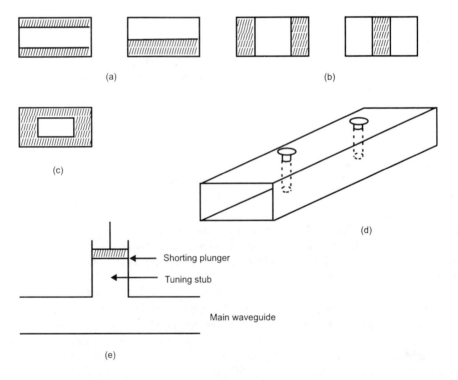

Figure 12.11 Waveguide impedance matching components: (a) capacitive waveguide irises; (b) inductive waveguide irises; (c) parallel resonant waveguide iris; (d) waveguide with two-screw tuner; and (e) stub tuner.

openings parallel to the short side of the guide. This type of obstacle is used for the equivalent of parallel inductance. Figure 12.11(c) shows a waveguide iris or aperture with a rectangular opening in the center of the guide with the long side of the opening parallel to the long side of the guide. This type of obstacle is used for the equivalent of a parallel resonant LC circuit. Irises are not easily adjustable and therefore normally are used to correct only permanent mismatches.

Another type of obstacle that can be used with waveguides for impedance matching is a cylindrical post extending into the waveguide from one of the broad sides. It can appear as either inductive or capacitive, depending on how far it extends into the waveguide. A short post appears capacitive, while a long post appears inductive.

Another type of obstacle that can be used when it is necessary to have adjustable matching elements is a screw extending into the waveguide from one of the broad sides of the guide. Again, the parallel impedance introduced by the screw depends on how far the screw extends into the guide.

It is common practice to use more than one screw for impedance-matching sections. These can be two-screw tuners or three-screw tuners. The spacing between screws typically is three-eighths guide wavelength at the center frequency in the waveguide band for the two-screw unit. A system of this type is shown in Figure 12.11(d).

The E-plane T also can be used in the same way as the adjustable TEM-mode transmission line shorted stub when it is provided with a sliding short-circuiting piston. A system of this type is shown in Figure 12.11(e).

12.10 WAVEGUIDE RESISTIVE LOADS AND ATTENUATORS

A common resistive termination for a waveguide is a length of epoxy-iron mixture fitted in at the end of the guide and tapered gradually with the point in the direction of the incoming wave. Such a termination absorbs the incoming waves and therefore does not cause reflections. Terminations of this type are not frequency sensitive if the length of the tapered section is greater than one-half the lowest frequency's guide wavelength.

A movable resistive vane can be used as a variable attenuator for a waveguide. In one type, the vane extends through a slot in the top of the waveguide. In another type, the vane is mounted on dielectric rods that extend through the sides of the guide.

12.11 WAVEGUIDE DIRECTIONAL COUPLERS

A waveguide directional coupler is illustrated in Figure 12.12. Waveguide directional couplers can be made by using a second waveguide parallel to the first with two holes providing coupling between them. The holes are spaced one-quarter guide

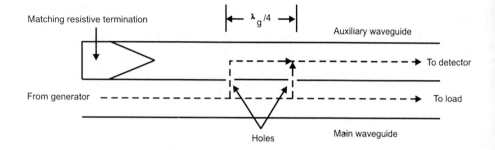

Figure 12.12 Two-hole waveguide directional coupler. (*After:* [2].)

wavelength apart. The operation and the use of waveguide directional couplers are similar to the operation and use of the TEM mode transmission line directional coupler. The couplers are used to measure simultaneously the forward and reflected power in the main waveguide. The amount of coupling depends on the size of the holes. A typical coupler has a coupling ratio of 20–40 dB and provides very low insertion loss.

12.12 WAVEGUIDE FERRITE ISOLATORS, CIRCULATORS, AND SWITCHES

In many cases at microwave frequencies, it is desirable to have only one-way transmission of signals. An example is a microwave generator where the amplitude and frequency could be affected by changes in the load impedance, a phenomenon that is termed frequency pulling when applied to frequency variation. The solution to that problem is to use an isolator between the generator and the load. A second example is a semiconductor device used for microwave amplification that is a two-terminal device, in which the input and the output would interfere unless some means of isolation are used. In that case, a circulator usually is used. Ferrites often are used in such devices with magnetic fields involved.

Ferrite isolators may be based on Faraday rotation, in which the direction of polarization is angularly rotated by the ferrite device. A waveguide isolator of this type is illustrated in Figure 12.13.

The isolator in Figure 12.13 uses a center section of a circular waveguide operating in the TE_{11} mode, with transition sections connected to standard rectangular waveguides operating in the TE_{10} mode. One of the transition sections is rotated 45 degrees with respect to the other. A small-diameter ferrite rod is mounted by means of a foam support in the center of the circular waveguide. A permanent magnet is placed about the circular waveguide and the ferrite rod. Two other important components placed in the circular waveguide are flat resistive attenuators, with one rotated 45 degrees with respect to the other, as shown.

Waveguides and Waveguide-Related Components | 359

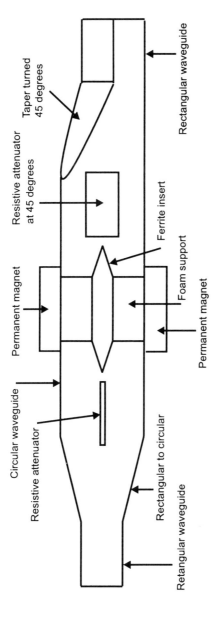

Figure 12.13 Faraday rotation isolator. (*After:* [2].)

The operation of a Faraday rotation isolator is as follows. A wave passing through the ferrite in the forward direction (to the right) has its plane of polarization rotated clockwise 45 degrees. It passes the resistive attenuator with low loss and exits the transition section with the desired direction of E-field. A typical total insertion loss in the forward direction for an X-band isolator of this type is in the range of 0.5–1.0 dB. A wave passing through the isolator in the reverse direction also is rotated in polarization in the clockwise direction. As a result, the E-field direction is such that the wave is largely absorbed by the resistive vane. It also cannot propagate in the input rectangular waveguide because of its direction of E-field and the dimensions of the waveguide at right angles to the E-field (the wave is in waveguide cutoff). A typical loss for the reverse is 20–30 dB.

This type of isolator finds many applications for peak power levels less than about 2 kW. The power limitation is the result of nonlinearities in the ferrite, resulting in polarization shifts departing from the ideal 45 degrees.

A second popular type of high-power waveguide isolator is the resonant absorption isolator, illustrated in Figure 12.14. It uses a section of a rectangular waveguide operating in the TE_{10} mode. A piece of ferrite material is placed about a quarter of the way from one side of the waveguide and halfway between its ends. A permanent magnet is used with the ferrite, as shown, with a much stronger field than in the Faraday rotation isolator. At the location of the ferrite, the magnetic field of the TE_{10} wave is strong and circularly polarized. The polarization is clockwise in one direction of propagation and counterclockwise in the other. Therefore, there is unaffected propagation in one direction but resonance and large absorption in the opposite direction. Maximum power capability for this type of isolator is limited

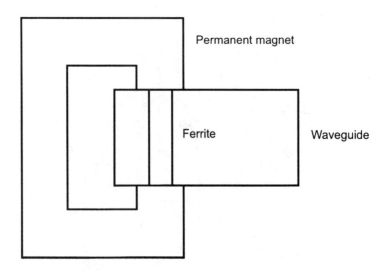

Figure 12.14 Resonance absorption isolator (end view). (*After:* [2].)

only by temperature rise that might bring the ferrite to its Curie point. Such systems can be built with peak power capabilities as high as 3,000 kW at S-band, 1,000 kW at C-band, and 300 kW at X-band.

There are waveguide, coaxial, and stripline versions of Y-junction ferrite circulators. Stripline versions were discussed in Chapter 11. Three-port, high-power, waveguide junction circulators and isolators are available in all standard waveguide sizes for WR28 through WR284. All circulators can be converted to isolators with the addition of a load. Peak power-handling capability depends on pressurization, duty cycle, pulse width, load mismatch, and altitude. Average power-handling capability can be increased by heat sinking, fins, or liquid cooling.

A waveguide can be placed at the junction of two or more intersecting waveguides to direct the flow of microwave energy. They can be used in any system where switchable isolation or duplexing is required. An SPDT switch is a junction circulator whose direction of circulation can be reversed on command by reversing the magnetic bias. That is achieved by replacing the permanent magnets found on a circulator with an electromagnetic source, such as an external electromagnet or an internal latch wire.

12.13 WAVEGUIDE DETECTORS AND MIXERS

A diode waveguide mount and detector are illustrated in Figure 12.15. Silicon-point contact diodes have been used for many years for microwave mixer and detector functions. More recently, Schottky-barrier diodes have been used for mixer functions because of their lower noise figure (below 6 dB at 10 GHz). In either case, it is necessary to use some form of diode mount to locate the diodes in the waveguide.

The detector in Figure 12.15 uses tuning for the waveguide, including a tuning plunger for providing the quarter guide wavelength location for the diode and a tuning screw. Figure 12.16 shows two types of waveguide mixers. If there is to be mixing in the waveguide, both the LO signal and the RF signal must be injected. One way to accomplish that is illustrated in Figure 12.16(a). The system includes a tuning screw in the LO line and a tuning plunger in the RF path. The diode is placed a quarter guide wavelength from the shorting plunger.

A balanced mixer configuration is illustrated in Figure 12.16(b). Here a magic T or hybrid T is used with two mixer diodes. This type of system is superior in performance to the single-ended mixer.

12.14 GAS-TUBE SWITCHES

Gas-tube switches are used in a number of different types of duplexers. These are sometimes called transmit/receive (TR) switches or anti-transmit/receive (ATR) switches. The gas-tube switch is basically a piece of waveguide with glass windows on each end. The tubes are filled with a gas mixture, such as hydrogen, argon,

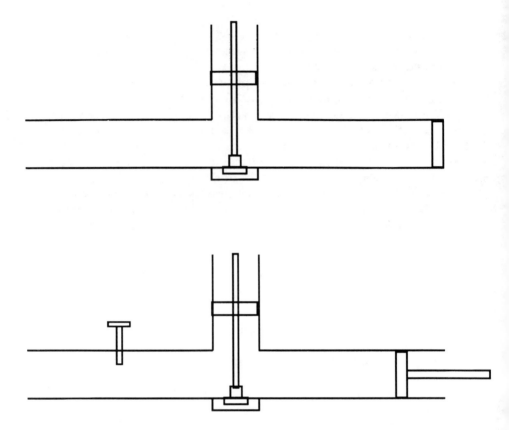

Figure 12.15 Diode waveguide mount and detector. (*After:* [2].)

water vapor, and ammonia, at low pressure. A pair of electrodes extending from the top and the bottom help to ionize the gas when high power is present.

At low power, the gas tube acts as an ordinary piece of waveguide, and the signal passes through with low insertion loss. When a high-power pulse arrives, the gas is ionized and becomes a poor conductor. The result is like placing a short circuit across the waveguide. Typical attenuations produced are greater than 60 dB. The switching action takes place rapidly (on the order of 10 ns). Quick deionization also is provided after the high-power pulse is ended.

12.15 DUPLEXERS

A duplexer is a circuit that is designed to allow the use of a common antenna for both transmit and receive functions. Figure 12.17 shows the case of a branch-guide duplexer for radar using TR and ATR switches. For simplicity of illustration, the

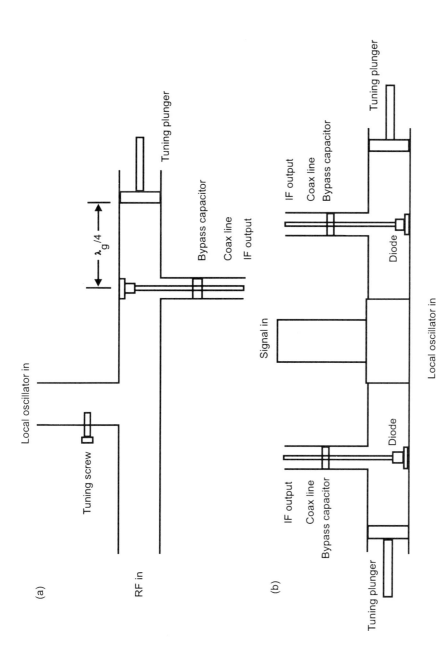

Figure 12.16 Two types of waveguide mixers: (a) single-ended waveguide diode mixer and (b) hybrid-T (magic-T) balanced waveguide mixer. (*After:* [2].)

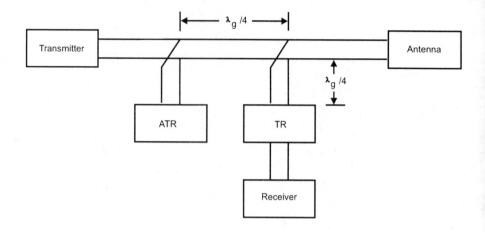

Figure 12.17 Branch-line type duplexer. (*After:* [3].)

waveguides are shown as two-wire lines. When the radar transmitter pulse is present, it travels directly to the antenna. The TR and ATR switches also fire, and short circuits are produced. These are transformed by the quarter guide wavelength lines to open circuits at the main line. When no transmit pulse is present, the ATR is an open circuit that is transformed to a short circuit at the transmitter port. It is further transformed by the quarter guide wavelength main line to an open circuit at the antenna port. The TR switch acts as a normal waveguide, passing the receive signal from the antenna to the receiver.

The branch-line duplexer is a relatively narrowband device because it relies on the length of the waveguides that connect the switches to the main waveguide. This type of duplexer generally has been replaced by the balanced duplexer in modern radars.

Figure 12.18 shows a balanced duplexer using dual TR tubes and two short-slot hybrid junctions. The transmit condition of operation is shown in Figure 12.18(a). The TR tubes are fired, producing a near short circuit at their interface to the waveguides.

The short-slot hybrid junction consists of two sections of waveguides joined along one of their narrow walls with a specially designed pair of slots cut in the common narrow wall to provide coupling between the two guides. The short-slot hybrid can be considered a broadband directional coupler with a 3-dB coupling ratio.

In the transmit condition, power is divided equally into each waveguide by the first short-slot hybrid junction. Both TR tubes break down and equally reflect the incident power out the antenna arm. The short-slot hybrid has the property that each time the energy passes through the slot in either direction, its phase is advanced 90 degrees. Any energy that leaks through the TR tubes is directed to the arm with the matched dummy load and not to the receiver.

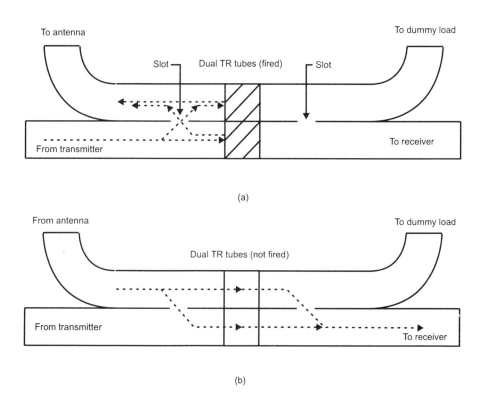

Figure 12.18 Balanced duplexer using dual TR tubes. (*After:* [3].)

Figure 12.18(b) shows the operation of the balance duplexer for the receive condition. The TR tubes are unfired, and the return signals pass through the duplexer and into the receiver. The power splits equally at the first junction, and because of the 90-degree phase advance on passing through the slot, the energy recombines in the receiving arm and not in the dummy-load line.

Other types of balance duplexers are sometimes used. One of these is a four ATR system. During transmission, the ATR tubes located in a mount between the two short-slot hybrids ionize and allow high power to pass to the antenna. During reception, the ATR tubes present a high impedance, which results in the return signal power being reflected to the receiver. This type of system has higher power-handling capability than the system in Figure 12.18, but it has less bandwidth.

12.16 CAVITY RESONATORS

Cavity resonators are used at microwave frequencies for much the same purposes as tuned LC circuits are used at the lower frequencies. Cavity resonators are characterized as having very high values of Q and therefore very narrow bandwidths.

Resonance occurs when the length of the resonator is some integer times the guide wavelength divided by 2. Figure 12.19 shows examples of cavity resonators. Figure 12.19(a) shows a halfwave waveguide cavity. The mode for that cavity is TE_{101}. The electric field is transverse to an axis in the length direction, as shown by the side view, and the variation of the electric field is one half-cycle, zero, and one half-cycle in the a, b, and l directions, respectively. The a dimension is the long dimension for the end view, the b dimension is the short dimension for the end view, and the l dimension is the long dimension for the side view.

Figure 12.19(b) shows a cylinder-type resonant cavity. The mode for this cavity is TM_{010}. Here, TM denotes that the magnetic field lies in planes transverse to the axis of the cylinder. The first and third subscripts denote that the variation of the magnetic field is zero with, respectively, radial direction and position along the axis. The second subscript indicates one half-cycle of variation in the field along a radial line passing from one edge of the cylinder to the other edge. The third number in the subscript indicates the number of half-cycles of electric field in the length dimension.

Figure 12.19(c) shows a reentrant-type resonant cavity. This type of cavity is frequently used in microwave tubes such as klystrons. Here the electric field is most intense in the gap or the region where the top and bottom sides are brought together to form a short space between walls. The magnetic field is most intense near the edges of that center section and falls off toward the outside edges of the cavity. When used with microwave tubes, an electron beam is sent through a hole in the center of the cavity where the top and bottom walls are close together. That couples energy into the cavity, or the cavity transmits energy to the beam for velocity modulation.

Exactly the same methods can be used for coupling to cavity resonators as are used to couple to waveguides. Thus, slots, loops, and probes are used when coupling power into or out of a cavity. Electron-beam coupling also is common for energy coupling when the cavity is used in microwave tubes.

Tuning of cavities is done by the same methods as used for impedance matching in waveguides. Adjustable screws or posts are the most often used methods. Another method of tuning a cavity is to have walls that can be moved in or out by mechanical means.

The Q of a cavity resonator can be very high. The 3-dB bandwidth of the cavity is the center frequency divided by the Q. Cavities such as those shown in Figure 12.19(a) and (b) have typical values for Q of about 24,000 if silver-plated copper walls are used. The Q for a reentrant cavity such as that shown in Figure 12.19(c) is typically about 4,000.

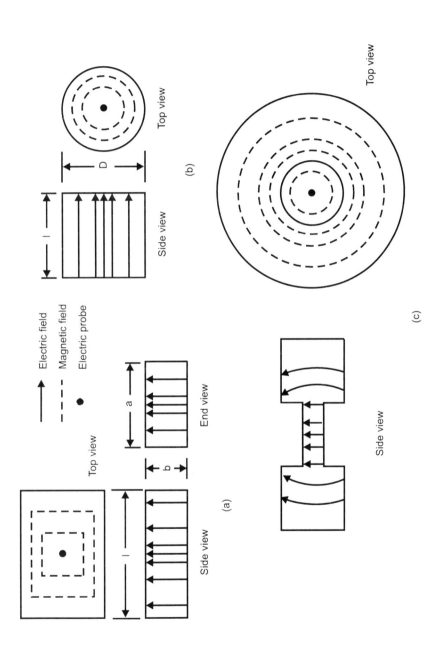

Figure 12.19 Examples of cavity resonators: (a) half-wave waveguide resonator; (b) cylindrical resonant cavity; and (c) resonant reentrant cavity. (*After:* [4].)

References

[1] Van Valkenburg, Mac E., ed., *Reference Data for Engineers*, 8th ed., SAMS, Carmel, IN: Prentice Hall Computer Publishing, 1993, pp. 30-1 to 30-31, "Waveguides and Resonators" by Tatsuo.
[2] Kennedy, George, *Electronic Communication Systems*, 3rd ed., New York: McGraw-Hill, 1985, pp. 288-347.
[3] Skolnik, Merrill I., *Introduction to Radar Systems*, 2nd ed., New York: McGraw-Hill, 1980, pp. 359-366.
[4] Terman, Frederick E., *Electronic and Radio Engineering*, 4th ed., New York: McGraw-Hill, 1955, pp. 127-165.

Selected Bibliography

Collin, R. E., *Field Theory of Guided Waves*, New York: McGraw-Hill Book, 1958.

Harvard University, Radio Research Laboratory, *Very High Frequency Techniques*, Vols. 1 and 2, New York: McGraw-Hill, 1947.

Marcuvitz, Nathan, *Waveguide Handbook*, MIT Radiation Laboratory Series, Vol. 10, New York: McGraw-Hill, 1951.

Monreno, T., *Microwave Transmission Design Data*, New York: Dover Publications, 1958.

Southworth, G. C., *Principles and Applications of Waveguide Transmission*, New York: Van Nostrand, 1960.

Young, Leo, "Synchronous Branch Guide Directional Couplers for Low and High Power Applications," *IRE Trans.*, Vol. PGMTT-10, Nov. 1962, pp. 459-475.

CHAPTER 13

Antennas

Much of the material presented in this chapter is quoted or adapted from [1–11].

The same antenna can be used as a transmit and as a receive antenna. A transmit antenna is a transducer that transforms electrical power delivered to the antenna into RF radiation. If the radiation from an antenna is uniform in all directions, it is called an isotropic antenna. Such an antenna is not possible in practice; however, it serves as a useful reference.

The theoretical isotropic antenna is taken as the reference, and the gain of the antenna in a given direction is a measure of how the power level in that direction compares with the power level that would exist if the isotropic antenna had been present. The gain can be either less than 1 or greater than 1. Expressed in decibels relative to isotropic, it can be either positive or negative. For example, a typical tracking radar might use a parabolic dish antenna that produces a 1 degree × 1 degree main beam. There are 41,300 square degrees in a spherical solid angle. The directional gain in the main beam would then be about 41,300, or 46.2 dBi. At angles other than the main beam, there will be sidelobes. A typical power gain in the sidelobe directions might be −10 dBi or less, depending on the design of the antenna.

The power gain of an antenna is less than the directional gain because of losses and radiation through the sidelobes. In a typical case, an antenna with a 1 degree × 1 degree beam will have a power gain of about 27,000, or 44 dBi. That assumes an antenna efficiency of about 65%, or about 1.85 dB loss. The typical relationship between antenna gain and beam angle is as follows:

$$G = 27{,}000/(\theta_{az} \cdot \theta_{el}) \qquad (13.1)$$

where

G = antenna gain as a number
θ_{az} = the azimuth beam angle in degrees
θ_{el} = the elevation beam angle in degrees

Beam angles ordinarily are measured and specified at the half power (−3 dB) points on the beam. The antenna gain expressed in dB with respect to isotropic is $10 \log_{10} G$.

For an example of the use of (13.1), assume the case of a 1.0 degree × 10 degree fan beam. Substituting values into (13.1), the gain would be

$$G = 27{,}000/(1 \cdot 10) = 2{,}700 = 34.3 \text{ dBi}$$

The effective radiated power (ERP) is defined as the product of the antenna gain and the radiated power. For example, if the numeric antenna gain is 10 and the radiated power is 200W, the ERP is $200 \cdot 10 = 2{,}000\text{W}$, or 33 dBw (63 dBm).

A receive antenna is a transducer that captures electromagnetic energy radiated from a distant transmitter antenna and converts it into electrical energy. For most antennas, the gain when an antenna is used as a receive antenna is the same as its gain when used as a transmit antenna. The effective capture area of a receive antenna is its gain times the area of an ideal isotropic antenna for the frequency of interest. The expression for the effective capture area of a receive antenna is given by

$$A_e = G_r \lambda^2 / 4\pi \tag{13.2}$$

where

A_e = effective capture area of an antenna (square meters)
λ = wavelength (meters)
G_r = receive directional antenna gain (numeric)

For an example of the use of (13.2), assume an antenna with a gain of 13 dBi or 20 numeric operating at a wavelength of 2m. Substituting values into (13.2), the effective capture area of the antenna would be

$$A_e = 20 \cdot 4/4\pi = 6.37 \text{ m}^2$$

In the case of aperture-type antennas, such as horn antennas and parabolic dish reflector antennas, the effective capture area of the antenna typically is about half the actual aperture area. For example, a parabolic dish antenna with a diameter of 60 ft (18.3m) has an aperture area of about 263 m^2 and an "effective" capture area of only about 132 m^2.

The assumption has been made that the polarization of the receiver antenna is the same as the polarization of the signal being received. If that is not the case, there is a polarization loss. If the signal being received has a vertical polarization (direction of the E-field is vertical) and the receive antenna polarization is also vertical, there is no polarization loss. If, on the other hand, the polarization of the receive antenna is horizontal, the polarization loss could be 30 dB or more. For cases where the polarization angle difference is greater than 0 degrees but less than 90 degrees, the polarization loss will be between 0 and 30 dB, depending on the angle difference.

If the polarization of the signal being received is circular (E-field rotates) and the polarization of the receiver antenna is linear (horizontal or vertical), there is a 3-dB polarization loss. If the polarization of the signal being received is right-hand-circular and the polarization of the receiver antenna is left-hand-circular, the polarization loss could be 30 dB or more.

13.1 MONOPOLE ANTENNAS

13.1.1 Thin-Wire Monopole Antennas

One of the most commonly used antennas is the quarter-wave monopole antenna. This type of antenna can be thought of as being derived from the coaxial transmission line and is illustrated in Figure 13.1(a). Resonance occurs when the length of the monopole center element is about a quarter of a wavelength at the frequency of interest.

The radiation pattern for a quarter-wave monopole antenna is as shown in Figure 13.1(b). For a vertical monopole, the polarization is linear and vertical. The gain pattern is uniform in the azimuth plane. The elevation pattern has a peak a few degrees above zero elevation angle. It then slowly drops to a lower value at 90 degrees. The exact pattern depends on how large and how conductive the ground plane is: the larger the ground plane, the lower the elevation angle of the peak in the radiation pattern.

Other patterns are produced when the monopole antenna has a height greater than a quarter-wavelength. An example is the case of a $5/8\lambda$-long monopole. The antenna pattern for this antenna is shown in Figure 13.1(c). The antenna gain in the horizontal direction is about 2 dB greater than that of the quarter-wave monopole. Such monopole antennas are frequently used for cellular telephones.

The peak antenna gain for a quarter-wave monopole is about 5 dBi. The exact antenna gain depends on elevation angle and losses in the antenna.

Figure 13.2(a) shows a quarter-wave monopole antenna with the ground plane at an angle of about 120 degrees from the center conductor. At that angle, the impedance at resonance has increased from about $37 + j0\Omega$ to about $50 + j0\Omega$.

The elevation pattern for this monopole antenna is shown in Figure 13.2(b). This type of monopole configuration is often used as a communica-

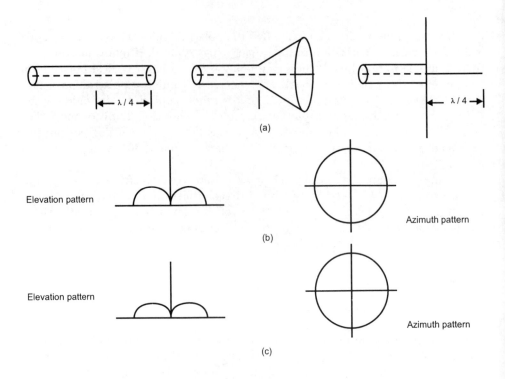

Figure 13.1 Monopole antennas: (a) monopole antenna being derived from the coax transmission line; (b) radiation and polarization for a quarter-wave monopole antenna; and (c) radiation pattern and polarization for a monopole antenna with 5/8λ height.

tion antenna at VHF and higher frequencies. It provides good near-horizon coverage, and the higher impedance facilitates proper and easy impedance matching between the antenna and a 50Ω coaxial cable.

13.1.2 Wideband Monopoles

Figure 13.2(c–e) shows "fat" monopoles. Each of these antennas is a wideband antenna. The reason for wide bandwidth is that the impedance for these monopoles remains fairly constant over a large range of frequencies.

13.1.3 Impedance of Monopole Antennas

Figure 13.3(a,b) shows impedance trends for a conical monopole antenna. In those plots, the resistances and reactances are given as a function of antenna length in electrical degrees and cone angle. The length, L, in electrical degrees is $L = 360h/\lambda$, where h is the height of the monopole and λ is the wavelength. Notice

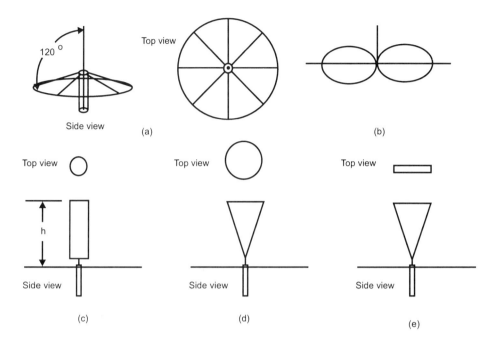

Figure 13.2 Example monopole antennas: (a) antenna with ground plane at 120 degrees from center conductor; (b) elevation radiation pattern for the antenna in (a); (c) fat cylinder monopole; (d) conical monopole; and (e) fan-shaped monopole.

that for cones with small cone angles there are vary large changes in both resistance and reactance as a function of antenna length. On the other hand, for cones with large cone angles, the variations are small, indicating a wider bandwidth capability.

Similar resistance and reactance plots can be provided for other shapes of monopoles, such as thin wires, fat cylinders, rectangular plates, and fan-shaped plates. The result would be the same general trend in impedance. Small-diameter monopoles may provide only 5% bandwidth, whereas large-diameter monopoles may provide 70–100% bandwidth.

13.1.4 Large-Size Monopole Antennas

At HF and MF frequencies, resonant monopole antennas are very large. For an example of the size of the quarter-wavelength monopole antenna, assume operation at 3 MHz. The wavelength at that frequency is 100m and a quarter-wavelength is 25m. Because of end effects, the required monopole length would be about 0.95 times that, or 23.75m. The ground plane should have a diameter of at least twice that, or 47.4m. An even larger ground diameter would be desirable for improving low-angle coverage.

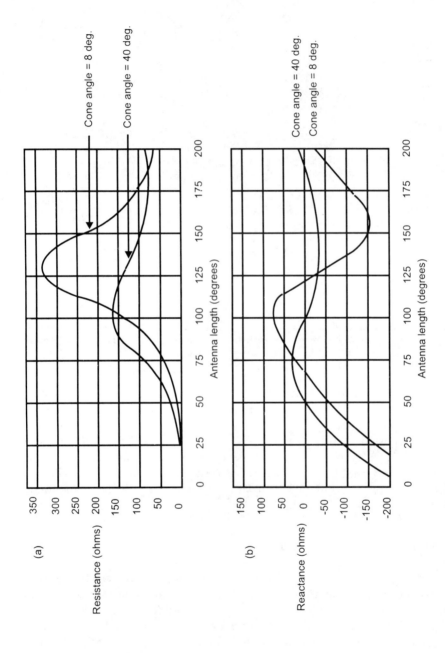

Figure 13.3 Resistance and reactance curves for conical monopole antennas: (a) resistance and (b) reactance. (*After:* [1].)

Ground planes for MF and lower frequency monopole transmitter antennas frequently are constructed over highly conductive ground. In addition, radial copper wires usually are used in the ground planes to reduce ground-plane losses.

13.1.5 Electrically Small Monopole Antennas

It is not practical to construct full-size resonant monopole antennas at low MF, LF, and VLF because of the very large wavelength that is involved. For example, LF has a frequency range of 30–300 kHz and a wavelength of 10,000–1,000m. A quarter-wavelength monopole at LF would have a height of 2,500–250m.

Antennas that are much less than a quarter-wavelength high are referred to as electrically small antennas. An electrically small antenna has a large capacitive reactance and a very small radiation resistance. The approximate radiation resistance for an electrically small monopole antenna is given by

$$R_r = 800(h/\lambda)^2 \qquad (13.3)$$

where

R_r = radiation resistance (ohms)
h = height of monopole (meters)
λ = wavelength (meters)

For an example of the use of (13.3), assume a monopole that is 50m high and operating at a wavelength of 1,000m. The antenna height is thus only 5% of a wavelength. Substituting values into (13.3), the approximate radiation resistance would be

$$R_r = 800(50/1,000)^2 = 2\Omega$$

One way to reduce the capacitive reactance of the electrically small antenna while at the same time improving the radiation resistance is to use capacitive or inductive loading. Examples of this are shown in Figure 13.4.

Figure 13.4(a) shows a typical LF antenna with flat-top loading consisting of a number of parallel wires connected between towers or masts, a vertical wire or set of wires that act as the monopole connected to the horizontal wires, and a series tuning inductance at the bottom.

Figure 13.4(b) shows a so-called umbrella-type loaded monopole in which conductors are extended at a large angle to the vertical using guy wires with insulators. The monopole antenna connects to the umbrella wires and has a tuning inductor at the bottom, that is, base loading. The same type of improvement is

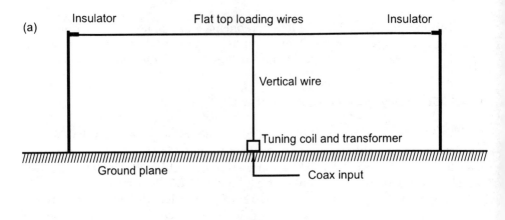

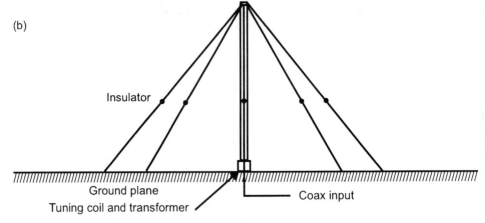

Figure 13.4 Examples of electrically small antennas for LF operation: (a) LF antenna with flat-top loading and inductive tuning and (b) umbrella-type loaded monopole.

provided with this approach as that provided by the system in Figure 13.4(a) without the need for two masts or towers.

The tuning coils and transformers needed for impedance matching at LF typically are very large. The reason is the need for very low ohmic resistance for the coil and the need for very high Q. Since we are dealing with series resonance, very large currents are involved in a high-power system.

The monopole antenna normally is fed by a coaxial transmission line and does not require a balun. At the lower frequencies, impedance matching usually is accomplished using lumped constant inductor and capacitor circuits. Stub matching can be used at higher frequencies.

13.2 DIPOLE ANTENNAS

13.2.1 Thin-Wire Dipole Antennas

Another commonly used antenna is the half-wave dipole antenna. The most common of this type is the thin-wire dipole. The thin-wire type of dipole antenna can be thought of as being derived from the two-wire transmission line shown in Figure 13.5(a).

Figure 13.5(b) shows a quarter-wave section of a two-wire line spread into a V-shape, which becomes a V-type dipole. This form of dipole has fairly narrow beams in the direction of the center line of the V.

At Figure 13.5(c) is shown the quarter-wave section of two-wire line spread 90 degrees on either side of the center line for the transmission line. This is the normal dipole configuration.

The antenna pattern for a half-wave dipole antenna is shown to the right of the dipole in Figure 13.5(c). The azimuth radiation pattern is omnidirectional in

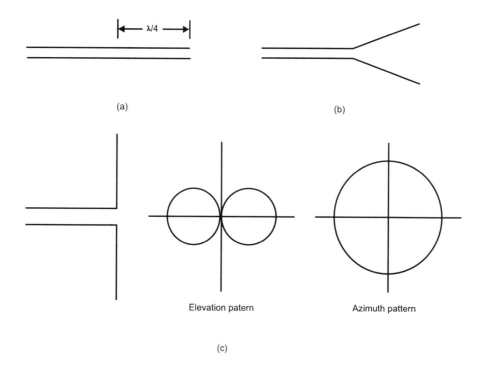

Figure 13.5 Dipole antennas derived from two-wire transmission line: (a) two-wire transmission line; (b) V-type antenna; and (c) dipole antenna.

the plane normal to the axis of the dipole. The elevation pattern is maximum normal to the dipole and varies as the cosine of the angle. The 3-dB elevation beam angle is about 78 degrees wide. The pattern has a deep null directly in line with the axis of the dipole. The maximum antenna gain for a half-wave dipole is about 2 dBi.

Other patterns result when the dipole antenna is other than a half-wave. In general, there are multiple lobes when the dipole is much larger than a half-wave. When shorter than a half-wave, the pattern is similar to that of the quarter-wave dipole.

Dipole antennas are fed by two-wire lines. When the transmission line to the antennas is coaxial, it is necessary to add a balun between the coax and the antenna because it is a balance (ungrounded) feed. A number of different types of baluns were discussed in Chapter 11.

13.2.2 Other Types of Dipole Antennas

Other types of dipole antennas are shown in Figure 13.6.

An electrically small dipole antenna, shown in Figure 13.6(a), has arms much less than a quarter-wavelength. It frequently is used at the lower frequencies, where it is not practical to use a full-size dipole. This type of antenna requires the use of impedance-matching circuits.

The antennas shown in Figures 13.6(b–e) are all so-called fat dipoles, used to provide broadband operation. They have essentially the same antenna patterns as thin-wire dipoles but different impedances.

Figure 13.6(f) shows the case of the folded dipole antenna. This antenna frequently is used as the feed element for television receiving antennas. A folded dipole antenna is made by connecting the ends of a regular antenna with a conductor that is spaced a short distance from the dipole. A transformer action is involved in the operation of the antenna. A typical system has the connecting conductor the same diameter as the dipole. When that is the case, the antenna has a radiation resistance about four times that of an ordinary dipole antenna. The resistance thus is about 292Ω at resonance, a decent value for matching to a 300Ω two-wire transmission line. Other radiation resistance values are possible with the folded dipole if different diameters are used for the two antenna elements. The folded dipole has a larger bandwidth than a simple thin-wire dipole, while having essentially the same antenna pattern. Polarization direction is again in line with the axis of the dipole.

13.2.3 Dipole Impedance

Figure 13.7(a,b) shows example resistive and reactive components of the antenna impedance as a function of the ratio of the dipole length to wavelength (L/λ) for

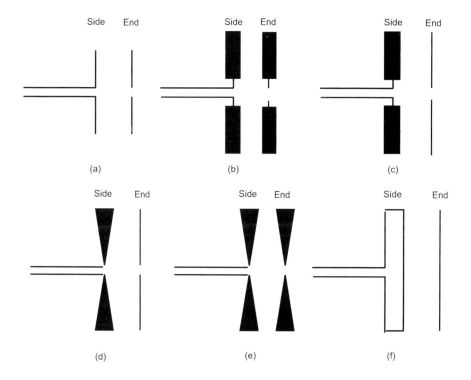

Figure 13.6 Other types of dipole antennas: (a) thin-wire electrically small dipole; (b) fat-cylinder dipole; (c) rectangular-plate dipole; (d) fan dipole; (e) conical dipole; and (f) folded dipole.

a thin cylindrical dipole antenna. The impedance behaves very much like that of an open-circuit transmission line, except that radiation resistance is involved. Like the two-wire transmission line, the impedance of the dipole antenna is high and capacitive when the distance from the end of the dipole to the feed point is small compared to a quarter-wavelength. As the dipole is made larger, the capacitive reactance becomes less. When the arms of dipole are each a quarter-wavelength long, the reactance becomes zero and the antenna is at resonance. In the case of the thin-wire dipole, the resistive component is about 73Ω. The exact value depends on the length-to-diameter ratio of the dipole. The resistive component is made up largely of the radiation resistance of the antenna, which accounts for the power radiated.

As the arms of the dipole are increased beyond a quarter-wavelength but less than a half-wavelength, the impedance becomes inductive and larger. The resistive component continues to increase with increasing dipole length-to-wavelength ratio until it reaches a peak when the length of the dipole is near a wavelength (arms one-half wavelength). With increasing L/λ, the resistance drops and the reactance switches from inductive to capacitive. It then falls again, reaching 0 when L/λ is

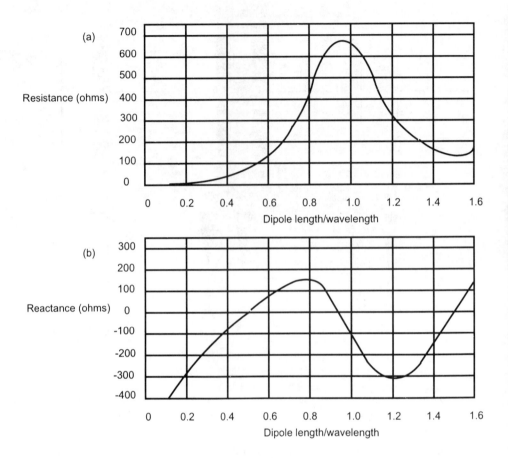

Figure 13.7 Resistance and reactance curves for a thin-wire dipole antenna: (a) resistance and (b) reactance. (*After:* [1].)

near 1.5. The resistance has dropped to a low value at that condition. We thus have another resonance. With larger L/λ the impedance switches to inductive, the resistance increases, and so on.

The thin-wire dipole has relatively narrow bandwidth, as small as 5%. The exact bandwidth depends on the length-to-diameter ratio: the smaller that ratio, the larger the bandwidth. On the other hand, as the dipole elements become large, the bandwidth can be quite large. Bandwidths near three-quarters of an octave are possible with wideband dipole antennas. The antenna gain, antenna patterns, and polarization characteristics are similar to those of the narrowband dipoles.

The peak antenna gain for a normal $\lambda/2$ dipole antenna is about 2 dB. Larger antenna gains are possible by using reflectors, directors, or both with the dipole. An example of a wideband UHF fan-type dipole with a corner reflector is shown in Figure 13.8. An example antenna of this type is useful over a frequency range of about 450–900 MHz. It has a gain in the range of about 8–12 dBi, depending on frequency.

The measured impedance characteristics for the antenna in Figure 13.8 are plotted on the Smith chart shown in Figure 13.9. For this case, the impedance of the Smith chart is normalized to 280Ω. It can be seen from this chart that the VSWR remains less than 2.5:1 over the full octave frequency range of 450–900 MHz. It is common practice to use a Smith chart in showing the impedance characteristic of antennas.

13.2.4 Dipole Current Distribution and Antenna Patterns for Different L/λ Ratios

Figure 13.10 illustrates current distributions on a thin-wire dipole for different L/λ ratios, where L is the full length of the dipole ($L = 2h$) and λ is the wavelength. This figure also shows corresponding antenna elevation patterns.

Figure 13.10(a) shows the case of $L < \lambda/2$. This is the electrically small antenna case. The currents at the ends of the dipole are zero. They increase toward the center but remain small compared to that of the half-wave resonant dipole. The elevation pattern is shown as circles on either side of the dipole with the −3 dB beamwidth equal to about 90 degrees.

Figure 13.10(b) shows the case of $L = \lambda/2$, that is, the resonant antenna case. The currents at the ends of the dipole are zero. They increase toward the center

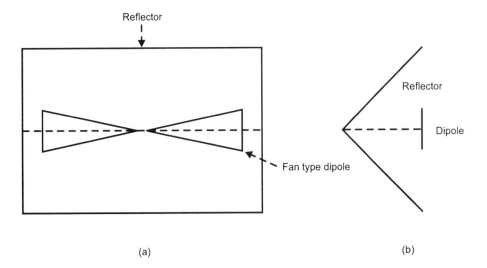

Figure 13.8 Fan dipole with corner reflector: (a) front view and (b) side view. (*After:* [3].)

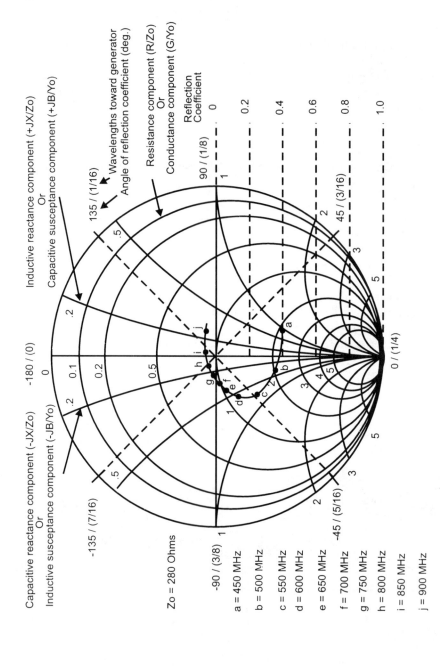

Figure 13.9 Impedance characteristics for a UHF corner reflector antenna. (*After*: [3].)

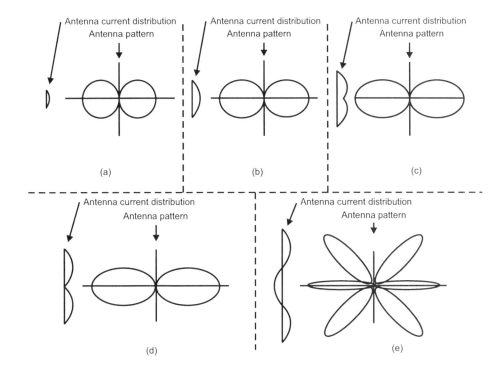

Figure 13.10 Current distribution and antenna patterns for different dipole sizes: (a) electrically small antenna ($L < 0.5$ wavelength); (b) resonant antenna ($L = 0.5$ wavelength); (c) $L = 0.75$ wavelength; (d) $L = 1$ wavelength; and (e) $L = 1.5$ wavelength.

in a sinusoidal fashion, reaching a maximum at the center of the antenna. The elevation pattern is shown as a fat ellipse on either side of the dipole, with the −3 dB beamwidth equal to about 78 degrees.

Figure 13.10(c) shows the case of $L = 3\lambda/4$. The currents at the ends of the dipole are zero. They increase as we move toward the center in a sine wave fashion, reaching a maximum and then dropping off to about 70% of the peak at the center of the antenna. The elevation pattern is shown as an ellipse on either side of the dipole, with the −3 dB beamwidth equal to about 60 degrees.

Figure 13.10(d) shows the case of $L = \lambda$. The currents at the ends of the dipole are zero. They also increase toward the center in a sine wave fashion, reaching a maximum and then dropping off to nearly zero at the center of the antenna. The elevation pattern is shown as an ellipse on either side of the dipole, with the −3 dB beamwidth equal to about 47 degrees.

Figure 13.10(e) shows the case of $L = 3\lambda/2$. The currents at the ends of the dipole are zero. They increase toward the center in a sine wave fashion, reaching a maximum and then dropping off to zero at a point one-third the way from the center to the ends and then increasing in the opposite direction to a maximum at the center of the antenna. The elevation pattern is shown as a six-lobe pattern.

13.2.5 Turnstile Antenna

Figure 13.11(a) shows the turnstile antenna configuration. A turnstile antenna is made up of two dipole antennas placed 90 degrees from each other and fed 90 degrees out of phase. A quadrature coupler can be used to provide the 90-degree quadrature outputs. This antenna frequently is used to transmit or receive circular polarization. In a direction normal to the two dipoles, the polarization is circular. In the plane of the dipoles, the polarization is linear with the E-field in the plane of the dipoles.

Figure 13.11(b,c) shows the azimuth and elevation patterns for the turnstile antenna. The azimuth pattern shows the individual patterns for the two dipole antennas plus the vector sum or superpositon of the two patterns. The elevation

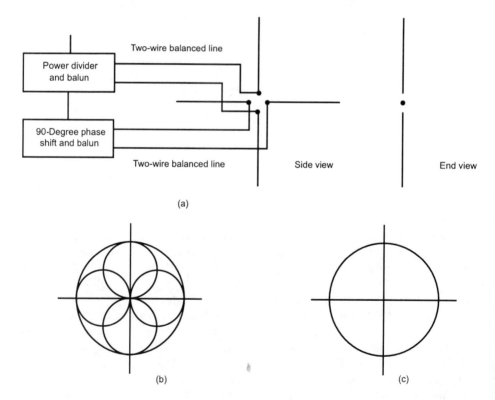

Figure 13.11 Turnstile antenna with antenna patterns: (a) turnstile antenna configuration; (b) azimuth pattern; and (c) elevation pattern.

pattern shows that the coverage is nearly isotropic. In this pattern, the polarization is shown to be circular. Circular polarization means that the direction of the electric field vector rotates in either the right-hand or left-hand direction, depending on the direction of the phase shift. If the two antennas have different effective radiated powers, the polarization is elliptical.

13.3 YAGI-UDA ANTENNAS

Figure 13.12 shows a Yagi-Uda antenna, or Yagi antenna as it is usually called. This antenna is an important example of a directional antenna that uses a dipole as its feed element. Often this dipole is a folded dipole. This type of antenna frequently is used as a receiving antenna for television. It also is used for many other applications where moderate directional gain is required.

On one side of the dipole are a number of director elements that are shorter than the feed dipole by 5–10%. The elements, which are spaced by 0.15–0.25 wavelength, are not dipoles but single rods that act as parasitic elements. The

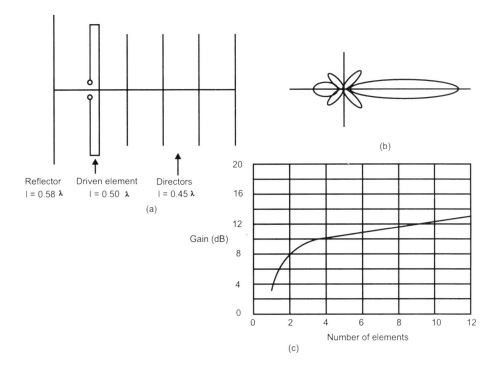

Figure 13.12 A Yagi-Uda, or Yagi, antenna: (a) example antenna configuration; (b) typical antenna pattern; and (c) gain characteristics for a Yagi antenna. (*After:* [4].)

number of elements depends on the desired gain for the antenna. On the other side of the dipole is a single reflector element. The length of that rod is 5–10% greater than that of the feed dipole. Again, the spacing from the dipole is in the range of 0.15–0.25 wavelength.

A typical antenna pattern for a Yagi antenna is shown in Figure 13.12(b). It has a main forward lobe with a typical peak antenna gain in the range of 10–12 dBi. It also has a back lobe with a peak gain of the order of 0–3 dBi. Thus, the typical front-to-back ratio is 9 to 10 dB.

A Yagi antenna is an end-fire traveling-wave antenna. Because the director array of elements is parasitic, the currents on the elements farther out from the driver have decreasing current amplitudes. Figure 13.12(c) shows how the gain of the antenna increases as the number of elements increases. We see that there is little value in increasing the number of director elements beyond about 10.

13.4 SLEEVE ANTENNAS

The addition of a sleeve to a monopole or dipole antenna can increase the bandwidth to more than an octave.

13.4.1 Sleeve Monopoles

Figure 13.13(a) illustrates a sleeve monopole configuration that has a 4:1 pattern bandwidth. The antenna is fed by a coaxial transmission line. The first sleeve monopole resonance occurs when $L + 1$ is approximately $\lambda/4$. The physical length

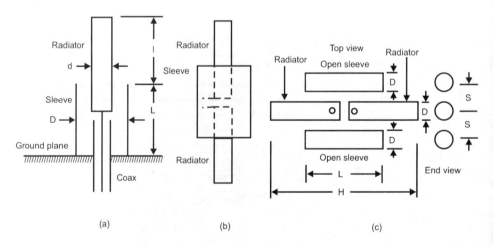

Figure 13.13 Sleeve antennas: (a) sleeve monopole; (b) sleeve dipole configuration; and (c) open-sleeve dipole antenna with reflector. (*After:* [5].)

is set by the low end of the frequency band to be covered. Thus, if we want to operate over a band of frequencies from 150 MHz to 600 MHz, the length $L + 1$ would have a value of about $2m/4 = 0.5m$.

The next requirement is to select the ratio $1/L$. It has been found experimentally that a value of $1/L$ of 2.25 yields optimum radiation patterns over a 4:1 band. The ratio D/d should be about 3.0. The VSWR is less than 8:1 for this 4:1 band. In most applications, that is too high, requiring a matching network for proper performance and to minimize mismatch losses.

13.4.2 Sleeve Dipoles

Two forms of sleeve dipole antennas are shown in Figure 13.13. The operation of these antennas is similar to that of sleeve monopoles. Figure 13.13(b) shows the case of a sleeve dipole with the two-wire transmission line input brought into the sleeve through a hole in the sleeve. Figure 13.13(c) shows the case of an open sleeve dipole antenna mounted above a reflector surface. In that case, the tubular sleeve is replaced by two parasitic conductors that simulate the sleeve. The length of the conductors is approximately one-half that of center-fed dipoles. Details for the feed for this open-sleeve dipole are shown in the figure. The antenna is fed by a coaxial cable through one support arm for the dipole. The outer conductor for the coax is connected to the left dipole element. The center conductor is connected to the right dipole element.

For an example design, assume an antenna designed to operate over the 225–400 MHz band. The dipole-to-reflector spacing, S_d, is chosen to be 0.29 λ at 400 MHz. All the dimensions required for the antenna expressed in wavelengths at the lowest frequency are as follows:

- $D = 0.026$;
- $H = 0.385$;
- $L = 0.216$;
- $S = 0.381$;
- $S_d = 0.163$.

These design values yield a VSWR that is less than 2.5 over the full operating range from 225 to 400 MHz.

13.5 LOOP ANTENNAS

Loop antennas seldom are used as transmit antennas. They are, however, used extensively as receive antennas at lower frequencies and sometimes in direction-finding systems. The two main types of loop antennas are air-core loop antennas and ferrite-core loop antennas.

13.5.1 Air-Core Loop Antennas

Figure 13.14(a) is a small air-core loop antenna with a few turns. This type of antenna sometimes is referred to as a magnetic dipole.

The antenna pattern for loop antennas is shown in Figure 13.14(c). The antenna pattern is similar to that of an electric dipole antenna, with the maximum gain in the plane of the loop and a deep null normal to the plane of the loop. The polarization is in the plane of the loop.

The electrically small loop antenna has a large inductive reactance and a small radiation resistance. It can be shown that when the perimeter of a circular loop antenna is less than about three-tenths of a wavelength, the approximate radiation resistance of this antenna is given by

$$R_r = 19{,}000 \ N^2 (D/\lambda)^4 \tag{13.4}$$

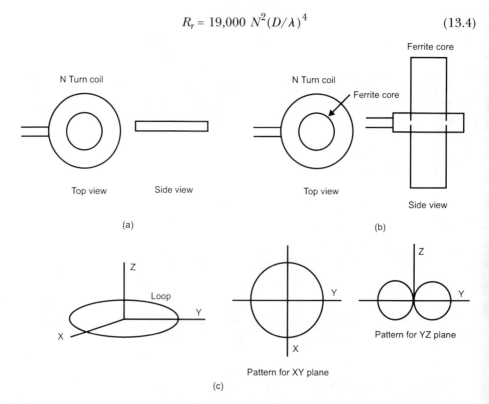

Figure 13.14 Loop antennas: (a) air-core loop antenna; (b) ferrite-core loop antenna; and (c) antenna pattern for loop antennas.

where

    D = loop diameter
    λ = wavelength
    N = number of loops or turns

For example, if $D = 0.1\lambda$ and $N = 4$, then

$$R_r = 19{,}000(16)(0.1)^4 = 30.8\,\Omega$$

Equation (13.4) shows that the radiation resistance of a loop antenna is increased by using a number of turns rather than just one. However, the total length of the wire involved must be less than about three-tenths of a wavelength.

The voltage induced in a small loop receiving antenna is given by

$$V = (2\pi/\lambda)\,nSE_\phi \sin\theta \qquad (13.5)$$

where

    n = number of turns
    S = surface area of loop
    E_ϕ = component of electric field in the plane of the loop
    θ = angle measured from the axis of the loop

Other terms are as defined for (13.4).

13.5.2 Ferrite-Core Loop Antennas

Another way to improve the radiation resistance of the loop antenna is to use a ferrite core for the loop. A ferrite-core multiturn loop is often called a loop-stick antenna. An antenna of this type is shown in Figure 13.14(b). Ferrite-core loop antennas are used for many AM broadcast receivers.

Antenna pattern and polarization for the ferrite-core loop antenna are the same as that of the air-core loop antenna. The radiation resistance of a coil of N turns wound on a ferrite core is given by

$$R_r = 19{,}000 N^2 \mu_{\text{eff}} (D/\lambda)^4 \qquad (13.6)$$

where μ_{eff} = effective permeability of the core material and other terms are as defined for (13.4). The inductive reactance term can be canceled using a series capacitor of equal reactance magnitude.

13.6 HELICAL ANTENNAS

Figure 13.15 shows a helical antenna operating in the axial mode of radiation. In this mode, the helix radiates as an end-fire, traveling-wave antenna with a single maximum along the axis of the helix and a phase velocity along the helix axis less than the speed of light. The helix is thus a slow wave structure. The radiation has circular polarization. The antenna may have either left-hand or right-hand circular polarization, depending on the sense of the winding. If the turns go counterclockwise toward the end of the antenna (in the direction of radiation), polarization is left-hand circular polarization and vice versa.

A helical antenna has a coaxial input with the outer conductor connected to a ground plane and the inner conductor connected to the helix. The ground-plane size is not critical but should be made wider than a half-wavelength at the lowest frequency of interest. The circumference is about a wavelength at the center frequency, and the antenna is effective with the circumference in the range of three-quarters wavelength to four-thirds wavelength. The ratio of the lowest operating frequency to the highest operating frequency thus is about 1.8 to 1.

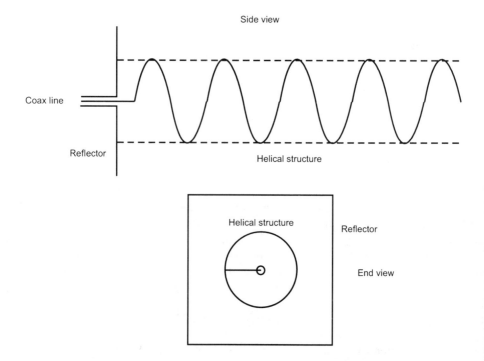

Figure 13.15 Axial-mode helical antenna.

The spacing between turns is about 0.21 wavelength, and the number of turns is approximately 12.

Equation (13.7) is an empirical formula for the half-power beamwidth for a helical antenna.

$$\theta_{hp} = 52/[(C/\lambda)(NS/\lambda)^{1/2}] \qquad (13.7)$$

where

C = circumference of helix
λ = wavelength
N = number of turns for helix
S = spacing between turns

For an example of the use of (13.7), assume that $C/\lambda = 1$, $S/\lambda = 0.21$, and $N = 12$. Substituting those values into (13.7) gives

$$\theta_{hp} = 52/[(1)(12 \cdot 0.21)^{1/2}] = 32.8 \text{ degrees}$$

The directional gain for this example antenna is

$$G = 41{,}253/\theta^2 \qquad (13.8)$$
$$G = 41{,}253/(32.8 \cdot 32.8) = 38.3 = 15.8 \text{ dBi}$$

The terminal impedance of a helical antenna in the axial mode is nearly purely resistive. An empirical formula for the input resistance for the helical antenna is given by

$$R_{in} = 140 \, C/\lambda \qquad (13.9)$$

where the terms are as previously defined. A typical helical antenna thus would have a center frequency input resistance of about 140Ω.

Helical antennas often are used in arrays with four or more antennas mounted on the same ground plane and separated from each other by about a wavelength. These are all fed in phase so that the powers add.

13.7 SPIRAL ANTENNAS

Frequently it is desirable to have the pattern and the impedance of an antenna remain constant over a wide range of frequencies, such as 10:1 or higher. An antenna of this type is often referred to as a frequency-independent antenna.

Actually, there is no such thing as a frequency-independent antenna. They always have a lower frequency limit set by the largest dimensions that are possible or desirable and an upper frequency limit set by the smallest dimensions that are practical. Frequency-independent antennas generally are of two types: spiral antennas and log periodic antennas. This section describes three types of spiral antennas.

13.7.1 Equiangular Spiral Antennas

Figure 13.16(a) illustrates a planar equiangular spiral antenna. The figure shows metal sheets mounted on a dielectric support, which usually is done by printed-circuit techniques. Two arms start at the center and spiral outward. The antenna is fed at the center by a coaxial feed line that is wound along one of the two antenna arms toward the feed points. The outer conductor of the coaxial feed line is connected at the feed location to one of the antenna arms, and the center conductor is connected to the other antenna arm.

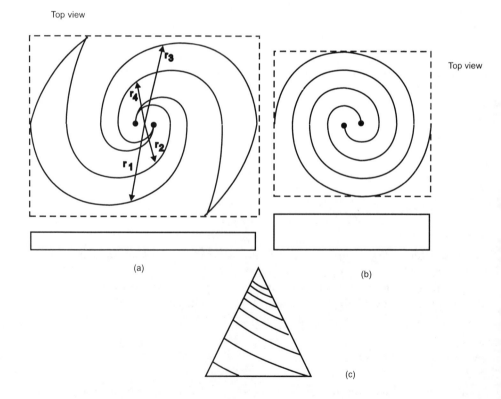

Figure 13.16 Spiral antennas: (a) planar equiangular spiral antenna; (b) archimedian spiral antenna; and (c) conical equiangular spiral antenna. (*After:* [6].)

The four edges of the metal antenna elements each have an equation for their curves. Edge 1 has its radius, r_1, given by (13.10). Edge 2 has its radius, r_2, given by (13.11). The other half of the antenna has edges that make the structure symmetric, that is, if one spiral arm were rotated one-half turn, it would coincide with the other arm. Edge 3 thus has its radius, r_3, given by (13.12), and edge 4 has its radius, r_4, give by (13.13).

$$r_1 = r_0 e^{a\phi} \quad (13.10)$$
$$r_2 = r_0 e^{a(\phi-\delta)} \quad (13.11)$$
$$r_3 = r_0 e^{a(\phi-\pi)} \quad (13.12)$$
$$r_4 = r_0 e^{a(\phi-\pi-\delta)} \quad (13.13)$$

A typical flare rate, a, is 0.22. The values of δ is $\pi/2$. The value of r_0 is the minimum radius of the antenna.

Spirals of one and one-half turns appear to be optimum. Bandwidths of 8:1 are typical; however, bandwidths as high as 20:1 can be obtained. A typical input impedance for the equiangular spiral antenna in Figure 13.16(a) is about 164Ω. The radiation pattern is bidirectional with two wide beams broadside to the plane of the antenna. The half-power beamwidth is approximately 90 degrees. The polarization of the radiation is close to circular over wide angles, out to as far as 70 degrees from broadside. The sense of the polarization is determined by the sense of the flare of the spiral. The spiral of Figure 13.16 radiates in the right-hand sense for directions out of the page and radiates in the left-hand sense for opposite propagation directions.

13.7.2 Archimedean Spiral Antennas

Figure 13.16(b) shows the Archimedean planar spiral antenna. This antenna is usually constructed using printed circuit techniques. The equations for the two spirals are given by

$$r = r_0 \phi \quad (13.14)$$
$$r = r_0(\phi - \pi) \quad (13.15)$$

The properties of the Archimedean spiral antenna are similar to those of the equiangular planar spiral antenna. It is a circularly polarized antenna. A single main beam can be obtained by placing a cylindrical cavity on one side of the spiral, thus forming a cavity-backed Archimedean spiral antenna. Commercially available antennas of this type have a nearly 90-degree half-power beamwidth, a 2:1 VSWR, a 1.1 axial ratio of polarization on boresight, over a 10:1 bandwidth. The circumference of such an antenna is roughly equal to a wavelength at the lowest frequency of operation. In spiral antennas, most of the radiation comes from the region of the structure where the circumference is about one wavelength. Thus, as the frequency is changed, a different part of the spiral supports the majority of the current. This feature is responsible for the broadband performance.

13.7.3 Conical Spiral Antennas

Nonplanar types of spiral antennas are possible, for example, the conical equiangular serial antenna in Figure 13.16(c). With that antenna, a single beam is produced off the tip of the cone. The typical front-to-back ratio of radiation is about 15 dB. A typical antenna has a half-power beam angle of about 80 degrees. The polarization is circular; however, the ellipticity does increase with off-axis angle. The antenna impedance is typically abut 165Ω. The beamwidth is controlled by the cone angle and by the length of the cone.

13.8 LOG-PERIODIC ANTENNAS

Log-periodic (LP) antennas make up a second class of so-called frequency-independent antennas. These antennas are not really frequency independent, but they do have very large bandwidth capability. A 10-to-1 frequency range is typical for this class of antennas. This section discusses a number of different types of LP antennas.

13.8.1 Log-Periodic Dipole Array

The LP dipole array (LPDA) is a popular broadband antenna that is simple in construction, low cost, and lightweight. This antenna is shown in Figure 13.17.

The LPDA antenna is a series-fed array of parallel wire or thin rod dipoles of successively increasing lengths outward from the feed point at the apex. In the single plane version of this type of antenna, the interconnecting feed lines cross over between adjacent elements. The enclosed angle, α, bounds the dipole lengths.

One common way to construct an LPDA is to use a two-plane version of the antenna in which the two feed conductors are parallel and closely spaced, with monopole arms alternating in direction. A coaxial transmission line is run through the inside of one of the feed conductors. The outer conductor of the coax is attached to that conductor at the apex, and the inner conductor of the coax is connected to the other conductor of the LPDA transmission line at the apex. This type of construction is illustrated in Figure 13.17(b).

The scale factor, τ, for the LPDA is given by

$$\tau = R_{n+1}/R_n < 1 \qquad (13.16)$$

$$\tau = L_{n+1}/L_n \qquad (13.17)$$

$$\tau = d_{n+1}/d_n \qquad (13.18)$$

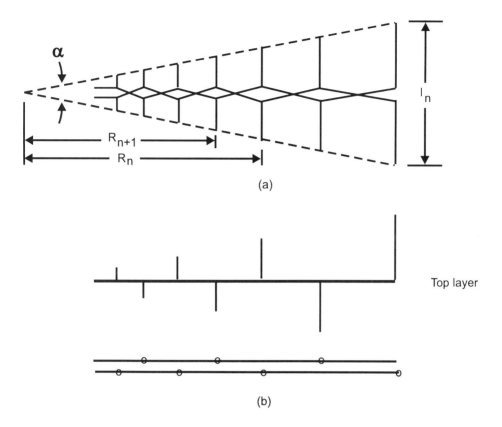

Figure 13.17 LPDAs: (a) LPDA configuration and (b) two-layer construction for LPDA. (*After:* [7].)

The ratio of successive element positions equals the ratio of successive dipole lengths and the ratio of successive dipole spacings. The spacing factor for the LPDA is defined as

$$\sigma = d_n/2L_n \tag{13.19}$$

The apex angle, α, is given by

$$\alpha = 2\tan^{-1}[(1-\tau)/4\sigma] \tag{13.20}$$

There is an active region for the LPDA where the few dipoles near the one that is a half-wavelength long support much more current than do the other radiating elements. The longer dipole behind the most active dipole behaves as a reflector, and the shorter dipole in front of the most active dipole acts as a director. The radiation is thus off the apex.

The pattern, gain, and impedance of an LPDA depend on the design parameters τ and σ. Figure 13.18 shows a plot of the antenna gain and optimum values of the scale factor, τ, and the spacing factor, σ. Note that there is an optimum combination for a given antenna gain, as indicated by the gain lines in the plot.

For an example of the use of Figure 13.18, assume the need for an antenna to cover the frequency range 54 MHz–216 MHz with a gain of 8.5 dB. From the figure, the optimum value of τ is about 0.822, and the optimum value of σ is about 0.149. Substituting values into (3.20) gives the apex angle, α:

$$\alpha = 2 \tan^{-1}[(1 - 0.822)/4 \cdot 0.149)]$$
$$\alpha = 2 \tan^{-1}(0.30) = 33.25 \text{ degrees}$$

The length of the longest dipole is near a half-wavelength at 54 MHz. That would be 2.78m. The shortest radiating dipole should have a length of about a half wavelength at 216 MHz. That would be about 0.69m. The distance from the apex to the longest element is given by

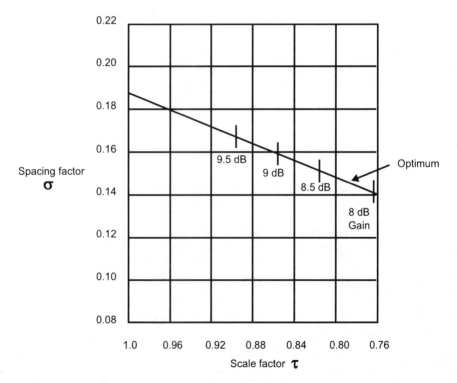

Figure 13.18 Antenna gain and optimum values of the scale factor and the spacing factor for an LPDA. (*After:* [7].)

$$R = 1.39/\tan 16.6 \text{ degrees} = 4.65\text{m}$$

The element lengths are found from the longest element by using (13.17), which is repeated here:

$$\tau = L_{n+1}/L_n \qquad (13.17)$$
$$L_{n+1} = \tau L_n$$

For example,

$$L_2 = (0.822)(2.78) = 2.28\text{m}$$

and

$$L_3 = (0.822)(2.28) = 1.88\text{m}$$

Completing this process yields

$L_4 = 1.54$m
$L_5 = 1.27$m
$L_6 = 1.04$m
$L_7 = 0.856$m
$L_8 = 0.704$m
$L_9 = 0.578$m
$L_{10} = 0.475$m

The element spacing is given by $2\sigma L_n = 0.298\ L_n$. Using the above element lengths, the spacings are

$d_1 = 0.828$m
$d_2 = 0.679$m
$d_3 = 0.560$m
$d_4 = 0.459$m
$d_5 = 0.378$m
$d_6 = 0.310$m
$d_7 = 0.255$m
$d_8 = 0.210$m
$d_9 = 0.172$m

13.8.2 Trapezoidal-Toothed Log-Periodic Antennas

Figure 13.19 shows construction details for two types of a trapezoidal-toothed LP antenna. Figure 13.19(a) shows the sheet metal-type antenna, while Figure 13.19(b) is shown the wire-type antenna.

Wedge angles (ψ) range from about 20 degrees to as much as 80 degrees. This sheet-metal version of the trapezoidal-toothed LP antenna often is used at microwave frequencies. This type of antenna is unidirectional, having a single main beam in the direction of the apex. A typical E-plane beam angle is about 65 degrees for apex angles (α) in the range of 20 to 60 degrees. The H-plane beam angle is determined largely by the wedge angle, ψ, and ranges from about 100 degrees, with ψ = 20 degrees, to 60 degrees, with ψ = 90 degrees. Over this wedge-angle range, the gain of the antenna typically is in the range of 8-10 dB with the smallest gain for the smallest wedge angle. The front-to-back ratio is greatest for small wedge angles. It is about 20 dB for a wedge angle of 30 degrees and about 10 dB for a wedge angle of 60 degrees. The polarization for this antenna is linear and is in the same direction as the teeth of the antenna.

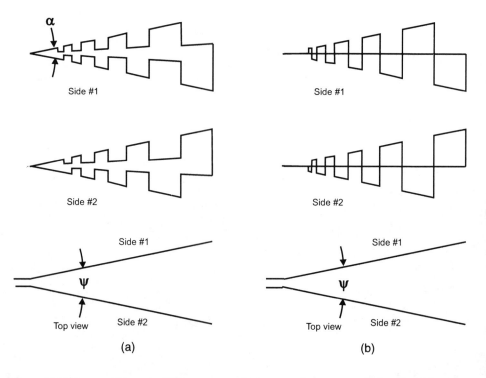

Figure 13.19 Trapezoidal toothed LP antennas: (a) sheet metal-type trapezoidal-toothed LP antenna and (b) wire-type trapezoidal-toothed LP antenna. (*After:* [7].)

The scale factor, t, for the trapezoidal-toothed LP antenna is given by (13.21), (13.22), and (13.25), the same as (13.16), (3.17), and (13.18).

$$\tau = R_{n+1}/R_n < 1 \qquad (13.21)$$

$$\tau = L_{n+1}/L_n \qquad (13.22)$$

$$\tau = d_{n+1}/d_n \qquad (13.23)$$

Thus, the ratio of successive element positions equals the ratio of successive element lengths and the ratio of successive element spacings. Because of the way trapezoidal-toothed LP antennas are made, a typical value for τ is much smaller than is the case for LPDA antennas. Typical values used are about 0.65.

There is an active region for the trapezoidal-toothed LP antenna where radiation takes place. That region is where elements are near a quarter-wavelength long. The longer element behind the most active element behaves as a reflector, and the shorter element in front of the most active element acts as a director. The radiation is thus off the apex.

Measurements for a trapezoidal-toothed LP antenna with a scale factor, τ, of 0.65, an apex angle, α, of 60 degrees, and a wedge angle, ψ, of 45 degrees have yielded E- and H-plane half-power beamwidths of 66 degrees, a gain of 9.2 dB, and a front-to-back ratio of 12.3 dB. The average input impedance has been measured as 110Ω with a VSWR of 1.45 over a 10:1 band. The radiation is linear polarized.

The usable bandwidth of the trapezoidal-toothed LP antenna is 10:1 or greater, which is one of the main advantages for this antenna.

One practical way to feed the trapezoidal-toothed LP antenna is to use a coaxial wideband balun that converts a 50Ω coaxial to a 110Ω balanced two-wire line over a distance of greater than a half-wavelength at the lowest frequency of operation. That usually is done with the balun located halfway between the two sides of the wedge and the two-wire line connected to the antenna at the apex.

Figure 13.19(b) shows construction details for a wire version of a trapezoidal-toothed LP antenna. This antenna is similar in many ways to the sheet metal-type trapezoidal-toothed LP antenna. It has the advantages of being lower in weight and with lower wind resistance than the sheet-metal version. This wire version of the trapezoidal-toothed LP antenna often is used at HF, VHF, and UHF frequencies.

Wedge angles (ψ) range from about 20 degrees to as much as 80 degrees. A typical E-plane beam angle is about 65 degrees for a values in the range of 20–60 degrees. The H-plane beam angle is determined largely by the wedge angle, ψ, and ranges from about 100 degrees, with $\psi = 20$ degrees, to 60 degrees, with $\psi = 90$ degrees. Over this wedge-angle range, the gain of the antenna typically is in the range of 8 to 10 dB, with the smallest gain for the smallest wedge angle. The front-to-back ratio is greatest for small wedge angles. It is about 20 dB for a wedge angle of 30 degrees and about 10 dB for a wedge angle of 60 degrees. The polarization for this antenna is linear and is in the same direction as the teeth of the antenna.

As in the case of the sheet-metal version, there is an active region for the wire trapezoidal-toothed LP antenna where radiation takes place. This region is where elements are near a quarter-wavelength long. The longer element behind the most active element behaves as a reflector, and the shorter element in front of the most active element acts as a director. The radiation is thus off the apex.

The performances for the two types of trapezoidal-toothed LP antennas are nearly the same.

13.8.3 Triangular-Toothed Log-Periodic Antennas

Figure 13.20 shows two types of triangular-toothed LP antennas. The performance of these types of antenna is similar to that of the trapezoidal-toothed LP antenna. Again, the typical value of the scale factor, τ, is about 0.65, the typical value of the apex angle, α, is about 60 degrees, and the typical value for the wedge angle, ψ, is about 45 degrees.

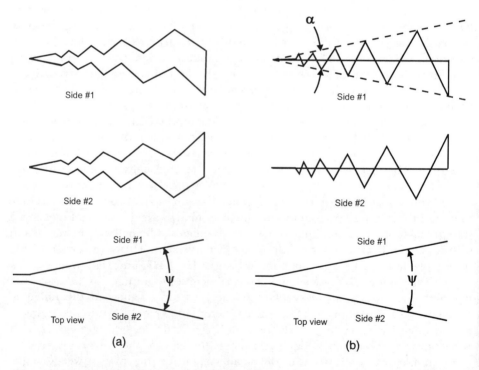

Figure 13.20 Triangular-toothed LP antennas: (a) sheet metal-type triangular-toothed LP antenna and (b) wire-type triangular-toothed LP antenna.

Figure 13.20(b) shows the wire version of the triangular-toothed LP antenna. The performance for this type of antenna is similar to that of its corresponding sheet-metal version, and design parameters are as given in the preceding paragraph.

13.9 SLOT ANTENNAS

13.9.1 Open-Slot Antennas

Figure 13.21(a) shows a slot antenna in a metallic plane. The slot has a length of about 0.5 wavelength and a width of about 0.1 wavelength. Many of the properties of the slot antenna can be deduced from the properties of the complementary metallic strip dipole antenna by the use of Babinet's equivalence principle. The E-plane pattern of the slot and the H-plane pattern of the dipole are omnidirectional, while the slot H-plane pattern is the same as the dipole E-plane pattern.

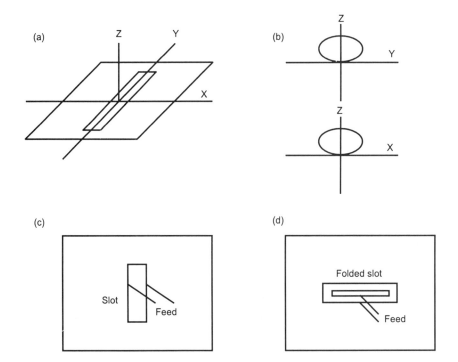

Figure 13.21 Slot antennas: (a) slot antenna in a metallic plane; (b) principal plane field diagrams for slot antenna in a metallic plane; (c) offset feed for lowering slot impedance; and (d) folded slot for lowering slot impedance. (*After:* [8].)

The principal plane field diagram for the slot in Figure 13.21(a) is shown in Figure 13.21(b).

Slots can be excited by a transmission line connected across the slot, by an energized cavity placed behind it, or by a waveguide. In the case of feeding by a transmission line, it is usually necessary to provide some type of impedance matching. Resonance occurs at the frequency for which the slot length is near a half-wavelength. The resistance of the center-fed resonant slot antenna is about 363Ω. Two methods for lowering slot resistance are shown in Figure 13.21(c,d).

It is possible to transform the slot impedance by feeding the slot at some point off center, an approach shown in Figure 13.21(c). The closer the feed point is to the side of the slot, the lower the impedance will be. That is just opposite of what we would have with the metallic-strip dipole antenna. In the dipole case, the closer the feed is to the ends of the dipole, the higher the antenna impedance will be.

Figure 13.21(d) shows the case of a folded slot fed by a 75Ω coaxial line. By folding the slot, the impedance of the slot is reduced by a factor of 4, from about 363Ω to about 90Ω. Again, that is just opposite of the case of a folded dipole, where folding increases the impedance by a factor of 4.

13.9.2 Cavity-Backed Rectangular-Slot Antennas

Cavity-backed slot antennas often are used where surface-mounted antennas are needed with radiation in only one hemisphere. Such an antenna is shown in Figure 13.22(a).

One way to achieve a fairly wideband cavity-backed rectangular-slot antenna is to use a T-fed slot antenna, like the one shown in Figure 13.22. The input to the antenna is by means of a coaxial cable. The center conductor of this line is connected to the T-shaped element located in the bottom of the cavity. The length of the slot is about 0.62 wavelength, the width about 0.20 wavelength, and the depth of the slot about 0.20 wavelength. The diameter of the T-feed element, positioned in the center of the cavity, is about 0.073 wavelength. Measurements show that this type of antenna has a VSWR less than 2:1 over a frequency range of about an octave. Other types of wideband feed systems also are used with cavity-backed slot antennas.

Figure 13.22(b) shows a typical antenna pattern for a cavity-backed slot antenna on a ground plane. The larger the ground plane, the wider is the pattern in the xz-plane. The pattern is narrower in the yz-plane than in the xz-plane.

13.9.3 Waveguide-Fed Slot Antennas

Many radar and other microwave systems use waveguide-fed slot antennas. These systems often involve arrays of many slot antennas. Systems of this kind are shown in Figure 13.23.

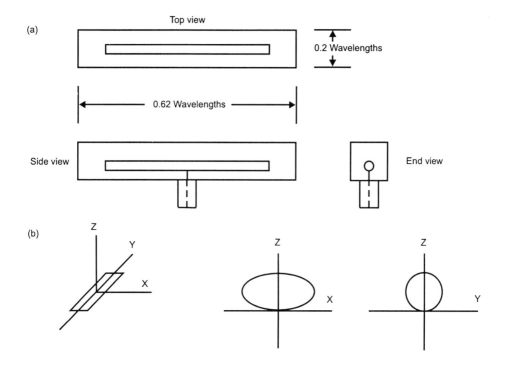

Figure 13.22 Cavity-backed rectangular slot antenna with T-feed: (a) antenna configuration and (b) antenna patterns. (*After:* [8].)

Figure 13.23(a) shows a section of rectangular waveguide with longitudinal slots in the top wall of the waveguide. The slots are positioned at points one-half guide wavelength along the guide, with alternate slots on either side of the center line. With the slots positioned in this manner, the slots are in phase as needed for the array. The longitudinal slots are in position to interrupt current flow for the dominant TE_{10} mode and are, therefore, good radiators. The slots are each a half-wavelength long. The selection of the distance from the center line often is done experimentally so the desired radiation is provided by all slots.

Figure 13.23(b) shows a second type of waveguide-fed slot array antenna, one that has slots on the side wall of the waveguide. Each slot is slanted with respect to the vertical. Again, they are positioned so they are one-half guide wavelength apart. Every other slot is in a different direction, so the slots are in phase. Each of these slots is a good radiator because each blocks current flow in the side wall of the waveguide. The larger the angle between the slot and the vertical, the larger the radiation is. For small angles, it is necessary for the slot to cover not only the side wall but also part of the top and bottom walls if it is to be a half-wavelength

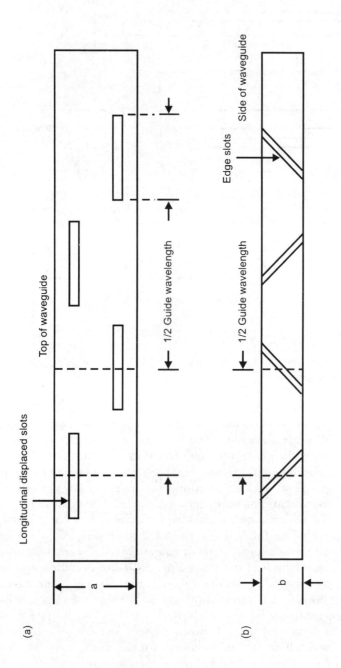

Figure 13.23 Waveguide-fed slot antenna array: (a) top-wall waveguide-fed slot antenna array and (b) side-wall waveguide-fed slot antenna array.

slot. That is because the side wall of the guide normally is much less than a half-wavelength high.

The polarization for the slots is at right angles to the long axis of the slots. With small slant angles from vertical, as is normally the case, there is considerable mutual coupling between the slots, which must be taken into account. The slant angle that will be used for each slot in an array usually is selected experimentally so the desired radiation is provided by all slots.

The choice of whether to use a side-wall or a top-wall slot antenna is based mainly on the ease of fabrication for the type of system used.

13.10 NOTCH ANTENNAS

A notch antenna, shown in Figure 13.24, is a broadband radiator that is often used with aircraft. For example, the edge of a wing or a rudder can be used for a notch antenna.

Figure 13.24(a) shows the notch antenna configuration. The depth of the notch is a quarter-wavelength. The feed points are across the notch. The exact impedance depends on the distance that the feed is placed from the edge of the ground plane. It is greatest at the edge and decreases as we approach the end of the notch.

Figure 13.24(b) shows the radiation patterns for a notch antenna where the ground plane is large with respect to a wavelength.

13.11 HORN ANTENNAS

Horn antennas are one of the more important types of antennas for use with waveguides. Horn antennas often are used as feeds for parabolic dish antennas. They are used separately where the gain requirements are not too high and also as standards in measuring performance of other antennas.

Figure 13.25 show three types of rectangular horn antennas.

The pyramidal horn and the conical horn emit pencil-like beams that have high directivity in both horizontal and vertical planes. Fan-shaped beams result when one dimension of the horn mouth is much smaller than the other. The sectoral horns formed by flaring in only one dimension exhibit this behavior.

In Figure 13.25, the dimension of the long width of the rectangular horn is a, and the dimension of the short width is b, just as in the case of the rectangular waveguide. Not labeled in the figure, the horn length from mouth to apex is L. The diameter of the mouth of the conical horn is d. Formulas for beam widths and power gains applicable to optimum horns of various types are given in Table 13.1. Directive gain is given by

$$\text{Directive gain} = 4\pi A_0 k/\lambda^2 \qquad (13.24)$$

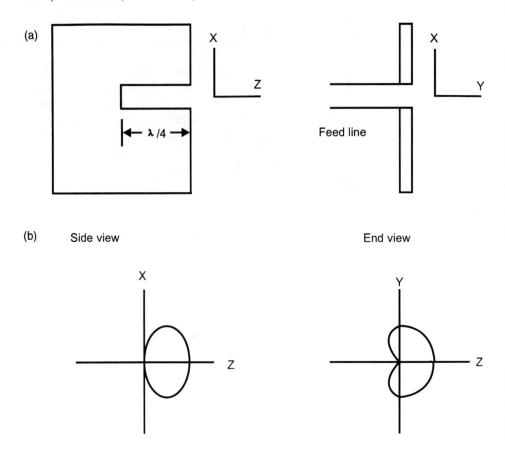

Figure 13.24 Notch antenna configuration and antenna patterns: (a) antenna configuration and (b) antenna radiation patterns. (*After:* [8].)

For an example of the use of Table 13.1, assume the design of a pyramidal horn in which the property that is optimized is the gain. Further assume a wavelength of 10 cm and a length, L, of 30 cm. The required values for dimensions a and b are

$$a = (3L\lambda)^{0.5} = (3 \cdot 30 \cdot 10)^{0.5} = 30 \text{ cm}$$
$$b = 0.81a = 0.81 \cdot 30 = 24.3 \text{ cm}$$

The value of A_0 is

$$a \cdot b = 30 \cdot 24.3 = 729 \text{ cm}^2$$

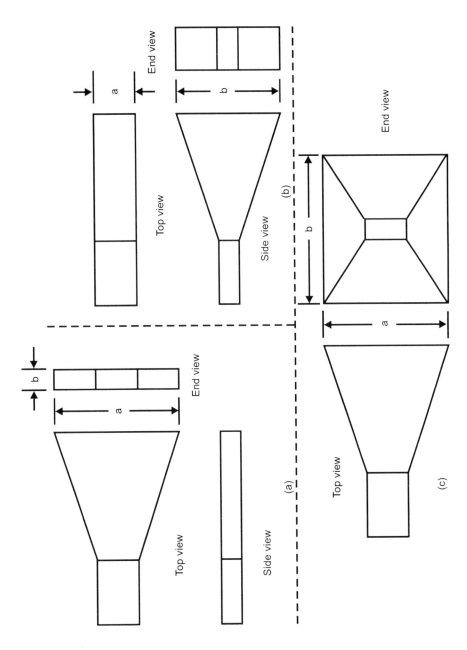

Figure 13.25 Three types of rectangular horn antennas: (a) H-plane sectoral horn; (b) E-plane sectoral horn; and (c) pyramidal horn.

Table 13.1
Formulas for Optimum Horns

Horn Type	Optimized Property	Optimum Properties	HP Beam Widths (degrees) H-Plane	E-Plane	Values of k for (13.24)
Pyramidal	Gain	$a = (3L\lambda)^{0.5}$ $b = 0.81a$ $G = 15.3\ L/\lambda$	$80\lambda/a$	$53\lambda/b$	0.50
H-plane sectoral	Beam width in H-plane	$a = (3L\lambda)^{0.5}$	$80\lambda/a$	$51\lambda/b$	0.63
E-plane sectoral	Beam width in E-plane	$a = (2L\lambda)^{0.5}$	$68\lambda/a$	$53\lambda/b$	0.65
Conical	Gain	$a = (2.8L\lambda)^{0.5}$	$70\lambda/d$	$60\lambda/d$	0.52

Source: [9].

The half-power beam widths and the directional gain are

H-plane beam width = $80\lambda/a$ = 80 · 10/30 = 26.7 degrees
E-plane beam width = $53\lambda/b$ = 53 · 10/24.3 = 21.8 degrees
Directive gain = $4\pi A_0 k/\lambda^2$ = 4 · π · 729 · 0.50/100 = 45.8 = 16.6 dB

13.12 LENS ANTENNAS

Figure 13.26 shows three types of lens antennas: dielectric lens antenna, Luneburg lens antenna, and a metal-plate lens antenna.

13.12.1 Dielectric Lens Antenna

Figure 13.26(a) shows the design and operation of a dielectric lens antenna. The objective of the dielectric lens is to convert a spherical wavefront from a point source such as a horn to a planar wavefront. To do that, the dielectric lens is made thick near the center and progressively thinner toward the edges of the lens. The speed of the RF wave is less in the dielectric than in air, being reduced by the square root of the relative permittivity of the dielectric. With the correct thickness, the total delay for the wave in traveling from the point source to the back of the lens is the same over all parts of the lens. The result is a planar wavefront.

The gain and the beam angle for a dielectric lens antenna depend on the area of the lens and the wavelength. The antenna gain is given by (13.24), with k equal to approximately 0.65, and A_0 is the area of the lens.

While the dielectric lens can be made with ordinary dielectric, it is possible to use a much lower weight artificial dielectric that consists of conducting small rods or spheres embedded in a low-weight dielectric such as styrofoam.

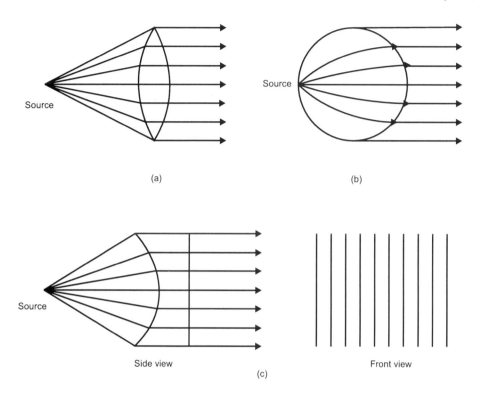

Figure 13.26 Lens antennas: (a) dielectric lens antenna; (b) Luneburg lens antenna; and (c) metal-plate lens antenna.

13.12.2 Luneburg Lens Antenna

A special case of a dielectric lens is a Luneburg lens, which is shown in Figure 13.26(b). A Luneberg lens is spherical in shape. The index of refraction (the square root of the dielectric constant) is a maximum at the center of the sphere, where it is equal to 1.414 and decreases to a value of 1.0 on the periphery. A practical Luneburg lens can be constructed from a large number of spherical shells, each of constant index of refraction. Discrete changes in index of refraction approximate a continuous variation. In one example of a Luneburg lens, 10 concentric spherical shells are arranged one within the other. The dielectric constant of the individual shells varies from 1.1 to 2.0 in increments of 0.1. The Luneburg lens focuses on a point source on the periphery of the sphere. It is possible to use many such point sources or receiver antennas simultaneously with a single Luneburg lens.

13.12.3 Metallic-Plate Lens Antenna

Figure 13.26(c) shows the design and operation of a metal-plate lens. The objective of this lens is to convert a spherical wavefront from a point source such as a horn to a planar wavefront. To do that, the metal-plate lens is thin near the center and progressively thicker toward the edges of the lens. The phase velocity of the RF wave is greater in a metal-plate lens than in air. By selection of the correct thickness, the total delay for the wave in traveling from the point source to the back of the lens is the same over all parts of the lens. The result is a planar wavefront.

The metal-plate lens consists of conducting strips that are placed parallel to the electric field of the wave that approaches the lens and are spaced slightly in excess of a half wavelength. To the incident wave, such a structure behaves like a large number of waveguides in parallel.

One weakness of this type of lens antenna is that its focusing action is frequency sensitive. That is because the phase velocity in a waveguide depends on the frequency. The gain and the beam angle for a metal-plate antenna depend on the area of the lens and the wavelength. The antenna gain is given by (13.24).

13.13 ANTENNA ARRAYS

An antenna array is made up of a number of individual antennas, such as monopoles, dipoles, helical antennas, slots, and open-ended waveguide. These are positioned and phased in such a way as to provide a higher gain antenna of the desired beam shape and other desired characteristics.

13.13.1 End-Fire Line Antenna Arrays

Figure 13.27(a) illustrates a two-element end-fire line array. The array might be made up of two monopole antennas, each with omnidirectional azimuth patterns. These antenna elements are spaced one-quarter wavelength apart and are fed 90 degrees out of phase. If the phase of the right-hand element lags the phase of the left-hand element, the two radiated signals will be in phase, and their powers will add in the right-hand direction, as shown in the antenna pattern for the array. In the left-hand direction, the two radiated signals will be 180 degrees out of phase. If their amplitudes are equal, their fields will add to zero. At other azimuth angles, the resulting radiated power will be the vector addition of the two signals. The resulting pattern shown in Figure 13.27(b) is known as a cardioid pattern.

The typical maximum spacing of elements is $3\lambda/8$ for end-fire line arrays where many antenna elements are used. There is a progressive phase difference between adjacent antennas equal in cycles to the spacing between the antennas, expressed in wavelengths.

Directivity of the individual antennas of an end-fire array is taken into account by multiplying the array pattern obtained by postulating isotropic radiators by the actual directional characteristic of the antennas used. For example, assume that the array gain using isotropic radiators in the main beam is 10 and the element

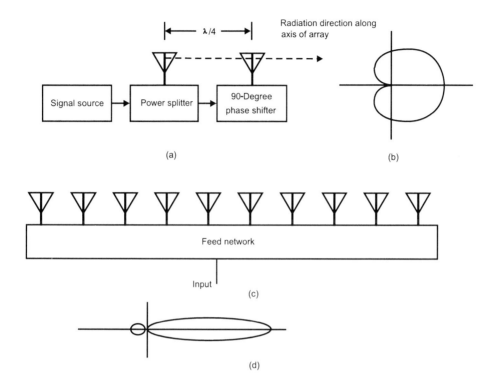

Figure 13.27 End-fire line array antennas: (a) simple two-element line array antenna; (b) azimuth antenna pattern for the two-element line array antenna; (c) example of 10-element end-fire line array; and (d) azimuth antenna pattern for 10-element end-fire line array.

gain in that direction is 3. The total antenna gain in the main beam, then, is 30 (14.8 dB).

The following equations apply to the end-fire line array with isotropic elements, where L is the length of the array between the centers of the end antennas:

Width of the major lobe between nulls is

$$\text{Width} = 115/(L/2\lambda)^{0.5} \text{ degrees} \quad (13.25)$$

Directive gain (numeric) = $4L/\lambda$

For an example of the use of (13.25) and (13.26), assume the use of 20 isotropic antennas in a line spaced by $3\lambda/8$. The value of L is 7.125λ. Substituting values into (13.25) and (13.26) gives

$$\text{Width} = 115/(7.13/2)^{0.5} = 60.9 \text{ degrees} \quad (13.26)$$

Directive gain = $4 \cdot 7.13 = 28.5 = 14.5$ dB

If each antenna element of the array had a gain in the desired direction of 3 dB, the total directive gain would be 17.5 dB. Figure 13.27(c) illustrates that array, while the resulting end-fire azimuth antenna pattern is shown in Figure 13.27(d).

13.13.2 Broadside Line Antenna Arrays

Figure 13.28(a) shows a broadside line antenna array. The typical antenna element spacing is $\lambda/2$ and all elements are fed in phase. If the elements are assumed to radiate uniformly in all directions, the azimuth antenna pattern might be as shown in Figure 13.28(b).

Equations (13.27), (13.28), and (13.29) apply to the broadside line arrays with isotropic elements, where L is the length of the array between the centers of the end antennas.

The width of the major lobe between nulls is given by

$$w_1 = 115/(L/\lambda) \text{ degrees} \tag{13.27}$$

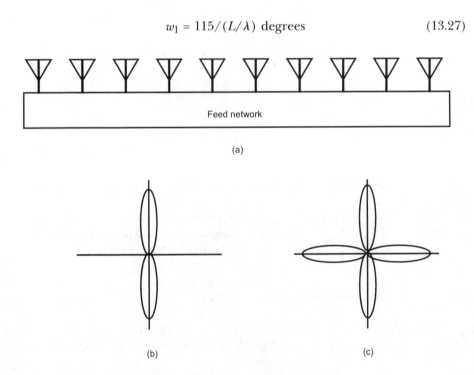

Figure 13.28 Broadside line arrays: (a) antenna configuration; (b) azimuth antenna pattern for element spacing of one-half wavelengths and isotropic radiation by elements; and (c) azimuth antenna pattern for element spacing of one wavelength and isotropic radiation by elements.

The width of the major lobe between half-power points is given by

$$w_2 = 51/(L/\lambda) \text{ degrees} \tag{13.28}$$

The directive gain (numeric) is given by

$$G_d = 2L/\lambda \tag{13.29}$$

For an example of the use of (13.27), (13.28), and (13.29), assume the use of 41 isotropic antennas in a line spaced by $\lambda/2$. The value of L is 20λ. Substituting values into (13.27) gives the width of the major lobe between nulls:

$$w_1 = 115/20 = 5.75 \text{ degrees}$$

Substituting values into (13.28) gives the width of the major lobe between half-power points:

$$w_2 = 51/20 = 2.55 \text{ degrees}$$

Substituting values into (13.29) gives the directive gain for isotropic radiators:

$$G_d = 2 \cdot 40 = 80 = 19 \text{ dB}$$

If the antenna elements are directional antennas with a gain of 3 dB in the desired broadside direction, the directive gain would total 22 dB.

Figure 13.28(c) shows the case of a line antenna with elements spaced by one wavelength. With that spacing, there will be major lobes in line with the antenna as well as broadside to the antenna. These lobes are called grating lobes. As the spacing is increased beyond one wavelength, the grating lobes occur at smaller angles from the main lobe. The angle for grating lobes is given by

$$\sin \theta = \pm n\lambda/d \tag{13.30}$$

where

θ = the angle from broadside
n = the number of the grating lobe
λ = wavelength
d = element spacing

For an example of the use of (13.30), assume the following:

$n = 1$ (i.e., the first lobe)

$d = 2\lambda$

$\sin \theta = \pm n\lambda / d$

$\sin \theta = \pm 1 \cdot \lambda / 2\lambda = \pm 0.5$

$\theta = \pm 30$ degrees

When $d = \lambda$, $\sin \theta = \pm 1$, $\theta = \pm 90$ degrees, which is the case shown in Figure 13.28(c).

Sometimes high-gain elements, such as helical antennas, are used in line arrays. In those cases, the spacing often is more than a wavelength and grating lobes will exist. They may not be of large amplitude, however, if the element pattern has low gain at those angles where there are grating lobes. Again, the antenna pattern will be the product of the array pattern and the element pattern.

13.14 PLANAR ARRAYS

Figure 13.29(a) shows an example of a planar antenna array. The antenna pattern for the array is shown in Figure 13.29(b). This type of array is a two-dimensional array with a "height" and a "width." The example array is an 8 × 4 element array,

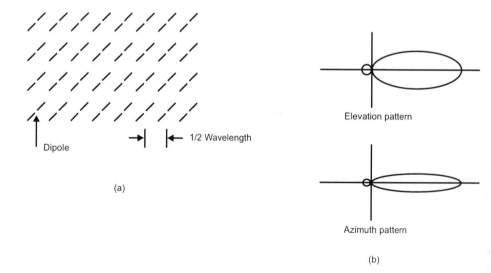

Figure 13.29 Planar array antenna: (a) antenna configuration and (b) antenna patterns.

which has a total of 32 elements. This might be viewed as four line arrays, with each line array having eight elements. Each element in the line array typically is spaced by $\lambda/2$, and each line array is spaced by $\lambda/2$.

Examples of antenna elements for the planar array are dipoles with reflector surfaces behind the dipoles, waveguides with slots, and open-ended waveguides. With dipoles having reflectors, the gain for each element may be about 4 dB. With 32 elements, the gain of the planar array would 19 dB.

A 100 × 100 element array using 4-dB gain elements would have a gain of about 44 dB and a beam angle of 1 degree × 1 degree. This antenna can be used as a reference in estimating the number of elements in a given direction needed for a given azimuth or elevation beam angle. This requirement is given by

$$n = 100/\theta \quad (13.31)$$

where θ is the desired azimuth or elevation −3 dB beamwidth in degrees.

For example, a 2-degree azimuth × 20-degree elevation fan-shaped beam would require $100/2 = 50$ antenna elements in the horizontal direction by $100/20 = 5$ elements in the vertical direction, or a total of $50 \cdot 5 = 250$ antenna elements.

13.15 SCANNING METHODS

13.15.1 Mechanically Scanned Arrays

There are a number of ways to point the high-gain small-width beams of array antennas. One way is to point the beam by mechanically positioning the array in the desired direction. In the case of a scanning system, the array can be rotated at the desired rate by mechanical means, provided that the desired rotational rate is not too high. Maximum possible rotation rates depend on the size of the antenna. A typical maximum rate for a microwave antenna array might be on the order of one rotation per second.

Combinations often are used in which the array antenna is scanned or moved in one direction by mechanical means and in the other direction by either phase scan or frequency scan. These latter two scanning methods have the advantage of very rapid scan and positioning of the beams.

13.15.1.1 Phase-Scanned Arrays

Figure 13.30 shows an example of a phase-scanned line array. This system uses four antennas, with each antenna input having a variable phase shifter. The phase shifters are fed with equal amplitude signals of the same phase. This type of feed system is know as a corporate feed. Other possible feed systems include an end-fed series phase-shift arrangement and a center-fed series phase-shift arrangement.

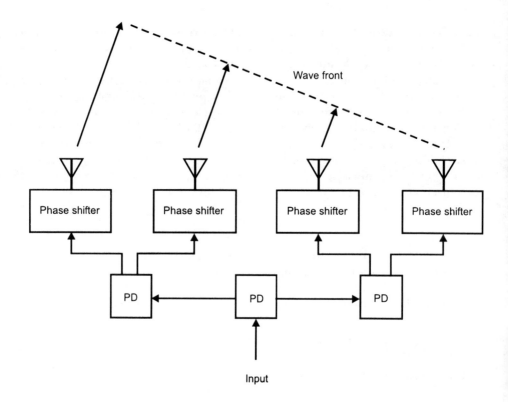

Figure 13.30 Phase-scanned line array antenna.

Figure 13.30 also shows how the beam angle can be changed by a change in the phase angle of the individual phase shifters. When the phase shift is progressively larger from left to right, the beam is shifted to the left. When the phase shift is progressively larger from right to left, the beam is shifted to the right. The wave front shown in Figure 13.30 is the line where the signals from each antenna have the same phase. The beam angle is normal to that line.

A planar array with phase-shift volumetric scan in two angular coordinates can be used for some military applications. Such a system requires a phase-shifter for each antenna and is costly because of the large numbers of antennas and phase shifters involved.

A phased-array antenna can form a number of simultaneous beams in the receive condition. A separate set of phase shifters is required for each beam formed.

13.15.1.2 Frequency-Scanned Arrays

A convenient method for achieving scanning in a line array is to use frequency scan. Such a system is shown in Figure 13.31. The array is fed from one side with

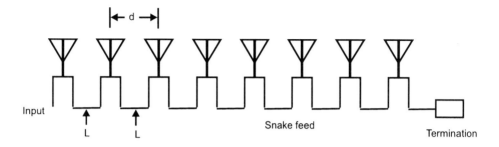

Figure 13.31 Frequency-scanned line array.

what is sometimes termed a snake, serpentine, or sinuous feed. The antennas are spaced by a distance d, which may be about a half-wavelength for low-gain antenna elements. The antennas tap off the snake line using suitable couplers. The connecting line between the antennas has a length, L, that is much greater than d. At the center frequency of operation, the length L is some multiple, m, of 360 degrees phase shift, such that the elements of the array all have the same phase. The antenna beam in this case is pointed normal to the line array. If the frequency is either increased or decreased, the phase shift between elements is changed. It is no longer a multiple of 360 degrees but some different phase angle, and the antenna elements have a progressive phase shift. That allows the antenna beam to be pointed either to the right or to the left, depending on which way the frequency is shifted.

The direction of main beam pointing is given by

$$\sin \theta = (c/v)(L/d)[1 - (f_0/f)] \qquad (13.32)$$

where

c = velocity of light
v = velocity of wave in the snake line
L = length of the snake line between antennas
d = spacing between antennas
f_0 = center frequency
f = actual signal frequency

For an example of the use of (13.32), assume the following:

$v = 0.8c$
$L = 6.4d$
$f = 1.1 f_0$

Substituting values into (13.32) gives

$$\sin \theta = (c/0.8c)(6.4\ d/d)[1 - (f_0/1.1f_0)]$$
$$\sin \theta = 1.25 \cdot 6.4\ (1 - 0.91) = 0.72$$
$$\theta = 46.05 \text{ degrees}$$

It is possible with a frequency-scanned array to use a number of frequencies simultaneously to provide multiple beams. That can be done to permit a higher scan rate than would be possible with a single-beam antenna. The U.S. Navy AN/SPS-48 radar, which is used on many naval vessels, uses a mechanically and frequency-scanned antenna of this type.

Frequency scanning can also be used in a planar array. In one type of system, the horizontal line arrays are frequency-scanned arrays that use delay lines between elements. Many of these lines are stacked one above another. Each line has a phase shifter in the input to the line, and all lines are fed in parallel using suitable power dividers. Frequency change is used to steer the beam in the azimuth direction, and phase change is used to steer the beam in elevation. This type of antenna is sometimes called a phase-frequency array. Another type of system uses frequency scanning in one direction and mechanical scanning in the other direction.

13.15.2 Arrays With Space Feeds

The feed systems for antenna arrays can be complex and costly. They usually involve many sections of coaxial line or waveguide and many coax or waveguide power dividers and combiners. An alternative method is to use so-called space feeds.

Figure 13.32 shows two examples of arrays with space feeds. Figure 13.32(a) is a lens array. With this system, a primary feed antenna, such as a horn, is used to transmit the signal to an array of small receiver horns, each connected to transmission lines containing phase shifters. The output of the phase shifters is sent to an array of transmitting horn antennas. That array can be either a line array or a two-dimensional planar array. The beam is pointed by adjusting the phase of all the phase shifters.

Figure 13.32(b) shows a so-called reflectarray antenna, in which an offset primary feed horn is used to direct energy to the array of receive horns. The output of the receive horns is sent to transmission lines that include phase shifters and short circuits for reflection. The reflected waves are sent back through the phase shifters and thence to the same array of horns that now act as transmit horns. The beam angle is thus pointed by means of the phase shifters.

13.16 FLAT-PLATE REFLECTOR TYPE ANTENNAS

Many types of reflector antennas are used with RF systems. Two of these are discussed here.

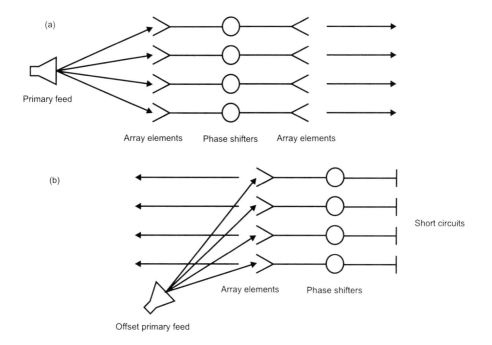

Figure 13.32 Arrays with space feeds: (a) lens array and (b) reflectarray.

13.16.1 Half-Wave Dipole Antennas With Reflectors

When a reflector such as a copper sheet is placed close to a half-wave dipole antenna, a unidirectional radiation pattern is obtained with an antenna gain of about 4 dB. With a quarter-wave separation between the dipole and the reflector and with other dipoles spaced by about a half-wavelength, the impedance of the dipole is about 153Ω. That is quite different for the impedance of a single thin dipole in free space, which has an impedance at resonance of about 73Ω.

13.16.2 Corner Reflector Antennas

The directivity of an antenna-reflector combination can be increased by bending the reflector to form a corner. A directional gain of about 12 dB is obtained from a dipole antenna with a 90-degree corner and the feed spaced from the corner by 0.25λ. A 60-degree corner reflector can provide a directive gain of about 14.5 dB. The dimensions of the reflector should be about one wavelength by one wavelength.

13.17 PARABOLIC REFLECTOR ANTENNAS

Parabolic reflectors are the most commonly used high-gain antennas for microwave frequencies. Figure 13.33 shows a parabolic reflector antenna with a circular aperture. When used for transmitting, the signal is radiated from a horn or other feed located at the focus point, as shown. When used for receiving, the parabolic dish collimates a planar wavefront to the focal point, at which a horn antenna is located.

Spherical wavefronts are shown before reflection as the wave travels to the reflector surface. The surface is so designed that it converts the spherical wavefront signal to a planar wavefront signal. That requires that the distance from the focus to the reflector and then to the planar wave front be the same regardless of position of reflection.

The width and the shape of the major lobe of the radiation pattern of a parabola depend on the size and shape of the mouth of the parabola and the variation of field intensity over the aperture defined by the mouth. If the shape of the mouth is circular and the field distribution across the mouth is uniform, the width of the main lobe between half-power points is 58 degrees/(D/λ), where D is the diameter of the circular mouth. The width of the main lobe between nulls is 140 degrees/(D/λ). If the shape of the mouth is rectangular and the field distribution is uniform, the width of the main lobe between half-power points is 51 degrees/(D/λ), and the width between nulls is 115 degrees/(D/λ). If the field distribution for the rectangular aperture is sinusoidal along D, the width of the main lobe between half-power points is 68/(D/λ), and the width between nulls is 182.5/(D/λ).

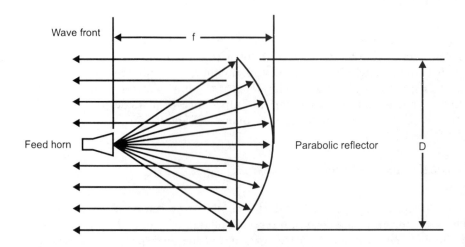

Figure 13.33 Parabolic reflector with a circular aperture and a front feed. (*After:* [10].)

The field pattern of a parabolic antenna ordinarily possesses minor sidelobes, just like other types of antennas. These sidelobes can be minimized by tapering the field distribution across the aperture of the parabola so that the field intensity is maximum at the center and minimum at the edges.

The directive gain of the parabolic reflector antenna is given by (13.33), where A_0 is the actual area of the mouth of the antenna, and k is a correction factor that accounts for the nonuniform distribution of energy across the aperture due to tapering of the field. In a practical system, the value of k is on the order of 0.5 to 0.7.

$$\text{Directive gain} = 4\pi A_0 k / \lambda^2 \qquad (13.33)$$

Feed systems for parabolic reflectors include horn antennas, dipole plus reflector antennas, Yagi antennas, and LP antennas. One important class of feeds is a monopulse feed system that consists of a group of horn antennas rather than a single horn. Such an arrangement allows a radar to measure more precisely the angular position of a target in azimuth and elevation on a single pulse. When transmitting, all four antennas radiate in a sum mode. When receiving, a sum channel is formed using all four antennas, an azimuth difference channel is formed, and an elevation difference channel is formed.

In many applications, only a section of a parabola is used, a situation sometimes referred to as a truncated paraboloid. When this shape is used, the resulting beam is a fan-shaped beam rather than a pencil beam as provided by a circular aperture.

Figure 13.34(a) shows a parabolic reflector antenna with an offset feed. This is a very important type of antenna system for communication systems as well as for radar. This type of antenna is normally used for a fan shaped beam.

Figure 13.34(b) shows a so-called hoghorn antenna, which frequently is used in microwave relay systems. It consists of a parabolic reflector joined by a pyramidal horn.

Figure 13.35 shows a parabolic reflector with a Cassegrain feed. The primary feed is located behind the parabolic reflector. The signal from the feed is reflected by a hyperbolic subreflector; it then travels to the parabolic reflector, where it is reflected as a plane wave. This type of antenna has the advantage of permitting very short feed lines between the transmitter or receiver and the antenna. The advantage of short feed lines is reduced line loss, which improves the effective radiated power in a transmitter and reduces the system noise figure in a receiver. That is important for high-microwave and millimeter-wave systems, where the attenuation in waveguide can be large.

The use of Cassegrain feeds also permits a system with a much larger focal length than is possible with a front-feed system. Large focal length is desirable when dealing with short pulses. A disadvantage of this type of feed is that the subreflector causes aperture blocking, which reduces the efficiency of the antenna and increases the sidelobes.

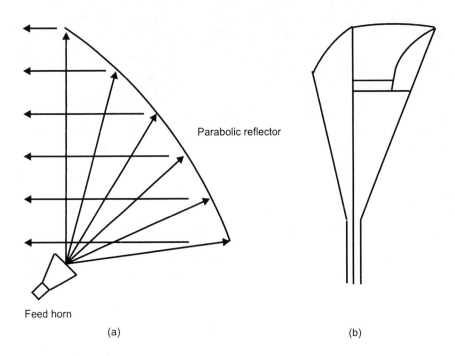

Figure 13.34 Parabolic reflector antenna with an offset feed: (a) system using a small horn feed and (b) hoghorn antenna. (*After:* [2].)

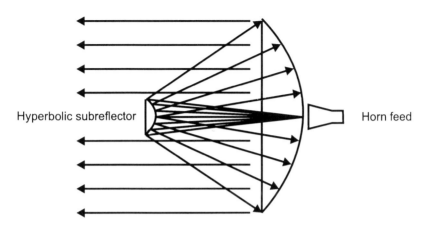

Figure 13.35 Parabolic reflector antenna with a Cassegrain feed. (*After:* [10].)

References

[1] Jasik, Henry, ed., *Antenna Engineering Handbook,* New York: McGraw-Hill, 1961, Chap. 3.
[2] Kennedy, George, *Electronic Communication Systems,* 3rd ed., New York: McGraw-Hill, 1985, Chap. 9.
[3] Jasik, pp. 24-28 to 24-29.
[4] Stutzman, Warren L., and Gary A. Thiele, *Antenna Theory And Design,* New York: John Wiley & Sons, 1981, pp. 220–226.
[5] Ibid., pp. 278–281.
[6] Ibid., pp. 281–287.
[7] Ibid., pp. 287–303.
[8] Jasik, Chapter 8.
[9] Terman, Fredric E., *Electronic and Radio Engineering,* 4th ed., New York: McGraw-Hill, 1955, p. 915.
[10] Skolnik, Merrill I., *Introduction to Radar Systems,* 2nd ed., New York: McGraw-Hill, 1980, Chapter 7.
[11] Van Valkenburg, Mac E., ed., *Reference Data for Engineers,* 8th ed., SAMS, Carmel, IN: Prentice Hall Computer Publishing, 1993, pp. 32-1 to 32-56, "Antennas" by Robert C. Hansen.

CHAPTER 14

Lumped Constant Components and Circuits

This chapter discusses resistors, inductors, and capacitors, so-called lumped constant components. The dimensions of these components generally are small compared to a wavelength. This chapter also discusses a number of circuits that use those components, including series resonant circuits, parallel resonant circuits, impedance-matching networks, and LC filters. Other lumped constant components and circuits are discussed in subsequent chapters.

14.1 CONDUCTORS AND SKIN EFFECT

The following is based on [1].

One of the important differences between LF systems and HF systems is that at high frequencies the current in a conductor is not uniform over the cross-section. Inductive reactance effects at high frequencies cause the current to flow only in the outer surface of the conductor. In computing resistance of a conductor at RF, it is common practice to first compute the skin depth. The skin depth is that distance below the surface of the conductor where the current density has diminished to $1/e$ its value at the surface, where e is the base of the natural logarithm (2.718). The thickness of the conductor is assumed to be several times the skin depth at the lowest frequency of interest to minimize ohmic or conductor losses. This value also applies to the thickness of microstrip or stripline conductors. The acceptable standard is ≥ 3 for δ low-loss microstrip or strip transmission line.

In computing resistance of a round conductor, the conductor is simulated by a cylindrical shell of the same surface shape but a thickness equal to the skin depth, with uniform current density equal to that which exists at the surface of the actual conductor.

The skin depth is given by

$$\delta = (\lambda/\pi\sigma\mu c)^{1/2} \qquad (14.1)$$

where

δ = skin depth (centimeters)
λ = free-space wavelength (centimeters)
σ = conductivity (siemens per centimeter) = $1/\rho$
ρ = resistivity (ohm-centimeters)
μ = permeability of conductor = $4\pi \times 10^{-9} \mu_r$ in henry/cm
μ_r = relative permeability of conductor material = 1.0
c = speed of light = 3×10^{10} cm per second

For numerical computations,

$$\delta = (6.61/f^{1/2})k_1 \text{ cm} \qquad (14.2)$$

where

$$k_1 = [(1/\mu_r)\rho/\rho_c]^{1/2} \qquad (14.3)$$

and

f = frequency (hertz)
ρ_c = resistivity of copper (1.724×10^{-6} Ω-cm)

At microwave and higher frequencies, the skin depth is very small and the resistance of even the best conductors may be high unless large dimensions are used.

14.2 RF RESISTORS

Resistors are used in many RF circuits. Applications include transistor biasing, loads for wideband amplifiers, gain control for receivers, and signal attenuators. Resistors used in RF circuits can have a number of shapes, with cylindrical being the most common at the lower frequencies. Resistors usually are made of carbon. At the higher frequencies, they may be in the form of carbon compound deposits on insulator surfaces or tantalum nitride (TaN) or nichrome (NiCr) for thin-film circuits. Platinum- or palladium-based compounds are used for thick-film resistors.

All resistors must have conducting leads connected to each end. At the higher frequencies, the length of the leads must be kept as short as possible. With chip resistors, mounting-pad size is determined by placement accuracy, part dimensional tolerance, and attachment method.

14.3 INDUCTORS AND INDUCTIVE REACTANCE

The approximate inductance for a straight short wire is given by

$$L = (\mu_0 l / 2\pi)[\ln(4l/d)] \tag{14.4}$$

where

L = inductance (henrys)
l = wire length (centimeters)
d = wire diameter (centimeters)
ln = natural logarithm

For an example of the use of (14.4), assume a wire with a length of 3 cm and a diameter of 0.2 cm. Substituting values into (14.4), the inductance, L, would be

$$L = 2 \times 10^{-9} \, l[\ln(4l/d)]$$
$$L = 2 \times 10^{-9} \times 3.0[\ln(12/0.2)]$$
$$= 24.5 \times 10^{-9} \text{ H}$$
$$= 24.5 \text{ nH}$$

The inductive reactance is given by

$$X_L = \omega L = 2\pi f L \tag{14.5}$$

where

f = frequency (hertz)
L = inductance (henrys)

At a frequency of 200 MHz, a wire with a diameter of 0.2 cm and a length of 3 cm would have a reactance of 30.8Ω. We thus see the reason that resistor or capacitor leads must be kept as short as possible at RF frequencies to avoid having an undesirably large inductance in series with the resistance or capacitance.

An air-core solenoid made of thin wire has a much greater inductance for the same length than the straight thin wire. At RF, such solenoids often have only one turn or, at most, a few turns.

The approximate inductance for an air-core solenoid of a number of turns is given by

$$L = n^2[r^2/(9r + 10l)] \qquad (14.6)$$

where

L = inductance (microhenrys)

n = number of turns

r = solenoid radius (inches)

l = solenoid length (inches)

For an example of the use of (14.6), assume a solenoid with a radius of 0.25 in, a length of 0.5 in, and 5 turns. Substituting those values into (14.6), the inductance would be

$$L = 25[0.0625/(2.25 + 5)] = 0.216 \; \mu\text{H}$$

The inductive reactance for this solenoid at a frequency of 200 MHz would be 270.8 Ω.

Figure 14.1 shows the flat square spiral inductor. This device frequently is used with printed transmission line circuits, that is, microstrip.

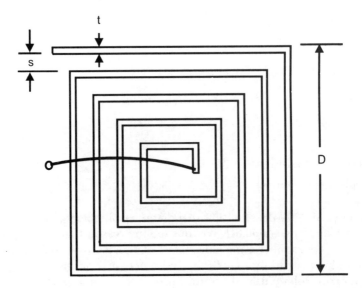

Figure 14.1 Flat square spiral inductor.

The approximate inductance for the flat square spiral inductance is given by

$$L = 8.5 \times 10^{-3} \times D \times N^{1.7} \tag{14.7}$$

where

L = inductance (microhenrys)
D = maximum outside dimension in cm
$N = (D/2)/(w + s)$
s = space between lines (centimeters)
w = width of lines (centimeters)

Another important type of inductor used with RF circuits is a toroidal inductor, which is shown in Figure 14.2. It consists of a few turns of wire with a powdered iron or ferrite core. The core adds greatly to the inductance of the coil at VHF and lower frequencies.

The toroidal inductor with a ferrite or powdered iron core and a square cross-section has an inductance that is given by

$$L = (N^2 \mu_r \mu_0 \, t/2\pi) \ln(d_2/d_1) \tag{14.8}$$

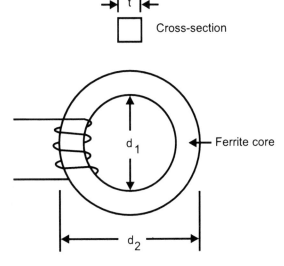

Figure 14.2 Toroidal inductor.

where

L = inductance (henrys)
N = number of turns
μ_r = relative permeability of core
μ_0 = permeability of free space = $4\pi \times 10^{-7}$ H/m
t = toroid thickness (meters)
d_2 = toroid outer diameter (meters)
d_1 = toroid inner diameter (meters)

For an example of the use of (14.8), assume the following:

$N = 4$
$\mu_r = 1{,}900$
$\mu_0 = 4\pi \times 10^{-7}$ H/m
$t = 0.0005$m
$d_2 = 0.02$m
$d_1 = 0.01$m
$L = (N^2 \mu_r \mu_0 \, t/2\pi) \ln(d_2/d_1)$
$L = (16 \times 1{,}900 \times 4\pi \times 10^{-7} \times 0.005/2\pi) \ln(0.02/0.01)$
$L = 2.10 \times 10^{-5}$ H = $21.0 \, \mu$H

The current in an inductor lags the applied voltage by 90 degrees. The inductive reactance, which is the voltage divided by the current, thus has a phase angle of +90 degrees.

The total inductance for two or more inductors in series is the sum of the individual inductances. The total inductance for two or more inductors in parallel is the reciprocal of the sum of the inductive susceptances, where the inductive susceptance is the reciprocal of the inductive reactance ($-jB_L = 1/+jX_L$). Unlike resistors, no power is dissipated in pure inductors. Energy is stored in the magnetic field and is transferred back to the conductor when the field strength is returned to zero.

The impedance of an RL circuit containing both resistance and inductance in series is given by

$$Z = R + jX_L \qquad (14.9)$$

where

j = operator indicating a phase angle of 90 degrees
Z = impedance
R = resistance
X_L = inductive reactance = $2\pi f L = \omega L$
f = frequency
ω = radian frequency

The operator j is also equal to the square root of -1.

14.4 RF CHOKES

An RF choke is another name for an inductor that is used to pass dc current but that blocks ac or RF current. Lumped-element chokes have capacitance between windings and are parallel resonant circuits at some frequency. This resonant frequency normally is chosen to be above the highest frequency with which we want to operate. This component is used in many RF circuits. In microstrip or stripline, chokes normally are printed sections of high-impedance transmission line a quarter-wavelength long at the center frequency of interest.

14.5 CAPACITORS AND CAPACITIVE REACTANCE

A capacitor typically consists of two conducting surfaces or plates separated by a dielectric insulation material that permits the storage of energy in the electric field between the plates. This component is illustrated in Figure 14.3. The dielectric prevents current flow when the applied voltage is constant, but a time-varying voltage produces a current proportional to the rate of voltage change.

The current in a capacitor is given by

$$I = C\, dv/dt \qquad (14.10)$$

where C is the capacitance measured in farads.
The capacitance, C, of the parallel plate structure is given by

$$C = \epsilon A/d = \epsilon_0 \epsilon_r A/d \qquad (14.11)$$

where

ϵ = absolute permittivity of the dielectric = $\epsilon_0 \epsilon_r$
A = area of parallel plates or conductors
d = spacing of plates or conductors
ϵ_0 = permittivity of free space = 8.854×10^{-12} F/m
ϵ_r = relative permittivity or dielectric constant of dielectric medium

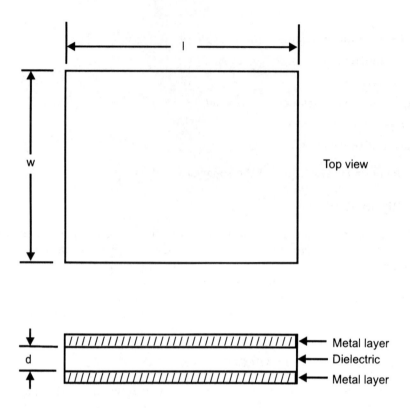

Figure 14.3 Illustration of a two-plate capacitor.

A typical dielectric such as mica has a permittivity of 5×10^{-11} F/m. Thus, the relative permittivity or dielectric constant of mica is 5.65. Materials used for printed circuits, microstrip, and stripline circuits have dielectric constants in the range of 2.3 to 2.6. FR-4 is about $\epsilon_r = 4.5$. Microstrip is also fabricated on alumina ($\epsilon_r = 9.6 - 10.0$) and beryllia ($\epsilon_r = 6.0$). The relative permittivity for fused silica is about $\epsilon_r = 2.1$.

For an example of the use of (14.11), assume the following:

Dielectric constant = 2.5
Area = 1 cm^2
Spacing = 0.1 cm

Substituting those values into (14.11) gives

$$C = 2.5 \times 8.854 \times 10^{-12} \times 1/0.1 = 2.21 \times 10^{-12} \text{F}$$
$$C = 2.21 \text{ pF}$$

In many cases, parallel plates are used to provide larger capacitance. In this case, the total capacitance is given by

$$C = (N-1)\epsilon_0 \epsilon_r A/d \qquad (14.12)$$

where N = number of plates, and the other terms are as defined for (14.11).

There are many other possible shapes and types for capacitors, including mechanically adjustable and voltage-variable capacitors. One form of solid-state diode has a capacitance that can be changed by changing the dc voltage across the diode. Such a voltage-variable capacitance diode is called a varactor.

Table 14.1 lists some of the more important types of available fixed-value capacitors with capacitance range, maximum voltage, and other characteristics. Comments about each type of capacitor follow.

Mica dielectric capacitors have excellent RF characteristics. Tubular, ceramic dielectric capacitors are available in very low values. Disk ceramic dielectric capacitors are small, inexpensive, and popular. Chip ceramic dielectric capacitors are used frequently in HF circuits. Mylar dielectric capacitors also are inexpensive and popular. Polystyrene dielectric capacitors are high quality but large. Polycarbonate dielectric capacitors are high quality. Glass and porcelain dielectric capacitors are good quality and inexpensive and have good long-term stability. Tantalum dielectric capacitors have high capacitance with acceptable leakage. They are polarized, meaning that there is a plus and a minus voltage assignment for terminals. They are small, have low inductance, and are popular. Electrolytic capacitors are used mainly for power supply filters. They are polarized. Oil-filled capacitors are used for high-voltage filters. They are large, have long life, and are not polarized.

Table 14.1
Characteristics for Fixed-Value Capacitors

Type	Capacitance Range	Maximum Voltage	Accuracy	Temperature Stability
Mica	1 pf–0.01 uf	100–600	Good	Good
Tubular	0.5–100 pf	100–600	Selectable	
Ceramic disk and chip	10 pf–1 uf	50–1,000		Poor
Ceramic mylar	0.001–10 uf	50–600	Good	Poor
Polystyrene	10 pf–0.01 uf	100–600	Good	
Polycarbonate	100 pf–10 uf	50–400	Good	Good
Glass	10–1,000 pf	100–600	Good	
Porcelain	100 pf–0.1 uf	50–400	Good	Good
Tantalum	0.1–500 uf	6–100	Poor	Poor
Electrolytic	0.1 uf–0.2 f	3–600	Poor	Poor
Oil	0.1–20 uf	200–10k		

The capacitive reactance is given by

$$X_C = 1/\omega C \tag{14.13}$$

where

X_C = capacitive reactance (ohms)
$\omega = 2\pi \times f$
f = frequency (hertz)
C = capacitance (farads)

The total capacitance for two or more capacitors connected in series is the reciprocal of the sum of the reciprocals of the individual capacitors. The total capacitance for two or more capacitors connected in parallel is the sum of capacitance for the individual capacitors. Capacitive susceptance is the reciprocal of the capacitive reactance. In other words, capacitances in parallel are like resistors or inductors in series and capacitances in series are like resistors or inductors in parallel.

Unlike resistors, no power is dissipated in pure capacitors. In a real capacitor, however, there is some loss or power dissipation due to a small conductance between the plates or the sides of the capacitor or the finite resistivity of the plate material itself.

The impedance of a series RC circuit containing both resistance and capacitance is given by

$$Z = R - jX_C \tag{4.14}$$

where

Z = impedance (voltage/current)
R = resistance
jX_C = capacitive reactance

The j in the capacitive reactance expression is the operator that indicates a phase angle of 90 degrees. The $-j$ indicates a phase angle of -90 degrees.

If a dc voltage is applied across a circuit consisting of a resistor and a capacitor connected in series, the voltage across the capacitor changes slowly with a rise time of RC. It changes by 63% in one RC time. The time it takes to go from 10% to 90% of its final value is 2.2 RC. That usually is referred to as the rise time of a pulse in an electronic circuit.

Some of the preceding material is based on [2].

14.6 SERIES RESONANT RLC CIRCUITS

Some of the material in the following three sections is quoted or adapted from [3].

A series resonant RLC circuit is made up of one or more resistors, one or more capacitors, and one or more inductors, all connected in series. Figure 14.4 shows impedance characteristics for a series resonant RLC circuit. The Q in this case is 10, where $Q = X/R$ at resonance. At resonance, the impedance is a minimum, and the impedance angle is zero. Below resonance, the circuit becomes capacitive and the impedance angle is negative. Above resonance, the impedance becomes inductive, and the impedance angle is positive.

The input impedance of the series resonant circuit is given by

$$Z = V/I = R + j(\omega L - 1/\omega C) \tag{14.15}$$

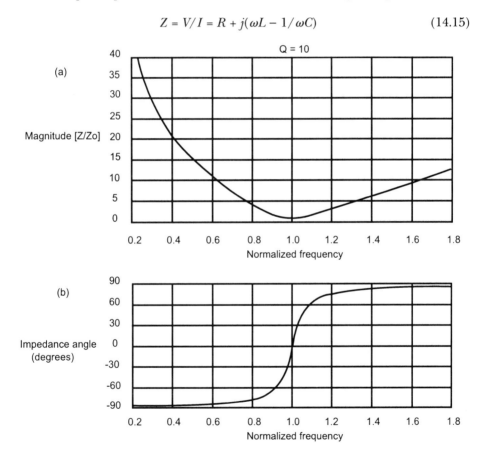

Figure 14.4 Series resonant RLC circuit impedance characteristics: (a) magnitude of normalized impedance as a function of normalized frequency and (b) phase angle of normalized impedance as a function of normalized frequency. (*After:* [3].)

where terms are as previously defined.

A large Q means that the resonant peak for Z will be narrow, while a small Q will produce a wide resonant peak. It can be shown that the −3 dB bandwidth, BW_{3dB}, is equal to the center or resonant frequency, f_0, divided by the Q, as shown in (14.16):

$$BW_{3dB} = f_0/Q \tag{14.16}$$

At resonance, the voltage across the capacitor is Q times the applied voltage. That can be important from the point of view of possible voltage breakdown for the capacitor.

14.7 PARALLEL RESONANT RLC CIRCUITS

A parallel resonant RLC circuit consists of a resistor, an inductor, and a capacitor, all in parallel. Examples of impedance characteristics for such a circuit are shown in Figure 14.5. Notice that the impedance is maximum at resonance and falls off on either side of resonance. The impedance angle is positive below resonance, zero at resonance, and negative above resonance.

Parallel resonance occurs when the input voltage and current are in phase and $B_L = B_C$ or $1/\omega L = \omega C$.

The Q of the parallel circuit is given by

$$Q = \omega_0 C/G = \omega_0 CR = R/\omega_0 L \tag{14.17}$$

$$G = \text{conductance} = 1/R \tag{14.18}$$

where ω_0 = resonant radian frequency.

The admittance for the circuit is

$$Y = G + j(\omega C - 1/\omega L) \tag{14.19}$$

where $Y = 1/Z$. The resonance peak is narrow for high values of Q and wide for small values of Q. Again, the 3-dB bandwidth is equal to the resonant center frequency divided by Q. It can be shown that, at resonance, the current in the capacitor is Q times the input current to the resonant circuit.

The following are examples of calculations for the impedance of a parallel resonant circuit:

At $f/f_0 = 0.5$ and $Q = 10$
Normalized $G = 1$
Normalized $B_C = 0.5Q = 5$
Normalized $B_L = Q/0.5 = 20$

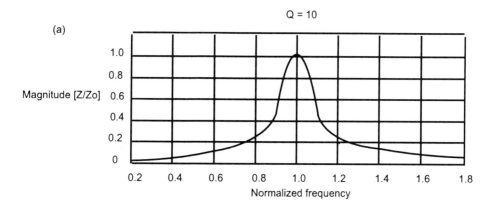

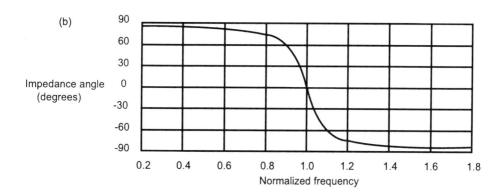

Figure 14.5 Impedance characteristics of parallel resonant RLC circuits: (a) normalized impedance as a function of normalized impedance and (b) phase angle of normalized impedance as a function of normalized frequency. (*After:* [3].)

$$Y/Y_0 = 1 + j5 - j20 = 1 - j15 = 15.03 \angle -86.19 \text{ degrees}$$
$$Z/Z_0 = 1/(Y/Y_0) = 0.0665 \angle 86.19 \text{ degrees}$$

For an example of component calculations, assume a Q of 10, a frequency of 100 MHz, and a resistance of 1,000Ω.

$$Q = \omega CR = R/\omega L \qquad (14.20)$$

$$C = 10/(6.28 \times 10^8 \times 1,000) = 15.9 \text{ pF}$$
$$L = 1,000/(10 \times 6.28 \times 10^8) = 0.159 \text{ } \mu\text{H}$$

Parallel resonant circuits are used for many applications in RF circuits. One important application is for selecting the desired frequency spectrum for amplifiers, oscillators, filters, and other RF components.

14.8 COMPLEX RESONANT CIRCUITS

In many practical cases, resonant circuits are more complex than simple series resonant or parallel resonant circuits. Four examples of such complex circuits are shown in Figure 14.6: (a) a circuit with a resistor in series with the inductor; (b) a circuit with a resistor in series with the capacitor; (c) a transformer-like circuit with two capacitors and the load resistor connected in parallel with one of them; and (d) a second transformer-like circuit but with two inductors and the load resistor connected in parallel with one of them. Figure 14.6 and the analyses that follow show how these circuits are converted to simple parallel resonant RLC circuits.

Circuit analysis can be used to calculate the equivalent circuits shown in Figure 14.6. Alternatively, it is possible to use a Smith chart for circuit analysis of complex circuits. Following is a brief discussion of a method of analyzing complex resonant circuits using Q values and formulas for converting from parallel to series equivalent circuits and from series to parallel equivalent circuits. The circuit in

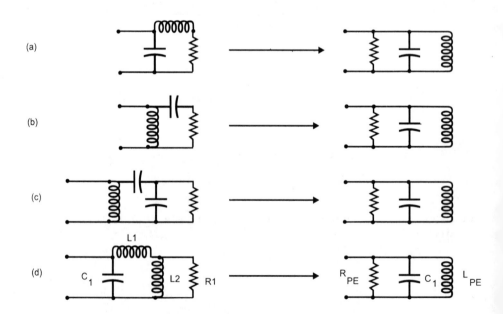

Figure 14.6 Resonant circuits and circuit transformation. (*After:* [3].)

Figure 14.6(d) is used as the circuit that is converted to a simple parallel resonant RLC circuit. The assumed impedance values for the circuit are as follows:

$$R_1 = 50\,\Omega$$
$$X_{L1} = 183\,\Omega$$
$$X_{L2} = 125\,\Omega$$
$$X_C = 209\,\Omega$$

The normalized values of impedance based on 50 Ω are as follows:

$R_1 = 1$
$X_{L1} = 3.66$
$X_{L2} = 2.5$
$X_C = 4.19$

The Q of the parallel and the series circuits are as follows:

$$\text{Parallel } Q = Q_P = R_P/X_P \qquad (14.21)$$

$$\text{Series } Q = Q_S = X_S/R_S \qquad (14.22)$$

The subscript P stands for parallel, and the subscript S for series. The exact formulas for conversion to equivalent circuit values are as follows:

$$R_{PE} = R_S (1 + Q_S^2) \qquad (14.23a)$$
$$X_{PE} = X_S(Q_S^2 + 1)/Q_S^2 \qquad (14.23b)$$

and

$$R_{SE} = R_P/(1 + Q_P^2) \qquad (14.24a)$$
$$X_{SE} = X_P[Q_P^2/(Q_P^2 + 1)] \qquad (14.24b)$$

The approximate formulas for conversion to equivalent circuits are as follows. These formulas assume $Q \gg 1$.

$$R_{PE} = R_S Q_S^2 \qquad (14.25a)$$
$$X_{PE} = X_S \qquad (14.25b)$$

and

$$R_{SE} = R_P/Q_P^2 \qquad (14.26a)$$
$$X_{SE} = X_P \qquad (14.26b)$$

The exact circuits are next applied to the circuit in Figure 14.6(d). First, we convert the parallel circuits L_2 and R_1 to a series circuit. The value of Q_P is found as follows:

$$\text{Parallel } Q = Q_P = R_P/X_P = 1/2.5 = 0.4$$

The series equivalent circuit values are found as follows:

$$R_{SE} = R_P/(1 + Q_P^2) = 1/(1 + 0.16) = 0.862$$
$$X_{SE} = X_P[Q_P^2/(Q_P^2 + 1)] = 2.5[0.16/(0.16 + 1)] = 0.345$$

Adding X_{L1} to those values, we have

$$Z = 0.862 + j\,3.66 + j\,0.345 = 0.862 + j\,4.0$$

We now convert this series RL circuit to a parallel RL circuit. The value of Q_S is found as follows:

$$\text{Series } Q = Q_S = X_S/R_S = 4.0/0.862 = 4.64$$

The parallel equivalent circuit values are found as follows:

$$R_{PE} = R_S(1 + Q_S^2) = 0.862(1 + 21.5) = 19.4$$
$$X_{PE} = X_S(Q_S^2 + 1)/Q_S^2 = 4.0[(21.5 + 1)/21.5] = 4.19$$

The value of X_C was given as (4.19), so reactance values for the equivalent parallel inductance and capacitance are the same. The total circuit Q is $19.4/4.19 = 4.6$.

Now we convert the normalized impedance to component values, assuming a radian frequency ω of 10^8 rps (frequency = 15.92 MHz), $Z_0 = 50\Omega$, and $Y_0 = 0.02$S.

$R_1 = 50\Omega$

$L_1 = 183.0 \times 10^{-8}$ H $= 1.83\ \mu$H

$L_2 = 125 \times 10^{-8}$ H $= 1.25\ \mu$H

$C_1 = 4.8 \times 10^{-11}$ F $= 48$ pF

$R_{PE} = 19.4 \times 50 = 970\Omega$

$X_{PE} = 4.19 \times 50 = 208.3\Omega$

$L_{PE} = 208.3 \times 10^{-8}$ H $= 2.08\ \mu$H

$Q = R_P/X_{C1} = 970/208.3 = 4.66$

$B = 15.92/4.66 = 3.42$ MHz

14.9 THE USE OF THE SMITH CHART FOR CIRCUIT ANALYSIS

In Chapter 11, the Smith chart was used in transmission line analysis and design. Here, it is used in lumped constant circuit analysis and design.

Figure 14.7 shows an example of the use of the Smith chart for circuit analysis. This example also performs the analysis for the circuit in Figure 14.6(d).

The starting point will be R_1. We will assume that this resistor has a normalized impedance R/Z_0 based on 50Ω of 1. It is thus located on the Smith chart at the center of the chart at point 1. The normalized conductance is also 1 and is located at the same point since the conductance is simply 1 over the resistance. To this conductance, we add the inductive admittance $-jB_{L2}$. We will assume that this normalized admittance is $-j\,0.4$, which is equal to $1/j\,X_{L2} = -j/2.5 = -j\,0.4$. Next, we travel along the 1.0 conductance circle to the $-j\,0.4$ arc. The normalized admittance at that point is $1 - j\,0.4$, located at point 2.

Next, we convert to a series impedance. That is done so that this impedance can be added to the impedance of L_1. The conversion is done by traveling along the straight line through the center of the chart to point 3, which is located the same distance from the center as point 2. Recall that to change from an admittance chart to an impedance chart, we simply move a quarter-wavelength on an SWR circle. That is equivalent to transferring through the center of the chart on a straight line to a point that is an equal distance from the center of the chart. The normalized impedance at this point is about $0.85 + j\,0.35$.

L_1 has a normalized impedance of $+j\,3.66$. To add that value to $+j\,0.34$, travel along the $R = 0.85$ circle to point 4. The normalized impedance at that point is $0.85 + j\,4.0$.

This impedance is now converted to an admittance so it may be added to the admittance of C. This is done by moving along a straight line through the center of the chart to point 5. This point is located the same distance from the center of the chart as point 4. The normalized admittance, $G_3 - j\,B_{L4}$, at this point is $0.051 - j\,0.24$.

The next move is along the $G = 0.051$ circle to point 6 by adding a capacitive susceptance B_{C1} of $+j\,0.24$. The admittance at point 6 is $0.051 + j\,0$.

Next, again convert this admittance to an impedance. Again, move on a straight line through the center of the chart to point 7, which also is located the same distance from the center of the chart as point 6. The normalized impedance at that point is $19.4 + j\,0$. The real impedance based on a normalized resistance of 50Ω is $R_P = 50 \times 19.4 + j\,0 = 970\,\Omega$. This completes the graphical analysis and design.

Now, convert the normalized impedance to component values, assuming a radian frequency, ω, of 10^8 rps (frequency = 15.92 MHz), $Z_0 = 50\,\Omega$, and $Y_0 = 0.02\,\text{S}$.

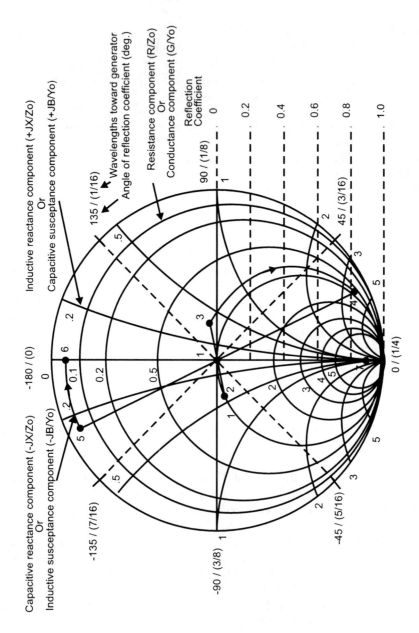

Figure 14.7 Example of the use of the Smith chart for circuit analysis.

$R_1 = 50\,\Omega$

$L_1 = 183.0 \times 10^{-8}\,\text{H} = 1.83\,\mu\text{H}$

$L_2 = 125 \times 10^{-8}\,\text{H} = 1.25\,\mu\text{H}$

$C_1 = 4.8 \times 10^{-11}\,\text{F} = 48\,\text{pF}$

$R_{PE} = 19.4 \times 50 = 970\,\Omega$

$X_{PE} = 4.19 \times 50 = 208.3\,\Omega$

$L_{PE} = 208.3 \times 10^{-8}\,\text{H} = 2.08\,\mu\text{H}$

$Q = R_P/X_{C1} = 970/208.3 = 4.66$

$B = 15.92/4.66 = 3.42\,\text{MHz}$

14.10 S-PARAMETERS

Another way to represent a complex circuit is by means of scattering parameters, or S-parameters. With this approach, parameters are measured by placing the system under test in an S-parameter test set with an associated network analyzer and signal source. The S-parameter test set is a mounting and switching arrangement with directional couplers and line terminations that allow for measurement of voltage reflection coefficients and network power-gain measurements. The signal source usually has a 50 Ω output impedance, and transmission lines and terminations also are 50 Ω. This test system is illustrated in Figure 14.8 for two types of test conditions.

In the circuit in Figure 14.8(a), directional couplers are used to measure the forward and reflected signal voltages, thus providing input and output reflection coefficients. The system under test uses switches, so that for the measurement of the input reflection coefficient (reflected signal voltage/forward signal voltage), the output of the two-way directional coupler is connected to the input of the system under test; and for measurement of output reflection coefficients, the output of the two-way directional coupler is connected to the output of the system under test. In each case, the system under test is terminated in Z_0. The input voltage reflection coefficient is called S_{11}, and the output voltage reflection coefficient is called S_{22}. These parameters have both magnitude and phase angle. They can be plotted on a Smith chart to obtain the input and output circuit impedances. Figure 14.8(b) shows the setup for measuring S_{21} and S_{12}. S_{21} is the output signal power over the input signal power, as measured with the two directional couplers when the signal is fed to the input of the system under test. S_{12} is the output signal power over the input signal power measured the same way when the signal is fed to the output of the system under test. In each case, the system under test is terminated in Z_0. These parameters have both magnitude and phase angle. Figure 14.9 illustrates the use of the Smith chart in converting S-parameters to input or output impedances. In this case, S_{11} is plotted. It is assumed to have a value of 0.5 at an angle of −140 degrees. It is plotted by constructing a radial line from the

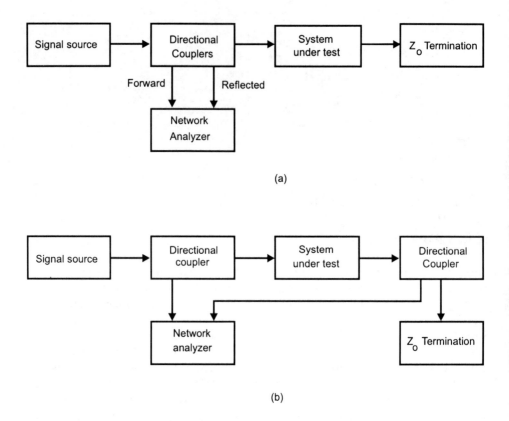

Figure 14.8 Illustrations of setups for measuring S-parameters: (a) setup for measuring S_{11} and S_{22} and (b) setup for measuring S_{21} and S_{12}.

center of the chart to −140 degrees, as shown on the angle of the reflection coefficient circle. A reflection coefficient circle is also constructed using the radial-scaled parameters for a voltage reflection coefficient of 0.5. The intersection of the radial angle line and the reflection coefficient circle gives us the S_{11} location. The corresponding input impedance is read from the Smith chart as $(0.37 - j\,0.32)\,Z_0$. If Z_0 is 50Ω, the input impedance would be $18.5 - j\,16\,\Omega$. In many cases, the network analyzer used to measure the S-parameters includes a Smith chart CRT display. The impedance values thus can be read directly from this chart. The magnitude and the angle also can be given by meter displays. The information can then be plotted on paper versions of the Smith chart.

14.11 IMPEDANCE MATCHING USING LC CIRCUITS

Impedance matching is important in RF systems. For example, the maximum power gain of an RF amplifier can be realized only if the input of the amplifier is impedance

Lumped Constant Components and Circuits | 445

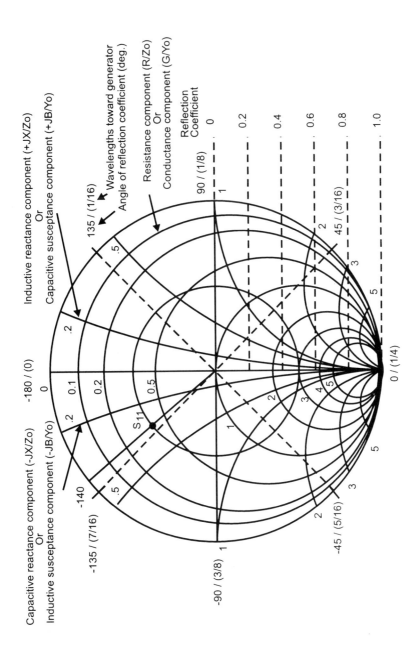

Figure 14.9 Example of the use of the Smith chart in converting S-parameters to impedances.

matched to the source impedance and the output of the amplifier is impedance matched to the load impedance. The same is true when matching the impedance of an antenna to the transmission line connected to it, so there will be no reflections.

With impedance-matching circuits, the objective is to move from the source impedance to the complex conjugate of the load impedance. The complex conjugate impedance has the same magnitudes for resistance and reactance components as the load impedance but the opposite sign for the reactance component. Thus, if the load impedance is $4.0 + j\,1.0$, the complex conjugate impedance is $4.0 - j\,1.0$.

Figure 14.10 shows four examples of discrete impedance-matching networks.

The circuit in Figure 14.10(a) shows a single L-section for output matching. It shows a transistor as the input or source impedance with dc current fed through an RF choke (RFC). This choke blocks RF signals and, if it is designed correctly, does not usually enter into the analysis. A low-impedance coupling capacitor, C_1, couples RF signals from the transistor to the L-section network made up of L and C_2. Again, this coupling capacitor does not enter into the analysis of the impedance-matching circuit. The load resistor, R_L, is connected across C_2. This type of circuit is a wideband matching circuit and is not intended to be a resonant circuit. It is used in wideband amplifiers and other wideband systems. The single L-section has the advantage of simplicity. Its disadvantage is that it is not practical for all types of impedance combinations because of limited degrees of freedom in component choice.

The circuit in Figure 14.10(b) shows double L-sections for output impedance matching. Again, the source impedance is the output impedance of a transistor. The two L-sections are L_1 and C_2 and L_2 and C_3. The load impedance is R_L. This type of circuit frequently is used with power amplifiers.

Another matching circuit sometimes used for output matching is the T-section shown in Figure 14.10(c). This network is used when a higher Q is required.

The circuit in Figure 14.10(d) is a pi network. This circuit frequently is used with power amplifiers and for impedance-matching of antennas.

Each of the circuits shown in Figure 14.10 also can be used as input matching circuits for transistors and other devices. Both the input and the output of transistor and tube amplifiers need to have impedance-matching networks if maximum gain and maximum power output are to be realized.

14.12 IMPEDANCE-MATCHING DESIGN USING THE SMITH CHART

The design of impedance matching circuits frequently is done using S-parameters and the Smith chart. Example designs for L, T, and pi networks are shown in Figures 14.11, 14.12, and 14.13. Each of these designs is discussed briefly in the following paragraphs.

Figure 14.11 shows an example where the goal is to match the output of a transistor amplifier to a 50Ω transmission line using an L-section LC network. The

Lumped Constant Components and Circuits | 447

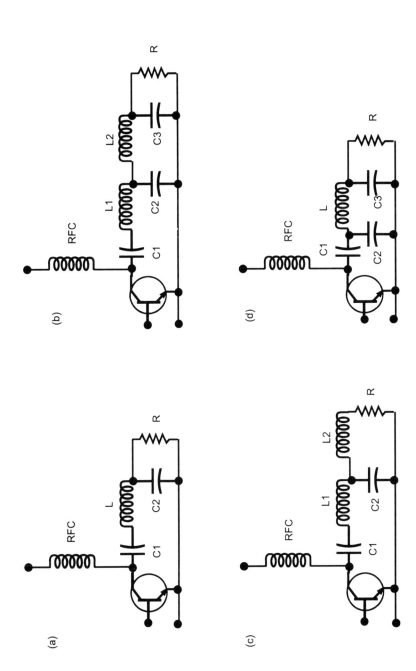

Figure 14.10 Examples of discrete impedance-matching networks: (a) single L-section; (b) double L-sections; (c) T-section; and (d) pi-section.

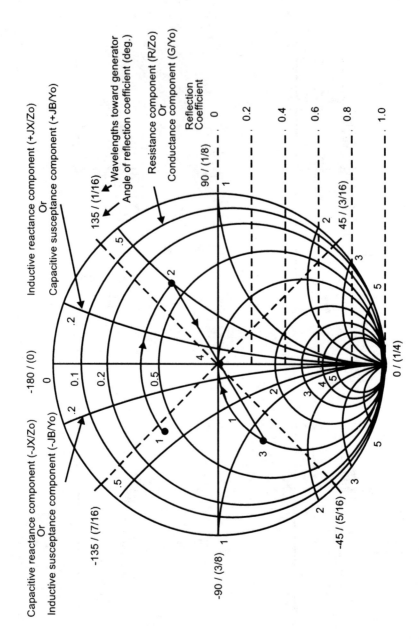

Figure 14.11 Example L-section impedance matching design using the Smith chart.

Lumped Constant Components and Circuits | 449

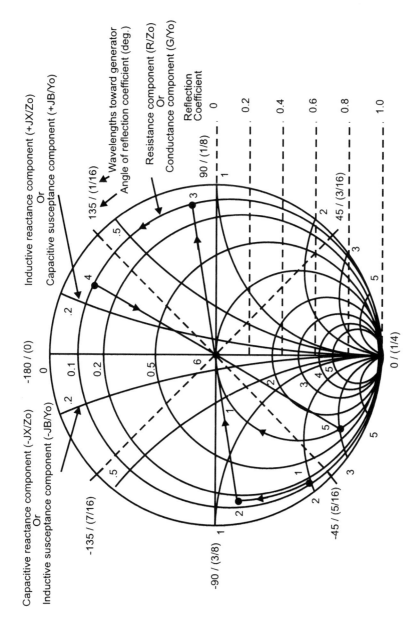

Figure 14.12 Example T-section impedance matching design using the Smith chart.

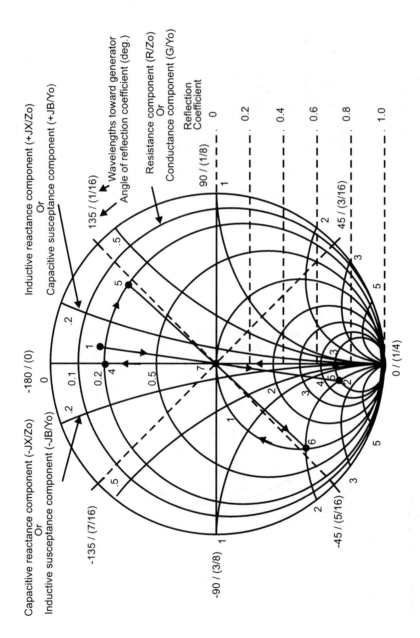

Figure 14.13 Example pi-section impedance-matching design using the Smith chart.

normalized transistor output impedance is assumed to be $0.4 - j\,0.4$ based on $50\,\Omega$, which is shown as point 1.

The selected starting point is point 1, located at $0.4 - j\,0.4$ on the Smith chart. First, move to the right on the 0.4 resistance circle to point 2. In doing so, series inductance is being added. This point is chosen so an equal distance on a straight line through the center will transform to the 1.0 resistance circle. The inductive impedance required to move from point 1 to point 2 is $j\,0.9$. The impedance at point 2 is $0.4 + j\,0.5$.

Next, convert to an admittance chart by transferring an equal radial distance to point 3. This point has an admittance of $1 - j\,1.2$. To provide an impedance match to the $50\,\Omega$ load, a capacitive susceptance of $j\,1.2$ must be added. That moves the point clockwise on the 1.0 conductance circle to point 4, which has an admittance of $1 + j\,0$, or an impedance of $1 + j\,0$, thus obtaining a perfect impedance match at that frequency.

Component values then can be determined by using the values of X and B that were selected. For example, if the operating radian frequency, ω, is 18^8 rps,

$$L_1 = Z_0 \times X_{L1}/\omega = 50 \times 0.9/10^8 = 0.45\ \mu H$$
$$C_2 = 1/Z_0 \times B_{C2}/\omega = 0.02 \times 1.2/10^8 = 240\ pF$$

Figure 14.12 shows an example where the output of a transistor amplifier is to be matched to a $50\,\Omega$ transmission line using a T-section LC network. The normalized transistor output impedance is assumed to be $0.15 - j\,2.0$ based on $50\,\Omega$ and is shown as point 1.

The starting point is point 1, located at $0.15 - j\,2.0$ on the Smith chart. Next, move clockwise on the 0.15 resistance circle to point 2. This point is chosen so we will be in good position to reach the $R = 1$ circle after two transfers. To reach point 2, 0.78 normalized inductive reactance must be added.

Now, convert to an admittance chart and transfer through the center an equal radial distance to point 3 on the 0.1 admittance circle. The admittance at this point is $0.1 + j\,0.8$. Now, move counterclockwise to point 4. This requires a B_C value of $-j\,0.5$.

Next, convert to an impedance chart and transfer through the center to point 5. This point is on the $R = 1$ circle. Finally, add a series inductive impedance of $+ j\,3.0$. This represents a move clockwise to point 6, which is at an impedance of $1 + j\,0$, creating an impedance match.

Component values then can be determined by using the values of X and B that were selected. For example, if the operating radian frequency, ω, is 18^8 rps,

$$L_1 = Z_0 \times X_{L1}/\omega = 50 \times 0.78/10^8 = 0.39\ \mu H$$
$$C_2 = 1/Z_0 \times B_{C2}/\omega = 0.02 \times 0.5/10^8 = 100\ pF$$
$$L_2 = Z_0 \times X_{L2}/\omega = 50 \times 3.0/10^8 = 1.5\ \mu H$$

Figure 14.13 shows an example of matching the output of a transistor amplifier to a 50Ω transmission line using a pi-section LC network. The normalized transistor output impedance is assumed to be $0.17 + j\,0.07$ in a 50Ω system, which is shown as point 1.

The selected starting point is point 1, located at $0.17 + j\,0.07$ on the Smith chart. The first step is to convert to an admittance chart and transfer through the center to point 2. The admittance at this point is $5.0 - j\,2.0$. Next, travel clockwise to point 3, which is reached by adding $-j\,2.0$ capacitive susceptance.

Now, convert to an impedance chart by transferring through the center to point 4. The impedance at this point is $0.2 + j\,0.0$. Then, add a series inductive reactance of $j\,0.4$. to move clockwise to point 5. The impedance at this point is $0.2 + j\,0.4$.

Next, convert to an admittance chart by transferring through the center to point 6. The admittance at this point is $1.0 - j\,2.0$. Finally, move clockwise on the unity conductance circle to point 7, where the admittance is $1 + j\,0$ and the impedance is also $1 + j\,0$, forming a perfect impedance match.

Component values then can be determined by using the values of X and B that were selected. For example, if the operating radian frequency, ω, is again 18^8 rps,

$$C_2 = 1/Z_0 \times B_{C2}/\omega = 0.02 \times 2.0/10^8 = 400 \text{ pF}$$
$$L_1 = Z_0 \times X_{L1}/\omega = 50 \times 0.4/10^8 = 0.2 \text{ }\mu\text{H}$$
$$C_3 = 1/Z_0 \times B_{C3}/\omega = 0.02 \times 2.0/10^8 = 400 \text{ pF}$$

14.13 LC FILTERS

Some of the following material is derived from [4].

LC filters are used extensively in communication and radar systems. These filters sometimes are referred to as lumped constant or lumped-element filters. There are four main types of LC filters: bandpass filters, band-reject filters, lowpass filters, and highpass filters. Typical attenuation versus frequency plots for these filter types are shown in Figure 14.14.

In the case of the bandpass filter response shown in Figure 14.14(a). the attenuation or loss is a minimum at the center of design frequency and increases on either side. The difference between the top of the plot and the first dashed line is the insertion loss over the passband. A typical loss might be 0.5–1.0 at VHF. The next dashed line indicates the −3 dB points on the attenuation curve. The difference between the two vertical lines is the −3 dB bandwidth. This may be as small as 1% of the center frequency or as wide as an octave, depending on the design of the filter. The plot is in decibels and shows a rapid increase in attenuation for the filter on either side of the −3 dB points. The slope of the curves depends on the number of sections or elements used in the filter.

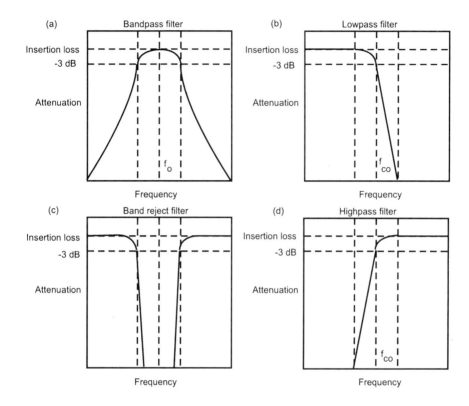

Figure 14.14 Attenuation versus frequency for LC filters: (a) bandpass filter; (b) lowpass filter; (c) band-reject filter; and (d) highpass filter. (*After:* [4].)

The typical frequency response of the lowpass filter is shown in Figure 14.14(b). Attenuation is a minimum from dc to near some desired critical frequency. As the frequency approaches that frequency, the attenuation begins to increase. At the cutoff frequency, the added loss over insertion loss is 3 dB. Above that frequency, the attenuation increases rapidly with the rate of increase, depending on the number of filter stages used and the ripple level for Chebychev designs.

A Chebychev filter is a filter that has an intentional small ripple in the passband. That allows use of designs with rapid increase in attenuation with change in frequency on either side of the −3 dB passband. The other class of filter, known as a Butterworth filter, has a maximally flat response within the passband.

The typical response for the band-reject filter is shown in Figure 14.14(c). The response in this case is the opposite of that of the bandpass filter. The attenuation in this case is low at all frequencies except near a desired center frequency. As the frequency approaches the desired frequency, the attenuation becomes large and is maximum at the center frequency. Again, the second horizontal dashed line indicates the points on the curve where the attenuation in addition to the insertion loss has increased to 3 dB. The bandwidth of the band-reject filter is indicated by the vertical dashed lines. The bandwidth may be small or large, depending on the frequency to reject, the slope of attenuation versus frequency response, acceptable in-band loss, the ripple level, and so on.

The typical frequency response of the highpass filter is shown in Figure 14.14(d). In this case, the attenuation for the filter is very large from dc to a frequency near a selected cutoff frequency. Above that frequency, the attenuation decreases and becomes low for higher frequencies. Again, the slope of the response curve and the cutoff frequency depend on the design of the filter and the number of filter stages used.

The detailed design for multiple-stage LC filters is complex and beyond the scope of this book.

Figure 14.15 shows four examples of bandpass filter circuits made by K&L Microwave Inc., a Dover Technologies Company, 408 Cole Circle Salisbury, Maryland 21801. K&L uses these four circuits plus at least four other circuits for bandpass filters to meet design requirements. Obviously, the circuits shown are simplified and in general show only a fraction of the number of stages used (typically in the range of four to nine).

Figure 14.15(a) shows a two-stage resonant ladder bandpass filter. Each stage includes a series resonant LC circuit followed by a parallel resonant LC circuit. The output is from a tap between the two resonant circuits. A six-stage filter would contain six of these resonant circuit pairs. This circuit is used in HF wideband applications. Figure 14.15(b) shows a three-stage capacitively coupled "tank" circuits bandpass filter. The term tank circuit refers to a parallel resonant LC circuit. The output tap is between the coupling capacitor and the tank circuit. This circuit is an excellent structure for narrowband-use applications.

The circuit in Figure 14.15(c) is a cascade of highpass and lowpass filters. The lowpass filter is made up of an inductor followed by a capacitor with the output tap between the two components. The highpass filter is made up of a capacitor followed by an inductor with the output tap between the two components. A typical bandpass filter circuit might include four to six lowpass stages and four to six highpass stages. This circuit is well suited for bandwidths approaching an octave bandwidth or wider.

The circuit in Figure 14.15(d) shows a two-stage narrowband symmetrical Chebychev bandpass filter structure. Each stage consists of a parallel resonant tank circuit followed by a second parallel resonant tank circuit with the output tap located between the two resonant circuits. A typical bandpass filter may have four to eight such stages.

Lumped Constant Components and Circuits | 455

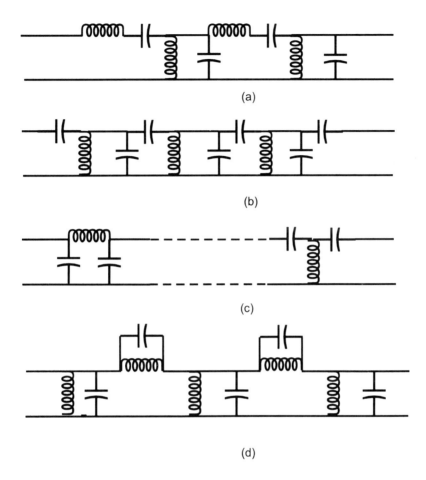

Figure 14.15 Examples of LC bandpass filters: (a) resonant ladder used in wideband applications; (b) capacitive coupled tank circuits; (c) cascade of highpass and lowpass filters; and (d) narrowband symmetrical Chebychev structure. (*After:* [4].)

References

[1] Van Valkenburg, Mac E., ed., *Reference Data for Engineers*, 8th ed., SAMS, Carmel, IN: Prentice Hall Computer Publishing, 1993, Chapter 6.
[2] Ibid., Chap. 5.
[3] Krauss, Herbert L., Charles W. Bostian, and Frederick H. Raab, *Solid State Radio Engineering*, New York: John Wiley & Sons, 1980, Chapter 3 and Appendix 3.1.
[4] K&L Microwave and RF Components Catalog, pp. 4–12.

CHAPTER 15

RF Transformer Devices and Circuits

This chapter discusses RF transformer devices and circuits. Topics include conventional transformers, core material for RF transformers, IF amplifier transformers, HF wideband conventional transformers, transmission-line transformers, and power combiners and splitters. Some of the information presented is based on [1–3].

15.1 CONVENTIONAL TRANSFORMERS

Figure 15.1 shows a conventional two-winding transformer with windings on a ferrite core. In an ideal transformer, the following relationships apply:

$$\text{Turns ratio} = a = N_1/N_2 = E_1/E_2 = I_2/I_1 \tag{15.1}$$

$$R_1 = a^2 R_2 \tag{15.2}$$

In practice in a real transformer, winding loss, leakage inductance, and magnetizing inductance must be taken into account. That may be done using the equivalent circuits in Figure 15.2.

Nomenclature for these circuits is as follows:

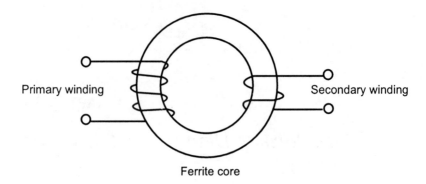

Figure 15.1 Conventional transformer with two windings.

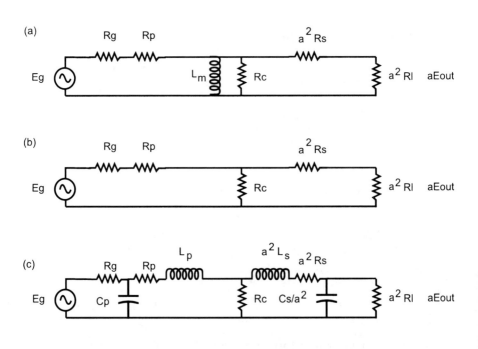

Figure 15.2 Equivalent circuits for conventional two-winding transformer: (a) low-frequencies equivalent circuit; (b) intermediate-frequencies equivalent circuit; and (c) high-frequencies equivalent circuit.

C_p = primary equivalent shunt capacitance
C_s = secondary equivalent shunt capacitance
E_g = rms generator voltage
E_{out} = rms output voltage
L_m = magnetizing inductance
L_p = primary leakage inductance
L_s = secondary leakage inductance
R_c = core-loss equivalent shunt resistance
R_g = generator resistance
R_L = load resistance
R_p = primary winding resistance
R_s = secondary winding resistance

The LF response of a conventional transformer is degraded by a shunt susceptance that appears across the primary winding. The magnetizing inductance, L_m, of the transformer is given by

$$L_m = 4\pi n^2 \mu_r \mu_0 A_e / L_e \qquad (15.3)$$

where

n = number of turns in the primary winding
μ_r = relative permeability of the core
μ_0 = permeability of vacuum = $4\pi \times 10^{-7}$ H/m
A_e = effective area of the core (square meters)
L_e = average magnetic path length, (meters)

For an example of the use of (15.3), assume the following:

$n = 4$
$\mu_r = 100$
$\mu_0 = 4\pi \times 10^{-7}$ H/m
$A_e = 2 \text{ cm}^2 = 2 \times 10^{-4} \text{ m}^2$
$L_e = 6 \text{ cm} = 0.06\text{m}$

Substituting those values into (15.3), we have

$$L_m = 4\pi \times 16 \times 100 \times 4\pi \times 10^{-7} \times 2 \times 10^{-4}/6 \times 0.06$$
$$L_m = 8.4 \times 10^{-5} \text{ H} = 84 \ \mu\text{H}$$

The inductive shunt susceptance depends on the frequency. For example, at 200 kHz, the inductive reactance for L_m would be 105.5Ω and the susceptance would be 0.0095S. The LF cutoff would be at 200 kHz if the parallel resistance is also 105.5Ω.

The equivalent circuit shown in Figure 15.2(b) is for the case where the magnetizing inductive reactance is high, and the primary and secondary inductive reactances are low. This is the normal operating frequency band for the transformer.

The equivalent circuit shown in Figure 15.2(c) is for the case where the magnetizing inductive reactance is high, and the primary and secondary inductive reactances are also high. This is the upper cutoff band for the transformer.

Figure 15.3 shows the case of a conventional autotransformer. In this case, there is only one winding on the core instead of two. The input to the transformer is between one side of the coil and a tap point located between the ends of the coil. In the Figure 15.3, the tap is located at the center, which makes the primary winding have one-half the turns of the secondary winding. In the ideal case, the output has twice the voltage of the input and half the current. The output impedance is four times the input impedance. As we will see later, many of the RF transformers used are of this type. Basically, RF autotransformers can be compared to their LF counterpart, except that, with increasing frequency, leakage reactance becomes an important factor.

15.2 MAGNETIC CORE MATERIAL FOR RF TRANSFORMERS

Some type of magnetic core is required for the lower frequencies and for wideband RF transformers to extend coverage at the low end of the frequency band. Powdered

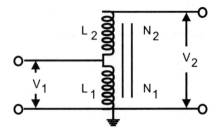

Figure 15.3 A conventional autotransformer.

iron often is used at the lower frequencies, while ferrites are the most common magnetic materials used for the higher frequency RF transformers.

Two main types of ferrites are used: nickel-manganese compositions and nickel-zinc compositions. Nickel-manganese compositions have higher permeabilities than nickel-zinc ferrites but larger losses at the higher frequencies. For that reason, nickel-zinc ferrites usually are used for the higher frequencies. One disadvantage of the nickel-zinc ferrites is that their Curie points can be as low as 130°C. The Curie point is the temperature at which damage may be done to the magnetic material by heat. Nickel-zinc ferrites can be manufactured only with relative permeability (μ_r) of less than about 1,000.

Because high-permeability ferrites saturate easier than low-permeability ones, it is good design practice to limit their maximum flux densities as shown in Table 15.1.

With RF transformers, either the primary or the secondary can be used for B_{max} calculations, but the 50Ω side (if applicable) commonly is used for convenience and standardization. A general formula for calculating flux density for a ferrite core is given by

$$B = V_{rms} \times 10^8 / 4.44 \, fnA \tag{15.4}$$

where

B = flux density (gauss)
V_{rms} = rms voltage on the winding
f = frequency (Hz)
A = core cross-sectional area (cm^2)
n = number of turns

For an example of the use of (15.4), assume the following:

Table 15.1
Maximum Flux Densities for High-Permeability Ferrites

μ_r	B_{max} (G/cm^2)
400–800	40–60
100–400	60–90
<100	90–120

$V_{rms} = 50V$

$f = 2.0$ MHz

$A = 1.0$ cm^2

$n = 4$

Substituting those values into (15.4) gives

$$B = 50 \times 10^8/(4.44 \times 2 \times 10^6 \times 4 \times 1)$$
$$B = 140.7G$$

Equation (15.4) can be modified to show the required core area for a given flux density, as shown next:

$$A = V_{rms} \times 10^8/4.44 \, fnB \tag{15.5}$$

where the terms are as defined for (15.4).

15.3 TUNED TRANSFORMERS

Tuned transformers, as used in IF amplifiers, are often constructed with adjustable rod-shaped cores, which can be moved in and out of the windings to change the value of the inductance, thereby changing the resonant frequency of the tuned circuit.

Figure 15.4 shows a transformer circuit with one side tuned. The circuit provides an alternative way to attain impedance matching. It can provide isolation between input and output circuits and introduce a phase reversal, if desired. The left-side coil, L_1, is the primary winding of the transformer, and the right-side coil, L_2, is the secondary winding. The coupling coefficient, k, is equal to the mutual inductance, M, divided by the square root of the product of L_1 and L_2, as shown in (15.6).

$$k = M/(L_1L_2)^{1/2} \tag{15.6}$$

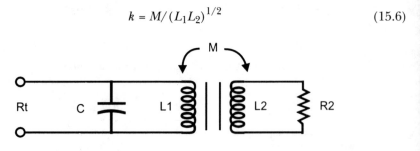

Figure 15.4 Single-tuned transformer circuit.

Figure 15.5 shows the case of a double-tuned transformer. These circuits have been used extensively in receiver IF amplifier stages as the primary means of filtering signals. They permit flexibility in adjusting the shape of the selectivity curve. Although they now are being supplanted by ceramic, crystal, and surface acoustic wave filters, they still are used where widely different impedance levels must be matched and in some FM discriminators. Both single-tuned and the double-tuned transformer circuits are packaged inside small metal cans or shields. In some cases, the transformers are designed for operation at 455 kHz, as used in standard AM IF systems. In other cases, they are designed for operation at 10.7 MHz, as used in FM IF systems.

Figure 15.5(a) shows the double-tuned transformer circuit. Figure 15.5(b) shows the frequency response of the double-tuned transformer for three values of the transformer coupling coefficient, k. Those values are the critical coupling, the below-critical coupling, and the above-critical coupling. For k equal to or less than k_c, the response curve has a single peak. For k greater than k_c, the resonant curve

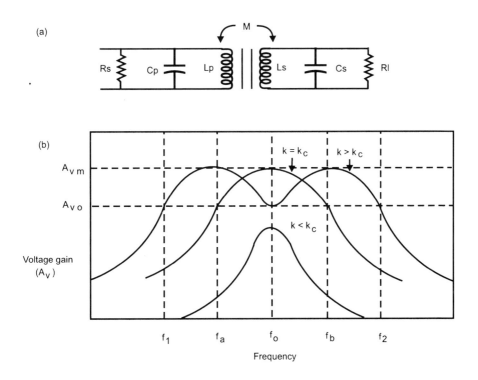

Figure 15.5 Double-tuned transformer: (a) transformer equivalent circuit and (b) frequency response of double-tuned transformer. (*After:* [1].)

has two peaks, as shown in Figure 15.5(b). This condition is called the overcoupled case. The ratio of the peak gain to the gain at the valley at f_0 is given approximately by $0.5(kQ + 1/kQ)$ and is controlled by the choice of k and Q. The bandwidth of the circuit often is defined as $f_2 - f_1$, as shown in the figure.

For maximum power transfer at resonance, $R = (\omega_0 M)^2/R$, or $\omega_0 M = R$. The circuit Q at resonance is defined by $Q = \omega_0 L/R$. It can be shown that the coefficient of coupling for the maximum power transfer condition (called critical coupling) is $k_c = 1/Q$.

It should be noted that if unequal impedances are to be matched on the two sides of the circuit, the primary or the secondary coil can be tapped.

Much of the foregoing discussion of tuned transformers is based on [1].

15.4 HIGH-FREQUENCY WIDEBAND CONVENTIONAL RF TRANSFORMERS

High-frequency (HF), wideband RF transformers are used for the following functions:

- Impedance transformation;
- Baluns;
- Phase inversion;
- Power combining for push-pull amplifiers;
- Hybrid combiners for multiple power sources.

There are two main types of HF, wideband RF transformers: conventional transformers and transmission-line transformers. The conventional transformer is usually inferior in performance to a transmission-line transformer. The difference is mainly in the power-handling capability, loss factor, and bandwidth. The conventional transformer, however, can be constructed for a wider range of impedance ratios than the transmission-line transformer.

Figure 15.6 shows a popular, conventional RF transformer that is often used for HF and VHF applications. This transformer uses only one turn for the primary winding and two or more turns for the secondary. Its construction is two metal tubes with a connection between the tubes on one end, making a U-shaped single turn. The second set of wires are threaded through the two tubes to form a continuous multiturn winding. That results in a tight coupling between the two windings with relatively low mutual winding capacitance, thereby allowing its use at very low impedance levels. This transformer has the disadvantage that only integer-squared impedance ratios, such as 4:1, 9:1, and 16:1, are possible.

The bandwidth for the one-turn transformer is largely determined by the impedance ratio used. For example, a 9:1 impedance ratio transformer can have a bandwidth of up to about 60 MHz, a 25:1 transformer can have a bandwidth of

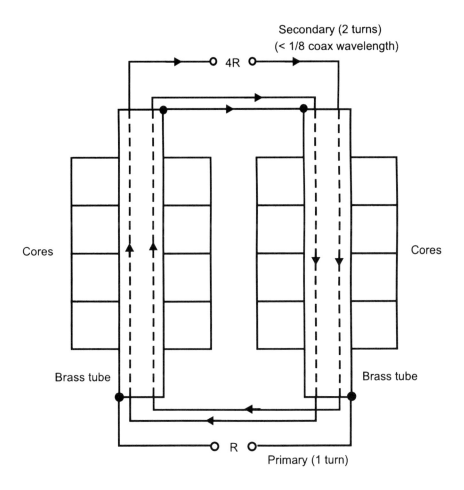

Figure 15.6 Conventional RF transformer using brass tubes. (*After:* [1].)

up to about 30 MHz, and a 36:1 unit has a usable bandwidth of only about 15–20 MHz.

Figure 15.7 shows another variation of the conventional HF transformer. With this transformer, a U-shaped section of semirigid coax cable has two straight sections of semirigid coaxial cable soldered to it. The center wires of the coax cables are connected so they make two loops, as shown. This arrangement is mounted in a ferrite core consisting of an E and an I core joined. The transformer so produced has a 1:4 impedance ratio. The outer conductors at their open ends provide terminals for the single-turn winding, and the two coaxial center conductor ends provide terminals for the two-turn winding.

A 1:9 impedance ratio transformer of this type is shown in Figure 15.8. It is made with two U-shaped sections of semirigid coax cable plus two straight sections

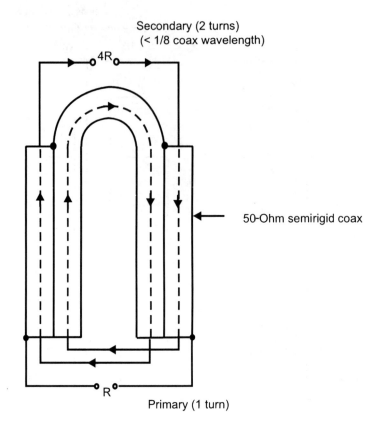

Figure 15.7 HF 4:1 transformer using coax cables. (*After:* [2].)

of semirigid coax cable soldered together to form a single U-shaped structure. The center wires of the coax cables are connected in series to form three loops. This concept can be extended to form four loop systems for a 1:16 impedance ratio transformer.

The types of transformers shown in Figures 15.6, 15.7, and 15.8 may or may not use ferrite cores to extend the LF coverage. At the higher frequencies, the core is of no value and would not be used. Transformers of the type shown in Figures 15.7 and 15.8 are usable up to about 300 MHz. The HF is limited by the fact that the total length of the high-impedance winding must be kept below about one-eighth of a wavelength at the highest frequency of operation to avoid major resonances. For operation at 200 MHz, the physical length of a U-shaped 4:1 unit is limited to about 3.5 cm. The length of a U-shaped 9:1 unit is limited to about 2.5 cm.

This type of transformer, having a typical length of about 3 cm, can be operated at frequencies as low as 3–10 MHz when a ferrite core is used. Without that core,

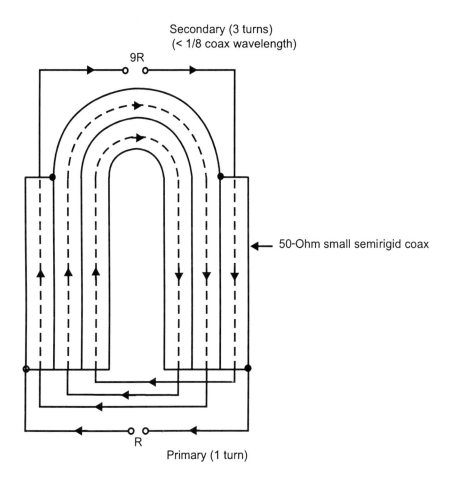

Figure 15.8 HF 9:1 transformer using coax cables. (*After:* [2].)

the LF limit is about 100 MHz. The transformer has a typical power rating of 200–300W.

Another type of VHF transformer using semirigid coax is shown in Figure 15.9(a). This type of transformer often is used at frequencies as high as 1,000 MHz. No core is used at the higher frequencies. The operation of this transformer is as follows. A single-loop path is formed starting at terminal A and following lines 1, 2, 3, 4, and 5 to terminal B. A two-loop path is formed starting at terminal C and following lines 6, 7, 8, 5, 4, 3, 2, 1, 9, and 10 to terminal D. We thus have a 2:1 voltage transformation or a 4:1 impedance transformation transformer.

A 1:1 balun transformer of this type is shown in Figure 15.9(b). This balun transformer is used to convert from a coax unbalanced line to a balanced two-wire line. There are no loops in this case. The voltage out is equal to the voltage in.

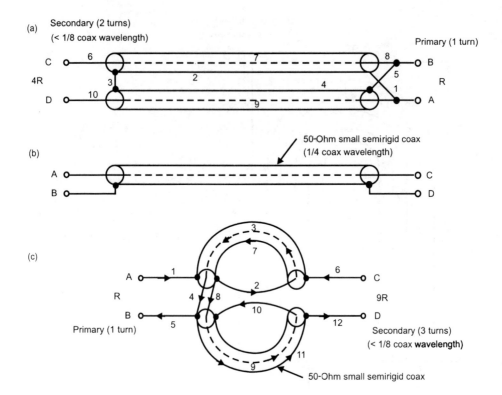

Figure 15.9 Coaxial transformers: (a) 4:1 coaxial transformer; (b) 1:1 coaxial balun transformer; and (c) 9:1 coaxial transformer.

The length of this type of transformer (balun) must be one-quarter guide wavelength in the coax. The coaxial cable impedance is $Zo\Omega$ in and out. Because of the one-quarter wavelength property, this type of transformer is inherently a narrowband device, providing no isolation between the balanced and the unbalanced ends.

Figure 15.9(c) shows the case of a 3:1 voltage transformer or a 9:1 impedance transformer of this type. A single-loop path is formed starting at A and following lines 1, 2, 3, 4, and 5 to B. A three-loop path is formed starting at C and following lines 6, 7, 8, 9, 10, 11, and 12 to D.

15.5 TRANSMISSION-LINE TRANSFORMERS

A number of RF applications require wideband transformers covering an octave or more, for example, hybrid combiners for wideband power amplifiers, hybrid power splitters, baluns, and impedance transformers. The type of transformers used are known as transmission-line transformers because they use transmission lines of either twisted-pair type or coaxial type with the characteristic impedance of the lines being an important part of the design. These transformers are not to be confused with the coaxial transformers discussed in Figure 15.4. Concepts for transmission-line transformers are shown in Figure 15.10.

To understand the bandwidth limitations inherent in conventional transformers, consider the conventional autotransformer shown in Figure 15.10(a). This unit is center tapped on the input to provide a two-to-one stepup in voltage and a four-to-one stepup in impedance. LF performance is degraded by the shunt susceptance of the windings, while HF performance is degraded by the series reactance of the windings.

Figure 15.10(b) shows the autotransformer rearranged into a primary winding and a secondary winding with connection between windings. The impedance and voltage transformations are the same as in Figure 15.10(a). The two separate windings can be replaced by the two conductors in a single transmission-line, as shown in Figure 15.10(c). There, the transformer consists of a magnetic core with a two-wire, twisted-pair transmission line wound around the core. At low frequencies, the

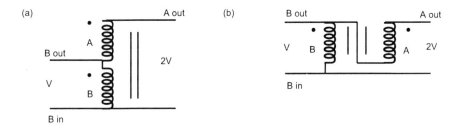

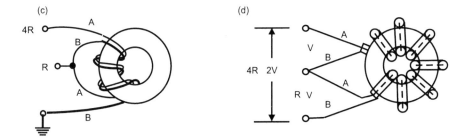

Figure 15.10 Concepts for transmission-line transformers: (a) autotransformer; (b) transmission-line transformer; (c) twisted-pair transmission-line transformer; and (d) coax transmission-line transformer. (*After:* [1].)

shunt susceptance degrades performance, as in the conventional autotransformer. At high frequencies, the circuit behaves as a transmission line, thereby greatly extending the HF limit for the transformer.

Notice the way in which the transmission lines are connected to provide the 4:1 impedance transformation. In the case of the twisted-pair transmission line in Figure 15.10(c), the two wires are labeled A and B. After looping around the core a number of times, the end of wire B is connected to the beginning of wire A. The low-impedance winding is the B-in to B-out connection. The high-impedance winding is the B-in to A-out connection. This winding has twice the path length and twice the number of turns as the low-impedance winding. Thus, the circuit is a 1:4 impedance ratio transformer.

Figure 15.10(d) illustrates a magnetic core with a coaxial transmission line wound around the core. The outer conductor is labeled B, and the inner conductor is labeled A. After looping around the core a number of times, the end of the outer conductor B is connected to the beginning of inner conductor A. The low-impedance winding is the B-in to B-out connection. The high-impedance winding is the B-in to A-out connection. This winding has twice the path length and twice the number of turns as the low-impedance winding. The circuit also behaves as a 1:4 impedance ratio transformer.

At low frequencies, the shunt susceptance degrades performance, as in the conventional autotransformer. At high frequencies, the circuit behaves as a transmission line, thereby greatly extending the HF limit for the transformer.

A number of possible transmission-line transformer configurations provide different impedance transformations. Four such circuits are shown in Figure 15.11: a balun with a 1:1 transformation ratio, a transformer with a 4:1 impedance ratio, a two-transformer combination with a 9:1 impedance ratio, and a three-transformer combination with a 16:1 impedance ratio. A second approach to providing a 16:1 impedance ratio is to use two 4:1 impedance transformers in series.

As in the case of the conventional transformer, the LF performance of the transmission-line transformer is determined by the ferrite core. With transmission-line transformers, the characteristic impedance of the transmission lines must be correct to take advantage of the optimum performance. At high frequencies, the series reactance combines with the interwinding capacitance, and the circuit behaves as a transmission line, greatly extending the HF response. The power transferred from the input to the output is coupled, not through the magnetic core except at very low frequencies, but rather through the dielectric medium separating the line conductors. It follows that a relatively small cross-section core can operate unsaturated at very high power levels.

The characteristic impedance of the line should be the geometric mean of the input and output impedances. For example, if the input impedance for the transmitter is 12.5Ω and the output impedance is 50Ω, the line characteristic impedance should be

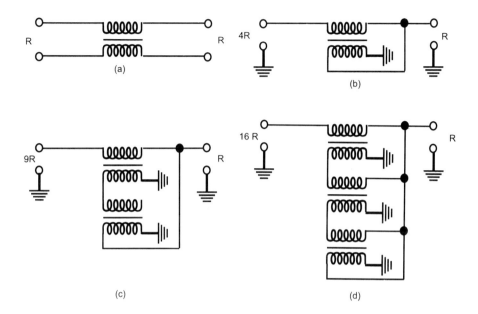

Figure 15.11 Transmission-line transformer configurations: (a) balun (1:1); (b) 4:1 transformer; (c) 9:1 transformer; and (d) 16:1 transformer. (*After:* [1].)

$$Z_0 = (R_{in} \times R_{out})^{1/2} \qquad (15.7)$$

$$Z_0 = (12.5 \times 50)^{1/2} = 25\Omega$$

15.6 POWER COMBINERS AND SPLITTERS

Hybrid coupler circuits can be used to combine two or more power amplifiers when a larger output is needed than can be obtained from a single amplifier. Direct parallel operation of the devices is usually unsatisfactory because the currents do not always divide equally among the devices.

Figure 15.12 shows hybrid couplers using both conventional transformers and transmission-line transformers. The basic hybrid combiner using a conventional center-tapped transformer, shown in Figure 15.12(a), is used to combine the outputs of two power amplifiers to achieve higher power.

The two amplifiers in Figure 15.12(a) are driven 180 degrees out of phase (push-pull) with respect to each other. The transformed sum of the two output currents is provided to the load R_o. The difference in the two output currents flows through the resistor R_h, provided for isolation purposes.

Figure 15.12(b) shows a hybrid splitter that uses a conventional center-tapped transformer. This type of circuit is used to drive two loads from a single source.

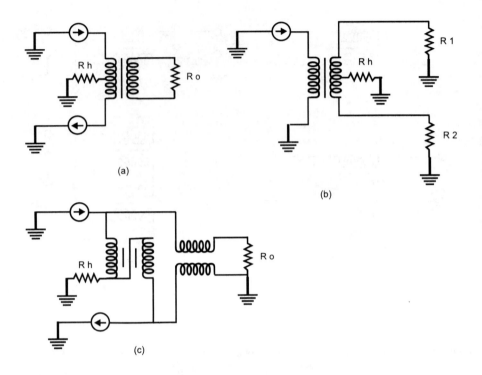

Figure 15.12 Hybrid couplers using transformers: (a) basic hybrid combiner; (b) hybrid splitter; and (c) broadband hybrid combiner. (*After:* [1].)

The circuit in Figure 15.12(c) is a wideband hybrid combiner that uses transmission-line transformers. By using many such combiners, it is possible to provide RF solid-state power amplifiers with power levels of more than a thousand watts at VHF and UHF frequencies.

References

[1] Krauss, Herbert L., Charles W. Bostian, and Frederick H. Raab, *Solid State Radio Engineering*, New York: John Wiley & Sons, 1980, pp. 371–382.

[2] Dye, Norm, and Helge Granberg, *Radio Frequency Transistors, Principles and Practical Applications*, Boston: Butterworth-Heinemann, 1993, Chap. 10.
Gottlieb, Irving M., *Solid-State High-Frequency Power*, Reston, VA: Reston Publishing Co., 1982.

[3] Van Valkenburg, Mac E., ed., *Reference Data for Engineers*, 8th ed., SAMS, Carmel, IN: Prentice Hall Computer Publishing, 1993, Chap. 13.

CHAPTER 16

Piezoelectric, Ferrimagnetic, and Acoustic Devices and Circuits

This chapter discusses piezoelectric, ferrimagnetic, and acoustic devices and circuits. Topics include quartz crystal oscillators, quartz crystal filters, ceramic filters, dielectric resonator oscillators, yttrium iron garnet (YIG) filters, YIG oscillators, surface acoustic wave (SAW) delay lines, and bulk acoustic wave (BAW) delay lines.

16.1 QUARTZ CRYSTAL RESONATORS AND OSCILLATORS

Some of the material presented in this section is quoted or adapted from [1–3].

Quartz crystals are piezoelectric devices. The term piezoelectric refers to the property of a material in which mechanical deformation along one crystal axis results in an electrical potential or electric field along another axis. Conversely, an applied voltage deforms and produces mechanical stress in the crystal.

Quartz crystals are used extensively as the frequency reference or frequency controlling element for oscillators used in RF transmitters and receivers. Figure 16.1(a) shows the symbol for the quartz crystal. Figure 16.1(b) shows the equivalent circuit for the crystal. We see that the crystal acts like a complex RLC resonant circuit.

Figure 16.2 shows an example of a terminal impedance characteristic for a quartz crystal. Notice that the crystal has two resonance frequencies. The first or lower frequency resonance is a series resonance in which the impedance is resistive and low. The second or higher frequency resonance is a parallel resonance in which the impedance is also resistive but very high. Between the two resonance frequencies, the impedance is inductive and has a positive phase angle near 90 degrees. Above the parallel resonance frequency, the impedance is capacitive and has a negative phase angle near −90 degrees.

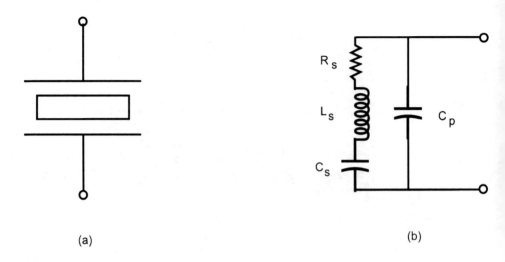

Figure 16.1 Quartz crystal: (a) crystal symbol and (b) equivalent circuit for a quartz crystal. (*After:* [2].)

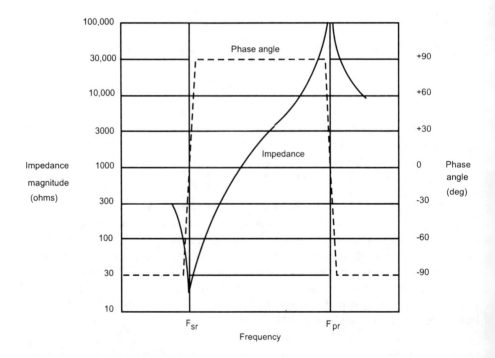

Figure 16.2 Example of a terminal impedance for a crystal (fundamental mode). (*After:* [2].)

Capacitors can be used for frequency correction or adjustment in conjunction with quartz crystals. The capacitors can be connected in either parallel or series and as either mechanically adjustable capacitors or electronically variable varactor diodes. Varactors also can be used for voltage-controlled frequency modulation of a crystal oscillator.

A quartz crystal can vibrate in a number of mechanical modes. The mode with the lowest resonance frequency is called the fundamental mode. Higher-order modes are called overtones or harmonics. Crystals intended for oscillator frequencies up to about 15 MHz normally operate in the fundamental mode and can be used in either series or parallel resonance. Above 15 MHz, an overtone normally is utilized. For that condition, operation is always with in-series resonant mode. The highest frequency that can be obtained with an overtone crystal is about 180 MHz.

Quartz crystals have a number of different sizes and shapes and normally are encapsulated in metal holders. Many crystals for surface-mount construction are plastic encapsulated. Their size depends on operating frequency. A typical holder for a 5-MHz crystal has the dimensions 0.75 in × 0.75 in × 0.35 in. A typical size holder for a 15-MHz crystal has the dimensions 0.36 in × 0.43 in × 0.18 in, due to a smaller-size crystal being employed. For some crystals, pins are used as terminals. In other cases, flexible wire leads are provided. CINOX offers a complete line of TO-5 packaged crystals from 10 MHz to 180 MHz. Table 16.1 lists several characteristics of these crystals.

A temperature-compensated crystal oscillator is one to which a network has been added to vary the crystal's load impedance over temperature. The network typically consists of a thermistor-resistor network driving a varactor diode in series with the crystal. If the elements of the thermistor-resistor network are chosen properly, the output shift caused by the varactor diode will cancel almost exactly the crystal's temperature characteristic. The net result is a very small variation of crystal oscillator frequency with changes in temperature. For example, the frequency variation over a frequency range from −55°C to +95°C is less than 0.5 ppm.

An oven-controlled crystal oscillator is one that is operated inside a temperature-controlled oven. With such a system, it is possible to achieve extremely small

Table 16.1
TO-5 (HC-35) Packaged Crystals Made by CINOX

Crystal Part Number	Frequency Range (MHz)	Overtone	R_S (max) (Ω)	Operating Temperature (°C)	Frequency Tolerance (ppm)
PC3125-0	30.0–70.0	3	40	25	±5
PC5125-0	60.0–130.0	5	60	25	±5
PC7125-0	130.0–170.0	7	150	25	±5

frequency errors. Different oven and oscillator systems have different accuracies. The best unit reported by CINOX for operation in the 0–50°C range has a $\Delta f/f$ of 2×10^{-10}. Other units reported have $\Delta f/f$ values of 5×10^{-9}, 2×10^{-8}, and 1×10^{-7}. As always, better performance costs more.

Figures 16.3 and 16.4 show a number of examples of transistor oscillators using quartz crystals to determine the operating frequency. The circuit at Figure 16.3(a) is a Pierce oscillator. This is basically a common source Colpitts circuit with the crystal acting as an inductor and forming a resonant circuit with C_G, C_D, and the internal capacitances of the FET. Notice that in this case the crystal operates in the frequency range between the series resonance and the parallel resonance. The approach to designing a FET Pierce crystal oscillator often is a cut-and-try approach. Trim capacitors often are used for frequency adjustments.

Figure 16.3(b) shows a Miller oscillator. A Miller oscillator is similar to the tuned-input, tuned-output oscillator. Both the crystal and the output tank circuit (parallel RLC circuit) appear as inductive reactances at the oscillation frequency.

Figure 16.4(a) shows the Colpitts crystal oscillator with the crystal operating in the series resonant mode. At resonance, the crystal behaves as a small resistance. That serves as the necessary feedback from the collector of the transistor to the emitter at only the desired frequency. This circuit is particularly useful at higher frequencies, where series resonant, overtone crystals are normally used.

Figure 16.4(b) shows a second type of Colpitts crystal oscillator. The crystal is connected between the base and the ground. It also operates in a series resonant

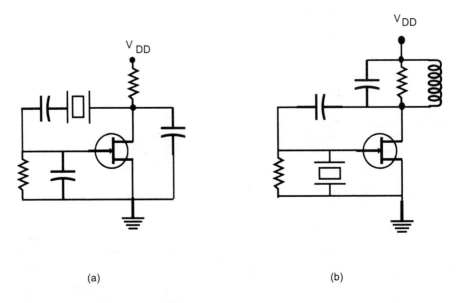

Figure 16.3 Examples of crystal oscillators: (a) Pierce crystal oscillator circuit and (b) Miller crystal oscillator circuit. (*After:* [2].)

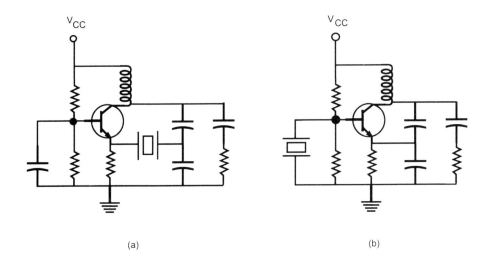

Figure 16.4 Two types of Colpitts crystal oscillators: (a) crystal in series resonant mode to provide feedback to emitter and (b) crystal in a series resonant mode to ground the base of the transistor. (*After:* [2].)

mode and grounds the transistor base at the operating frequency. In both of these Colpitts circuits, the transistor operates as a common-base amplifier, which requires that the base be ac grounded. Again, this type of circuit is useful at higher frequencies, where series resonant overtone crystals usually are utilized.

16.2 MONOLITHIC CRYSTAL FILTERS

Figure 16.5(a) illustrates a piezoelectric quartz crystal with three electrodes. The equivalent circuit for this filter is shown in Figure 16.5(b). Such a monolithic crystal can be used to fabricate a very high Q bandpass filter. The bandwidth of these filters is limited to a maximum of a few tenths of 1%. Their useful frequency range is about 5–350 MHz.

A typical frequency response of a four-pole monolithic crystal filter with center frequency at 75 MHz is shown in Figure 16.5(c). Quartz crystal filters have much higher Q values than ceramic filters and higher operating frequencies.

16.3 CERAMIC FILTERS

Some of the material presented in this section and Section 16.4 is quoted or adapted from [4–6].

478 | RF Systems, Components, and Circuits Handbook

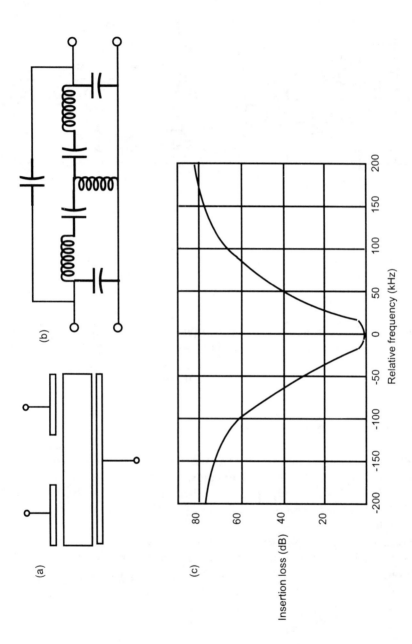

Figure 16.5 Characteristics for a monolithic crystal filter: (a) three-electrode monolithic crystal filter configuration; (b) approximate equivalent circuit; and (c) example frequency response for filter. (*After*: [5].)

Ceramic filters are made from piezoelectric ceramics. They are available with center frequencies ranging from a few kilohertz to more than 1.0 GHz. They have bandwidths ranging from 0.05–20%.

Figure 16.6 illustrates a ceramic disc resonator with two electrodes. Figure 16.6(a) shows the ceramic disk resonator configuration. Figure 16.6(b) shows the equivalent circuit for this single resonator. Figure 16.6(c) shows the impedance magnitude as a function of frequency. Ceramic resonators have a series resonant frequency in which the impedance is low and a parallel resonant frequency in which the impedance is very high.

A three-electrode piezoelectric ceramic resonator is shown in Figure 16.7(a). Figure 16.7(b) shows the approximate equivalent circuit for this three-electrode ceramic resonator. Figure 16.7(c) shows its approximate frequency response. This type of device can be used as a bandpass filter for IF amplifiers operating at 455 kHz. They also can be used for this purpose at 10.7 MHz.

16.4 DIELECTRIC RESONANT OSCILLATORS

16.4.1 Dielectric Resonator Description and Parameters

A dielectric resonant oscillator is a free-running oscillator stabilized by the insertion of a high-Q dielectric resonator into the circuit. This dielectric resonator acts in many ways like an air-filled metallic cavity resonator with the advantage that it is much smaller in size. That is due to the higher dielectric constant of the titanate material used, compared to that of air. The wave velocity in the cavity is reduced by approximately the square root of the dielectric constant. The type of resonators used have dielectric constants of about 38, so dimensions for the cavity are reduced by a factor of about 6.2. The material it is made of is usually a barium titanate-based material. The unloaded Q of such a resonator is about 7,000 at 4 GHz.

Figure 16.8 shows one configuration used for the dielectric resonator. Example dimensions are given for the resonator as a function of frequency.

The height of the resonator, H, is one-half wavelength at the resonant frequency. For example, at 10 GHz, the wavelength in free space is 3 cm. The wavelength in the dielectric is that value divided by the square root of the dielectric constant. If we assume that the dielectic constant is 38, the wavelength in the dielectic at 10 GHz is 0.49 cm. One-half wavelength would be 0.24 cm, or 0.095 in, as given in the table.

The diameter of the resonator is greater than one-half the cutoff wavelength. The cutoff wavelength in free space is given by

$$\lambda_0 = 2\pi r/(kr) \tag{16.1}$$

where

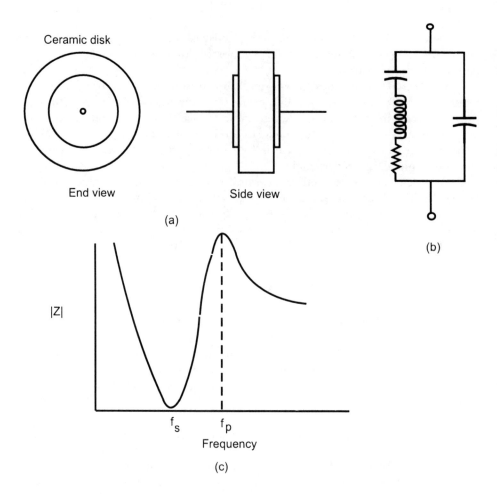

Figure 16.6 Ceramic disc resonator characteristics: (a) ceramic disk resonator configuration; (b) equivalent circuit; and (c) impedance magnitude versus frequency. (*After:* [5].)

r = radius of resonator

(kr) = solution of a Bessel function equation

For the $TE_{01\delta}$ mode, $(kr) = 3.83$. Thus, the cutoff wavelength for this mode is

$$\lambda_{c0} = 2\pi r/(kr) = 1.6\ r = 0.8\ D$$

For example, at 10 GHz and a dielectric constant of 38, the cutoff wavelength is 0.49 cm = 0.19 in and $D = 0.15$ in. The required diameter to be well above cutoff is about 0.22 in.

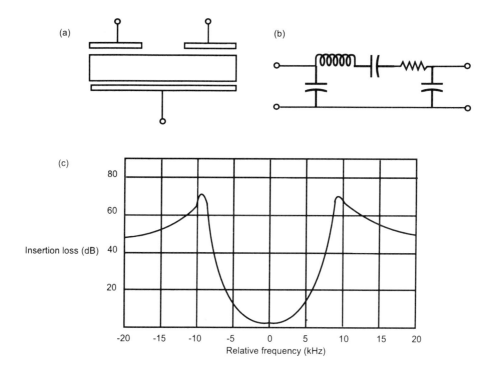

Figure 16.7 Characteristics for a ceramic IF amplifier filter: (a) three-electrode ceramic filter configuration; (b) approximate equivalent circuit; and (c) frequency response. (*After:* [5].)

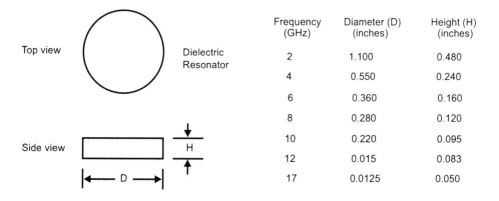

Figure 16.8 Dielectric resonator dimensions. (*After:* [6].)

16.4.2 Coupling Between a Dielectric Resonator and a Microstrip Line

With a metallic cavity, coupling to the cavity is by means of a probe or a loop inserted into the inside of the cavity. With a dielectric resonator with no metallic walls, coupling usually is done using a microstrip line near the edge of the resonator. This configuration is shown in Figure 16.9(a).

In the Figure 16.9, only one edge coupling is used. In other cases, two or more microstrip lines are used for multiple-edge coupling.

The dielectric resonator oscillator (DOR) configuration shown in Figure 16.9(b) uses a metallic enclosure and a pedestal or spacer for the resonator. The spacer is added to improve the loaded Q by optimizing the coupling. The lateral distance between the resonator and the microstrip conductor primarily determines the amount of coupling between the resonator and the microstrip line. The metallic shielding is required to minimize radiation losses and reduce unwanted stray coupling to adjacent circuitry.

The dielectric resonator placed adjacent to the microstrip line operates like a reaction cavity that reflects the RF energy at the resonant frequency. It is similar to an open circuit with a voltage maximum at the reference plane at the resonant frequency.

16.4.3 Mechanical and Electrical Tuning of Dielectric Resonators

It is possible to tune the dielectric resonator by providing mechanical tuning, electrical tuning, or a combination of both. Mechanical tuning involves use of a capacitance plate that can be moved with respect to the resonator. Electrical tuning involves the use of a voltage variable capacitor (varactor diode) coupled to the resonator with a microstrip line. Mechanical tuning is illustrated in Figure 16.9(b).

16.4.4 Examples of Dielectrically Stabilized Oscillators

Examples of dielectrically stabilized oscillators are produced by Anzac, a division of Adams-Russell Co. Inc., Burlington, MA. This company offers dielectric resonant oscillators from 2.65 GHz (dimensions 3.5 in × 1.92 in × 1.66 in) to 12 GHz (dimensions 1.88 in × 1.15 in × 0.77 in). Output powers are 10 dBm minimum. Mechanical tuning ranges from ±5 MHz at 2.65 GHz to ±15 MHz at 12 GHz. Harmonics typically are −25 dBc, and spurious signals typically are −90 dBc. Phase noise typically is −95 dBc/Hz at f_0 ±10 kHz. These units use SMA connectors for RF and solder feed-through for dc.

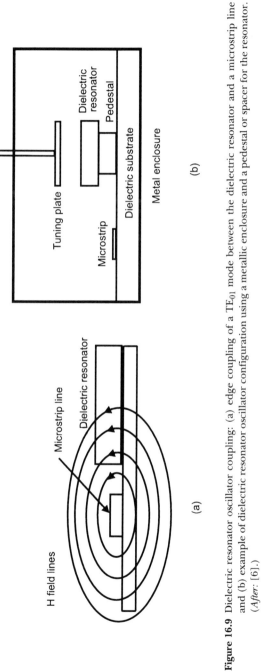

Figure 16.9 Dielectric resonator oscillator coupling: (a) edge coupling of a TE_{01} mode between the dielectric resonator and a microstrip line and (b) example of dielectric resonator oscillator configuration using a metallic enclosure and a pedestal or spacer for the resonator. (*After:* [6].)

16.5 YIG RESONATORS AND FILTERS

Some of the material presented in this section is quoted or adapted from [7].

16.5.1 Ferrimagnetic Resonance in Yttrium Iron Garnet Crystals

One of the most interesting types of crystals is the YIG crystal. This crystal is a magnetic insulator that resonates at a microwave frequency when magnetized by a suitable dc magnetic field. A unique feature of a YIG crystal is that for a spherical configuration the resonant frequency is related to only the direct magnetic field, not to its dimensions. YIG resonators are thus constructed as small, highly polished spheres with diameters between 1.0 mm and 2.0 mm. YIG crystals of this type are used in magnetic-field-controlled bandpass filters, bandstop filters, oscillators, limiters, discriminators, and numerous microwave systems. The useful frequency range for such systems is about 500 MHz to about 40 GHz.

The basic ferrimagnetic resonance phenomenon in a YIG crystal can be explained in terms of spinning electrons, which create a net magnetic moment in each molecule of the crystal, illustrated in Figure 16.10(a). Application of a biasing magnetic field causes the magnetic dipoles to align themselves in the direction of the magnetic field, thus producing a strong net magnetization. Any microwave magnetic field at right angles to the dc magnetic field results in precession of the magnetic dipoles around the biasing field. If the frequency of the microwave field coincides with the natural precessional frequency of the YIG crystal, strong interaction results.

For spherical resonators, the resonant frequency, f_0, is given by

$$f_0 = \gamma(H_0 + H_a) \tag{16.2}$$

where

γ = charge-mass ratio of an electron or the gyromagnetic ratio
   = 2.21×10^5 (rad/s)/(A/m)
H_0 = applied direct field (A/m)
H_a = internal crystal anistropy field

H_a usually is very small compared to H_0 and so is neglected in the following example.

For an example of the use of (16.2), assume $H_0 = 85.3$ kA/m. Thus, the resonant frequency is

$$f_0 = \gamma(H_0 + H_a)$$
$$f_0 = 2.21 \times 10^5 \times 85.3 \times 10^3 = 1.9 \times 10^{10} \text{ rad/s} = 3.0 \text{ GHz}$$

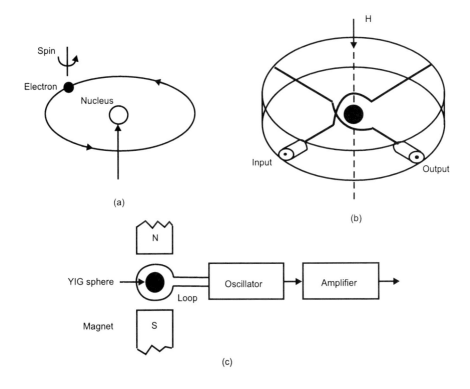

Figure 16.10 YIG crystal concepts and applications: (a) spin motion in YIG resonators; (b) YIG semiloop coupling structure; and (c) schematic diagram of YIG-tuned oscillator. (*After:* [7].)

Thus, as shown in (16.2), the resonant frequency is directly proportional to the applied direct magnetic field. The upper frequency limit is determined by the difficulty of establishing the required direct magnetic field with reasonable power supply specifications.

The unloaded Q for the YIG sphere is given by

$$Q_U = H_0/\Delta H \qquad (16.3)$$

where

H_0 = applied direct field (A/m)
ΔH = uniform mode line width = 40 A/m

For this example, the calculated unloaded Q is 4,600, which is as good as can be achieved with a conventional metallic cavity resonator at the same frequency. The advantage for the YIG resonator is that it can be tuned quickly by electrical means and over a large frequency range. Tuning time is not as fast as that of a varactor. Usually YIG systems require a heater to be built into the structure to maintain the YIG sphere at a constant temperature.

16.5.2 YIG Bandpass Filters

Figure 16.10(b) shows one version of a YIG bandpass filter using semiloop wires. The YIG sphere usually is mounted on a ceramic rod. The two semiloops are at right angles to each other and to the H_0 field. This field is supplied by a magnet, which is not shown in Figure 16.10.

In the absence of the direct magnetic field, the two circuits are decoupled and no transmission takes place between the input and the output terminals. When the magnetic field is applied, the microwave magnetic field at right angles to the direct field results in precession of the magnetic dipoles around the biasing field. The microwave magnetic field is applied by the input semiloop. Because of the precession of the magnetic dipoles around the biasing field, there is strong coupling to the output semiloop. That produces a bandpass filter structure with strong coupling in only one narrow frequency band. The center frequency can be shifted by changing the direct magnetic field by changing the input current.

Many other coupling structures can be used with YIG bandpass filters, including full-loop structures, stripline structures, coax line structures, and waveguide structures. There are single-sphere, two-sphere, and four-sphere filters.

16.5.3 YIG-Tuned Oscillators

Many possible oscillator configurations can use YIG resonators to control the frequency of oscillation. Figure 16.10(c) shows a diagram of a YIG-tuned oscillator. The magnet that supplies the direct field is shown. The loop about the YIG sphere is connected to an active device such as a BJT or FET oscillator or a Gunn-diode oscillator. (These types of oscillators are discussed in Chapter 17.)

16.6 SURFACE ACOUSTIC WAVE DELAY LINES

Some of the material presented in this section is quoted or adapted from [8,9].

16.6.1 Nondispersive Delay Lines

SAW delay lines are made using single large crystals of quartz, lithium niobate, lithium tantalate, and other piezoelectric materials. Such devices provide accurate wideband delays in a very small volume. That is because the velocity of the acoustic wave generated with electrical transducers is approximately 100,000 times slower than that of an electromagnetic wave. The useful frequency range for SAW delay lines is about 10 MHz to 1,600 MHz at this time. The upper frequency limitation and achievable bandwidths are due to fabrication limitations for the transducers used to launch the acoustic waves. Table 16.2 lists some SAW delay line material parameters.

The simplest SAW delay line configuration uses an input interdigital transducer (IDT) and an output IDT that are separated on the crystal surface by some distance that determines the amount of the time delay. This type of configuration is shown in Figure 16.11.

The SAW is launched by application of an RF signal to the interdigital electrode transducer on the left. In a nondispersive delay line, the fingers of the transducer on a given side are a full wavelength apart at the acoustic wave velocity. Adjacent fingers are a half-wavelength apart. The presence of the electric field between the

Table 16.2
SAW Delay Line Material Parameters

Material	Surface Wave Velocity	Temperature Dependency	Optimum Fractional Bandwidth
ST quartz	0.124 in/μs		0.1 to 0.5%
Lithium tantalate	0.129 in/μs	−23 ppm/°C	4 to 9%
YZ lithium niobate	0.134 in/μs	−94 ppm/°C	7 to 30%
Lithium niobate	0.153 in/μs	−72 ppm/°C	15 to 67%

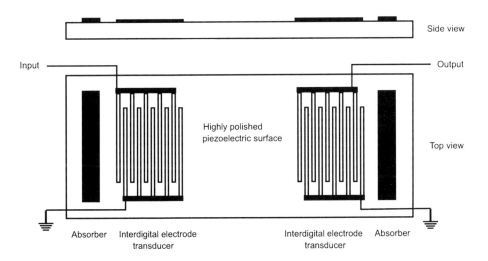

Figure 16.11 SAW in-line transversal filter or delay line structure. (*After:* [9].)

fingers causes mechanical stress in the piezoelectric crystal surface, and an acoustic wave is launched in both directions. The wave moving to the left is absorbed by the left absorber element, thus preventing reflections. The wave moving to the right travels across the highly polished piezoelectric surface to the receiving transducer. The acoustic wave is accompanied by an electric field. As the wave passes under the receiving transducer fingers, the electric field induces a delayed signal voltage, which is then fed out of the system. The acoustic wave that reaches the absorber element at the right is absorbed, thus preventing reflections.

The term nondispersive delay line means that the delay is constant and independent of frequency.

A dispersive delay line, on the other hand, means that the delay is a function of the frequency. Nondispersive delay lines are used in oscillators, frequency discriminator circuits, filters, and other signal processing applications. In filter configurations, the finger length usually is not uniform. The spacing determines the wavelength of the acoustic wave that is preferentially excited, the finger overlap determines the strength of each source of the waves and determines the shape of the filter response, and the number of fingers determines the bandwidth. An approximate equation for bandwidth for the SAW filter is

$$BW = f/N$$

where

f = center frequency

N = number of finger-pairs

For example, if 20 finger pairs are used, the bandwidth of the filter is about 5%. Five finger pairs are shown in each of the transducers shown in Figure 16.11. The launching transducer usually has only a few fingers compared to the receiving transducer, which may contain many fingers.

The insertion loss of a SAW bandpass filter is in the range of 10–30 dB, 6 dB of which is due to the bidirectional transducers. That is somewhat larger than the loss of ceramic or crystal filters. However, the wider bandwidth capability and the ability to shape the transfer function of the filter make the SAW filter attractive for many receiver applications. The filters are now used in televisions, cellular telephones, radars, and so on, for highly selective filtering.

Nondispersive IDTs are used when relatively narrow bandwidths (<30%) are required, while dispersive IDTs allow the implementation of very large (<67%) fractional bandwidths with relatively low loss.

Although most nondispersive delay lines are less than 20 μs in length, it is possible to produce delays of up to 150 μs in an area smaller than 2 in × 2 in. A minimum delay of 250 ns is recommended, even for very short delay lines.

Table 16.3 shows performance parameters for nondispersive delay lines made by Sawtek Inc., Orlando, FL.

16.6.2 Tapped Delay Lines

The design of the basic delay line can be extended to include many output transducers at different locations along the piezoelectric substrate, each having a different delay to realize a tapped delay line designed for a particular PSK code. It is possible to implement correlators for biphase, quadriphase, and MSK.

A PSK device can serve as an expander that elongates a short impulse into a coded waveform with uniform amplitude over its time duration and as a compressor that shortens the coded signal from an expander in a time-reversed code sequence to a short impulse with low sidelobes. The PSK device is most often used as a compressor because the phase-encoded signal can be generated easily by digital means. Table 16.4 shows performance parameters for Sawtek Inc. PSK delay lines.

SAW PSK correlators are widely used in spread spectrum communication systems, phase-coded radars, radio data links, communication modems, navigation and identification systems, and range differencing surveillance systems. These devices offer the advantages of compact size, real-time processing, and asynchronous operation.

16.6.3 Dispersive Delay Lines

SAW dispersive delay lines currently are used most extensively in pulse compression radars. They provide a high degree of flexibility in the implementation of different types of waveforms, which makes them suitable for the optimization of particular radar applications.

Table 16.3
Performance Parameters for Nondispersive Delay Lines

Parameters	Values
Center frequency	10 to 1,600 MHz
Fractional bandwidth	2 to 40%
Delay	0.25 to 150 μs
Insertion loss	10 to 35 dB
Amplitude ripple	0.1 to 1.0 dB
Phase ripple	1 to 10 degrees
Group delay ripple	10 to 250 ns
Triple-transit suppression	30 to 60 dB
Spurious suppression	40 to 70 dB

Table 16.4
Performance Parameters for PSK Delay Lines

Parameters	Values
Center frequency	10 to 800 MHz
Fractional bandwidth	2 to 40%
Insertion loss	20 to 50 dB
Maximum length	<40 μs
Chip rate	<200 MHz
Triple-transit suppression	30 to 60 dB
Spurious suppression	40 to 70 dB
Sidelobe level degradation from theoretical	<1 to 3 dB
Processing gain degradation from theoretical	<1 to 3 dB

The basic concepts of a pulse compression system are shown in Figure 16.12. A short impulse is applied to the SAW expander to produce an elongated frequency-modulated signal at the output of the device. The coding sense can be chosen to be "up," with high frequencies having longer delays than low frequencies, or "down," with high frequencies having shorter delays than low frequencies. After the amplified expanded signal reflects from the target, it is received and fed into a SAW compressor having the complex conjugate frequency characteristic, or simply the reverse impulse time response, of the expander. The delays encountered by the different frequencies are opposite to the delays in the expander network, resulting in all frequencies being compressed in time.

There are a number of ways to implement SAW dispersive delay lines. Two such systems are shown in Figure 16.13: the in-line device with a broadband transducer and the in-line device with two dispersive transducers.

Figure 16.14 shows the reflective array compressor. This compressor is used when long delays (up to 120 μs) or large-bandwidth time delay product products (>1,000) are needed. Shallow grooves etched in the delay path result in SAW reflections to form a delay that depends on the frequency. The reflective array compressor usually is more complex and costly than its IDT counterparts. It is also more susceptible to temperature effects and suffers from higher insertion loss than conventional IDT designs. On the other hand, it is more tolerant to manufacturing defects, allows higher BT products to be implemented, and offers a significant reduction in size.

Table 16.5 shows performance parameters for Sawtek dispersive delay lines. Parameters are presented for both expanders and compressors.

16.7 SURFACE ACOUSTIC WAVE DELAY LINE OSCILLATORS

SAW oscillators are available from Andersen Laboratories, Bloomfield, CT, at operating frequencies from 100 MHz to 2.6 GHz. These oscillators have low phase

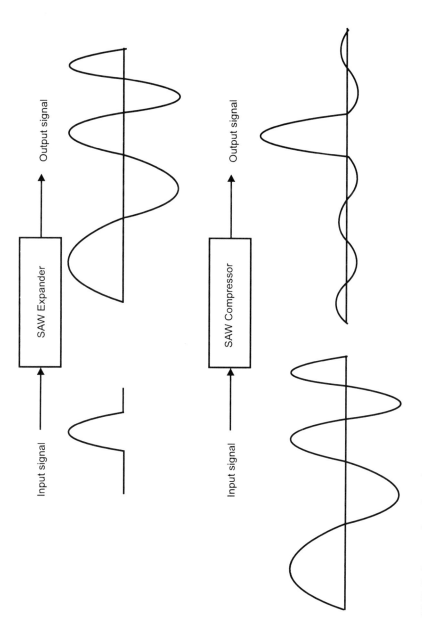

Figure 16.12 Basic operation of a pulse compression system. (*After:* [9].)

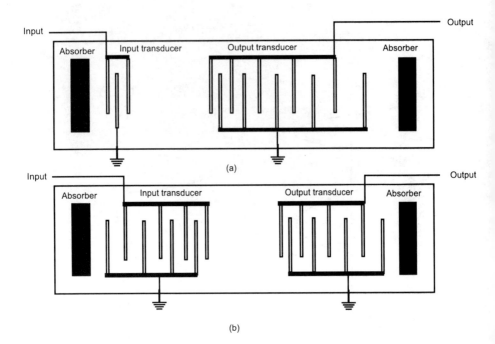

Figure 16.13 Dispersive delay lines: (a) in-line device with one dispersive transducers and (b) in-line device with two dispersive transducers. (*After:* [9].)

noise, compact size, good frequency stability, and rugged construction. They can be either fixed-frequency oscillators or voltage-controlled oscillators. Typical output power is about 10 mW. Figure 16.15 is a block diagram of a SAW delay line VCO. This oscillator uses a SAW device as the frequency-controlling element in the feedback loop of an amplifier. In the circuit shown, the output of a transistor amplifier is fed to a power divider or coupler. Part of the output power is fed to a buffer amplifier, and part is fed to a SAW delay line. The buffer amplifier isolates the oscillator from the load and provides the needed power output.

The delay line is specifically designed to set the fundamental operating frequency of the loop and provide the necessary phase noise characteristics for VCO requirements. The overall gain of the loop is ≥1. The phase shift around the loop is an integral number of 2π radians. The oscillator supports a comb of output frequencies spaced by $1/\tau$, where τ is the delay time of the SAW delay line. The delay line has a line frequency passband that selects the desired frequency line.

By introducing a predictable voltage-controlled phase shift into the feedback loop of the amplifier, the frequency of oscillation can be varied (pulled) from the center frequency over some specified operating range.

Table 16.6 shows performance parameters for standard product oscillators made by Andersen Laboratories. Oscillator specifications depend on each individual application and vary widely. Table 16.6 should be used only as a guide.

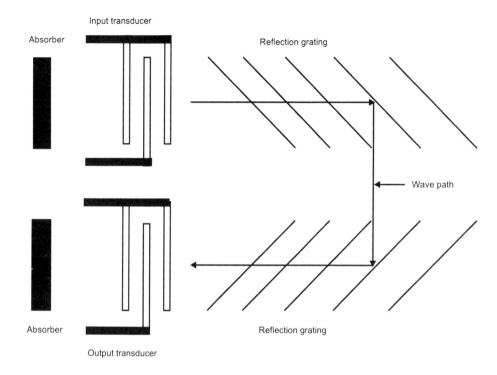

Figure 16.14 Reflective array compressor. (*After:* [9].)

The frequency stability for the SAW oscillator is approximately 12.5 ppm for the temperature range of 0–50°C. That can be improved by limiting the temperature range or by incorporating a heater. The packaged size for a standard Andersen Laboratories SAW oscillator is only about 1 in × 1.5 in × 0.25 in in the dual in-line package (DIP) configuration and slightly larger for other configurations.

16.8 BULK ACOUSTIC WAVE DELAY LINES

BAW delay lines are another important class of crystal delay line. A basic BAW delay line comprises two piezoelectric transducers bonded to a low-velocity medium such as quartz. The time delay is determined by the acoustic velocity and the length of the path.

Figure 16.16 shows some BAW delay line configurations. The simplest geometry, illustrated in Figure 16.16(a), is that of a rectangular bar with piezoelectric transducers bonded to each end. An electric signal is applied to the input transducer and subsequently converted to an acoustic signal. The acoustic signal is then transmitted through the crystal to the output transducer, where it is detected and converted back to an electrical signal.

Table 16.5
Performance Parameters for Dispersive Delay Lines

Parameters	Values
Expander	
Center frequency	20 to 1,000 MHz
Fractional bandwidth	2 to 67%
Pulse length	0.25 to 120 μs
Coding type	LFM, NLFM, MNLFM
Amplitude ripple of expanded pulse	±0.25 dB to ±0.5 dB
I/O impedance	50Ω
VSWR	1.5:1
Compressor	
Compressed pulse width (τ)	3 to 1,000 ns
Close in time sidelobes (t < 6τ)	<30 dB to <40 dB
Far out time sidelobes (t > 6τ)	<40 dB to <45 dB
Mismatch loss	0.1 to 2 dB
Signal-to-noise improvement	10 dB to 35 dB
I/O impedance	50Ω
VSWR	1.5:1

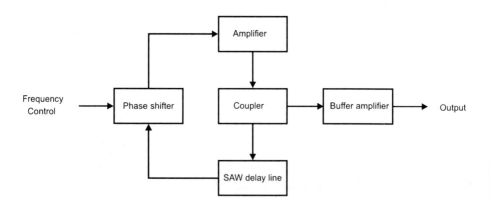

Figure 16.15 Block diagram of SAW VCO.

It is possible to provide long delays by using multiple reflection type crystals. Figure 16.16(b) shows the case of a single reflection crystal. There are other configurations that have large numbers of bounces.

The delay media used by Andersen Laboratories for BAW delay lines is fused quartz, crystalline quartz, or glass. Fused quartz is the delay medium most frequently used for standard BAW delay lines. The acoustic loss constant for fused quartz is

Table 16.6
Performance Parameters for SAW Oscillators

Parameters	Values $f < 1{,}300$ MHz	Values $f > 1{,}300$ MHz
Operating frequency	100 to 1,300 MHz	1,300 to 2,300 MHz
Tuning range	up to 1,000 kHz	up to 1,500 kHz
Modulation bandwidth	up to 500 kHz	up to 500 kHz
Tuning voltage	0 to 12V	0 to 12V
Output power	+10 dBm nominal	+10 dBm nominal
Power variation over temperature	±1.5 dB Max.	±2.0 dB Max.
Spurious outputs		
Harmonic	−30 dBc maximum	−30 dBc maximum
Nonharmonic	−60 dBc	−60 dBc maximum
Frequency accuracy at 25°C	±20 ppm	±20 ppm

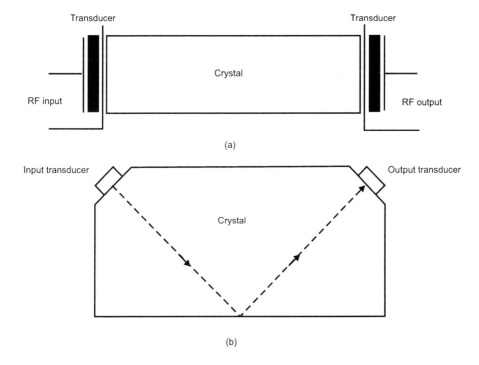

Figure 16.16 BAW delay line configurations: (a) schematic of rectangular bar BAW delay line and (b) reflection type BAW delay line.

low, making possible large delay lines with reasonable insertion loss. By using multiple internal reflections, delays up to 5,000 μs can be achieved with a single device. The wave velocity in fused quartz is about 4.3 μs/in in the compression mode, which is normally used for BAW delay lines.

Table 16.7 shows typical performance parameters for standard BAW delay lines made by Andersen Laboratories.

Teledyne Microwave, Mountain View, CA, produces BAW delay devices that operate at microwave frequencies. The RF input is through an impedance-matching network. The output of this network is a thin-film acoustic transducer. The delay crystal material is usually single-crystal sapphire or quartz. At the ends of the rod are thin-film acoustic transducers for converting from acoustic energy to electrical energy and vice versa. An impedance-matching network is used at the output.

This transducer is made up of three layers: a metalized counterelectrode, a piezoelectric layer, and a metalized top electrode. The counterelectrode is composed of Cr-Au composite metalization, which acts as the ground plane of the transducer. The piezoelectric film is a sputtered ZnO thin-film that converts electromagnetic energy to acoustic energy. The top electrode is also a Cr-Au composite metalization.

Delay devices using bulk acoustic waves can be built for operation in the 300 MHz to 18 GHz range. Bandwidths in excess of one octave have been achieved, but the narrower the bandwidth, the lower the insertion loss.

Time delay between 100 ns and 30 μs are achievable with a single crystal acting as a two-port device. The longer time delays can be realized only at lower frequencies since acoustic propagation losses increase with frequency and insertion losses become prohibitive at higher frequencies for long delays. For example, a delay device with 0.5 μs delay with a center frequency at 16 GHz and a bandwidth of 1 GHz will exhibit an insertion loss of about 60 dB. That loss can be overcome by amplification.

Table 16.7
Performance Parameters for LF BAW Delay Lines

Parameters	Values
Delay	0.25 to 5,000 μs
Center frequency	5 to 150 MHz
3-dB bandwidth	60%
Insertion loss	5 dB to 60 dB
Feed-through	Suppressed to >50 dB
Triple transit	Suppressed to >40 dB
Random signal	Suppressed to >50 dB
Cross-talk	Suppressed to >60 dB

References

[1] Kinsman, R. G., *Crystal Filters, Design, Manufacture, and Application*, New York: John Wiley & Sons, 1987.
[2] Krauss, Herbert L., Charles W. Bostian, and Frederick H. Raab, *Solid State Radio Engineering*, New York: John Wiley & Sons, 1980, pp.151–161.
[3] Matthys, R. J., *Crystal Oscillator Circuits*, New York: Wiley Interscience, 1983.
[4] Kajfez, Darko, and P. Guillon, *Dielectric Resonators*, Norwood, MA: Artech House, 1986.
[5] Krauss, Herbert L., Charles W. Bostian, and Frederick P. Raab, pp. 273–279.
[6] ANZAC RF and Microwave Signal Processing Components Catalog, Adams-Russell Co., Burlington, MA, pp. 455–459.
[7] Helszajn, J., *YIG Resonators and Filters*, New York: John Wiley & Sons, 1985.
[8] Kennedy, George, *Electronic Communication Systems*, 3rd ed., New York: McGraw-Hill, 1985, pp. 391–393.
[9] Skolnik, Merrill I., *Introduction to Radar Systems*, 2nd ed., New York: McGraw-Hill, 1980, pp. 420–434.

CHAPTER 17

Semiconductor Diodes and Their Circuits

This chapter discusses semiconductor diodes and circuits. Topics include semiconductor materials, "ordinary" junction diodes, zener diodes, Schottky-barrier diodes, PIN diodes, varactor diodes, step-recovery diodes, tunnel diodes, Gunn-effect diodes, IMPATT diodes, LEDs, IR laser diodes, and IR photodiodes.

Some of the information presented here is adapted from [1,2].

17.1 SEMICONDUCTOR MATERIALS

Some of the following material is quoted or adapted from [1].

Semiconductors are materials that have electrical conductivities intermediate between those of metals and insulators, that is, they are "semi" conductors. These materials are found in column IV and neighboring columns of the periodic table. The column IV materials of interest are silicon (Si) and germanium (Ge), with Si being the most important, and are referred to as elemental semiconductors. The elements in column III of interest are boron (B), aluminum (Al), gallium (Ga), and indium (In). These elements are used as doping materials for Si or other semiconductor materials for producing p-type material. The elements in column V of interest are phosphorus (P), arsenic (As), and antimony (Sb). These elements are used as doping materials for Si or other semiconductor materials for producing n-type materials.

Compounds of materials from columns III and V make up part of the intermetallic or compound semiconductors. These III-V semiconductors include AlP, AlAs, AlSb, GaP, GaAs, GaSb, InP, InAs, InSb. Of these, gallium arsenide (GaAs) is the most important and has been used for microwave and higher frequency FET devices and Gunn diodes. GaAs has also been used for Schottky-barrier diodes, tunnel diodes, varactors, and step-recovery diodes. The elements in column II that are of interest are zinc (Zn) and cadmium (Cd). The elements in column VI that are of interest are sulfur (S), selenium (Se), and tellurium (Te). Semiconductor compounds that use elements from columns II and VI are ZnS, SnSe, ZnTe, CdS, CdSe, and CdTe. Some of the compound semiconductors are used in LEDs.

17.2 "ORDINARY" JUNCTION DIODES

Figure 17.1 shows an ordinary junction diode. This diode consists of n- and p-doped Si with two terminals. The current-voltage (I-V) characteristics for the diode are shown in the figure. When the p-side of the diode is positive with respect to the n-side, the resistance of the diode is low. In the case of a Si diode, it takes about 0.7V to turn on the diode in the forward direction.

Two diodes in series would require about 1.4V to turn on the diodes.

When the p-side of the diode is negative with respect to the n-side, the resistance of the diode is at first very high and little current flows. At some high negative voltage, reverse breakdown takes place, and the reverse diode resistance becomes small. The mechanism for such high voltage breakdown is known as *avalanche breakdown*.

17.3 ZENER DIODES

A Zener diode consists of n- and p-doped Si with two terminals. The current-voltage characteristics for this diode look very much like the I-V characteristics for the

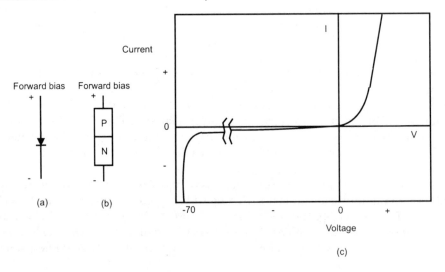

Figure 17.1 Ordinary junction diode: (a) symbol; (b) diode configuration; and (c) current-voltage characteristics. (*After:* [1].)

ordinary diode except that the reverse bias breakdown occurs at much lower voltages. Typical breakdown voltages for zener diodes are more like 5–20V rather than 70V or more for an ordinary diode. Zener breakdown occurs in heavily doped junctions in which the transition from p to n material is abrupt. The mechanism involved is called tunneling.

The Zener diode finds application as a voltage reference. A resistor usually is placed in series with the diode to limit the current. The reverse voltage across the diode is fixed by the design of the diode.

It should be pointed out ordinary diodes occasionally are used as voltage references in bias circuits for transistor amplifiers. In that case, the diode is forward biased, and the voltage drop across each diode is again about 0.7V. Two or more diodes sometimes are connected in series to increase the voltage of the reference.

17.4 SCHOTTKY-BARRIER DIODES

Much of the remainder of this chapter is quoted or adapted from [2].

The Schottky-barrier diode is an extension of the oldest semiconductor device of them all, the point-contact diode. With the Schottky-barrier diode, the metal-semiconductor interface is a surface rather than a point contact. Like the point-contact diode, the Schottky-barrier diode has no minority carriers in the reverse-bias condition. Thus, the delay present in junction diodes, due to hole-electron recombination time, is absent. Because of the larger contact area between the metal and the semiconductor, the Schottky-barrier diode has much lower resistance and lower noise than the point-contact diode. That makes this diode desirable for microwave and higher frequency applications, where ordinary junction diodes are not effective.

The most commonly used semiconductor materials for Schottky-barrier diodes are N-type Si and N-type GaAs. GaAs has the lower noise and the higher operating frequency limits. On the other hand, Si is easier to fabricate and is consequently used at X-band and lower frequencies in preference to GaAs. The metal at the interface with the semiconductor is often a thin layer of titanium surrounded by gold for protection and low ohmic resistance.

Schottky-barrier diodes can be used at frequencies as high as 100 GHz. They are used as detectors and mixers. The noise figures for mixers using these diodes are as low as 4 dB at 2 GHz and 15 dB at 100 GHz.

17.5 PIN DIODES

Figure 17.2 shows the PIN diode, which consists of a narrow layer of p-type semiconductor separated from an equally narrow layer of n-type material by a somewhat thicker region of intrinsic semiconductor material, thus the term PIN. PIN diodes often use lightly doped n-type semiconductor material rather than intrinsic material. PIN diodes usually are made of Si, although GaAs is sometimes used.

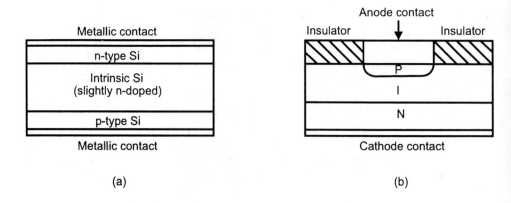

Figure 17.2 PIN diode: (a) schematic diagram and (b) planar PIN diode.

The PIN diode is used for microwave power switching, attenuating, limiting, and modulation. At microwave frequencies, the PIN diode acts as a variable resistance. The simplified equivalent circuits and the resistance variation with reverse and forward bias are shown in Figure 17.3(a,b). With reverse (negative) bias, the resistance to microwave energy typically is about 5,000–10,000Ω. When the diode is forward biased, the positive-bias resistance to microwave energy is typically 1–10Ω. If the PIN diode is placed across a waveguide, a 50Ω coaxial line or other transmission medium, it does not significantly load the line when negatively biased. When positively biased, however, it presents a near short circuit across the line and thus produces reflections on the line.

Figure 17.3(c) is a schematic diagram of a series-mounted PIN diode switch for a coaxial or microstrip line. The dc bias is fed into the diode using an RF choke, and the signal is injected using a coupling capacitor. The output is across an RF choke to ground.

Figure 17.3(d) is a schematic diagram for a shunt-mounted PIN diode switch for a coax line. Again, the dc bias is fed in to the diode using an RF choke, and the signal is fed in using a coupling capacitor. In this case, no output choke is used, and the PIN diode is connected directly to ground. The output is fed out using a coupling capacitor.

PIN diodes can be used in parallel or series, as desired. Individual diodes can handle up to about 200 kW peak or 200W average. Several diodes in parallel can handle as much a 1 MW peak. Switching times are in the range of 1–40 ns, depending on the power levels used.

17.6 VARACTOR DIODES

Almost every semiconductor diode has a junction capacitance that varies with the applied reverse bias. If the diode is manufactured to have suitable microwave

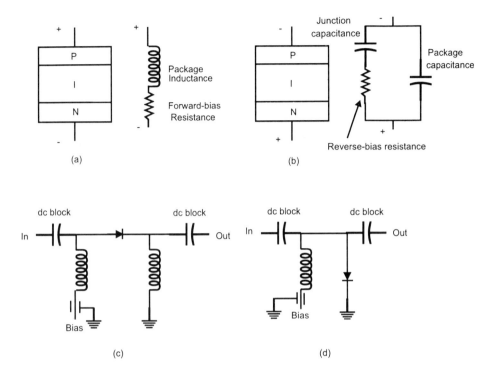

Figure 17.3 PIN diode simplified equivalent circuits and example switches: (a) forward bias condition; (b) reverse bias condition; (c) series-mounted switch; and (d) shunt-mounted switch.

characteristics, it is called a varactor diode. Varactor diodes are made of Si or GaAs. GaAs has the advantage of higher maximum operating frequency. Figure 17.4 shows the characteristics of varactor diodes, including the current versus voltage characteristics and the junction capacitance versus voltage characteristics. The bias voltage region of interest for a varactor diode is between just above the avalanche breakdown point and zero volts. For typical Si varactors, the minimum capacitance is about 1 pF and the maximum capacitance is about 25 pF.

Varactors find application as voltage-variable capacitors for frequency modulation and oscillator tuning. They also are used in frequency multipliers. Because snap-off varactor diodes multiply by high factors with better efficiency than ordinary varactor diodes, they are used where possible. GaAs varactors often are used at the higher frequencies. A varactor multiplier of this type can have an efficiency for a 60-GHz doubler of greater than 50%.

The maximum output power for the varactor diode multipliers ranges from more than 10W at 2 GHz to about 25 mW at 100 GHz. Tripler efficiencies range from 70% at 2 GHz to about 40% at 36 GHz; however, that is with proper design, including idlers to reflect fundamental and second harmonic power back to the input. Otherwise, 33% efficiency is more typical.

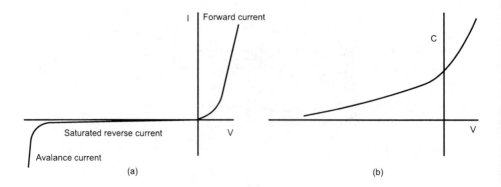

Figure 17.4 Varactor diode characteristics: (a) current versus voltage and (b) capacitance versus voltage. (*After:* [2].)

One of the current applications for multiplier chains is to provide a low-power signal to phase-lock a Gunn or IMPATT diode oscillator. (These devices are discussed later.)

17.7 STEP-RECOVERY DIODES

A step-recovery diode is a Si or GaAs p-n junction diode with construction similar to that of a varactor diode. It stores charge when conducting in the forward direction. When reverse bias is applied, the diode briefly discharges the stored energy in the form of a sharp pulse, an impulse, that is rich in harmonics. The duration of the pulse typically is only 100 to 1,000 ps, depending on diode design.

Step-recovery diodes are not available for frequencies above about 20 GHz, whereas varactors can be used well above 100 GHz. Step-recovery diodes are available for powers in excess of 50W at 300 MHz, 10W at 2 GHz, and 1W at 10 GHz. Multiplication ratios up to 12 commonly are available. Efficiency can be in excess of 50% for triplers at frequencies up to 1 GHz. The efficiency drops to about 15% for a times-5 multiplier with an output frequency of 12 GHz.

17.8 MICROWAVE TUNNEL DIODES AND CIRCUITS

Figure 17.5 shows the voltage-current characteristics for a Ge junction tunnel diode. This diode differs from the ordinary junction diode in that the semiconductor material is heavily doped, perhaps 1,000 times that of an ordinary rectifier diode. That permits a depletion layer so thin that tunneling can occur easily.

In the voltage region from A to B, there is a region of negative resistance. That means that this device can be used as an oscillator. The tunnel diode oscillator found use early after its development, but it no longer is used extensively because other negative-resistance semiconductor devices now provide higher output power.

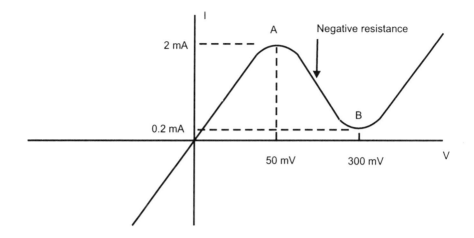

Figure 17.5 Tunnel diode voltage-current characteristics. (*After:* [2].)

The tunnel diode also can be used as a microwave amplifier when used with a circulator. A circuit of this type is shown in Figure 17.6. The tunnel diode amplifier (TDA) is a low-noise system. Reasons for that are that a tunnel diode is a low-resistance device, and the operating current is low. Tunnel diode amplifiers can be fairly broadband systems. They are small and simple, and they can operate at microwave frequencies with upper limits in excess of 50 GHz.

17.9 MICROWAVE GUNN DIODES AND CIRCUITS

Figure 17.7 shows an epitaxial GaAs Gunn diode. A negative resistance is provided by such a device if the voltage gradient across the slice of GaAs is in excess of about 3,300 V/cm.

Oscillations then occur if the slice is connected to a suitably tuned circuit. Proper doping profile is also required.

The Gunn effect is a bulk property of semiconductors and does not depend on either junction or contact properties. It occurs only in n-type materials, so it is associated only with electrons and not holes. GaAs is one of the few materials for which the Gunn effect works. In this material with n-type doping, there is an empty energy band higher in energy than the highest filled or partially filled band. The forbidden energy gap is small. In this diode with its very high voltage gradient, electrons acquire enough energy to be transferred to the higher energy band, in which they are much less mobile. Thus, the current has been reduced as a result of voltage rise. This voltage region therefore is a region of negative resistance. Eventually with increasing voltage, the voltage becomes high enough to remove electrons from the higher energy, lower mobility band so the current will increase with voltage once again.

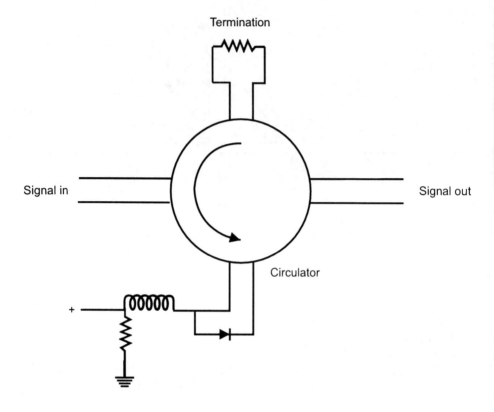

Figure 17.6 Tunnel-diode amplifier with circulator. (*After:* [2].)

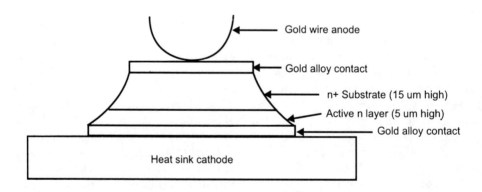

Figure 17.7 Epitaxial GaAs Gunn diode. (*After:* [2].)

A second phenomenon with Gunn diodes is important: the formation of bunches of electrons. Negative-resistance domains are formed that are less conductive. These travel toward the positive anode at a speed of about 10^7 cm/s in practice. These traveling domains may be thought of as low-conductivity, high-electron-transfer regions corresponding to a negative pulse of voltage. When they arrive at the positive end of the GaAS slice, a pulse is received by the associated resonant circuit, and oscillations take place. It is actually this arrival of pulses at the anode, rather than the negative resistance proper, that is responsible for oscillations in Gunn diode oscillators.

A typical Gunn oscillator uses a coaxial cavity. The Gunn diode is located between the end of the center conductor of the cavity and the end of the cavity. The bias for the diode is fed through the other end of the cavity using a bypass capacitor and a bias feed-through capacitor. A sliding short plunger is used to mechanically tune the half-wave cavity. A coaxial output is connected to a capacitive probe that enters the cavity. A tuning screw is used for coupling adjustment. This type of system is shown in Figure 17.8.

Gunn diodes are available for operation in the frequency range from 4 to beyond 100 GHz. A typical X-band Gunn diode oscillator requires a 9V dc bias and an operating current of 950 mA. The output RF power is 300 mW in the frequency range of 8–12.4 GHz. The efficiency, therefore, is only about 3.5%. A typical Gunn diode oscillator operating in the 26.5–40 GHz band produces about 250 mW with an efficiency of 2.5%. Gunn diode oscillators are used as low- and medium-power oscillators in microwave receivers. Higher power Gunn oscillators are used in a wide variety of frequency-modulated transmitters. Other applications currently include police radar, CW doppler radar, burglar alarms, and aircraft rate-of-climb indicators.

17.10 MICROWAVE IMPATT DIODES

IMPATT diodes, another important type of microwave oscillator, are also called avalanche diodes. The schematic diagram for this type of diode is shown in

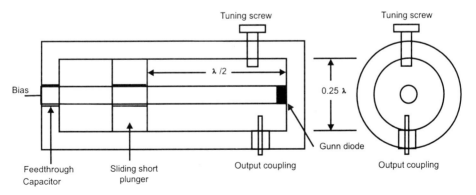

Figure 17.8 Layout of Gunn diode oscillator. (*After:* [2].)

Figure 17.9. The IMPATT diode is reverse biased. An extremely high voltage gradient (on the order of 400 kV/cm) is applied to the IMPATT diode. This causes a flow of minority carriers (electrons in this case) across the junction.

Figure 17.10 shows the dc voltage at the avalanche threshold. Let us assume the existence of oscillations and therefore an RF voltage added to the dc voltage. Avalanche takes place over a period of time such that the current pulse maximum at the junction occurs at the instant when the RF voltage across the diode is zero and going negative. A 90-degree phase difference between voltage and current has thus been obtained, as shown in Figure 7.10. The current pulse then drifts toward the cathode. The thickness of the drift region is selected such that the delay in reaching the cathode adds another 90-degree phase shift and thus a total phase shift between current pulse and voltage of 180 degrees. This is one type of negative resistance.

IMPATT diodes usually are made of Si; they also may be made of GaAs. IMPATT diodes are essentially narrowband devices because the thickness of the drift region is critical to the operation of the device.

Commercial IMPATT diodes currently are produced over the frequency range of about 4–200 GHz. The maximum power per diode varies from nearly 20W near 4 GHz to about 50 mW at the HF end. Above 20 GHz, this type of diode produces higher CW power output than any other semiconductor device.

17.11 SEMICONDUCTOR IR LASER DIODES

A GaAlAs diode of the type shown in Figure 17.11 is capable of producing laser action. Depending on its precise chemical composition, it is capable of producing an output with wavelength in the range of 0.75 to 0.9 μm with 0.85 μm being typical. This is in the near-infrared region. The attenuation in a fiber optic cable at this wavelength is about 2.5 dB/km.

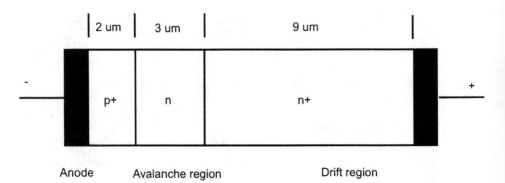

Figure 17.9 IMPATT diode schematic diagram. (*After:* [2].)

Semiconductor Diodes and Their Circuits | 509

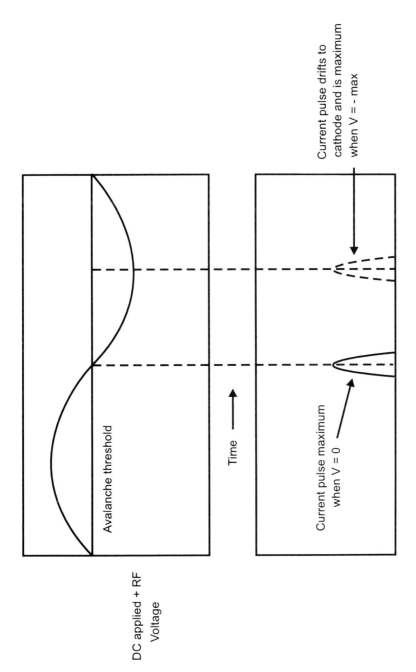

Figure 17.10 Current and voltage characteristics of IMPATT diodes. (*After:* [2].)

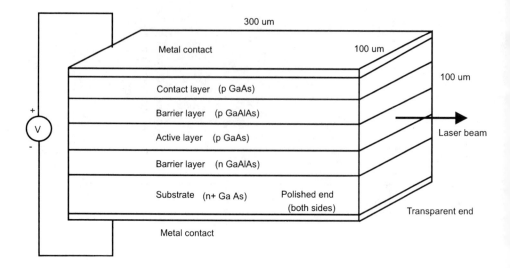

Figure 17.11 GaAlAs laser diode (0.75 to 0.9 μm wavelength). (*After:* [2].)

The GaAlAs laser is forward biased to turn it on. Electrons and holes originating in the GaAlAs layers cross the heterojunctions (junctions between dissimilar semiconductor materials, GaAlAs and GaAs in this case) and give off their excess recombination energy in the form of light. The heterojunctions are opaque, and the active region is constrained by them to the p-layer of GaAs. This layer is only a few micrometers thick. The two ends of the slice are highly polished, so that reinforcing reflection takes place between them, as in other lasers, and a continuous beam is emitted in the direction shown. The laser is capable of powers in excess of 1W.

Figure 17.12 shows an InGaAs phosphide laser. This semiconductor laser, a more recent development than the GaAlAs device, was developed to produce laser outputs at wavelengths longer than those the GaAs laser is capable of producing. The operation of this laser is similar to that of the GaAlAs laser. Output wavelengths in this case are at 1.3 μm or 1.55 μm. That permits the laser to take advantage of low-attenuation windows in the transmission spectrum of optic fibers. The attenuation in a fiber optic cable at 1.3 μm is 0.4 dB/km. The attenuation in a fiber optic cable at 1.55 μm is an even lower 0.25 dB/km.

17.12 LIGHT-EMITTING DIODES

The construction of an LED is similar to that of a laser diode, but the structure is simpler. There are no polished ends, and laser action does not take place. Consequently, the power output is much lower, and a much wider beam of light results. The light is not monochromatic. A small lens is often used with the diode.

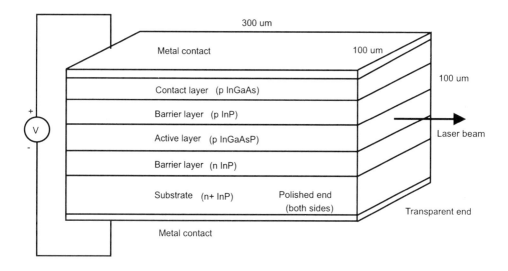

Figure 17.12 InGaAs phosphide laser diode (1.2 to 1.6 μm wavelength). (*After:* [2].)

In the operation of LEDs, electrons and holes are injected across heterojunctions, and light energy is given off during recombination. The materials used are the same as for the corresponding laser diodes.

17.13 IR PHOTODIODES

Photodiodes are used in a variety of applications, including optical communications, picosecond-pulse detection, and optical fiber characterization. A number of different types of photodiodes can be used. One type is the GaAlAs/GaAs PIN diode, which has a typical rise time of 50 ps. This photodiode has a 10-dB bandwidth of greater than 7.0 GHz and a quantum efficiency of nearly 65%. It also features extremely low capacitance and extremely low dark current.

The photodiode structure consists of a photosensitive GaAs layer, overlayed by a transparent GaAlAs window layer, grown on a semi-insulating substrate. The use of a semi-insulating GaAs substrate significantly reduces parasitic capacitance. The photodiode chip is passivated with a dielectric coating to ensure high reliability and is protected in the detector housing by a sapphire window. An SMA-type connector is incorporated into the diode for ease in coupling to 50Ω systems.

Semiconductor types used for PIN-type photodetectors include Ge, Si, GaAlAs, and InGaAs.

One problem with the PIN photodiode is that it is not as sensitive as we would like it to be. No gain takes place in the device. A single photon of light cannot create more than one hole-electron pair. This problem is overcome by the use of the avalanche photodiode (APD). An APD is reverse biased close to breakdown using a voltage in the range of 100–500V. A light quantum impinging on the diode causes a hole-electron pair to be created. With the high voltage used, avalanche multiplication can take place, as in the IMPATT diode. A typical APD is 10 to 150 times more sensitive than a PIN photodiode. Response time also is much shorter. The materials used for the APD are the same as those used for the PIN diode.

Figure 17.13 is an illustration of an APD.

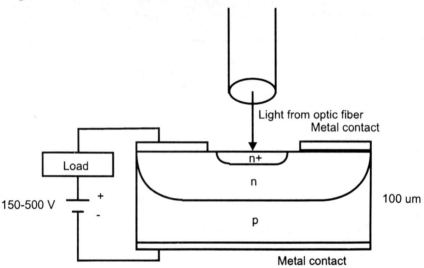

Figure 17.13 APD schematic. (*After:* [2].)

References

[1] Van Valkenburg, Mac E., ed., *Reference Data for Engineers*, 8th ed., SAMS, Carmel, IN: Prentice Hall Computer Publishing, 1993, Chap. 18, "Semiconductors and Transistors" by Ben G. Streetman.

[2] Kennedy, George, *Electronic Communication Systems*, 3rd ed., New York: McGraw-Hill, 1985, pp. 393–451.

CHAPTER 18

Bipolar and Field-Effect Transistors and Their Circuits

18.1 BIPOLAR JUNCTION TRANSISTORS

Figure 18.1 shows the geometry, representation, and the symbol for an NPN BJT.

Figure 18.1(a) shows the cross-section of a planar BJT. This drawing shows a central volume of n-type doped Si called the emitter, a p-type Si layer around the emitter volume called the base, and an n-type volume of doped Si called the collector. Metallic leads are connected to each of these semiconductor sections to form the transistor.

Figure 18.1(b) is another representation of an NPN BJT, showing the n-type emitter, the p-type base, and the n-type collector stacked one above the other, with leads exiting from each region. This representation shows the junctions between the elements.

The symbol for the NPN BJT used in schematic diagrams is shown in Figure 18.1(c). The arrow points in the direction of conventional current flow.

The operation of the Si NPN transistor is as follows. This device acts as a current-controlled valve, controlling the flow of electrons between the emitter and the collector. The collector is biased very positive with respect to the base. The combination of base and collector thus acts like a back-biased pn junction diode.

The base may be biased positive or negative with respect to the emitter, depending on whether the device is turned on or off. The combination of base and emitter also acts as a junction diode. If the base is biased positive with respect to the emitter, electrons are allowed to flow from the emitter to the base by the process of diffusion. When entering the base, the electrons are minority carriers (holes being the majority carriers in p-type material). Some of the carriers exit the base as base current. The largest numbers of them, however, see the very high

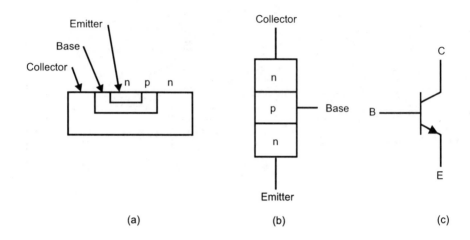

Figure 18.1 NPN BJT: (a) cross-section of a planar NPN BJT; (b) elements of an NPN BJT; and (c) symbol for the NPN BJT.

positive field at the collector-base junction and are swept into the collector region. There they exit the collector through the collector lead. The more positive the base bias, the larger the base current and the larger the collector current. Thus, the device is a current amplifier, since a small change in base current produces a large change in collector current.

Figure 18.2 shows similar illustrations for the PNP-type BJT. Figure 18.2(a) shows the cross-section of a planar BJT. The figure illustrates a central p-type volume of doped Si called the emitter, an n-type Si layer around the emitter volume called the base, and a p-type volume of doped Si called the collector. Metallic leads are connected to each of these semiconductor sections to form the transistor.

Another representation of a PNP BJT is shown in Figure 18.2(b). Illustrated are the p-type emitter, the n-type base, and the p-type collector stacked one above the other, with leads exiting each region. The figure shows clearly the junctions between the elements.

Figure 18.2(c) is the symbol for the PNP BJT used in schematic diagrams. As in Figure 18.1, the arrow points in the direction of conventional current flow.

The operation of the Si PNP transistor is as follows. This device acts as a current-controlled valve, controlling the flow of holes between the collector and the emitter. This time, the collector is biased very negative with respect to the base. The combination of base and collector acts like a back-biased pn junction diode.

The base may be biased negative or positive with respect to the emitter, depending on whether the device is turned on or off. The combination of base and emitter also acts as a junction diode. If the base is biased negative with respect to the emitter, holes are allowed to flow from the emitter to the base by the process of diffusion. When entering the base, the holes are minority carriers (electrons

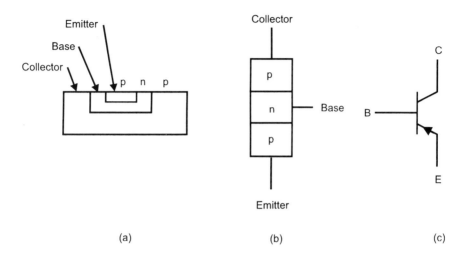

Figure 18.2 PNP BJT: (a) cross-section of a planar PNP BJT; (b) elements of a PNP BJT; and (c) symbol for the PNP BJT.

being the majority of carriers in n-type material). Some of these carriers exit the base as base current with electrons from the base combining with the holes. The largest number of the holes, however, see the very high negative field at the collector-base junction and are swept into the collector region. There they exit the collector through the collector lead (they combine with electrons, so there is a flow of electrons to the collector). The more negative the base bias, the larger the base current and the larger the collector current. Hence, this device is also a current amplifier, since a small change in base current produces a large change in collector current.

Currently both types of BJTs are used in electronic systems. NPN devices are the most frequently used of the two types. UHF and higher frequency devices are of the NPN type because of the higher mobility of electrons as majority carriers, which means improved HF power gain. PNP-type transistors are used primarily in land-mobile communications equipment requiring a positive ground system. The following discussions of amplifier configurations apply only to NPN transistors.

18.2 BJT AMPLIFIER CONFIGURATIONS

There are three main types of BJT amplifier configurations: the common emitter, the common base, and the common collector amplifiers.

18.2.1 Common-Emitter Amplifier

The common-emitter amplifier is the configuration most often used at VHF and lower frequencies. It is also frequently used at UHF and microwave frequencies.

Advantages of this configuration over the common-base amplifier include higher voltage gain, higher current gain, higher power gain, and higher input impedance. It usually has the lowest noise figure, good stability, and is relatively easy to match input and output impedances to 50Ω or 75Ω, as applicable. Figure 18.3 shows two schematic diagrams for this type of circuit (both for NPN transistors).

The circuit in Figure 18.3(a) uses resistors between the collector power supply and the collector and between the base bias power supply and the base. Signals are coupled in and out using coupling capacitors. A resistive load is used. This circuit is typical for a wideband, lower frequency, small-signal amplifier.

The circuit in Figure 18.3(b) uses RFCs between the collector power supply and the collector and between the base bias power supply and the base. Signals are coupled in and out using coupling capacitors. A resistive load also is used. This circuit is more typical for a narrowband, HF, small-signal amplifier.

There are other possible circuit variations for the NPN BJT common-emitter amplifier. One important one is the circuit that uses a transformer in place of the RFC or resistor that connects between the collector and the collector power supply. This amplifier configuration is shown in Figure 18.4.

The transformer may be either tuned or untuned. Transformers sometimes are also used as inputs to the base. One end of the transformer secondary is connected to the base bias supply. Such transformers also may be either tuned or untuned, depending on the application. These transformers can be used for

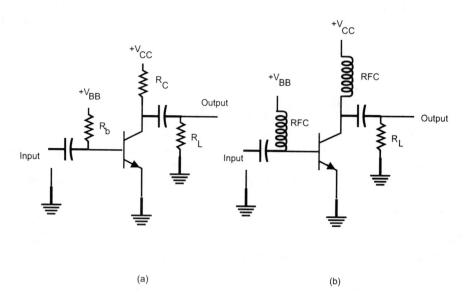

Figure 18.3 BJT common-emitter amplifiers: (a) amplifier using a resistor connected to V_{cc} and (b) amplifier using an RFC connected to V_{cc}.

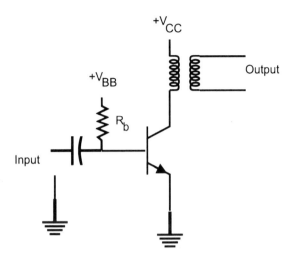

Figure 18.4 BJT common-emitter amplifier using a transformer connected to V_{cc}.

impedance matching and filtering. Other types of impedance matching, filtering, and coupling used with common-emitter amplifiers are discussed later.

Figure 18.5 shows a typical common-emitter characteristics plot for the amplifier in Figure 18.3(a). This plot shows the assumed collector current for the transistor as a function of base current and collector-to-emitter voltage. This plot shows typical dc and ac load lines with a Q point corresponding to the bias conditions. The supply voltage is +8V; the collector resistance, R_C, is 1 kΩ; the load resistance, R_L, is 1 kΩ; and the base bias current is 40 μA. The value of h_{FE} (current gain) for this transistor is assumed to be 100, so the collector current is 4 mA, and the dc collector voltage is 4V.

The dc load line is constructed by locating a first point where there is no current through the resistor ($I_C = 0$ and $V_{CE} = V_{CC}$), and a second point where the current is a maximum and all the voltage is across the resistor ($I_C = 8$ mA and $V_{CE} = 0$). The dc load line is a straight line drawn between those points.

The ac load line for this case is not the same as the dc load line, because the collector sees R_C in parallel with R_L. This parallel resistance is 500Ω. Thus, we have a 500Ω ac load line passing through the Q point. With no ac signal input, the transistor is biased at point Q. With an ac signal present, the operating point moves up and down the ac load line above and below the Q point. If the input signal goes positive so that the base current increases to 60 μA, the operating point is moved to $V_{CE} = 3$V and $I_C = 6$ mA. If the input signal goes negative so that the base current decreases to 20 μA, the operating point is moved to $V_{CE} = 5$V and $I_C = 2$ mA.

Figure 18.6 is a plot of the typical common-emitter characteristics plot for the amplifier in Figure 18.3 with an RFC connected to the collector for dc bias

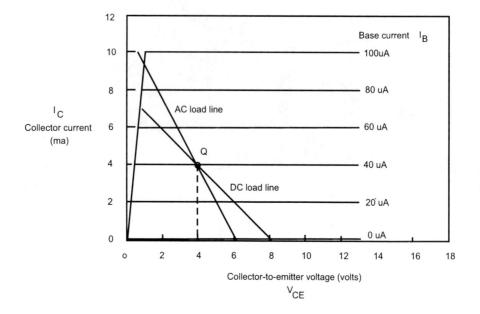

Figure 18.5 Plot of typical common-emitter characteristics (resistor between collector and power supply).

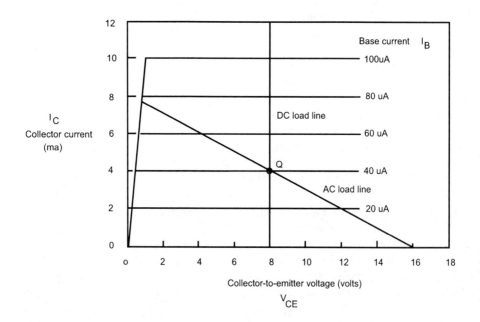

Figure 18.6 Plot of typical common-emitter characteristics (RFC between collector and power supply).

current. The collector is capacitively coupled to a 1-kΩ load resistor. Typical dc and ac load lines are shown, with point Q corresponding to the bias conditions.

The supply voltage is again +8V and the bias base current is 40 μA. The assumed value of h_{FE} for this transistor is again 100, so the collector current is 4 mA, and the dc collector voltage is 8V.

The dc load line is constructed by locating a first point where there is no current through the choke ($I_C = 0$ and $V_{CE} = V_{CC}$), and a second point where the current is a maximum and no voltage is across the choke ($I_C = 8$ mA and $V_{CE} = V_{CC}$). The dc load line is a straight line drawn between those points. In this case, the dc load line is a vertical line.

The ac load line is constructed by locating a first point at the Q point and a second point where the current is a maximum and no the voltage is across the transistor ($I_C = 8$ mA and $V_{CE} = 0$). The dc load line is a straight line drawn between those points and extending to the point where the collector current is zero.

With no input ac signal, we are at point Q. With the presence of an ac signal, movement is along the ac load line above and below the Q point. If the input signal goes positive so that the base current increases to 60 μA, the operating point is moved to $V_{CE} = 4$V and $I_C = 6$ mA. If the input signal goes negative so that the base current decreases to 20 μA, the operating point is moved to $V_{CE} = 12$V and $I_C = 2$ mA.

The current gain, β, of the transistor is not constant with frequency but changes, as shown in Figure 18.7. Over much of the frequency range, it decreases inversely with frequency. Key frequencies that are shown are f_β, where the power gain is down by 3 dB; f_t, which is known as the gain-bandwidth product; and f_1, where β reaches a value of 1.0. The value of h_{fe} or β changes widely from device to device of the same type.

Figure 18.8 shows ac equivalent circuits for the BJT common-emitter amplifier. The figure illustrates that the transistor is a fairly complex electrical circuit consisting of a current generator, resistors, and capacitors. Figure 18.8(a) shows the midband equivalent circuit, where no capacitors are involved. The circuit in Figure 18.8(b) shows the HF equivalent circuit, with two capacitors. C_π typically is in the range of 30–500 pf. C_μ typically is in the range of 1–20 pf.

The following midband design equations are for the BJT common-emitter amplifier.

Current gain:

$$A_i = h_{fe} = \beta = i_C/i_B \qquad (18.1)$$

Input resistance at the base:

$$r_i = r_\pi = (\beta + 1)(25 \text{ mV}/I_E) \qquad (18.2)$$

A small-current symbol (I) indicates ac; a large-current symbol (I) indicates dc.

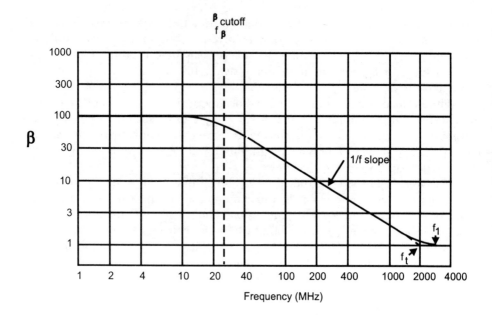

Figure 18.7 Characteristics of an NPN common-emitter amplifier current gain. (*After:* [1].)

Transconductance for the transistor:

$$g_m = I_E/(25 \text{ mV}) = 1/r_e \qquad (18.3)$$

where I_E = dc emitter current.

Output resistance for the amplifier:

$$r_o = R_L \parallel r_{ce} \qquad (18.4)$$

where R_L = load resistance seen by collector, and $\parallel$ indicates in parallel.

Voltage gain for the amplifier:

$$A_v = -g_m r_o \qquad (18.5)$$

The negative sign indicates a 180-degree phase shift between input and output.

Power gain for the amplifier:

$$A_p = g_m^2 r_o r_\pi \qquad (18.6)$$

For an example of the use of (18.1) to (18.6) in the evaluation of a common-emitter amplifier, assume the following:

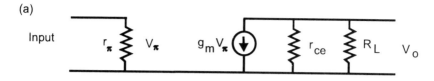

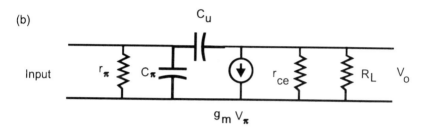

Figure 18.8 Equivalent circuit for BJT common-emitter amplifier: (a) midband equivalent circuit and (b) HF equivalent circuit. (*After:* [2].)

$$h_{fe} = \beta = 100$$
$$r_{ce} = 10 \text{ k}\Omega$$
$$I_E = 2.5 \text{ mA}$$
$$R_L = 1 \text{ k}\Omega$$

Then,

$$g_m = I_E/(25 \text{ mV}) \qquad (18.7)$$
$$g_m = 2.5 \text{ mA}/(25 \text{ mV}) = 0.1\text{S}$$

$$r_i = r_\pi = (\beta + 1)(25 \text{ mV}/I_E) \qquad (18.8)$$
$$r_i = r_\pi = (101)(25 \text{ mV}/2.5 \text{ mA}) = 1010 \Omega$$

$$r_o = R_L \parallel r_{ce} \qquad (18.9)$$
$$r_o = 1,000 \times 10,000/11,000 = 909 \Omega$$

$$A_v = -g_m r_o \qquad (18.10)$$
$$A_v = -0.1 \times 909 = -90.9$$

$$A_p = g_m^2 r_o r_\pi \tag{18.11}$$
$$A_p = 0.01 \times 909 \times 1{,}010 = 9{,}181 = 39.6 \text{ dB}$$

At the higher frequencies, the equivalent circuit also involves the inductance of the device input and output conductors. It is common practice for higher frequency applications to use S-parameters to characterize the devices.

18.2.2 Common-Base Amplifier

The common-base amplifier finds extensive use at the higher frequencies, such as UHF and microwave. That is because of higher cutoff frequency capability for this configuration and reduced coupling between output and input. Reduced coupling provides improved stability, thereby reducing the tendency for unwanted oscillations. Figure 18.9 is a schematic diagram of a common-base NPN BJT amplifier.

The base in a common-base amplifier is connected directly to ground or at least to RF ground. The signal input is through a coupling capacitor to an RFC that is connected at one end to the emitter and at the other end to a negative emitter bias power supply. The collector is connected through an RFC to the

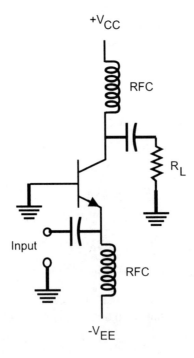

Figure 18.9 Schematic diagram of a common-base NPN BJT amplifier.

collector power supply. The signal output is through a coupling capacitor to a resistive load. The common-base configuration is the one used primarily in the design of microwave oscillators.

The following design equations are for the common-base amplifier at mid-band.

Current gain:

$$A_i = \alpha = i_C/i_E \quad (18.12)$$

Input resistance at the base:

$$r_i = (25 \text{ mV}/I_E) \quad (18.13)$$

Transconductance for the transistor:

$$g_m = I_E/(25 \text{ mV}) \quad (18.14)$$

Output resistance for the amplifier:

$$r_o = R_L \parallel r_{ce} \quad (18.15)$$

Voltage gain for the amplifier:

$$A_v = g_m r_o \quad (18.16)$$

Power gain for the amplifier:

$$A_p = \alpha^2 g_m r_o \quad (18.17)$$

For an example of the use of (18.12) to (18.17) in the evaluation of a common-base amplifier, assume the following:

$$\alpha = 1$$
$$r_{ce} = 10 \text{ k}\Omega$$
$$I_E = 2.5 \text{ mA}$$
$$R_L = 1 \text{ k}\Omega$$

Then,

$$g_m = I_E/(25 \text{ mV})$$
$$g_m = 2.5 \text{ mA}/(25 \text{ mV}) = 0.1\text{S} \quad (18.18)$$

$$r_i = (25 \text{ mV}/I_E)$$
$$r_i = (25 \text{ mV}/2.5 \text{ mA}) = 10\Omega \quad (18.19)$$

$$r_o = R_L \parallel r_{ce}$$
$$r_o = 909\Omega \quad (18.20)$$

$$A_v = \alpha g_m r_o$$
$$A_v = 0.1 \times 909 = 90.9 \quad (18.21)$$

$$A_p = \alpha^2 g_m r_o$$
$$A_p = 0.1 \times 909 = 90.9 = 19.6 \text{ dB} \quad (18.22)$$

18.2.3 Common-Collector Amplifier

Figure 18.10 is a schematic diagram of a common-collector amplifier.

The circuit in Figure 18.10 is usually called an emitter-follower. In this type of circuit, the output voltage follows the input voltage with an offset of about 0.7V. The emitter-follower is a feedback amplifier with a gain of about 1. The input resistance is high, as indicated by (18.23) and (18.24).

Input resistance:

$$r_\pi = (\beta + 1)25 \text{ mV}/I_E \quad (18.23)$$
$$r_i = r_\pi + (\beta + 1)R_L \quad (18.24)$$

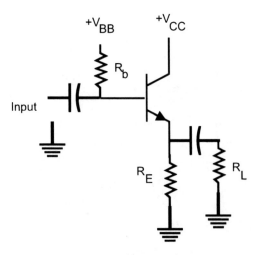

Figure 18.10 A common-collector amplifier.

Other design equations are as follows:
Output resistance:

$$r_o = R_L \parallel [(R_1 + r_\pi)/(\beta + 1)] \tag{18.25}$$

where R_1 is the signal source resistance.
Voltage gain:

$$A_v = (\beta + 1) R_L / r_i \tag{18.26}$$

Current gain:

$$A_i = \beta + 1 \tag{18.27}$$

For an example of the use of (18.23) to (18.27) in finding the performance characteristics of a common-collector amplifier, assume the following:

$$\beta = 50$$
$$I_E = 10 \text{ mA}$$
$$R_1 = 50 \Omega$$
$$R_L = 200 \Omega$$

Then,

$$r_\pi = (\beta + 1) 25 \text{ mV}/I_E$$
$$r_\pi = (50 + 1) 25 \text{ mV}/10 \text{ mA} = 127.5 \Omega \tag{18.28}$$

$$r_i = r_\pi + (\beta + 1) R_L$$
$$r_i = 127.5 + (51 \times 200) = 10{,}327 \Omega \tag{18.29}$$

$$r_o = R_L \parallel [(R_1 + r_\pi)/(\beta + 1)]$$
$$r_o = 200 \parallel [(50 + 127.5)/51] = 3.4 \Omega \tag{18.30}$$

$$A_v = (\beta + 1) R_L / r_i$$
$$A_v = 51 \times 200/10{,}327 = 0.99 \tag{18.31}$$

$$A_i = \beta + 1$$
$$A_i = 50 + 1 = 51 \tag{18.32}$$

18.3 FIELD EFFECT TRANSISTORS AND CIRCUITS

A number of different types of FETs are used in RF systems, including junction FETs (JFETs), metal-semiconductor FETs (MESFETs), and metal-oxide semiconductor FETs (MOSFETs).

18.3.1 Junction Field-Effect Transistors

An n-channel, JFET is illustrated in Figure 18.11. This device is made of Si with n- and p-type doping. The elements include the source, the gates, and the drain. An n-channel JFET has an n+ source, p+ gates, an n+ collector, and an n conducting channel connecting the source to the drain. The transistor acts as a voltage-controlled resistor, with the resistance of the conducting channel depending on the gate-to-source voltage, V_{GS}.

Under all conditions of bias, a depletion region forms near the junction of a JFET. The depletion region effectively reduces the width of the n-channel between source and drain and increases the resistance of the channel. The stronger the reverse bias, the larger the depletion region is and the greater the resistance of the channel. When the field becomes sufficiently strong, the channel is completely pinched off and no current flows.

Figure 18.12 is a schematic diagram of a common-source n-channel JFET amplifier. The gate is connected to a negative voltage by R_G, and the drain is connected to a positive voltage by R_D. The output is to a resistive load by means of a coupling capacitor connected to the drain.

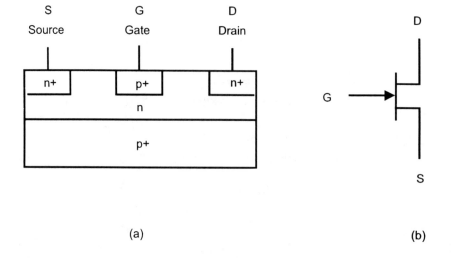

Figure 18.11 JFET: (a) geometry and (b) symbol. (*After:* [3].)

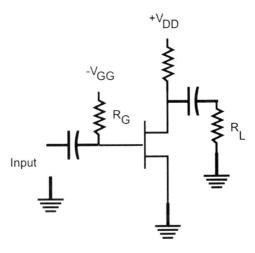

Figure 18.12 Schematic diagram of a common-source JFET amplifier.

The following midband design equations are for JFET common-source amplifiers:

Current gain:

$$A_i = {>}10^7 \tag{18.33}$$

Input resistance at the gate:

$$r_i = r_{gs} \tag{18.34}$$

Transconductance for the transistor: given by the manufacturer's data sheet.
Output resistance for the amplifier:

$$r_o = R_L \parallel r_{ds} \tag{18.35}$$

Voltage gain for the amplifier:

$$A_v = -g_m r_o \tag{18.36}$$

Typical small-signal parameters for JFETs are as follows:

$$g_m = 0.1 \text{ to } 10 \text{ mA/V}$$
$$r_{gs} = 10^9 \, \Omega$$
$$r_{ds} = 0.01 \text{ to } 1 \text{ M}\Omega$$
$$C_{gs} = 1 \text{ to } 10 \text{ pF}$$
$$C_{gd} = 1 \text{ to } 10 \text{ pF}$$

For an example of the performance of a JFET common-source amplifier, assume the following:

$$V_{DD} = 20\text{V} \qquad g_m = 0.005\text{S}$$
$$I_D = 5 \text{ mA} \qquad R_L = 2 \text{ k}\Omega$$
$$R_E = 200\Omega \qquad r_{ds} = 100 \text{ k}\Omega$$

Then,

$$A_v = -g_m r_o = -0.005 \times 1{,}961 = -9.8 = 19.8 \text{ dB}$$

The voltage gain of JFET amplifiers tends to be less than that of BJT amplifiers.

Figure 18.13 is a schematic diagram of a common-drain JFET amplifier, also known as a source-follower.

The following midband design equations are for JFET common-drain amplifiers:

Current gain:

$$A_i = {>}10^7 \qquad (18.37)$$

Input resistance at the base:

$$r_i = r_{gs} \qquad (18.38)$$

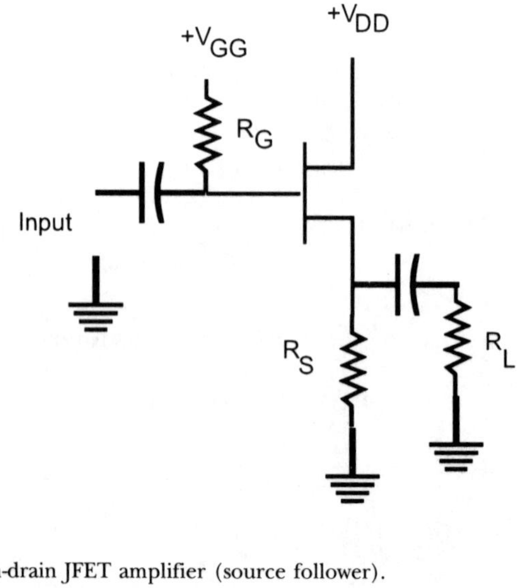

Figure 18.13 Common-drain JFET amplifier (source follower).

The transconductance for the transistor, which depends on the design of the JFET, usually is given by the manufacturer's data sheet. It depends on the design of the JFET.

Output resistance for the amplifier:

$$r_o = R_L \parallel 1/g_m \qquad (18.39)$$

Voltage gain for the amplifier:

$$A_v = R_L/(1/g_m + R_L) \qquad (18.40)$$

The JFET is not a good HF device, largely because it has the problem of charge storage in the gate and it does not use high-mobility materials. It also is not a high-power device and therefore is used primarily for low-power LF applications, where high input resistance is desirable.

18.3.2 Metal-Semiconductor Field-Effect Transistors

Figure 18.14 is an illustration of a MESFET. The MESFET diagram looks very much like the JFET diagram. The difference is that the PN junction-type diode used as the gate is replaced by an HF Schottky-barrier diode. That eliminates the problem of charge storage, one of the main limitations for the JFET for HF operation.

The MESFET uses an n-type GaAs semiconductor channel, which connects a source contact region and a drain contact region. Between those regions is a narrow gate consisting of a metallic contact placed on the surface of the semiconductor channel and forming a Schottky-barrier diode. Associated with this interface is a depletion region that has a width controlled by the gate voltage.

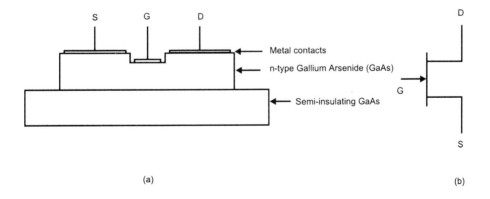

Figure 18.14 MESFET: (a) geometry and (b) symbol. (*After:* [4].)

GaAs is used with this device rather than Si because of its much higher mobility. That, coupled with the use of a Schottky-barrier gate with a length of only about 1 μm, allows its use as a microwave amplifier with very good operating characteristics.

The operation of the MESFET is similar to that of the JFET. Both devices are depletion-mode devices in which the depletion region cross-section near the gate is determined by the voltage at the gate diode. As the gate-to-source voltage is made more negative, the cross-section of the conducting channel decreases. At some negative voltage, the conducting channel is completely pinched off. A typical pinch-off voltage is about −1.5V. At saturation, the output resistance is about 400–500Ω.

The main type of amplifier configuration used with the MESFET is the common-source amplifier. Common-gate designs also are used. The voltage gain for the MESFET amplifier is approximately equal to $g_m r_o$. Thus, for an output resistance of 400Ω and a g_m of 60 mS, the voltage gain would be 24. There could be a fairly large power gain for the device because of the very high input resistance of the device.

The MESFET has been the workhorse of the microwave industry for many years. It is used as the active device for both low-noise amplifiers and power amplifiers. It is also used for oscillators, mixers, and other microwave circuits. It can be used in monolithic circuitry as well as in single devices. For more information on MESFETs, see [5].

18.3.3 Metal-Oxide Semiconductor Field Effect Transistors

An n-channel enhancement-mode MOSFET is illustrated in Figure 18.15. The symbol for this transistor shows a broken line between the source and the drain.

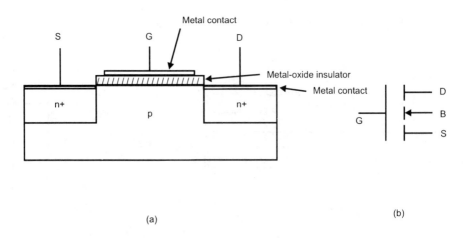

Figure 18.15 N-channel enhancement-mode MOSFET: (a) geometry and (b) symbol. (*After:* [3].)

That indicates that the device is normally cut off. (Sometimes the symbol for a MOSFET does not use the broken line, for reasons of simplicity.)

The channel current for a MOSFET is controlled by a voltage at a gate electrode that is isolated from the channel by an insulator. For that reason, the resulting device also is called an insulated-gate FET (IGFET). In the most common configuration, a metal-oxide layer is grown or deposited on the semiconductor surface, and the metal gate electrode is deposited onto that oxide layer.

In an insulated-gate or enhancement-mode MOSFET the source and drain regions may be n+ type material. These regions are separated by p-type material, with the insulated gate bridging over this connecting material. When a positive voltage is applied to the gate, electrons in the p-type material are attracted to the surface below the insulator, forming a connecting near-surface region of n-type material between the source and the drain. The larger or more positive the gate-to-source voltage, the lower the resistance of this region between the source and the drain. It takes some minimum gate voltage, called the threshold voltage, to turn this device on. A typical value may be in the range of 2–4V, depending on the device. Near the threshold voltage, the transistor is very nonlinear. At some higher voltages, the device becomes more nearly linear, and the value of forward transconductance, g_m, becomes larger.

The only amplifier configuration that is used extensively for the MOSFET is the common-source amplifier. This circuit is shown in Figure 18.16. The gate normally is biased a few volts positive with respect to the source, so the device will operate in the more linear and higher gain region of its operating characteristics.

An n-channel depletion/enhancement-mode MOSFET is illustrated in Figure 18.17. The solid line in the symbol between the source and the drain indicates that the device normally is conducting.

The channel current is controlled by a voltage at a gate electrode that is isolated from the channel by an insulator. In the most common configuration, a metal-oxide layer is grown or deposited on the semiconductor surface, and the metal gate electrode is deposited onto that oxide layer.

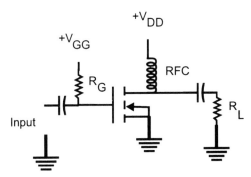

Figure 18.16 Common-source MOSFET amplifier.

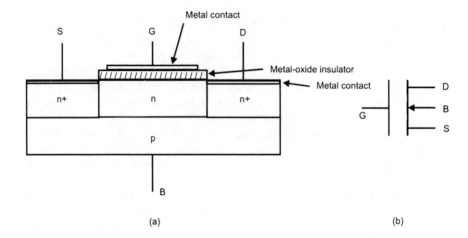

Figure 18.17 N-channel depletion/enhancement-mode MOSFET: (a) geometry and (b) symbol.

In a depletion/enhancement-mode MOSFET, the source and drain regions may be n+ type material. These regions are separated by n-type material, with the insulated gate bridging over the connecting material. When the gate-to-source voltage is zero, there is a moderate conductance between the source and the drain. When a negative voltage is applied to the gate, the number of electrons in the n-type channel between source and drain are reduced at the surface below the insulator. The resistance of the conduction channel thus is increased. The more negative the gate-to-source voltage, the higher the resistance of the region between the source and the drain. When a positive voltage is applied to the gate, the number of electrons in the n-type channel between source and drain are increased at the surface below the insulator. The resistance of the conduction channel is thus decreased. The more positive the gate-to-source voltage, the lower the resistance of the region between the source and the drain.

Figure 18.18 shows a common-source amplifier using an n-channel depletion/enhancement-mode MOSFET. The gate is connected to ground using a resistor. The drain is connected to a positive drain power supply, using an RFC. The output is to a resistive load via a coupling capacitor.

18.4 COMPARISON OF FET AND BJT AMPLIFIERS

As indicated earlier, there are only two types of BJTs: the NPN and the PNP. UHF and higher frequency devices are of the NPN type because of the higher mobility of electrons as majority carriers, which means higher f_t and improved HF power gain. PNP-type transistors are used primarily in land-mobile communications equipment requiring a positive ground system. They also may be used in circuits involving complementary symmetry operation, as in a push-pull amplifier.

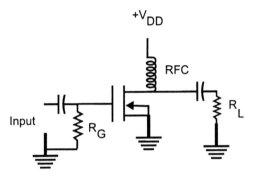

Figure 18.18 Depletion/enhancement type MOSFET amplifier.

Far more types of FETs are commercially available for RF amplifier use. Many of these devices have been around only a fairly short time, whereas BJTs have been around quite a while. The first 1-GHz BJTs were Ge, came to market around 1965 from Texas Instruments, and cost about $300 each. One such device was the TIX3024 transistor. It took five or six of them for a 20-dB gain, 1- to 2-GHz amplifier.

There are applications where BJTs clearly are better than FETs, and there will continue to be many systems that use this type of device. On the other hand, there clearly are many applications where FETs are superior to BJTs. That is particularly true for power applications and for microwave and millimeter-wave small-signal amplifiers.

FETs have the advantage over BJTs at the higher frequencies because they are able to use GaAs rather than Si. GaAs has a higher mobility than Si and higher peak electron velocities. These two features result in faster transit time and lower power dissipation. Therefore, they have higher frequency capability, higher gain, lower noise figure, and usually better efficiency.

MESFETs are used at all frequencies, from a few cycles per second to 40 GHz or more. At the lower frequencies, these transistors use Si as the device material. At microwave frequencies, these devices use GaAs. Several watts per transistor are available up to 15 GHz. Power outputs of a few hundred milliwatts are available from single GaAs FETs at 30 GHz. Noise figures below 0.3 dB are attainable at 4 GHz for low-noise GaAs FETs (e.g., the NEC32584 device). They are as low as 1.4 dB at 20 GHz.

MOSFET power transistors have relatively high CW output power at frequencies up to about 1.5 GHz. Output CW power from the M/A-COM type DU28200M transistor is reported to be 200W at 175 MHz. The M/A-COM type UHF2815OJ transistor is reported to provide an output CW power of 150W at 500 MHz, while the M/A-COM type LF40100 transistor is reported to provide an output CW power of 100W at 1,000 MHz.

The same RF design practices, such as grounding, filtering, bypassing, and creating a good circuitboard layout apply equally to circuits using any of these devices. Precautions must be taken with each type of device to prevent damage or destruction of the device. FETs are highly sensitive to gate rupture, which can be caused by excessive dc potentials or transients between the gate and the source. Of particular concern is static electricity. Technicians handling or testing the devices must be careful to be properly grounded along with the device.

A weak spot with BJTs is the possibility of thermal runaway. The main reason for thermal runaway with BJTs is the increasing h_{FE}. Care must be taken in the design of the bias circuit to make sure that does not happen. Self-biasing techniques often are used that provide negative feedback designed to prevent the problem.

18.5 HIGH ELECTRON MOBILITY TRANSISTORS AND HETEROJUNCTION BIPOLAR TRANSISTORS

High electron mobility transistors (HEMTs) and heterojunction bipolar transistors (HBTs) are important recent developments in microwave and millimeter-wave transistors. These devices make use of heterojunctions for their operation. The heterojunctions are formed between semiconductors of different compositions and bandgaps, for example, GaAs/AlGaAs and InGaAs/InP. That is unlike conventional transistors, which use junctions between like materials. These relatively new types of devices offer significant improvements for low-noise amplifiers and microwave power amplifiers.

Some of the information presented in this section is adapted from [6].

18.5.1 High Electron Mobility Transistors

Figure 18.19 shows a schematic cross section of an HEMT structure using GaAs and AlGaAs. The conventional HEMT is similar to a GaAs MESFET. As seen in

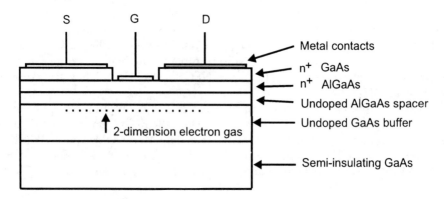

Figure 18.19 An HEMT device. (*After:* [6].)

Figure 18.19, the HEMT has two ohmic contacts (source and drain) and a Schottky gate that modulates the flow of current in the channel between the two contacts. The difference between the two types of devices and the key to the HEMT's improved performance is in the underlying semiconductor material. The HEMT has superior electron transport properties and much higher sheet charge density than the MESFET because of a two-dimension electron gas layer that is formed in a thin layer between the AlGaAs and the undoped GaAs layers. This is shown in Figure 18.19.

HEMPTs have demonstrated unprecedented noise performance at cryogenic temperatures and good microwave and millimeter-wave noise and power performance at room temperature at frequencies up to 60 GHz. Typical noise figures at 12 GHz for commercially available low-noise HEMTs are about 1.0 dB. The best reported noise figures at the same frequency are about half that, or about 0.5 dB.

Figure 18.20 shows a comparison of HEMT and MESFET room-temperature noise performance up to about 60 GHz. This figure is based on the best reported noise figures for the two devices and clearly shows the advantages provided by the HEMT device.

In addition to lower noise figure, HEMTs also have several characteristics that make them more attractive for low-noise applications. They are easier to provide impedance matching, and they have a larger gain-bandwidth product.

18.5.2 Heterojunction Bipolar Transistors

The cross-section of a basic n-p-n HBT is shown in Figure 18.21. The n-type emitter is formed in the wideband gap AlGaAs while the p-type base is formed in the lower-band gap GaAs. The n-type collector is also formed in GaAs.

The GaAs MESFET is currently the most widely used microwave device for the amplification of microwave signals. HBTs are also well suited to amplification of large microwave signals. Laboratory results have shown that in the future HBT amplifiers will have a significant advantage over MESFETs for power amplification in terms of output power, compactness, efficiency, and novel circuit usage such as complementary amplifiers. HBTs have been used for frequencies ranging from 3 to 60 GHz. At 3 GHz, up to 1W of CW output power has been obtained with 61% efficiency. The highest output power obtained at microwave frequencies is 5.3W of CW of output power at 8 GHz with 33% efficiency. Improvements are expected in the future.

18.6 DC BIAS CIRCUITS FOR BJT AMPLIFIERS

Good design of the bias circuit for a BJT requires that these two deficiencies of the transistor be overcome: (1) that the transistor is temperature sensitive and (2) that the parameters of the transistor are subject to process variations. The bias circuit shown in Figure 18.22 is often used to solve these problems. In the circuit

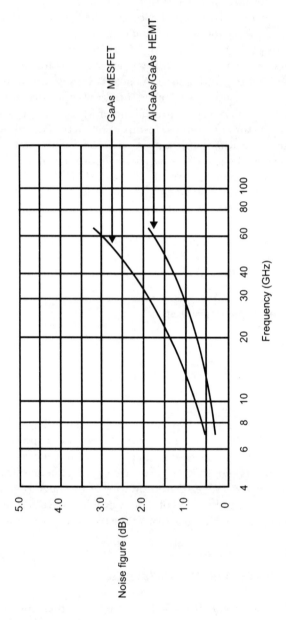

Figure 18.20 Comparison of noise figures for GaAs MESFET and AlGaAs/GaAs HEMT devices. (*After:* [6].)

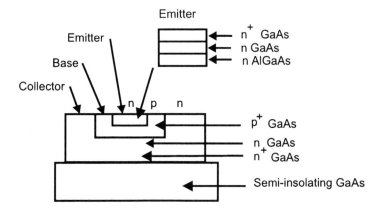

Figure 18.21 An HBT structure. (*After:* [6].)

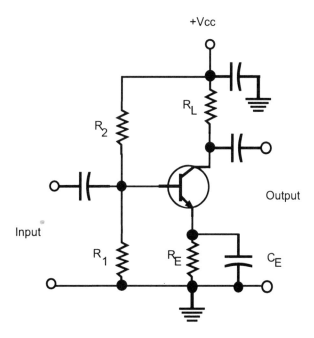

Figure 18.22 Self-bias circuit for common-emitter amplifier.

in Figure 18.22, the transistor is shown with a feedback resistor, R_E, and two base resistors, R_1 and R_2. It is called a self-bias circuit because only one supply voltage is required. This same circuit can also be used for MESFET common-source amplifiers and usually is done that way unless separate gate bias is supplied.

The value of V_{BB} and R_B are given by (18.41) and (18.42).

$$V_{BB} = R_1 V_{CC}/(R_1 + R_2) \qquad (18.41)$$

$$R_B = R_1 R_2/(R_1 + R_2) \qquad (18.42)$$

The value of R_E usually is chosen such that there is a 1–3V drop across it with no RF signal. Thus, if the quiescent operating point for the circuit is 4 ma and a 2V drop is used, the value of R_E would be 500Ω.

The value of R_1 is chosen to be less than $\beta_{min} R_E/5$. Thus, with a transistor minimum β of 50 and $R_E = 500Ω$, R_1 would be

$$50 \times 500/5 = 5 \text{ K}\Omega$$

The voltage at the base is the voltage drop across $R_E + 0.7V$. For our example, that would be 2.7V. The current in R_1 thus would be about 2.7/5K = 0.54 ma. If the value of V_{CC} is 20V, the voltage drop across R_2 is 17.3V.

The current through R_2 is the current through R_1 plus the base current. Assuming a β of 50, the base current would be 0.08 ma. The value of R_2 is thus 17.3/0.62 = 27.9 KΩ. The value of R_B is thus 8.5 KΩ.

With this method of bias, there tends to be only a small change in Q-point current for a large change in β or a large change in temperature. Part or all of the resistance, R_E, can be bypassed with a capacitor to avoid negative feedback for ac signals. That allows the gain of the amplifier to be high and the noise figure to be low. For the example shown, all the emitter resistance is bypassed by C_E for maximum gain.

High-power BJT RF amplifiers sometimes use a clamping diode plus adjustable series resistance to establish the desired base bias voltage for a common-emitter amplifier with the emitter grounded. This dc voltage is fed through an RFC to the base circuit. The diode can be connected mechanically to the transistor heat sink to perform a temperature-compensating function.

More sophisticated bias sources can be made that use integrated circuit voltage regulators. Transistor source followers can be used with these IC regulators to provide the needed current boost and to lower the source impedance.

18.7 BIAS CIRCUITS FOR FET AMPLIFIERS

Low-power LF FET amplifiers can use simple voltage dividers for bias circuits. A circuit of this kind is shown in Figure 18.23 for a small-signal enhancement-mode MOSFET amplifier. The gate in this case is biased positive with respect to the source. A typical threshold voltage would be in the range of 2–4V. For the case of a common-source amplifier with class A operation, the bias voltage would need to be a few volts positive with respect to the threshold voltage.

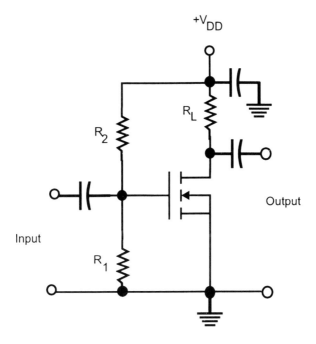

Figure 18.23 A dc bias circuit for enhancement-mode MOSFET.

18.8 STABILITY WITH BJT AND FET AMPLIFIERS

One of the important problems with high-gain amplifiers is the problem of unwanted oscillations. One approach used to help improve stability and prevent oscillations is to use shielding. It is common practice to place amplifiers in small metal containers with feed-through capacitors and chokes for dc inputs and coaxial connectors and cables for RF inputs and outputs. Radar-absorbing material may be attached to the inside of the cover to reduce radiative coupling effects. A shielding system of this type is shown in Figure 18.24.

Another way to improve stability is to use neutralizing circuits. The goal is to provide a negative feedback signal that is equal in amplitude but opposite in phase to the positive feedback signal that is causing the unwanted oscillations.

18.9 IMPEDANCE MATCHING

There are two main methods for providing impedance matching for transistor amplifiers and other RF circuits: with transformers and with L, T, and π LC circuits. The transformer approach is useful mainly at 500 MHz and lower frequencies. The L, T, and π LC circuits method is useful at all RF frequencies, including UHF and microwave. The types of components used with the LC circuits are lumped constant

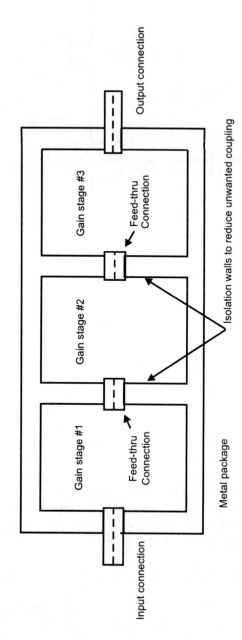

Figure 18.24 Shielding concepts for RF amplifiers.

circuits for frequencies below about 500 MHz. Above 500 MHz, it is difficult to realize discrete inductors and capacitors. It then becomes necessary to use microstrip or stripline techniques to realize the required matching circuits.

Methods of impedance matching with LC circuits were discussed in detail in Chapter 14. Methods of impedance matching with transformers were discussed in detail in chapter 15.

18.10 DESIGN METHODS WITH S-PARAMETERS

Some of the following material is quoted or adapted from [7].

Many RF transistors for HF and higher frequencies are characterized using S-parameters. Packaged amplifier and other RF products also may be characterized by S-parameters.

S-parameters can be measured quickly and accurately using test equipment called network analyzers with S-parameter test sets added. Automated S-parameter test equipment is available from Hewlett-Packard Company and Wiltron Corporation and are used by the manufacturers of RF components. Less costly nonautomated S-parameter test equipment nearly always is available in industry for users of the RF components.

S-parameters are simply the coefficients of the incident and reflected voltage waves. S_{11} is the input voltage reflection coefficient and is defined as reflected-wave voltage (b_1) divided by incident-wave voltage (a_1) when incident-wave voltage (a_2) is zero volts. S_{22} is the output voltage reflection coefficient and is defined as the reflected-wave voltage (b_2) divided by incident wave voltage (a_2) when incident-wave voltage (a_1) is zero volts. S_{21} is the forward voltage transmission coefficient and is defined as output-wave voltage (b_2) divided by the input-wave voltage (a_1) voltage when the incident-wave voltage (a_2) is zero volts. S_{12} is the reverse transmission coefficient and is defined as the output-wave voltage (b_1) divided by the incident- or input-wave voltage (a_2) when the incident wave (a_1) is zero volts.

The quantity $|S_{21}|^2$ is the power gain of the transistor at the specified bias conditions and frequency with 50Ω source and load terminations. When S_{21} is given in a data sheet expressed in decibels, it is referring to $10 \log|S_{21}|^2$ or $20 \log|S_{21}|$.

The term G_{tu} is the unilateral, that is, one-way transmission only, transducer power gain with $S_{12} = 0$. It is defined as the power delivered to the load divided by the power available from the source with $S_{12} = 0$. The assumption that S_{12} is close to zero is usually a good one for many high-quality transistors, and it usually is used in calculations of amplifier gain since it greatly simplifies calculations. The value of G_{tu} is given by (18.43).

$$G_{tu} = [1 - |\Gamma_S|^2]/|1 - S_{11}\Gamma_S|^2 \times |S_{21}|^2 \times [1 - |\Gamma_L|^2]/|1 - S_{22}\Gamma_L|^2 \quad (18.43)$$

where Γ_S is the source reflection coefficient, and Γ_L is the load reflection coefficient.

Equation (18.43) can be broken into three sources of gain. These are:

$$G_S = [1 - |\Gamma_S|^2]/|1 - S_{11}\Gamma_S|^2 \qquad (18.44)$$

$$G_0 = |S_{21}|^2 \qquad (18.45)$$

$$G_L = [1 - |\Gamma_L|^2]/|1 - S_{22}\Gamma_L|^2 \qquad (18.46)$$

G_S is the gain contribution achieved through input impedance matching, G_0 is the gain contribution of the transistor itself, and G_L is the gain contribution achieved by output impedance matching. The total gain for the three sources expressed in decibels is given by (18.47):

$$G_{tu} = G_S + G_0 + G_L \qquad (18.47)$$

where the terms are in decibels.

If the circuit design is narrowband and we want maximum power gain, all that is required is to set $\Gamma_S = S_{11}^*$ and $\Gamma_L = S_{22}^*$, where * indicates complex conjugate. If the circuit is broadband and we want a certain amount of gain across a band of frequencies, we can use circuits that compensate for the variations in gain with frequency for the device. That can be done with selective mismatching.

For an example of the use of S-parameters in the design of a small-signal RF amplifier, assume that the source and load impedances are each 50Ω. Also assume that interest is only in maximum gain at 1,000 MHz. The assumed transistor is the Motorola MRF 571. Our bias will be 6V at 50 mA. The manufacturer's data sheet shows that f_T is near its peak at 50 mA, and values of scattering parameters are given for this bias.

S-parameters at the desired frequency and bias points are as follows:

S_{11} = 0.60 at an angle of +156 degrees
S_{22} = 0.11 at an angle of −164 degrees
S_{12} = 0.09 at an angle of + 70 degrees
S_{21} = 4.40 at an angle of + 75 degrees

The required value of $\Gamma_s = S_{11}^* = 0.60$ at an angle of −156 degrees. The required value of $\Gamma_L = S_{22}^* = 0.11$ at an angle of +164 degrees.

The gains contributed by each of the gain sources are as follows:

$$G_S = [1 - |\Gamma_S|^2]/|1 - S_{11}\Gamma_S|^2 \qquad (18.48)$$
$$G_S = 0.64/0.41 = 1.56 = 1.93 \text{ dB}$$

$$G_0 = |S_{21}|^2 \qquad (18.49)$$
$$G_0 = 100 = 20.00 \text{ dB}$$

$$G_L = [1 - |\Gamma_L|^2]/|1 - S_{22}\Gamma_L|^2 \qquad (18.50)$$
$$G_L = 0.988/0.976 = 1.012 = 0.053 \text{ dB}$$

The total gain with conjugate matching is then

$$G_{tu} = G_S + G_0 + G_L \qquad (18.51)$$
$$G_{tu} = 1.93 + 20.00 + 0.05 = 21.98 \text{ dB}$$

From the preceding example, note that there is already a nearly perfect impedance match at the output without adding matching circuits. Therefore, we will select a design that uses only input matching.

Next, select the design of the impedance-matching sections using the Smith chart shown in Figure 18.25.

One approach to finding the impedance-matching components involves starting at the 50Ω input and moving to the complex conjugate value of S_{11}^*. That value is $S_{11}^* = 0.60$ at an angle of −156 degrees.

For the input, start at the center of the Smith chart at point 0 and use the chart as an admittance chart. Travel on the 1.0 conductance circle to point A. The length of that path is 1.63 units long, indicating a normalized capacitive susceptance component of $+j\,1.63$ and a real capacitive susceptance of 0.032S. Next, transfer through the center of the chart to point B, converting to an impedance chart. Then travel to point C, which is the S_{11}^* point. The length of that path is found to be 0.24 unit long, indicating a normalized inductive reactance component of $+j\,0.24$. The real inductive reactance is 50 times that value, or $+j\,12.0\Omega$.

Let us now find the component values for impedance matching at 1,000 MHz.

$X_L = 12.0\Omega$
$L = X_L/2\pi f = 12.0/(2\pi \times 1,000 \times 10^6) = 1.9$ nH
$B_C = 0.032$S
$C_1 = B_C/2\pi f = 0.032/(2\pi \times 1,000 \times 10^6) = 5.1$ pF

The optimum source reflection coefficient for minimum noise figure for a transistor amplifier can be given by a transistor data sheet. In general, that will be different from the S_{11}^* point for maximum gain. The gain available when the transistor is matched for low-noise figure general will in be less than would be the case if the amplifier using the device had been designed for maximum or near-maximum gain. That gain is termed associated gain and usually is supplied on data sheets for low-noise devices.

18.11 MANUFACTURERS' DATA SHEETS FOR TRANSISTORS

Manufacturers' data sheets normally are available for the RF transistors and packaged amplifiers of interest. In the data sheets, transistors are characterized by two

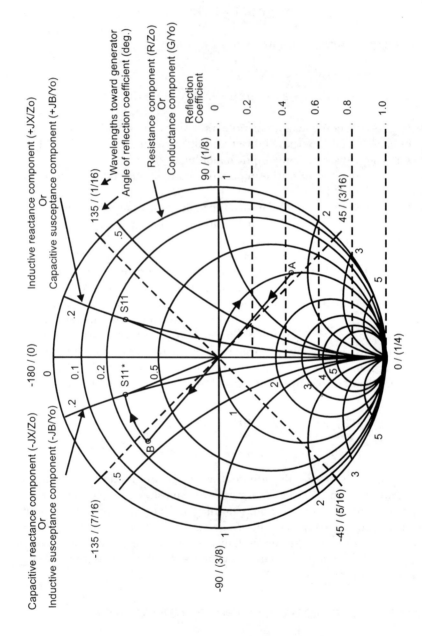

Figure 18.25 Example design of impedance matching circuits for narrowband transistor amplifier using S-parameters and Smith Chart.

types of parameters: DC and functional. The DC specifications consist of breakdown voltages, leakage currents, the DC beta (h_{FE}) for BJTs, and capacitances. The functional specifications cover gain, noise figure, Z_{in}, Z_{out}, S-parameters, current gain-bandwidth product, distortion, ruggedness, and so on. The data sheets also provide thermal characteristics for the transistors and packaged amplifiers. In providing those characteristics, the data sheets may provide minimum, typical, and maximum ratings for the various characteristics. In other cases, they provide only minimum and maximum values or ratings. Curves often are supplied for the various parameters as a function of frequency. In some cases, schematic diagrams of test fixtures used in measuring the various parameters are provided. Table 18.1 is an example of a typical set of manufacturers' data sheets for an RF power MOSFET. The manufacturer is M/A-COM Power Hybrids Operation, 1742 Crenshaw Blvd., Torrance, CA 90501.

Table 18.1
DU1260T N-Channel RF Power MOSFET

Features

DMOS structures
Lower capacitances for broadband operation
High saturated output power
Lower noise figure than bipolar devices
Specifically designed for 12V applications

Absolute Maximum Ratings

Parameter	Symbol	Rating	Units
Drain-source voltage	V_{ds}	40	V
Gate-source voltage	V_{gs}	20	V
Drain-source current	I_{ds}	24	A
Total device dissipation at 25°C	P_d	250	W
Junction temperature	T_j	200	°C
Storage temperature range	T_{stg}	−55 to +150	°C
Thermal resistance		0.7	°C/W

Electrical Characteristics at 25°C

Parameter	Symbol	Min	Max	Units
Drain-source breakdown voltage	B_{dss}	40		V
(Test condition V_{gs} = 0V, I_{ds} = 30.0 mA)				
Drain-source leakage current	I_{dss}		6	mA
(Test condition V_{ds} = 15.0V, V_{gs} = 0V)				
Gate-source leakage current	I_{gss}		6	uA
(Test condition V_{gs} = 20.0V, I_{ds} = 0 mA)				
Threshold voltage	V_{gs}(th)	2	6	V
(Test condition V_{ds} = 10.0V, I_d = 600.0 mA)				
Forward transconductance	G_m	3		
(Test condition V_{ds} = 10.0V, I_d = 6,000.0 mA)				
Input capacitance	C_{iss}		200	pF
(Test condition V_{ds} = 12.0V, f = 1 MHz)				
Output capacitance	C_{oss}		240	pF
(Test condition V_{ds} = 12.0V, f = 1 MHz)				
Reverse capacitance	C_{rss}		48	pF
(Test condition V_{ds} = 12.0V, f = 1 MHz)				
Power gain	Pg	8		dB
(Test condition V_{ds} = 12.0V, I_{dq} = 600 mA, f = 175.0 MHz, PO = 60.0W)				
Drain efficiency			60	%
(Test condition V_{ds} = 12.0V, I_{dq} = 600 mA, f = 175.0 MHz, P_O = 60.0W)				
Load mismatch tolerance	VSWR 30:1			
(Test condition V_{ds} = 12.0V, I_{dq} = 600 mA, f = 175.0 MHz, P_O = 60.0 W)				

Typical Device Impedance

Frequency (MHz)	Z_{in} (Ω)	Z_{ol} (Ω)
30	4.5–j8.0	4.6–j7.9
100	1.4–j4.0	1.4–j8.0
175	1.0–j0.5	1.0–j0.5

V_{ds} = 12.0V, I_{dq} = 600 mA, P_O = 60.0W

Z_{in} is the series equivalent input impedance of the device from gate to source.
Z_{ol} is the optimum series equivalent load impedance as measured from drain to ground.

Drawings and Performance Curves Presented

1. Typical drain efficiency vs. frequency
2. Typical gain vs. frequency
3. Typical power output vs. power input
4. Typical power output vs. supply voltage
5. Package outline
6. Test fixture

References

[1] Dye, Norm, and Helge Granberg, *Radio Frequency Transistors, Principles and Practical Applications*, Boston: Butterworth-Heinemann, 1993.
[2] Van Valkenburg, Mac E., ed., *Reference Data for Engineers*, 8th ed., SAMS, Carmel, IN: Prentice Hall Computer Publishing, 1993.
[3] Ibid., pp. 18-13–18-21.
[4] Kennedy, George, *Electronic Communication Systems*, 3rd ed., New York: McGraw-Hill, 1985.
[5] Golio, J. Michael, *Microwave MESFETs & HEMTs*, Boston: Artech House, 1991.
[6] Ali, Fazal, and Aditya Cupta, eds., *HEMTs and HBTs: Devices, Fabrication, and Circuits*, Boston: Artech House, 1991.
[7] Dye, Norm, and Helge Granberg, Chap. 13.

CHAPTER 19

High-Power Vacuum Tube Amplifiers and Oscillators

Vacuum tubes are used for nearly all high-power transmitter systems including power grid tubes, klystrons, helix-type TWTs, CCTWTs, crossed-field amplifiers (CFAs), magnetrons, gyrotrons oscillators, and gyrotron amplifiers.

19.1 GRID TUBES

Much of the information in this section is adapted from [1].

19.1.1 Triode, Tetrode, and Pentode Vacuum Tubes

Grid-type vacuum tubes find extensive use as high-power amplifiers and modulators for frequencies below about 4 GHz. Their main application is at VHF and lower frequencies. Types of power grid tubes used include triodes, tetrodes, and pentodes. The triode is shown in Figure 19.1(a). Elements of the triode include the thermionic cathode, which emits electrons; the control grid, which controls the flow of electrons; and the anode, which collects most of the electrons. If the grid of a triode is biased to a sufficiently negative voltage, no current flows. As the grid is made less negative, more current flows to the anode. When the grid becomes positive with respect to the cathode, both the grid and the anode draw current.

In operation as an RF power amplifier, the triode must be either neutralized or operated in the common-grid mode. If that is not done, internal capacitance between the anode and the grid produces positive feedback that may cause oscillation.

Very large triodes may be used at the lower frequencies in high-power applications. As frequency is increased, however, the triodes must become smaller. This

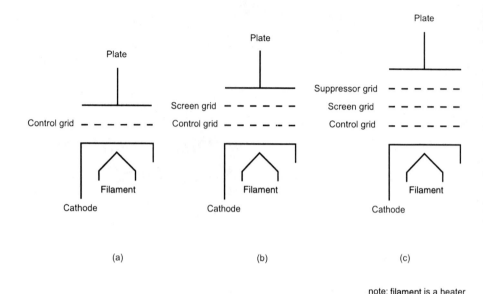

Figure 19.1 Power grid tubes: (a) triode; (b) tetrode; and (c) pentode.

limitation is the result of reduced available transit time as frequency is increased. The transit time, the time it takes an electron to move from the cathode to the control grid, must be much less than one-half the period of the RF cycle. With short available transit time, the space between the cathode and the control grid must be made small, and the power handling capability of all elements thus is reduced.

Small, planar triodes are used at UHF and microwave frequencies up to about 4 GHz. The anode supply voltage typically is 1,000V or more.

Small cylindrical triodes are used mainly at VHF and UHF where CW power of a few hundred watts or pulse power of tens of kilowatts is required. Modern triodes use beam-forming cathodes and control-grid geometry to allow simplicity of design and the circuit advantages of a triode with the gain of a tetrode.

The tetrode is illustrated in Figure 19.1(b). It has the same three elements as the triode plus a screen grid between the control grid and the anode. The presence of the screen grid greatly reduces the capacitance between the anode and the control grid and makes neutralization completely unnecessary or at least easy to accomplish.

Tetrode tubes are available in various sizes. Applications are in radio broadcasting (AM and FM), television broadcasting (VHF and UHF), communications, radar, navigational aids, among others. Uses are in RF power generation, modulation (AM), and switching for pulse service.

The pentode is shown in Figure 19.1(c). It has the same elements as the tetrode plus an additional suppressor grid to control secondary electrons. Modern tetrodes accomplish this control in other ways and have almost completely displaced pentodes.

There is a limit to the power capability of the tubes that is determined by power dissipation of the anode and the grids, which depends on the type of cooling that is provided. The lower power triodes usually use fins and air cooling. In some cases, this will be forced air cooling. The higher power tubes usually use water cooling.

Some important terms used with grid-type vacuum tubes follow:

μ = amplification factor with current held constant ($\Delta E_b/\Delta E_{c1}$)
R_p = dynamic plate resistance ($\Delta E_b/\Delta I_b$)
G_m = transconductance ($\Delta I_b/\Delta E_{c1}$)
E_b = total instantaneous plate voltage
E_{c1} = total instantaneous control grid voltage
I_b = total instantaneous plate current

The value of μ may be in the range of 20–200, depending on the deign of the tube. In a triode, plate current increases with increasing plate voltage. In a tetrode and a pentode, the plate current is nearly independent of plate voltage. Figure 19.2 shows the characteristics of a typical tetrode plate current. Plate current is plotted as a function of plate voltage and control-grid voltage for a fixed screen voltage of 300V.

Figure 19.3 is a drawing of the radial beam power tetrode, type 4CX15000A. The figure shows the dimensions of the tube and identifies the location of the elements. This tube is typical of high-power air-cooled tubes.

19.1.2 Grid-Type Vacuum Tube Amplifiers and Modulated Amplifiers

Figure 19.4 is a schematic diagram of a tetrode RF power amplifier. This circuit is an example of a low-VHF, HF, MF, or lower frequency system that uses lumped constant inductors and capacitors. Higher frequency VHF or UHF amplifier circuits use cavity resonators rather than LC circuits.

The filament or heater circuit and the cathode circuit for the amplifier in Figure 19.4 are shown at the lower part of the diagram. The cathode is ac coupled to ground. It is also dc coupled to ground through the center tap of the filament transformer. The grid bias is injected through an RFC, as is the plate bias current. Bypass capacitors also are incorporated to decouple the RF signals from the power supplies. The RF input to the grid circuit is supplied by a tuned transformer with series tuning on the primary and parallel tuning on the secondary. The RF output

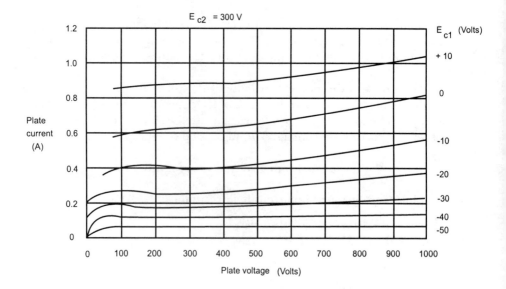

Figure 19.2 Characteristics of a typical tetrode plate current. (*After:* [1].)

is coupled from the plate using a series capacitor followed by an impedance-matching circuit. The circuit also shows the use of a neutralization capacitor. With neutralization, feedback is provided from the plate circuit to the grid circuit in such a way that other feedback signals are canceled. That is done to prevent unwanted oscillations.

Most of the unwanted oscillations in RF power amplifiers using grid-type tubes fall into the following three categories:

- Oscillations at VHF from about 40 to 200 MHz, regardless of the operating frequency of the amplifier;
- Self-oscillation on the fundamental frequency of the amplifier;
- Oscillation at a low RF below the normal frequency of the amplifier.

LF oscillations usually involve the RFCs. These may be eliminated by changes in the positions or orientations of the chokes. Oscillations near the fundamental frequency involve the normal resonant circuits. These are eliminated by effective design and the use of the neutralization circuit. Oscillations at VHF usually are referred to as parasitic oscillations. These may not be eliminated by the simple neutralization circuit. Solutions to this problem may require the addition of an RFC or an RFC plus a resistor in parallel that are located between the plate and the bias choke or output circuit.

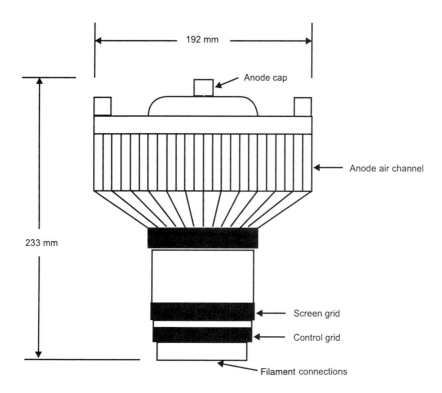

Figure 19.3 Drawing of the type-4CH15000A radial beam power tetrode. (*After:* [1].)

Figure 19.5 shows one version of the plate-modulated power amplifier that is frequently used for AM applications. In this instance, triodes are used rather than tetrodes. Either type of tube may be used.

The RF signal is fed to the grid of the upper triode using a single-tuned transformer. The tube is operated as a class C amplifier. It has its cathode connected to ground and its grid biased negatively. The output short pulses of current at the RF frequency are fed to a single-tuned RF transformer, which, by the fly-wheel effect, converts the pulse-type signal to a sine wave signal.

The audio or other modulating signal is fed to the grid of the lower triode amplifier. That amplifier modulates the voltage applied to the RF amplifier, thereby amplitude modulating the RF signal.

A more common plate-modulated amplifier is one that uses a modulation transformer in the plate voltage supply and a class B amplifier to drive the transformer. All the sideband power must be supplied by the audio or other LF amplifier that drives the transformer. Thus, the amplifier is a large, high-power system for high-power modulators.

Figure 19.6 shows grid modulation of a class C RF amplifier. In this case, the modulating signal is applied to a transformer that is in series with the input RF

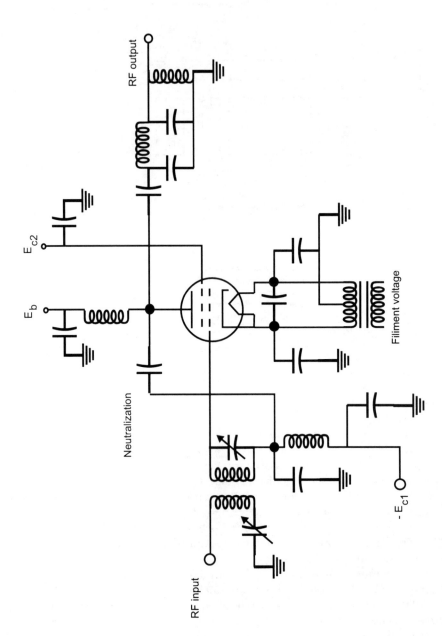

Figure 19.4 Schematic diagram of a tetrode RF power amplifier. (*After*: [1].)

High-Power Vacuum Tube Amplifiers and Oscillators | 555

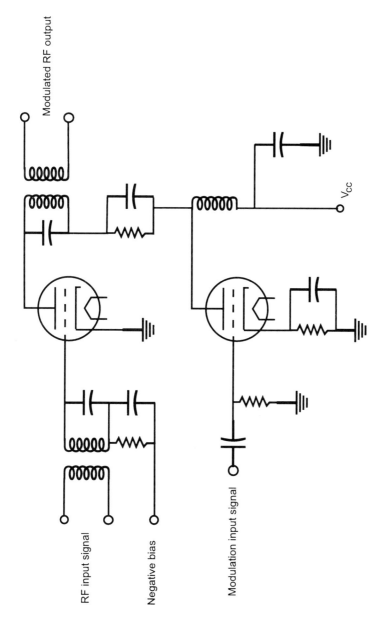

Figure 19.5 Classic plate-modulated class C power amplifier circuit for AM. (*After:* [1].)

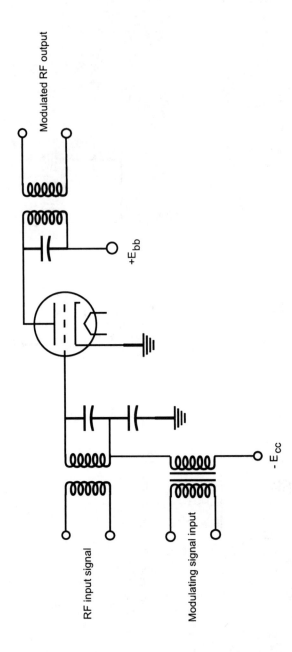

Figure 19.6 Grid-modulated class C amplifier for AM. (*After:* [1].)

carrier signal transformer, and the combination is applied to the grid of the triode power amplifier. Again, a single-tuned transformer is used at the output of the triode amplifier to convert the pulsed signals to modulated sine waves. This type of circuit avoids the problems of supplying high modulating power to the plate circuit.

19.1.3 The Use of Cavities as Resonators for High-Power, High-Frequency Triodes and Tetrodes

Cavities frequently are used as resonators for high-power triode or tetrode amplifiers that operate at VHF and higher frequencies. The type of cavities most often used are coaxial cavities. These may be either quarter-wavelength cavities or half-wavelength cavities.

Figure 19.7 shows the case of the quarter-wave cavity with a tetrode amplifier. An RFC connected to a coupling line brings dc plate bias into the tube. The RF input is to the base terminals. The RF output is coupled to a coax line using a small coupling loop. Tuning is accomplished by means of an adjustable shorting deck. Additional tuning is performed by means of tuning capacitors, which may include a movable tuning-capacitor plate.

In a quarter-wavelength cavity, the short circuit at the end of the cavity is reflected as an open circuit. Thus, it behaves like a parallel resonant circuit at the position of the plate.

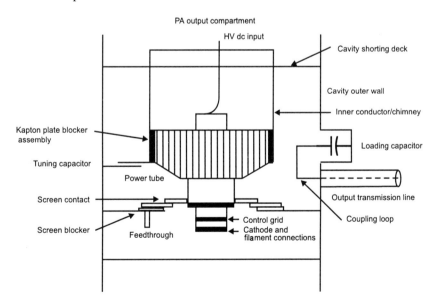

Figure 19.7 Example VHF quarter-wavelength cavity for high-power tetrodes. (*After:* [1].)

The equivalent circuit for the quarter-wavelength cavity amplifier with capacitive coupling is shown in Figure 19.8. The cavity acts as a high-Q parallel resonant circuit.

Half-wavelength power cavities also are used for tube amplifiers. In this case, the transmission line cavity is open circuited at the far end rather than being short circuited. The open circuit is reflected as an open circuit at the location of the plate, again acting as a parallel resonant circuit. Needless to say, this type of cavity is large at the lower frequencies. For example, at 88 MHz the length of the cavity above the plate would be 67 in.

19.2 MICROWAVE TUBES AND CIRCUITS

19.2.1 Introduction to Microwave Tubes

Grid-type tubes such as triodes and tetrodes cannot be used at high microwave frequencies because of the problem of transit-time limitations. Transit time is the time it takes for an electron to travel from the cathode to the control grid and needs to be much less than the time for a half-cycle of the RF signal. Microwave tubes, on the other hand, do not have this problem. As a matter of fact, some microwave tubes, such as the klystron and the TWT, use transit-time effects to an advantage. Instead of using electron-density modulation, as in the case of triodes and tetrodes, they use velocity modulation of the electron beam as a means of producing electron bunching.

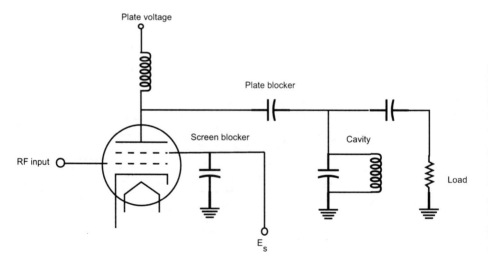

Figure 19.8 Equivalent circuit for a quarter-wavelength cavity amplifier with capacitive coupling. (*After:* [1].)

Table 19.1, which is adapted from information in [2], is a comparison of the operating characteristics for the four main types of high-power microwave tubes: the klystron, the CCTWT, the magnetron, and the CFA. Other important types of microwave tubes are not shown in the table: lower power, helix-type TWTs; low-power, two-cavity klystron oscillators; reflex klystron oscillators; high-power gyrotron oscillators; extended interaction devices; and fast-wave devices.

The maximum average or CW power capability of microwave tubes is shown in Figure 19.9 as a function of frequency. Also included on the plot are the maximum CW power capabilities for solid-state devices and for triodes and tetrodes.

19.2.2 Multiple-Cavity Klystron Amplifiers

A three-cavity klystron is shown in Figure 19.10. Elements of this tube include an electron gun, which forms a thin pencil-like electron beam; an input resonant cavity

Table 19.1
Comparison of Operating Characteristic for Medium- to High-Power Microwave Tubes

| *Linear Beam Tubes (LBT)* | | |
Characteristic	Klystron	CCTWT
Application	Amplifier	Amplifier
Frequency	UHF to Ka-band	UHF to Ka-band
Maximum peak power	5 MW at L-band	240 kW at UHF
Maximum average power	1 MW at L-band	12 kW at L-band
	>10 kW at X-band	10 kW at X-band
Cathode voltage for peak power	Up to 125 kV	Up to 42 kV
	(5 MW, L-band)	(0.2 MW, L-band)
Percentage bandwidth	1–10%	5–15% for CCTWT
		100% for helix TWT
Gain (dB)	30–65	30–65
Efficiency	Up to 65%	Up to 60%

| *Crossed-Field Tubes (CFTs)* | | |
	Magnetron	CFA
Application	Oscillator	Amplifier
Frequency	UHF to Ka-band	UHF to Ka-band
Maximum peak power	1 MW at L-band	5 MW at S-band
Maximum average power	1.2 kW at L-band	1.3 kW at L-band
	100 W at X-band	2 kW at X-band
Cathode voltage for peak power	Up to 60 kV	Up to 105 kV
	(1 MW, L-band)	(5 MW, L-band)
Percentage bandwidth	1–15%	5–15%
Gain (dB)		10–20
Efficiency	Up to 70%	Up to 80%

Source: [2].

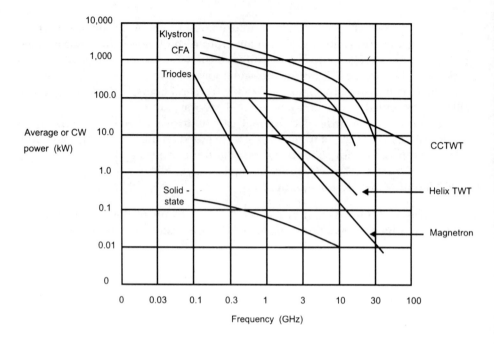

Figure 19.9 Average power for microwave power tubes. (*After:* [3].)

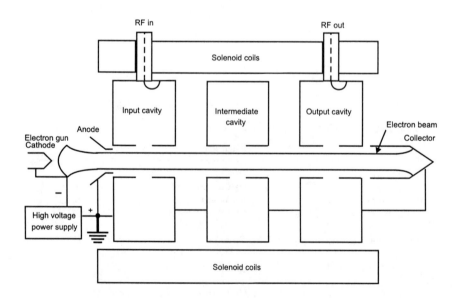

Figure 19.10 Three-cavity klystron amplifier. (*After:* [4].)

with openings to the beam; a similar intermediate resonant cavity tuned to a slightly higher frequency; an output resonant cavity tuned to the same frequency as the input cavity; and a collector for capturing the electrons. Electromagnets are used to keep the beam from spreading. Simple klystron amplifiers may use only two cavities (an input cavity and an output cavity). Very high gain cavities use more than three cavities.

The RF input signal field in the input cavity velocity modulates the electron beam. That causes the beam to become density modulated as the electrons move toward the collector. The intermediate cavity helps to further velocity modulate the beam. As the bunched electrons pass through the output cavity, they couple energy into the cavity. Suitable couplings are used at both the input cavity and the output cavity, so the cavities can be connected to the input and output transmission lines.

Klystrons are used for UHF television transmitters. Varian Associates produces tubes of this kind with output powers up to 65 kW CW. Examples of characteristics of this type of UHF-TV klystron amplifier are listed in Table 19.2. These characteristics are for the Varian VKP-7553S klystron, which is only 1 of 22 tubes of this kind listed by Varian in their sale brochure.

Many modern radar systems need combinations of high peak and average power, high gain, broad bandwidth, and long pulse capability in conjunction with linear phase characteristics and stable performance. These requirements can be satisfied by high power pulsed klystron and twystron amplifiers. Twystrons are a combination of a klystron and a TWT made by Varian. Somewhat larger bandwidths can be obtained with twystrons than with klystrons.

Table 19.3 lists characteristics of three types of high-power pulse klystrons made by Varian: types VA-963A and VKL-7796 high power; pulsed klystron

Table 19.2
Characteristics of Varian Associates VKP-7553S Klystron

Characteristics	Values
Operating frequency (MHz)	470–566
Output power (kW)	60
Gain (dB)	31
Tuning range (MHz)	96
1-dB bandwidth (MHz)	8
Number of cavities	5
Beam voltage (kV dc)	24.5
Beam current (A dc)	5.2
Heater voltage (V)	7.0
Heater current (A)	17.0
Weight (lb)	930 (includes both tube and magnet)
Cooling	Vapor/liquid/forced-air cooled

Table 19.3
Characteristics for Varian Types VA-963A and VKL-7796 High-Power, Pulsed Klystron Amplifiers and VA-145E High-Power, Pulsed Twystron

Characteristics	VA-936A	VKL-7796	VA-145E
Frequency (GHz)	1.25–1.35	1.29–1.36	2.9–3.1
Peak output power (MW)	5 MW	4 MW	2.5 MW
Typical gain (dB)	50	45	37
Typical efficiency (%)	40	43	35
1-dB bandwidth (MHz)	15	70	200
Duty cycle	0.002	0.075	0.002
Pulse length (ms)	3	130	7
Tunable in system	Yes	No	No
Beam voltage (kV dc)	130	112	117
Beam current (A dc)	101	85	80
Tube weight (lb)	150	600	140
Tube length (in)	60	95	43
Cooling	Liquid	Liquid	Liquid
Tube type	Klystron	Klystron	Twystron

amplifiers; and type VA-145E high-power, pulsed twystron. Varian also produces lower power klystrons for radars in the frequency range of 0.43–36 GHz. Typical lower power ranges are 50–100 kW.

Note that the VA-145E twystron has a 1.5-dB bandwidth of 200 MHz, whereas the klystron amplifiers have 1-dB bandwidths of only 15 and 70 MHz. The VA-145E twystron is used for applications involving chirp modulation, where FM is used in the form of linear sweep over the duration of a fairly long pulse. That permits the generation of high average power needed for long-range capability and, at the same time, short pulses needed for good range resolution. An example of an application is the long-range tracking radar at the Pacific Missile Test Center, Vandenberg, CA.

19.2.3 Helix-Type Traveling-Wave Tube Amplifiers

The basic helix TWT is illustrated in Figure 19.11. This tube uses an electron gun to form a thin pencil-type beam. Focusing magnets are used to keep the beam from spreading. The electrons in the beam are collected at the far end of the tube by a beam collector. A helix that is capable of propagating a slow wave is placed around the electron beam. The RF wave on this helix travels at about one-sixth the speed of light. The velocity of the electron stream is adjusted to be approximately the same as the axial phase velocity of the wave on the helix so there can be interaction between a signal on the slow-wave structure and the electrons in the beam.

The input signal on the helix causes velocity modulation of the electron beam. That, in turn, causes the electron beam to become density modulated as the

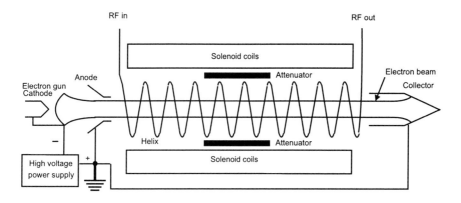

Figure 19.11 Basic helix type TWT. (*After:* [5].)

electrons move down the tube. The modulated electron beam induces waves on the helix. The mutual interaction continues along the length of the tube with the net result that dc energy is given up by the beam to the circuit as RF energy and the wave is thus amplified. It is necessary to provide a resistance midway in the helix to prevent backward-wave oscillations. The wave on the helix is attenuated by the resistive material that surrounds the helix. The signal is then reintroduced to the helix after the resistor by the density-modulated electron beam. The type of TWT that uses the helix slow-wave structure is characterized by wide bandwidth (about an octave). Gains of 20–50 dB are possible, depending on the TWT design. Some TWT amplifiers are designed to be low-noise, low-power front ends for microwave receivers. They have smaller gains than the higher power amplifiers.

Some helix-type TWTs use one or more grids to switch the electron beam on or off. In a radar, that is done to reduce the noise to the receiver when in the receive mode. Many TWTs use a more complex collector than that shown. By using a so-called depressed collector in which the voltage is more positive than the cathode but less than zero volts, it is possible to still collect the beam, increasing the overall efficiency of the tube.

TWTs are used extensively for microwave communication systems in ground, air, and satellite applications. The devices may be either low or high power.

Table 19.4 lists some characteristics for the Varian VTU-6291C1 conduction-cooled, Ku-band TWT used for communications.

Table 19.5 lists some characteristics of a Varian VTF-5121A8 1-kW pulsed TWT that operates in the 2.5–2.8 GHz range.

Table 19.6 lists some Varian millimeter-wave TWTs. These tubes have output powers less than 100W. They are small and lightweight.

TWTs are used in satellite transmitters. This type of transmitter has the advantage of better efficiency than solid-state transmitters of the same power. Typical efficiency for the TWT transmitter is about 60%, compared with about 30% for

Table 19.4
Characteristics of the Varian VTU-6291C1 Communication-Type TWT

Characteristics	Values
Frequency (GHz)	14–14.5
Minimum output power (W)	125
Gain (dB)	50
Helix voltage (kV)	6.9
Cathode current (mA)	0.15
Helix current with RF (mA)	1.5
Collector voltage (kV)	3.2
Cooling	Conduction
Input connector	SMA coax
Output connector	WR-75 waveguide
Dimensions, L × W × H (in)	14.5 × 3 × 2.8
Weight (lb)	5

Table 19.5
Characteristics of the Type VTF-5121A8 Pulsed TWT

Characteristics	Values
Frequency (GHz)	2.5–2.8
Minimum output power (kW)	1.0
Maximum drive power (dBm)	25
Grid pulse (V)	+160
Grid bias (V)	−200
Beam voltage (kV)	9
Collector voltage (kV)	6.4
Grid current (mA)	10
Beam current (Amps)	1.4
Beam duty cycle (%)	4
Maximum pulse width (ms)	125
Modulating element	Unigrid
Dimensions, L × W × H	21.5 × 2.5 × 3
Weight (lb)	7.5
Input connector	SMA coax
Output connector	TNC coax
Cooling	Conduction

the solid-state version. Much higher efficiency is important for a satellite system, where prime power must come from solar cells. Additional weight in orbit is costly.

The main type of microwave tube used for EW transmitters is the wideband helix-type TWT amplifier. Sophisticated ECM systems require ultrawide bandwidths, high efficiency, multistage depressed collectors, low voltage, and grid-controlled

Table 19.6
Characteristics of Several Varian Millimeter-Wave TWTs

Characteristics	VTK-6193D1	VTA-6193M2	VTE-6193G1
Frequency (GHz)	18–26.5	27.5–30.0	50–60
Minimum output power (W)	20	80	20
Gain (dB)	47	49	40
Helix voltage (kV)	12.5	12.5	14.0
Helix current (mA)	2.0	0.2	0.2
Collector voltage (kV)	2	5.6	6.3
Beam current (mA)	95	90	50
Modulating element	Focus electrode	Focus electrode	Focus electrode
Electrode on/off voltage (V)	−5/−700	−5/−700	−5/−700
Anode voltage (V)	None	+500	+500
Input RF connector	SMA	UG-599/U	UG-383U Mod
Output RF connector	SMA coax	UG-599/U	UG-383U Mod
Dimensions, L × W × H (in)	12.3 × 0.8 × 1.6	16 × 2.8 × 3.4	16 × 2.8 × 3.4
Weight (pounds)	4	7	7
Cooling	Conduction	Conduction	Conduction

electron guns. TWTs using these features range from MINI-TWT amplifiers to high-power, wide-bandwidth TWTs such as the Varian VTM-6292 family, which is well suited for active phased-array applications, driver amplifiers, and final output stages. An example of an ECM pod is the AN/ALQ-131 ECM pod, which contains a high-power ECM TWT amplifier. Table 19.7 lists some characteristics of the Varian VTM-6292F4 CW EW TWT.

Table 19.7
Characteristics of Varian VTM-6292F4 CW

Characteristics	Values
Frequency (GHz)	6.5–18
Minimum output power (W)	200
Minimum gain (dB)	30
Helix voltage (kV)	10.3
Cathode current (mA)	280
Helix current with RF (mA)	9
Collector voltage (kV)	5.1/3.0
Electron gun configuration	Focus electrode
Voltage for beam on/off (V)	−10/−1200
Input/output connectors	WRD650
Dimensions, L × W × H (in)	17 × 3 × 3.5
Weight (lb)	9

19.2.4 Coupled-Cavity Traveling-Wave Tube Amplifiers

Helix-type TWTs are limited in power capability by the heat-dissipation capability of the helix. For higher power output capability, it is necessary to use a different type of slow-wave structure. The best one to use is the coupled-cavity slow-wave structure. A TWT of this type is shown in Figure 19.12.

CCTWTs are linear amplifiers. They can be used over the frequency range from UHF to Ka-band. The maximum output peak power at UHF is about 240 kW. The maximum average power output is about 12 kW at L-band and 10 kW at X-band. The amplifier gain is in the range of 30–65 dB. The percentage bandwidth is in the range of 5–15%. That compares with 1–10% for the klystron. The efficiency can be as high as 60%. For the higher frequencies, it tends to be much lower. The cathode voltage for peak power can reach 42 kV.

19.2.5 Conventional Magnetrons

Figure 19.13 shows an important type of conventional magnetron oscillator. It consists of a cathode and an anode with openings to cavities. This tube is a *diode*, and it operates as an oscillator. It is mounted between the pole pieces of a magnet. The interaction between the magnetic field and the movement of electrons causes the electrons to move in a circular path. The path becomes modified by interaction with the RF fields associated with the cavities. The result is that electron spokes are formed that rotate. Energy is coupled out of one of the cavities to a waveguide or a coax.

Openings are provided from the cavities to the space between the anode and the cathode. An output is taken from one of the cavities. In this example, the output is coupled using a loop coupler and coax line.

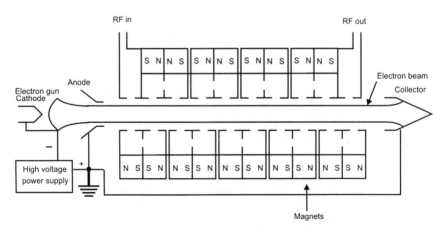

Figure 19.12 CCTWT. (*After:* [5].)

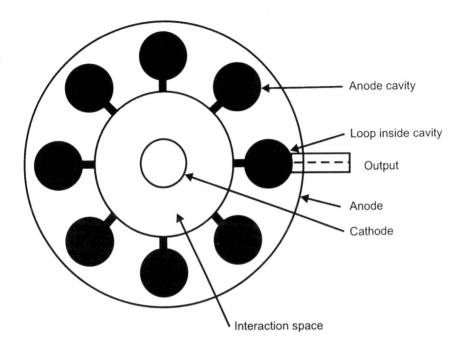

Figure 19.13 Conventional magnetron structure using eight cylindrical cavities in the anode. (*After:* [6,7].)

Magnetrons of this type having eight or more coupled cavities usually are strapped in such a way that there are two sets of conductors connected between every other anode section between the cavities. That is done to avoid mode jumping.

Figure 19.14 shows the case of the vane-type magnetron anode structure, which is used with very high frequency magnetrons. The configuration has the advantage that it does not require mode strapping.

19.2.6 Coaxial Cavity Magnetron

A different type of magnetron, a coaxial magnetron, is shown in Figure 19.15. The magnetic field is perpendicular to the page. It is seen that there is an integral coaxial cavity present in this magnetron. The Q of the cavity is much higher than the Qs of the various resonators, so it is the coaxial cavity that determines the operating frequency. The output from the cavity is through a waveguide window to a rectangular waveguide operating in the dominant TE_{10} mode.

The performance of the coaxial magnetron is better than that of earlier types of magnetrons. The enlarged anode area, compared with conventional magnetrons, permits better dissipation of heat and consequently smaller size for a given output power. Mean time between failure (MTBF) is considerably longer.

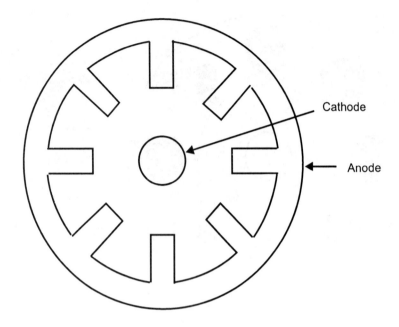

Figure 19.14 Vane-type magnetron anode configuration. (*After:* [6,7].)

An example of a coaxial magnetron is the SFD-341 mechanically tuned C-band coaxial magnetron. This tube is used extensively for shipboard and ground-based radars. It delivers a peak power of 250 kW with a 0.001 duty cycle. It operates over the frequency range of 5.45–5.825 GHz. The efficiency of the tube is in the range of 40–45%. The operating life of coaxial magnetron tubes can be 5,000 to 10,000 hr. That is a 5- to 20-fold improvement over conventional magnetrons.

Since most of the RF energy is stored in the TE011 mode cavity rather than in the resonator region, reliable broadband tuning of the magnetron can be accomplished by a noncontacting plunger in the cavity. Both the pushing figure (change in the frequency with a change in anode current) and the pulling figure (change in the frequency with a change in the phase of the load) are much less in a coaxial magnetron than in the conventional configuration.

Magnetron oscillators are low-cost, rugged devices. For those reasons, the magnetrons are used in home microwave ovens. This type of tube also is used for industrial heating and numerous other applications.

19.2.7 Amplitrons

Another crossed-field device of interest is shown in Figure 19.16: the CFA known as the amplitron. In its operation, an amplitron is a cross between a TWT and a

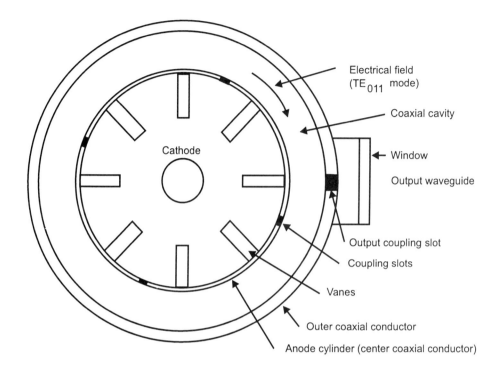

Figure 19.15 Cross-section of a coaxial cavity magnetron. (*After:* [6,7].)

magnetron. It uses a magnetic field normal to the surface of the page and a large cathode. It uses a slow-wave structure to provide a continuous interaction between the electron beam and a moving RF field. In practice, a vane-type slow-wave structure usually is used.

CFAs also find application in radar systems. These amplifiers have only limited gain (10–20 dB), so they must be driven by a fairly high-power driver stage, which could be a klystron or a CCTWT. This microwave tube is very much like a magnetron in appearance and operation. The big difference, of course, is that the CFA is a saturated amplifier, whereas the magnetron is an oscillator.

The main advantages of the CFA are its very high power capability and a very high efficiency. The maximum peak power is about 5 MW at S-band. The maximum average power is 1.3 kW at L-band and 2 kW at X-band. The efficiency can be up to 80%. The cathode voltage is up to 15 kV.

The bandwidth is 5–15%. The narrower the bandwidth, the higher the gain.

19.2.8 Gyrotron Oscillators and Amplifiers

The gyrotron oscillator is a microwave vacuum tube that operates on the basis of the interaction between an electron beam and microwave fields where coupling is

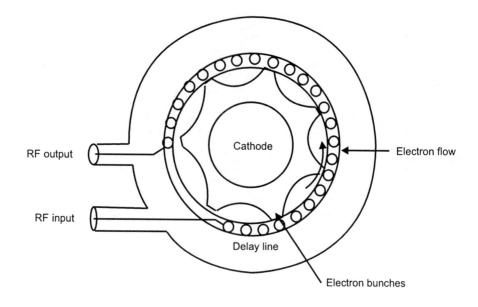

Figure 19.16 Operational concepts of the amplitron. (*After:* [8].)

achieved by the cyclotron resonance condition. This type of coupling allows the beam and microwave circuit dimensions to be large compared to a wavelength. Hence, the gyrotron avoids the power-density problems encountered in conventional klystrons and TWTs at millimeter wavelengths.

The device includes a cathode or electron gun, a beam collector area, an interaction cavity, and a circular waveguide of gradually varying diameter. Electrons are emitted at the cathode with small variations in speed. The electrons are then accelerated by an electric field and guided by a static magnetic field through the device. The nonuniform induction field causes the rotational speed of the electrons to increase. The linear velocity of the electrons, as a result, decreases. The interaction of the microwave field within the waveguide and the rotating (helical) electrons cause bunching similar to the bunching in a klystron. A decompression zone at the end of the device permits decompression and collection of the electrons.

The circuit area in the gyrotron devices is 100 times the circuit area in a klystron or conventional TWT. That permits gyro-devices to have 100 times the output power capability of conventional tubes.

Another interesting feature of the gyrotron is that the applied magnetic field is proportional to frequency. Superconducting magnets are required for operation at 50 GHz and higher.

Initially, gyro-devices were single-cavity oscillators, in which the entire interaction occurred in a single microwave cavity. However, the same basic interaction can be used with different variations, such as amplifiers, using several resonant cavities, or traveling-wave circuits. These variations are called gyro-klystrons and gyro-TWTs, respectively.

Varian gyrotrons have produced over 300 kW at 28 GHz, 200 kW at 60 GHz, 850 kW at 140 GHz, and 12 kW at 250 GHz. A wide range of both oscillators and gyro-TWTs are under development by Varian and others at the time of this writing.

19.2.9 Circuit Configurations for Microwave Tubes

All the vacuum-tube circuits discussed in this chapter need high-voltage power supplies, controls, and protection circuits, which must be tailored to the devices involved. These systems include sensors for measuring voltage, current, temperature, and standing-wave ratio. Means for automatic shut-off are provided in the event that a problem is detected. A number of different types of controls and displays are provided by these systems. X-ray shielding is an important safety feature required in very high voltage klystrons. Other required safety features are interlock systems that turn the systems off if a cabinet is opened while the systems are turned on.

Figure 19.17 shows two examples of high-power microwave transmitter systems that use TWTs. Figure 19.17(a) is a single-stage TWT transmitter and associated power supply and control system. The RF signal is fed into the TWT at a relatively low power level, since the TWT normally has very high gain. The output is then connected to the antenna circuit, which may consist of a duplexer circuit, monitoring circuits, and the antenna. The TWT in the figure is connected to a high-voltage power supply and a modulator. Inputs to this stage are the prime power and modulation pulses. The modulation pulses are used to turn the electron beam on or off. The third input is a control bus that provides the operating controls for the transmitter. Outputs from this system are monitor and status signals.

Figure 19.17(b) shows a high-power transmitter that uses an amplitron CFA as the output stage and a TWT as the driver for the amplifier. This system might be used for a high-power radar. The RF input signal to the TWT is first amplified to a level 10–15 dB below the final output. The output of the TWT is then fed to the input of the amplitron.

The output of the amplitron is connected to the antenna circuit, which may consist of a duplexer circuit, monitoring circuits, and the antenna. The TWT shown is connected to a high-voltage power supply and a modulator. Inputs to this stage are the prime power and modulation pulses. The modulation pulses are used to turn the TWT electron beam on or off. The third input is a control bus that provides the operating controls for the transmitter. Outputs from this system are monitor and status signals. A similar high-voltage power supply is used for the amplitron.

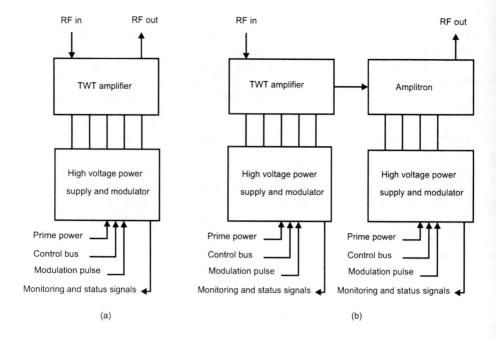

Figure 19.17 High-power microwave transmitters that use TWT amplifiers: (a) single-stage TWT transmitter and (b) high-power transmitter that uses an amplitron CFA as the output stage and a TWT as the driver stage for the amplitron.

References

[1] Whitaker, Jerry C., *Power Vacuum Tubes Handbook*, New York: Van Nostrand Reinhold, 1994.
[2] Sivan, L., *Microwave Tube Transmitters*, London: Chapman and Hall, 1994, p. 15.
[3] Ibid., p. 11.
[4] Ibid., pp. 125–129.
[5] Ibid., pp. 39–41.
[6] Kennedy, George, *Electronic Communication Systems*, 3rd ed., New York: McGraw-Hill, 1985, pp. 351–385.
[7] Skolnik, Merrill I., *Introduction to Radar Systems*, 2nd ed., New York: McGraw-Hill, 1980, pp. 192–216.
[8] Ibid., p. 196.

About the Author

Ferril Losee has been both a university educator and an RF engineer who has successfully practiced in industry. He has spent more than 43 years as an electrical engineer, with 22 years working in industry, 18 years as a university professor, and three years as a consultant and author.

Mr. Losee received his bachelor of science degree in electrical engineering from the University of Utah in 1953 and his master of science degree in electrical engineering from the University of Southern California in 1957. He spent six years (1953–1959) as an electrical engineer for Hughes Aircraft Company, where he was a design engineer and a project leader in areas of HF communications and space and satellite communications. He spent the next six years at the Aeronutronic Division of Philco-Ford working on space and satellite communications, missile defense penetration aids, ECM, and missile defense systems.

The author taught electrical engineering at Brigham Young University for 18 years (1965–1983). During the first 12 of those years, he served as chairman of the Electrical Engineering Department. During much of his 18 years at Brigham Young, Mr. Losee did substantial consulting and research for government and industry in areas of missile defense, reentry systems, ECM systems, and special-purpose antennas.

After early retirement in 1983, the author worked for SRS Technology for five years as a design engineer and project leader in the areas of radar and special-purpose X-ray systems. His final position was with EG&G Special Projects, where he worked for five years as an RF systems design engineer and project leader in the areas of instrumentation radar systems, radar simulation, ELINT systems, and air defense communication systems. He is currently a part-time consultant and author in the areas of communications, radar, and other RF systems.

Mr. Losee has been married to his sweetheart Dona for 44 years. Together, they are the parents of nine children, the grandparents of 26 children, and the great-grandparents of three children.

Index

Acquisition radar, 49
Active circuit, 297
Active region, 395, 399, 400, 510
Adaptive spot noise jammer, 64
ADC. *See* American digital cellular
Adder circuit, 213–14, 283
ADF. *See* Airborne direction finder
Admittance chart
    conversion, 310, 441, 451–52
Admittance coordinates, Smith
    chart, 304–6, 309–10
Advanced Mobile Phone Service, 22, 236–46
    calling procedures, 242–45
    control functions, 240–41
Aeronautical mobile satellite service, 9–10
Aeronautical mobile service, 7–9
AF modulator. *See* Audio frequency
    modulator
AGC. *See* Automatic gain control
Airborne direction finder, 52, 55
Airborne moving target indicator, 285
Airborne multiple-function radar, 46, 94
Airborne radar amplifier, 158–59
Airborne side-looking radar, 47–48
Airborne terrain-avoidance radar, 47
Airborne terrain-following radar, 47
Airborne weather-avoidance radar, 39, 44
Air cooling, 551
Air-core loop antenna, 388–89
Air-core solenoid, 427–28
Aircraft detection/tracking, 49

Aircraft-ground station link, 205–6
Aircraft navigation, 255
Aircraft radar cross-section, 263–64
Aircraft surveillance radar, 37–38, 93–94
Air dielectric, 291–92, 294
Air-filled cavity resonator, 479
Airport high-resolution vehicular
    monitoring, 41–43
Air-to-ground communications, 41, 222
Air traffic control radar beacon
    system, 60–61
Alerting procedure, 243
Altimeter, 44–45, 276–78
Aluminum, 353, 499
Aluminum gallium arsenide, 534–36
Alumispline cable, 294
AM. *See* Amplitude modulation
Amateur service, 9
American digital cellular, 246–54
Amospheric noise, 100–1
Amplifier gain, 5
Amplifiers
    applications, 143–44
    audio/low frequency, 154
    bipolar junction transistor, 535, 537–38
    chains, 154, 156, 159, 178, 186–88,
        209, 211, 218, 221, 236–37
    class A, 145–46, 165, 181, 208
    class AB, 145–46, 155–56
    class B, 145–46, 178, 181, 206, 553
    class C, 145–46, 177–80, 553, 555–56

Amplifiers (continued)
    collector-modulated, 178–80
    common-base, 522–24
    common-collector, 524–25
    common-emitter, 515–22
    efficiency formula, 145
    fiber optic system, 257
    frequency division multiplexing, 134
    front-end low-noise, 145–51
    grid-modulated, 178
    grid-type vacuum tube, 551–58
    gyrotron, 569–71
    intermediate frequency, 151–54
    low-noise, 99
    oscillation problem, 150
    plate-modulated, 177–78
    power, 154–59
    push-pull, 145, 154–55, 178, 180
    radio astronomy system, 68
    transmitter system, 133
Amplitron, 568–72
Amplitude modulation, 9, 86, 107, 153, 172
    citizens band radio, 231
    conventional, 116–18
    ground-to-air communication, 208–10
    loop antenna use, 389
    modulator types, 177–80
    power amplifier, 553, 555–56
    radio broadcasting, 550
    single-sideband, 223
Amplitude modulation detector, 195–97
Amplitude-phase keying, 140
AMPS. *See* Advanced Mobile Phone Service
AMTI. *See* Airborne moving target indicator
Analog delay line, 282–83
Analog modulation, 22, 219
Analog multiplexing, 27
Analog cellular telephone, 232
Analog-to-digital converter, 134–36, 283
AN/ALQ-131 system, 565
Andersen Laboratories, 490, 492–95
Angle measurement, 273–74
ANL. *See* Automatic noise limiter
Anode, 549–50, 566–68
Antenna arrays
    broadside line, 412–14
    end-fire line, 410–12
    planar, 414–15
Antenna gain/capture area, 5, 72–73

Antimony, 499
Anti-transmit/receive switch, 285, 361, 364–65
ANZAC, 151, 170, 482
APD. *See* Avalanche photodiode
Aperture antenna, 370
Aperture blocking, 421
APK. *See* Amplitude-phase keying
AR6A radio system, 29
Archimedean spiral antenna, 392–94
Arsenic, 499
A-scope display, 41
AT&T, 29, 31, 219, 230
ATCRBS. *See* Air traffic control radar beacon system
Atmospheric absorption, 76–78
Atmospheric-loss noise, 95, 102
Atmospheric refraction, 79–80
ATR. *See* Anti-transmit/receive switch
Attenuation, 12
    bandpass filter, 327, 329
    cellular telephone, 236
    coax cable, 294
    feed line, 421
    fiber optic, 31, 88–89, 256–57
    gas-tube, 362
    highpass filter, 326–27
    laser diode, 508, 510
    LC filter, 452–54
    radar, 39–40
    stripline, 300
    waveguide, 344–45, 350
    weather, 78–79, 214, 217
Attenuator, resistive vane as, 357
Audio frequency amplifier, 143–44, 154 208, 213
Audio frequency modulator, 178, 187, 189
Audio frequency transformer, 180
Automatic gain control, 153, 208–9, 213, 218, 231, 285. 426
Automatic noise limiter, 231
Automobile location, 255
Automobile speed monitoring, 275
Autotransformer, 460, 469–70
Avalanche breakdown, 500, 503
Avalanche diode, 162, 507–9
Avalanche photodiode, 257, 512
Average frequency counter, 278
Azimuth angle, 58–60, 264, 266–67, 286
Azimuth angle to target, 35, 41, 46
Azimuth difference channel, 286

Babinet's equivalence principle, 401

Backscatter, 89–90, 261
Balanced duplexer, 364–65
Balanced mixer, 133, 168–73, 196, 198, 209, 211, 221, 278, 361, 363
Balanced modulator, 119, 180–83, 185, 192, 196, 213
Balanced to unbalanced transformer, 206, 317–18, 378, 399, 464, 467–68, 471
Balanced transmission line, 291
Balun. *See* Balanced to unbalanced transformer
Bandpass filter, 121, 131, 133, 147, 151–52, 169–70, 181, 202–3, 206–9, 211, 213, 218–19, 221, 277–79, 327–29, 452–53, 455, 477, 479, 485–86
Band-reject filter, 452–54
Barium titanate, 479
Barker code, 129–30
Base, transistor, 513–14
Base loading, 375
Base station, 14
Base station-to-ground commuications, 208
Base station-to-mobile unit system, 91–92
BAW. *See* Bulk acoustic wave
Beam angle, 72, 369–70, 416
Beam collector, 562–63, 570
Beam-forming cathode, 550
Bessel function, 122, 480
Bias circuit, 146–47, 535, 537–39
Biasing, 155, 171, 180, 334, 426, 484, 486, 501, 513, 538
Binary phase-shift keying, 126, 138, 190–91, 202
Binary signal, 137–38
Biphase modulation, 171
Bipolar junction transistors
  common-base amplifier, 522–24
  common-collector amplifier, 524–25
  common-emitter amplifier, 515–22
  compared to FET, 532–34
  dc bias circuits, 535, 537–38
  NPN/PNP, 513–15
  power amplifier, 180, 206
  radio frequency amplifier, 145–46, 148, 153, 156, 165, 173–75
  stability, 539
  YIG-tuned oscillator, 486
Bistatic radar, 36, 50
BJT. *See* Bipolar junction transistors

Blind speed, 282
BNC cable connector, 295
Boltzmann's constant, 97–98
Boron, 499
BPSK. *See* Binary phase-shift keying
Branch-line duplexer, 364
Branch-line hybrid coupler, 319–20, 322
Breakdown voltage, 500–1, 503, 512
Brewser angle, 83–84
Broadband systems, 378, 394, 405–6, 472, 505
Broadcasting-satellite service, 12–13
Broadcasting service, 9, 11–12, 121, 123, 550
  frequency modulation, 211–14
Broadcast remote pickup, 21
Broadside-coupled directional coupler, 323–24
Broken line symbol, 530–31
B-scope display, 41
BT product, 490
Buffer amplifier, 143–44, 163, 178, 180–81, 205, 207, 209, 211, 213, 492
Bulk acoustic wave delay line, 473, 493–96
Burn-through range, 66
Business/personal communication systems
  49-MHz cordless telephone, 227–29
  900-MHz cordless telephone, 229–30
  citizens band radio, 231
  pager system, 230–31
  two-way radio, 232
Butterworth filter, 326, 329, 453
Bypass capacitor, 551

Cable television, 222, 257
C/A code. *See* Coarse/acquisition code
Cadmium, 500
Call completion, 244
Call origination, 250
Cancellation circuit, 280
Capacitive loading, 87, 375–76
Capacitive reactance, 379, 434
Capacitive susceptance, 434, 452
Capacitors
  bypass, 168, 551
  characteristics, 431–434
  lumped constant, 376
  series, 434
  quartz crystal, 475
  variable, 160
Capture area, 72–73, 76, 89, 370

Carbon, 426
Cardioid antenna, 55, 59
Carrier, single-sideband, 120–21
Carrier power, 117–18
Carson's rule, 122–23
Cascading, 154
Cassegrain feed, 421, 423
Cathode circuit, 551, 566
Cathode ray tube, 41, 52, 107, 444
Cavity-backed Archimedean spiral
    antenna, 393
Cavity-backed rectangular-slot
    antenna, 402–3
Cavity-backed slot antenna, 277
Cavity resonator, 365–67, 482, 551,
    557–58
C-band radar, 38–39, 46, 214, 217, 361
CB radio. *See* Citizens band radio
CCTWT. *See* Coupled-cavity traveling
    wave tube
CDI. *See* Course deviation indicator
CDMA. *See* Code division multiple access
CDVCC, 247, 250
Cell division, 23–25
Cell site, 23, 242–43
Cellular mobile telephone
        service, 22–25, 488
    Advanced Mobile Phone
        Service, 236–46
    digital systems, 246–54
    propagation, 234–36
    spatial frequency reuse, 232–34
Center-fed series system, 415
Center-tapped transformer, 471–72
Central switching center, 26–27
Ceramic dielectric capacitor, 433
Ceramic filter, 153, 183–84, 208, 231,
    463, 477–81
Ceramic resonator, 160
CFA. *See* Crossed-field amplifier
Channel designation, 242–43
Channel/path reconfiguration, 244
Chebychev filter, 326, 329, 453–54
Chip resistor, 426–27
Chirp modulation, 36, 46, 124, 129, 562
CINOX, 475–76
Circuit analysis
    complex circuit, 438–40
    Smith chart, 438, 441–43
Circular aperture, 420–21
Circular electron path, 566

Circular polarization, 73–74, 254, 371,
    384–85, 390, 393–94
Circular waveguide, 350–52
Circulator systems, 217–218, 221, 224, 270
    ferrite, 329–30, 358, 361
    tunnel-diode amplifier, 506
Citizens band radio, 12, 231
Civil marine radar, 39
Cladding, 30, 88
Clamping, 168, 538
Clapp oscillator, 160
Closed loop, 253
Clutter, 49, 263–67, 280
Coarse/acquisition code, 254
Coax diode mixer, 168
Coaxial cable
    characteristics, 27–28, 291–96
    dipole antenna, 394
    feed line, 392
    high-frequency transformer, 465–69
    monopole antenna, 376
    sleeve monopole, 386
    slot antenna, 402–3
    TE10 mode, 347–49
    termination, 315–16
    waveguide mixer, 168
Coaxial cavity magnetron, 567–69
Coaxial directional coupler, 316–17
Coaxial electromechanical switch, 331–33
Coaxial frequency multiplier, 165
CODEC, 250
Code division multiple access, 25, 232, 252
Coding, cellular digital, 247
Coherent detection, 198, 202–3, 272–73
Coherent integration, 105–7, 125
Coherent oscillator, 279–81, 283, 285
Coherent-phase signal, 138
Coherent quadrature phase-shift
    keying, 139
Coho. *See* Coherent oscillator
Collector, 513–17, 519, 522, 561–63, 570
Collector-modulated transistor
    amplifier, 178, 180
Colpitts oscillator, 160–61, 476–77
Combiner, 240, 464, 471–72
Combiner circuit, 156–57, 192
Combining signals, 283
COMINT. *See* Communication intelligence
    gathering systems
Common-base amplifier, 522–24
Common-collector amplifier, 524–25

Index | 579

Common-emitter amplifier, 515–22, 537–38
Common-gate amplifier, 530
Common-grid mode, 549
Common-source amplifier, 530–31, 537
Common-source junction field-effect
 transistor, 526–28
Communication intelligence gathering
 systems, 63
Communication satellite, 29–30
Communication systems
 defined, 3
 grid tube use, 550
 link analysis, 109–11
 propagation performance, 90–92
 range measurement, 125–26
 single-sideband modulation, 205–8
 transmitter power amplifier, 154–57
 traveling wave tube amplifier, 563–64
Complementary symmetry operation, 532
Complex conjugate, 446, 542–43
Complex resonant circuit, 438–440
Composition triple-beat distortion, 148
Comstar systems, 223
Conducting sphere, 261–62
Conductors
 loss, 298
 round, 425
 skin effect, 425–26
Conical monopole antenna, 372–74
Conical scan tracking, 273–74, 286
Conical spiral antenna, 392, 394
Connector, coaxial cable, 295–96
Continuous wave, 35–36
 amplifier chain, 154–55
 Doppler radar, 43
 modulation, 130–31
 omnidirectional system, 55
 polarization, 257
 pulsed, 115–16, 126–28
 radio beacon, 52
 range measurement, 269–70
 velocity measurement, 270–72, 274–76
Control circuit, 240
Control grid, 549–50
Controller, 23
Control station, 63
Conventional transformer, 457–60, 464–67
 bandwidth limitations, 469–70
 hybrid coupling, 471–72
Conversion gain, 174–75
Conversion loss, 170

Converter, 23
Copper, 295, 353, 375
Cordless telphone, 227–30
Corner reflector, 261, 380–82, 419
Corporate feed, 415–16
Correlator, phase-shift keying, 489
Counter, frequency, 278
Coupled-cavity magnetron, 566–67
Coupled-cavity traveling wave
 tube, 158, 279,559, 566, 569
Coupled-line directional coupler, 323–24
Coupled-line hybrid coupler, 323, 325
Coupled-tank bandpass filter, 454–55
Coupled-tuned output circuit, 180
Coupler/coupling, 202–3
 amplifier, 522, 557, 561, 569–70
 cable connector, 295–96
 cathode tube, 551
 critical, 463–64
 directional, 316–17, 357–58, 443–44
 hybrid, 319–23, 325, 354–56, 471–72
 quadrature, 384
 quarter-wave, 323–24
 resonator, 366, 482–83
 snake feed, 417
 tuned transformer, 462–64
 waveguide, 343–44, 348
 YIG bandpass filter, 486
Coupling capacitor, 199–200, 446, 502–3,
 516, 522–23, 526, 532
Course deviation indicator, 59
Critical coupling, 463–64
Critical frequency, 84
Crossed-field amplifier, 157–59, 559, 568–72
Cross-modulation distortion, 148, 175
CRT. See Cathode ray tube
Cryogenic temperature, 535
Crystal. See Quartz crystal filter; Quartz
 crystal oscillator
C-scope display, 41
CT-500 cellular telephone, 239
Curie point, 461
Current-controlled amplifier, 146
Cutoff wavelength, 338–42, 347, 350–52,
 360, 453–54, 479, 522
CW radar. See Continuous wave radar
Cylinder monopole, 373
Cylinder-type waveguide resonator, 366–67

Data collection system, 224
Data link network, 23
DBS receiver, 153

DC bias circuit, 535, 537–38
DC specifications, 545
Deception jammer, 64, 66, 68
Decibel use, 4–6
Decompression, electron, 570
Deemphasis, 123–24, 189, 209, 213
Delay-line canceler, 281–85
Delay lines, 136–37, 194, 287, 418
    bulk acoustic wave, 493–96
    surface acoustic wave, 486–93
Delay time, 36
Demodulators
    amplitude modulation, 195–96
    frequency modulation, 199–202
    phase, 202–3
    product, 196–99
    satellite system, 224
Demultiplex system, 224
Depletion/enhancement mode, 531–33
Depletion layer, 504, 526, 529
Depressed collector, 563–64
Detection probability, 107–8
Detectors
    envelope, 107, 195–97
    optical, 257–58
    synchronous, 107
    waveguide, 361–62
    See also Demodulators
DGPS. *See* Differential global positioning satellite
Dielectric, transmission line, 291–95, 297–98, 300–1, 353
    See also Air dielectric; Foam dielectric; Microfiber teflon fiberglass; Polyethylene; Solid dielectric; Teflon
Dielectrically stabilized oscillator, 482
Dielectric constant, 79, 87, 432
Dielectric lens antenna, 408–9
Dielectric-loss tangent, 298
Dielectric resonator oscillator, 160–61, 479–83
Differential amplifier, 202
Differential global positioning satellite, 255
Differential-pair oscillator, 160
Differential phase-shift signal, 138
Differential quadrature phase-shift keying, 139, 190, 246, 248, 252
Diffraction, 11, 15, 36, 80–82, 211
Digital cellular telephone
    Europe, 251–52

Japan, 251
    North America, 246–54
Digital delay line, 281–82
Digital filter bank, 284
Digital multiplexing, 27
Digital signal processing, 283–84
Digital speech interpolation, 251
Digital spread spectrum system, 229
Diode, 566
Diode envelope detector, 195–97, 199
Diode laser, 257
Diode mixer, 167–73, 196
Diode transformer, 199
DIP. *See* Dual in-line package
Diplexer, 270
Dipole antennas, 5, 254
    current distribution, 381, 383
    as feed system, 421
    impedance, 378–82
    receiving, 211
    sleeve, 386–87
    thin-wire, 377–78
    turnstile, 384–85
    types of, 378–79
    with reflector, 415
Direct frequency, 162, 211
Directional antenna, 385–86, 413
Directional coupler, 316–17, 357–58, 443–44
Directional gain, 369, 405, 408, 413, 419, 421
Direct reception, 9, 12
Direct sequence modulation, 126, 252
Dispersive delay line, 488–90, 494
Dispersive transducer, 490, 492
Distance-measuring equipment, 55–57, 59
Distortion, transistor, 148, 156, 175
Diversity reception, 17
D-layer, 84
DME. *See* Distance-measuring equipment
Dominant mode, 338–39, 344–46, 351, 403, 567
Doppler frequency, 35, 38, 198–99, 270–76, 278, 280, 285
Doppler navigation radar, 39, 44–45
DOR. *See* Dielectric resonator oscillator
Double-balanced diode mixer, 169–73, 202, 208
Double-sideband detector, 196, 198
Double-sideband
    modulation, 180–82, 213–14
    suppressed carrier, 118–19
Double-sleeve impedance matching, 315–16
Double-stub impedance matching, 315–16

Double-tuned transformer, 463–64
Down-conversion, 119, 153, 208, 213, 224, 248, 280–81
Downlink, satellite, 19–20, 30, 156, 223
DQPSK. *See* Differential quadrature phase-shift keying
DR11-40 system, 29
DR6-30-135 system, 29, 219
DR6-30 system, 29
Drift region, 508
DRO. *See* Dielectric resonator oscillator
DSB. *See* Double-sideband modulation
DSI. *See* Digital speech interpolation
DSS system. *See* Digital spread spectrum system
Dual conversion, 208
Dual in-line package, 493
Dual polarization, 29
Ducting, 80
Duplexers, 206, 209, 221, 224–25, 237, 240, 248, 281, 571
 balanced, 364–65
 with TR/ATR switch, 362, 364–65
Duplex system, cordless telephone, 227–29
Dynamic radar cross-section test range, 50

Early-warning radar, 37, 49
Earth, Doppler frequency for, 275
Earth exploration-satellite service, 12, 20
Earth station, 223–25
ECCM system. *See* Electronic counter-countermeasure system
Eccosorb, 150
ECM system. *See* Electronic countermeasure system
Edge-coupled directional coupler, 323–24
Effective radiated power, 72, 370
E-field. *See* Electrical field
EHF, 9
Eight-state phase modulation, 139
Eight-state phase-shift keying, 190, 192–93
E-layer, 84–85
Electrical field, 14, 73, 75, 371
Electrically small dipole antenna, 378–79, 381, 383
Electrically small loop antenna, 388–89
Electrically small monopole antenna, 375–76
Electrical tuning, 482
Electric coupling, 343–44
Electric flux, 343–44
Electric-plane horn antenna, 405, 407–8

Electrolytic capacitor, 433
Electromagnet, 561
Electromagnetic pulse, 129
Electromagnetic wave, 73–75
Electromechanical switch, 331–33
Electron-beam coupling, 366
Electron bunching, 558, 561, 570
Electron gun, 559, 562–63, 570
Electronically variable varactor, 475
Electronic counter-countermeasure system, 64
Electronic countermeasure system, 64–68
 power amplifier, 157–159
 radar performance, 112–14
 range gate, 287
 traveling wave tube, 564–65
Electronic intelligence system, 64, 153
Electronic warfare system, 35, 63–68, 564–65
Electrons, 484–85, 510–12, 515, 532
Elevation angle, 83, 60, 264, 266
Elevation angle error, 285
Elevation angle to target, 35, 46
Elevation difference channel, 285
ELINT. *See* Electronic intelligence system
Elliptical polarization, 73–74, 385
Emitter, 513–14, 522
Emitter follower, 154, 524
EMP. *See* Electromagnetic pulse
End-fed series system, 415
End-fire traveling-wave antenna, 386, 390
Enhancement mode, 530–33, 538–39
Enhancer, 23
Envelope detector, 107, 195–97
E-plane T junction, 357
Epoxy-iron, 357
Equiangular spiral antenna, 392–93
Equivalent circuit, 457–58, 474, 519–22, 558
ERP. *See* Effective radiated power
Error voltage, 189
ET-532 cordless telephone, 229
ET-909 cordless telephone, 230
E-TDMA. *See* Extended time-division multiple access
European digital cellular, 246, 251–52
Evanescent mode, 349
EW system. *See* Electronic warfare system
Excess path loss, 76–79
Exchange system, 27
Extended time-division multiple access, 251
External noise, 95–96, 145

FACCH, 247, 251

Fading, 14
Fall time, pulse, 127–28
False alarm probability, 107–9
Fan-shaped antenna, 60, 373, 380–81
Fan-shaped beam, 421
Faraday rotation isolator, 358–60
Fast Fourier transform, 284
Fat dipole antenna, 378–79
FCC. *See* Federal Communications Commission
FDM. *See* Frequency division multiplexing
FDMA. *See* Frequency division multiple access
Federal Communication Commission, 69
Federal Communications Commission, 7, 234
Feedback circuit, 177–78
Feedback resistor, 537
Feed systems, 415–23
Ferrimagnetic resonance, 484–86
Ferrite circulator, 329–30, 358, 361
Ferrite core, 461, 465–66, 470
Ferrite-core loop antenna, 388–89
Ferrite-core transformer, 171, 457–58
Ferrite isolator, 330–31, 358–60
Ferrite switch, 361
Ferrite toroidal-core transformer, 170
Ferrodisc, 330
FET. *See* Field-effect transistor
FFT. *See* Fast Fourier transform
Fiber optic cable, 27, 30–31, 88–89, 256–59
Field-effect transistor, 145–46, 154, 173–75, 180, 206, 219, 476, 486, 499
  bias circuit, 538–39
  compared to BJT, 532–34
  junction, 526–529
  metal-oxide semiconductor, 530–32
  metal-semiconductor, 529–30
  stability, 539
Filament circuit, 551
Filter bank, digital, 284
Filters
  balanced mixer, 119
  bandpass, 121, 131, 133, 147, 151–52, 169–70, 181, 202–3, 206–9, 211, 213, 218–19, 221, 277–79, 327–29, 452–53, 455, 477, 479, 485–86
  band-reject, 452–54
  Butterworth, 326, 329, 453
  ceramic, 153, 183–84, 208, 231, 463, 477–81
  Chebychev, 326, 329, 453–54

  double-tuned transformer, 463
  highpass, 326–27, 452, 454–55
  lowpass, 131, 181, 195–96, 198, 202–3, 325–26, 452–53, 455
  lumped constant, 183, 452–55
  matched, 130
  narrowband, 152
  quartz crystal, 183–84, 463, 477–78
  range resolution and, 278
  sideband, 183–84
  surface acoustic wave, 154, 183–84, 208, 463
Fire-control radar, 46, 159
Fixed-frequency oscillator, 160, 198, 492
Fixed-satellite service, 19
Fixed service, 12, 14–19
Fixed-value capacitor, 433
Flange, waveguide, 352
Flat-plate reflector antenna, 418–19
Flat square spiral inductor, 428–29
Flat-top loading, 375–76
F-layer, 84–85
Flexible waveguide, 346, 353
Flux density, ferrite core, 461–62
Fly-wheel action, 178, 553
FM. *See* Frequency modulation
FMCW. *See* Frequency-modulated continuous wave
Foam dielectric, 292
Focusing magnet, 562–63
Fog, 78–79
Folded dipole antenna, 378–79, 385
Folded-slot antenna, 401–2
Forward bias, 334, 501–4, 510, 513–14
Forward channel, 238, 240, 531
Forward error correction, 253
Foster-*See*ley discriminator, 199–200
Four-state modulation, 138–39
Four-thirds earth rule, 79–80
FPLMTS. *See* Future Public Land Mobile Telecommunications System
FPS-16 tracking radar, 273
Frame, time slot, 246
Free-space electrical field, 73–74
Free-space path loss, 75–76, 85–86, 88
Free-space propagation, 3, 8–9, 11, 14–15, 74–75, 211
Free-space wavelength, 341–42
Frequency allocation, 6–7
Frequency bands, 4–5
Frequency control, 492

Frequency conversion, 151–55, 181, 187, 206, 208
  See also Down-conversion; Up-conversion
Frequency diversity, 17
Frequency divider, 162
Frequency division multiple access, 224–25
  cordless telephone application, 232
Frequency-division multiplexing, 27, 124–25, 135, 186, 224–25
Frequency division multiplex transmitter, 133–34
Frequency-independent antenna, 391–92, 394
Frequency-modulated continuous wave, 36, 44, 131
  radar altimeter, 276–78
  radar range measurement, 269–70
  radar velocity measurement, 272
Frequency-modulated transmitter, 507
Frequency modulation, 9, 11, 21–22, 30
  broadcast systems, 211–14, 550
  characteristics, 121–24
  detector, 199–202
  ground-to-air communication, 208–10
  klystron amplifier, 562
  modulator, 186–89
  pulsed, 36
  satellite relay, 224–25
  transmitter, 240, 277
  two-way radio, 232
Frequency modulation detector, 199–202
Frequency modulation transmitter, 240, 277
Frequency multiplex, 219
Frequency multiplier, 154, 162–66, 187, 209, 211, 503–4
Frequency prescaler, 162
Frequency pulling, 358
Frequency reference control, 473
Frequency reuse, 23–25
Frequency-scanned array, 416–18
Frequency shift keying, 22, 137–38, 189
Frequency sweep modulation, 124
Frequency synthesizer, 162–63, 206, 208–9, 213, 232, 240
Fresnel reflection coefficient, 83
Front-end low amplifier, 143–51
Front-to-back ratio, 398–99
FSK. *See* Frequency shift keying
FT3/FT3C optical fiber, 31
FTX-180 optical fiber, 31

Functional specifications, 545
Fundamental mode, 475
Fused quartz, 494, 496
Future Public Land Mobile Telecommunications System, 234
Gain
  amplifier, 5, 154, 157–59, 519–20, 523, 527–30, 569
  antenna, 5, 71–73, 89, 370–71, 380–81, 385–86, 396
  delay line, 492
  directional, 369, 405, 408, 413, 419, 421
  homodyne receiver, 153
  impedance matching, 542–43
  intermediate frequency, 151
  planar array, 415
  traveling wave tube, 566
  ultra high frequency, 157
  See also Automatic gain control
Galactic noise, 101–3
Gallium, 499
Gallium aluminum arsenide, 508, 510–11
Gallium arsenide, 146, 151, 164, 499, 501, 503–8, 510–11, 529–30, 533–36
Gas-tube switch, 361, 364–65
Gate bias, 180, 537
Gate rupture, 534
Gate voltage, 529–32
Gating, 134–35, 184, 286–87
Generator, 128–29
Geostationary orbit satellite, 19, 222
Germanium, 499, 504, 511, 533
Glass, 257, 297, 433, 494
Glide-slope indicator, 58–59
Global positioning system, 60, 62–63, 254–56
Global System for Mobile Communications, 246, 251–52
GPS. *See* Global positioning system
Graded-index fiber, 31, 257
Grating lobe, 413
Grazing angle, 264, 266
GRI. *See* Group repetition interval
Grid-modulated class C amplifier, 178
Grid-type vacuum tube amplifier, 156–57
Grid-type vacuum tubes
  cavity resonator, 557–58
  power amplifier, 551–57
  triode/tetrode/pentode, 549–52, 557–58
Ground-based surveillance radar, 42, 49
Ground-based weather radar, 43–44
Ground clutter, 49

Ground mapping, 47
Ground noise, 104–5
Ground plane construction, 375, 390
Ground-station antenna, 221
Ground-to-air communication, 41, 208–10
Ground-to-ground microwave relay, 214–19
Ground wave differentiation, 52
Ground wave propagation, 11, 36, 86–87
Group repetition interval, 52
Groups, frequency channel, 186–87, 234
GSM. *See* Global System for Mobile Communications
Guided missile, 50
Gunn diode, 43, 159, 162, 164, 486, 499, 504–7
Gyro-klystron amplifier, 571
Gyro-traveling wave tube amplifier, 571
Gyrotron oscillator/amplifier, 569–71

Half-power beamwidth, 391, 393
Half-rate speech coding, 190, 251
Half-wave cavity, 558
Half-wave dipole antenna, 377–78, 419
Half-wave waveguide resonator, 366–67
Hand-held transceiver, 14
Handoff, 244, 250–51, 253
Harmonics, 148, 475, 504
HBT. *See* Heterojunction bipolar transistor
Heater circuit, 551
Height-finding radar, 41
Heliax cable, 294
Helical antenna, 390–91, 414
Helium, 68
Helix-type traveling wave tube, 159, 562–65
HEMT. *See* High electron mobility transistor
Heterojunction, 510
Heterojunction bipolar transistor, 534–35
Hewlett Packard Company, 162, 541
HF. *See* High frequency
H-field. *See* Magnetic field
High electron mobility transistor, 534–35
High frequency, 7, 9, 11, 14, 17, 20–21, 84–87, 91, 95, 145, 373, 399, 425, 454, 458, 529, 541, 551, 554
 characteristics, 36–37
 problems, 86
 single-sideband modulation, 205–8
High-frequency wideband transformer
 conventional, 464–68
 transmission line, 464, 468–71
High-gain amplifier, 150, 539

High-gain antenna, 14, 19, 87, 88, 103, 219, 221, 279, 414, 420
High-order mode, waveguide, 346–47
Highpass filter, 326–27, 452, 454–55
High-power impulse generator, 128–29
High-power microwave generator, 128–29
High-resolution radar, 41, 43, 47
Hoghorn antenna, 217, 421–22
Hole, waveguide wall, 343, 510–12, 515
Home-location register, 252
Homodyne receiver, 153
Hopping, propagation, 85
Horizontal polarization, 30, 73, 211, 371
Horn antenna, 29, 73, 217, 352, 370, 405–8, 418–19, 421
Hot-earth noise, 95, 102
H-plane horn antenna, 405, 407–8
HPM generator. *See* High-power microwave generator
Hughes Network Systems, Inc., 251
Humans, radar cross-section for, 263–264
Hybrid combiner, 464
Hybrid coupler, 319–23, 325, 471–72
Hybrid divider/combiner, 322–23
Hybrid junction, 172, 173, 285, 364–65
Hybrid ring, 354–56
Hybrid T-junction, 354–56, 361, 363
Hyperbolic subreflector, 421, 423

I and Q product detector, 198
ICAO. *See* International Civil Aviation Organization
ICBM. *See* Intercontinental ballistic missile
Idle, cellular call, 245
Idler circuit, 164–65, 503
IDT. *See* Input interdigital transducer
IFF. *See* Information friend or foe
IGFET. *See* Insulated-gate field-effect transistor
IIF. *See* Integration improvement factor
ILS. *See* Instrument landing system
Image frequency suppression, 151–52
Image rejection, 218
IMPATT diode, 162, 164, 504, 507–9
Impedance, 74, 291–92, 295, 297–99, 306–10, 349, 353, 372–73, 78–82, 391, 396, 399, 430–31, 434–35, 436–37, 439–40, 460, 470, 473
Impedance conducting state, 165
Impedance coordinates, Smith chart, 304–6

Impedance matching, 147–48, 163–64, 206, 209, 211, 303, 310–15, 338, 348, 353, 356–57, 402, 444–47, 517, 539–41, 542–43, 544
Impedance ratio, 464–66, 470
Impedance transformer, 318–19, 464, 470
Impulse generator, 128–29, 165
Impulse jammer, 129
Incident wave, 338
Indirect frequency, 162–63, 188
Indium, 499
Indium gallium arsenide, 510–11, 534
Inductive reactance, 303, 379–80, 388–89, 425, 427–31, 459–60
  total, 430
Inductive shunt susceptance, 460
Inductor, 165, 180, 431, 434
Industrial systems, 21
Information friend or foe, 41
Infrared-emitting diode, 257
Infrared laser, 50, 508–11
Infrared photodiode, 511–12
INMARSAT, 222
In-phase channel, 283
Input interdigital transducer, 487, 490
Input matching circuit, 446–47
Input/output system, 7, 240
Insertion loss, 4–5, 331–32, 358, 362, 453, 488
Instrument landing system, 56, 58–59
Insulated-gate field-effect transistor, 531
Integrated circuit, 153, 162, 538
Integration
  coherent, 105–7, 125
  false alarm probability, 109
  noncoherent, 107
  postdetector, 107
Integration improvement factor, 107
INTELSAT, 19, 223
Intensity, electromagnetic wave, 73
Intensity modulation, 257
Intercept probability, 107–8
Intercontinental ballistic missile, 37, 49
Interference, 95, 105, 125–27, 253
Interferometer, 68, 286
Intermediate frequency, 133, 135, 171, 183, 195–96, 202, 208–9, 213, 224 248, 458
Intermediate frequency amplifier, 143–44, 151–54, 224, 231, 275–78, 280–81, 283, 285–86, 462–63, 479, 481
Intermodulation distortion, 148, 175

Intermodulation product, 133, 150
International Civil Aviation Organization, 56, 60
International Telecommunication Union, 6–7, 36–37, 69
Interrogation, 41, 56, 60
Intersatellite service, 19
I/O system. *See* Input/output system
Ionospheric loss, 84–86
Ionospheric reflection propagation, 7–9, 11, 14, 17, 36, 84–86, 91, 205–6
Ionspheric scatter propagation, 87
I/O system. *See* Input/output system
Iris, waveguide, 356–57
Iron core, 461
IR laser. *See* Infrared laser
IS-54 digital cellular, 246–51
IS-95 digital cellular, 252–54
Isolators
  Faraday rotation, 358–60
  ferrite, 330–31, 358–60
  resonant absorption, 360–61
Isotropic antenna, 5, 71–72, 369–70, 413
ITU. *See* International Telecommunication Union

Jammer, 64–68, 129, 287
Jamming, 95, 112–14, 125, 254
JAN type number, 340
Japanese digital cellular, 246, 251
JDC. *See* Japanese digital cellular
JFET. *See* Junction field-effect transistor
Joint, waveguide, 343, 351–53
Joint Tactical Information Distribution System, 125–26
JTIDS. *See* Joint Tactical Information Distribution System
Junction diode, 500
Junction field-effect transistor, 160, 175, 526–29

K&L Microwave, Inc., 454–55
K-band radar, 39
Ka-band radar, 19, 39–40, 158, 566
Klystron amplifier, 156–59, 181, 225, 279, 569
  multiple-cavity, 559–62
  operating characteristics, 559
  velocity modulation, 558
Knife-edge diffraction, 80, 82
Ku-band radar, 19, 39, 214, 297

Land-based coax cable, 27
Land clutter, 265–66

Land mobile-satellite service, 20
Land transportation systems, 21
Laser diode, 508–11
Laser diode emitter, 257
Laser radar, 50
Laser speed detector, 43
Laser target designator, 50
L-band aircraft surveillance radar, 93–94
L-band radar, 38, 41, 49, 78, 158, 282, 566
LC. *See* Lumped-constant circuits;
        Lumped-constant filter;
        Lumped-constant inductor
LCD. *See* Liquid crystal display
LED. *See* Light-emitting diode
Left-hand circular polarization, 385, 390, 393
Lens antennas
    dielectric, 408–9
    Luneburg, 409
    metallic-plate, 409–10
LF. *See* Low frequency
Light-emitting diode, 231, 239, 257, 500,
        510–11
Lightning, 95, 101, 206, 209, 334–35
Limited space-charge accumulation, 162
Limiter, 219
Line antenna array, 410–14
Linear amplifier, 145, 566
Linear polarization, 73, 384, 398–400
Line loss, 298
Line-of-sight system, 14, 55, 101, 211
Liquid crystal display, 229–31
Lithium niobate/tantalate, 486
LNA. *See* Low noise amplifier
LO. *See* Local oscillator
Load conditions, 302–3
Load impedance, 307, 349, 446
Loading
    base, 375
    capacitive, 375–76
Load line, 517, 519
Load resistance, 426, 519
Lobe, grating, 413
Localizer, 59
Local oscillator, 119, 133–34, 151–52, 161–62,
        168–72, 174, 190, 206, 208–9, 213,
        218–19, 275–77, 285, 361
Lock coupling, 295
Logic circuit, 190, 240
Logic unit, 23
Log-periodic antennas
    dipole array, 394–97

    trapezoidal-toothed, 398–400
    triangular-toothed, 400–1
Log-periodic dipole array, 394–97
Long-range communication, 14, 16
Long-range radar, 38–39
Loop antennas, 387
    air-core, 388–89
    ferrite-core, 389
Loop coupling, 316–17, 366
Loops, open/closed, 253
Loop-stick antenna, 389
Loran navigational aids, 52, 54
Loss
    coax cable, 295
    conductor, 298
    delay line, 282, 496
    dielectric, 298
    excess path, 76–79
    feed line, 421
    free-space path, 75–76, 85–86
    ground-plane, 375
    ground-wave propagation, 86
    insertion, 4–5, 331–32, 358, 362, 453, 488
    ionospheric reflection, 84–86
    line, 298
    polarization, 371
    stripline, 300
    tropospheric scatter, 87–88
Lowest usable frequency, 84
Low frequency, 12, 20–21, 84, 86, 95, 375–76,
        425, 529
Low frequency amplifier, 143–44, 154
Low-frequency equivalent circuit, 458
Low frequency field-effect transistor, 538
Low noise amplifier, 99, 143–51, 206, 209,
        211, 218, 224, 236, 276, 285, 505,
        530, 535, 563
Low-noise receiver, 224
Lowpass filter, 131, 181, 195–96, 198, 202–3,
        325–26, 452–53, 455
LP antennas. *See* Log-periodic antennas
LPDA. *See* Log-periodic dipole array
LSA. *See* Limited space-charge accumulation
L-section impedance matching, 446–48, 452
LUF. *See* Lowest usable frequency
Lumped-constant circuits
    impedance matching, 147, 444–46
    series, 356–57, 365
    vacuum-tube amplifier, 551, 554
Lumped-constant filter, 183, 452–55
Lumped-constant inductor, 376

Lumped-element choke, 431
Luneburg lens antenna, 409

M/A-COM, Inc., 330, 332, 533, 545–47
Magic T junction, 172, 320–21, 354–56, 361, 363
Magnetic core, 460–62, 469–70
Magnetic coupling, 343–44
Magnetic dipole, 388–89
Magnetic field, 74
Magnetic flux, 341, 343
Magnetic transformer, 153
Magnetizing inductance, 459–60
Magnetrons, 69, 159, 280–81
    coaxial cavity, 567–69
    conventional, 566–67
    operation, 559
    vane-type anode, 567–68
Main-beam noise jammer, 64, 66
Main-beam repeater jammer, 66, 68
Man-made noise, 95–96, 105
Manufacturers' data sheets, 543, 545
MARISAT, 222
Maritime mobile-satellite service, 20
Maritime mobile service, 20
Marker beacon, 58–59
Mastergroup, 186
Matched filter, 36, 130
Matching circuit, 147
Maximum usable frequency, 84
Maxwell's equations, 337
MDS. *See* Minimum detectable signal
Mean time between failure, 567
Measurement, radar
    angle, 273–74
    range, 267–70
    velocity, 270–73
Mechanical filter, 183–84, 208
Mechanically adjustable capacitor, 433, 475
Mechanically scanned array, 415–18
Mechanical mode, quartz crystal, 475
Mechanical tuning, 160, 482, 568
Medium frequency, 9, 11, 14, 20–21, 36, 84, 95, 145, 151, 373, 375, 551, 554
Medium frequency receiver, 145
MESFET. *See* Metal-semiconductor field-effect transistor
Metallic cavity, 482
Metallic-plate lens antenna, 409–10
Metallic-strip dipole antenna, 401

Metal-oxide semiconductor field-effect transistor, 175, 526, 530–33, 538–39, 546–47
Metal-semiconductor field-effect transistor, 175, 526, 529–30, 533, 535–36
Meteor-burst scatter propagation, 87
Meteorological aids service, 20
Meteorological-satellite service, 20–21
MF. *See* Medium frequency
Mica dielectric capacitor, 432–33
Microfiber teflon fiberglass, 300
Microphone, 205, 207, 209, 211, 240
Microprocessor, 52, 254
Microstrip transmission line, 296–98, 428
    conductor, 425
    coupling, 482
    diode mixer, 168
    frequency multiplier, 165
    impedance matching, 541
    patch antenna, 254
    RF choke, 431
Microwave amplifier, 181, 505, 515, 522
Microwave applications
    coaxial cable, 295
    delay device, 496
    diodes, 501–2, 504–9
    double-sideband modulation, 181
    energy resistance, 502–3
    FET versus BJT, 533
    generators, 128–29
    heating systems, 69–70
    impedance matching, 539, 541
    landing system, 60
MESFET, 530
    mixers, 170–72
    passive circuit, 297
    satellite uplink, 92
Microwave oven, 568
Microwave relay, 14, 16–17, 27–29, 133, 135
    ground-to-ground, 214–19
    sixteen-quadrature modulation, 192
Microwave terrestrial radio. *See* Microwave relay
Microwave tubes, 158, 181, 558–59
    amplitron, 568–69
    circuit configuration, 571–72
    coupled-cavity TWT, 566
    gyrotron oscillator, 569–71
    helix-type TWT, 562–565
    magnetron, 566–68

Microwave tubes (continued)
  multiple-cavity klystron, 559–62
Military aircraft, 262
Military radar
  airborne, 46
  aircraft/missile defense, 49
  ground-based target, 49
  ground-wave propagation, 86
  ship-based, 49
Military weapon control, 39
Miller oscillator, 160, 476
Millimeter-wave FET versus BJT, 533
Millimeter-wave radar, 40
Millimeter-wave radiometer, 68–69
Millimeter-wave traveling wave tube, 563, 565
Minimum detectable signal, 258
Minimum shift keying, 126, 252, 489
MINI-traveling wave tube amplifier, 565
Minority carrier, 513
Mismatching, 542
Missile detection/tracking, 46, 49
Mixers
  Advanced Mobile Phone Service, 236
  balanced, 119, 133, 196, 198, 209, 211, 221, 278, 361, 363
  continuous wave, 270, 275–79
  diode, 167–73
  double-balanced, 202, 208
  down-converter, 213
  frequency division multiplexing, 134
  frequency synthesis, 162
  performance characteristics, 174
  quadrature phase-shift keying, 202–3
  transistor, 148, 173–75
  transmitter, 133, 219
  up-converter, 206, 237
  waveguide, 361, 363
MLS. See Microwave landing system
MMIC amplifier, 151
Mobile communications, 123, 156, 208
Mobile initialization sequence, 249
Mobile-initiated disconnect, 245
Mobile retune command, 244
Mobile-satellite service, 25
Mobile service, 7–9, 21–22
Mobile station, 222
Mobile telecommunication switching office, 23, 239, 240, 242–45, 250
Mobile telephone service, 22–25
Mode jumping, 567
Moderate range surveillance radar, 38

Modulation index, 116–17, 123, 187, 211
MOLNIYA, 222
Monolithic circuit, 530
Monolithic crystal filter, 477–78
Monopole antennas
  electrically small, 375–76
  impedance, 372–73
  large-size, 373–75
  sleeve, 386–87
  thin-wire, 371–72
  wideband, 372
Monopulse antenna, 254
Monopulse tracking, 274, 285–86
Monostatic radar, 36–37
MOSFET. See Metal-oxide semiconductor field-effect transistor
Motorola, 232, 246, 542
Mounting pad, 427
Moving platform, 284–85
Moving target indicator, 38, 49, 278–85
MSK. See Minimum shift keying
MTBF. See Mean time between failure
MTI. See Moving target indicator
MTR. See Monopulse tracking radar
MTSO. See Mobile telecommunication switching office
MUF. See Maximum usable frequency
Multimode fiber, 257
Multimode propagation, 30–31
Multipath, 14, 75, 82–84, 138, 236
Multiple access, 194, 221
Multiple-function radar, 94
Multiple reflection, 494, 496
Multiple-state system, 98–100
Multiplex demodulator, 213
Multiplexing, 27, 213
Multiplier chain, 503–4
Multiposition coaxial switch, 333
Multivibrator, 201–2
Mylar dielectric capacitor, 433

NAMPS. See Narrowband Advanced Mobile Phone Service
Narrowband adaptive spot noise jammer, 64
Narrowband Advanced Mobile Phone Service, 246
Narrowband frequency modulation, 123, 133, 208–10
Narrowband systems, 21, 152, 364, 380, 454–55, 468, 508
Natural noise, 95–96, 101–3
Navigational aids, 550

Navigation systems
  airborne direction finder, 52, 55
  air traffic control, 60–61
  distance-measuring equipment, 56–57
  instrument landing, 56, 58–59
  Loran, 52, 54
  microwave landing, 60
  multiple-function radar, 46
  NAVSTAR/GPS, 60, 62–63
  Omega, 50, 52–53
  power amplifier, 157–59
  radio beacon, 52, 55
  tactical air, 59
  transit, 60
  VHF omnidirectional range, 55–56
N-channel enhancement mode, 530–32
Near-surface region, 531
Negative resistance, 504–5, 507–8, 539
Negative resistance oscillator, 161–62
Net magnetization, 484–85
Neutralization capacitor, 552
Neutralizing circuit, 150, 178, 539
New-channel preparation, 244
Nichrome, 426
Nickel core, 461
Nodding height finder, 41
Noise
  atmospheric, 100–1
  external, 95–96, 145
  frequency modulation, 189
  galactic, 101–3
  ground, 104–5
  man-made, 95–96, 105
  multiple-state system, 98–100
  natural, 95–96, 101–3
  solar, 95, 102–4
  sources, 95–97
Noise cancellation, 172
Noise figure, 98–100, 145, 168, 170, 211, 361,
    501, 516, 533, 535, 538, 543
Noise immunity, 124
Noise power, 97–98
Noise temperature, 97–99
Noncoherent integration, 107
Noncoherent optical detector, 257–58
Nondispersive delay line, 486–89
North American digital cellular
    telephone, 246–51
Notch antenna, 405–6
NPN bipolar junction transistor, 155, 513–14,
    516, 522, 532

N-type cable connector, 295–96, 333
Nulls
  impulse generator, 129
  multipath, 83
  transmission line, 302–4
Null-to-null main lobe frequency
    spectrum, 126
Number, in decibels, 6
Nyquist sampling rate, 135

Octave bandwidth, 454
Octonary phase-shift keying, 139
Offset feed, 421–22
Ohmic resistance, 376, 501
Oil-filled capacitor, 433
Omega system navigational aid, 50, 52–53
Omnidirectional antenna, 59, 254
On-board jammer, 66
One-way attenuation, 78, 80, 82
One-way transmission, 358
On-off amplitude keying, 137
Open-circuit load, 302–3
Open-circuit stub, 356
Open-circuit termination, 315–16
Open-ended waveguide, 415
Open loop, 253
Open-sleeve dipole antenna, 386–87
Open-slot antenna, 401–2
Open-wire line, 28
OPSK. *See* Octonary phase-shift keying
Optical laser radar, 50
Optical region, 261–62
Origination, 243
OR service. *See* Aeronautical mobile service
Oscillations
  amplifier, 150
  grid-type tube, 552
  high-gain amplifier, 539
  parasitic, 552
Oscillators
  coherent, 279–81, 283, 285
  dielectric resonant, 160–61, 479–483
  gyrotron, 569–571
  local, 119, 133–34, 151–52, 161–62,
      168–72, 174, 190, 206, 208–9,
      213, 218–19, 275–77, 279, 281,
      285, 361
  negative resistance, 161–62
  quartz crystal, 134, 160, 162, 165–66, 178,
      180–81, 188, 196, 202, 205, 208–9
      211, 219, 473, 475–77
  surface acoustic wave, 490–93

Oscillators (continued)
    transistor feedback, 159–61
    tunnel diode, 504
    YIG-tuned, 486
Output impedance, 147
Output tap, 454–55
Oven-controlled crystal oscillator, 475–76
Overcoupling, 464
Over-the-horizon radar, 36, 87–88
Overtone, 475–76
Oxygen, 76

Pacific Missile Test Center, 562
Paging, 230–31, 242, 249–50
Palladium, 426
PAM. *See* Pulse amplitude modulation
Parabolic dish reflector antenna, 29, 46, 68, 71, 73, 88, 217, 219, 285, 369, 370, 405, 420–23
Parallel impedance, 356–57
Parallel PIN diode, 502
Parallel plate capacitance, 431–34, 475
Parallel resonant LC circuit, 454–55
Parallel resonant mode, 160, 473, 475, 479
Parallel resonant RLC circuit, 436–38
Parallel to series conversion, 438–40
Parasitic capacitance, 511
Parasitic oscillation, 552
Passive microwave circuit, 297
PCB. *See* Printed circuit board
PCM. *See* Pulse code modulation; Phase code modulation
P code. *See* Precision code
PCS. *See* Personal communications service
PDR. *See* Pulse Doppler radar
Peaks
    multipath, 83
    transmission line, 302–4
Peak-type detector, 196
Pedestal, 482–83
Penetration depth, guide wall, 343
Pentode grid tube, 549, 551
Perfect integrator, 107
Perfect reflector, 69
Personal communications service, 234
Personal communication systems.
    See Business/personal communication systems
Personal computer, 26
Phase code modulation, 52, 129–30, 252
Phase-comparison monopulse radar, 286
Phased-array antenna, 158

Phase detector, 107, 202–3, 280–81
Phase-frequency antenna array, 418
Phase inversion, 464
Phase-locked loop, 138, 162, 188–89, 198, 201, 232
Phase modulation, 36, 219
Phase-scanned array, 415–16
Phase-sensitive detector, 285
Phase-shifter switching system, 334
Phase-shift frequency modulation, 188–89
Phase shifting
    scanning, 415–16, 418
    sideband suppression, 183–85
    space feed, 418–19
Phase-shift keying, 189–92, 489–90
    binary, 138
    four-state/eight-state, 138–39
Phase velocity, 300–1
    metal-plate lens, 410
    waveguide, 338, 351
    wavelength, 340–41
pHemt devices, 39
Phosphorus, 499
Photoconductor, 257
Photodetector, 257
Photodiodes, 31, 257
    infrared, 511–12
Photomultiplier, 257
Phototransistor, 31, 257
Pickup truck, 263
Pierce oscillator, 160, 476
Piezoelectric ceramic resonator, 479, 481
Piezoelectric transducer, 493, 495
PIN diode, 333–34, 501–2
Pi-section impedance matching, 446–47, 450, 452
Planar antenna array, 414–15, 418
Planar wavefront, 408, 410, 420–21
Plan position indicator, 41, 60
PLARS. *See* Position Location and Reporting System
Plastic optical fiber, 257
Plate-modulated amplifier, 177–78, 553, 555
Platinum, 426
Plunger, magnetron cavity, 568
p-n junction, 164, 529
PNP bipolar junction transistor, 514–15, 532
POI. *See* Probability of intercept
Point-contact diode, 168
Pointing error, 273–74, 285
Point-to-point communications, 156

Polarization
   circular, 73–74, 371, 385, 390, 393
   diversity, 17
   elliptical, 73–74, 385
   horizontal, 30, 73, 211, 371
   linear, 73, 384, 389–400
   modulation, 257
   receive antenna, 371
   vertical, 11, 30, 73, 83, 371
Polycarbonate dielectric capacitor, 433
Polyethylene, 293, 295, 301, 353
Polystyrene dielectric capacitor, 433
Porcelain dielectric capacitor, 433
Position Location and Reporting System, 125
Positive feedback, 539
Postprocessing, 107
Power amplifier
   collector-modulated, 178, 180
   communication systems, 143–44, 154–57
   digital cellular, 248
   double-sideband modulator, 181
   FET/TWT, 219
   frequency modulation, 188
   grid tube, 549–58
   moving target indicator, 279
   pi network, 446
   plate-modulated, 178
   power combining, 156–57
   push-pull, 206
   radar system, 157–59
   satellite system, 225
   tetrode-tube, 211
   vacuum tube, 177–79
Power combiner, 464, 471–72
Power density, 74–75
Power divider, 169
Power level, 6
Power ratio, 6
Power reflection coefficient, 306
Power splitter, 202–3, 471–72
PPI. *See* Plan position indicator
PPL. *See* Phase-locked loop
PPM. *See* Pulse position modulation
Precision code, 254–55
Precursor jammer, 66
Preemphasis, 123–24, 189, 209, 211
Preorigination, 243
Prescaler, 162
PRF. *See* Pulse repetition frequency
PRI. *See* Pulse repetition interval
Printed circuit board, 295

Printed circuit, 392–93, 428, 432
   See also Microstrip transmission line
Probability of intercept, 107–8
Probe coupler, 316–17, 366
Product detector, 196–99
Propagation
   cellular telephone, 234–36
   ground wave, 11
   man-made noise, 105
   performance equations, 90–94
   radio frequency, 71
   waveguide, 338
   wavelength, 340–42
   See also Free-space propagation;
      Ionospheric reflection
      propagation; Tropospheric
      scatter propagation
Protection circuit, 218
PRR. *See* Pulse repetition rate
Pseudorandom noise code, 63, 126, 252–54
PSK. *See* Phase-shift keying
Public safety systems, 21
Pulse amplitude modulation, 125
Pulse broadening, 257
Pulse code modulation, 36, 125
   phase code, 52, 129–30
   time-division multiplex, 135–37, 192–96
Pulse compression, 36, 46, 129, 489–91
Pulsed continuous wave, 35, 126–28
   Omega system, 50
   waveform, 115–16
Pulsed frequency modulation, 36
Pulsed klystron amplifier, 561–62
Pulsed magnetron, 159
Pulse-Doppler radar, 49, 278–79
Pulsed traveling wave tube, 159, 563–64
Pulse position modulation, 125
Pulse-power amplifier, 158
Pulse radar
   range measurement, 268–69
   types, 35–36
   velocity measurement, 272–73
Pulse repetition frequency, 126–27, 278–79, 281–82
Pulse repetition interval, 126
Pulse repetition rate, 108, 127
Pulse width modulation, 125
Push-pull amplifier, 145, 154–55, 178, 180, 206, 464, 471–72, 532
PWM. *See* Pulse width modulation
Pyramidal horn antenna, 405–8, 421–22

QAM. *See* Quadrature amplitude modulation
QPSK. *See* Quadrature phase-shift keying
QSELP. *See* Quadrature sum-excited linear predictive
Quadrature amplitude modulation, 29, 63, 192, 194, 219
Quadrature channel, 283
Quadrature coupler, 384
Quadrature phase-shift keying, 126, 138–39, 190, 192, 202–3
Quadrature sum-excited linear predictive, 253
Quadruple diversity, 17
QUALCOMM, Inc., 252–53
Quantization, 278, 282
Quarter-wave cavity, 557–58
Quarter-wave coupled directiona coupler, 323–24
Quarter-wave monopole antenna, 371–72
Quarter-wave sleeve balun, 318
Quarter-wave stub bandpass filter, 327–28
Quarter-wave transformer, 310–13
Quartz crystal, 298, 496
  characteristics, 473–75
  delay media, 281, 486, 494
Quartz crystal amplifier, 178
Quartz crystal filter, 183–84, 463, 477
Quartz crystal oscillator, 134, 160, 162, 165–66, 178, 180–81, 188, 196, 202, 205, 208–9, 211, 219, 475–77
Quiescent operating point, 538

Radar, defined, 35
Radar absorbing material, 150, 262
Radar cross-section, 35, 41, 46, 278
  characteristics, 261–63
  clutter, 263–67
  target backscatter, 89–90
  target specifications, 263–64
  test ranges, 49–51
Radar systems
  concepts, 35–36
  crossed-field amplifier, 569
  electronic countermeasures, 112–14
  frequencies, 36–40
  grid tube use, 550
  klystron amplifier use, 561
  link analysis, 111–12
  power amplifier, 157–59
  propagation performance, 92–94
  types, 40–50
Radial velocity, 35
Radiation resistance, 388–89

Radio astronomy systems, 68
Radio beacon, 52, 55
Radio frequency, 3, 14, 35, 180, 431
Radio frequency choke, 446, 502–3, 516, 518–19, 522, 538, 552, 557
Radiometer, 68–69
RadioShack, 229–31, 239
Radiotelephone, 236–38
Rain, 78–79, 217, 266–68
RAM. *See* Radar absorbing material
Range capability, 9
Range gate, split, 286–87
Range-gate pulloff, 68, 287
Range-gate stealer, 68
Range measurement, 125–26, 267–70
Range-squared proportionality, 76
Range to target, 35, 41, 46
Ratio detector, 200–1
Rat race hybrid coupler, 320–22, 354–55
RATSCAT test range, 50
Rayleigh region, 261–62
RC circuit, 434
RC-coupled amplifier, 154
RCS. *See* Radar cross-section
Receive antenna, 72–73, 89, 369–71, 387, 420
Receive-only system, 63
Receivers
  global positioning satellite, 254
  homodyne, 153
  superheterodyne, 152–53
Rectangular-plate antenna, 373
Rectangular pulse, 127–28
Rectangular-slot antenna, 402–3
Rectangular waveguide, 338–47, 349
Reentrant-type waveguide resonator, 366–67
Reference oscillator, 196
Reference temperature, 98–99
Reflectarray antenna, 418–19
Reflection
  ionospheric, 83, 85
  loss due to, 85–86
  transmission line, 301–3, 306
  voltage/power, 306
  waveguide, 338
Reflective array compressor, 490, 493
Reflector antennas, 418–19
  corner, 419
  half-wave dipole, 419
  parabolic dish, 420–23
Refraction, 79–80, 85, 409
Release, cellular call, 245

Repeater, 23, 31, 258–59
Repeater jammer, 66, 68
Reply, 41
Resistive load, 302–3, 357, 516
Resistive switching mixer, 170–71
Resistivity, waveguide, 344
Resistor, 426–27, 434, 532
Resonance frequency, water/oxygen, 76
Resonance region, 261–62
Resonant absorption isolator, 360–61
Resonant circuits, 145
    complex, 438–40
    parallel, 436–38
    series, 435–36
Resonant dipole antenna, 381, 383
Resonant ladder bandpass filter, 454–55
Resonators
    cavity, 365–67, 559, 561
    ceramic disk, 479–80
    quartz crystal, 473–75
RETMA type number, 340
Reverse bias, 334, 501–5, 512–15, 526
Reverse channel, 238, 240
Reverse diode breakdown, 500–1
RF. *See* Radio frequency
RFC. *See* Radio frequency choke
RGPO. *See* Range-gate pulloff
Rho-theta system, 56
Ridged waveguide, 350
Right-hand circular
    polarization, 371, 385, 390, 393
Rise time, pulse, 127–28, 434
RL circuit, 430–31
RLC circuit, 435–38
Rod-shaped core, 462
Rolloff rate, 126
Rotary joint, 351–53
Rotating loop antenna, 52
R service. *See* Aeronautical mobile service
R/T. *See* Radiotelephone
Rubber, 353

SA. *See* Selective availability
SACC, 250
SACCH, 247
Sampling circuit, 134–35, 192, 194
SAT. *See* Supervisory audio tone
Satellite navigation systems
    NAVSTAR/GPS, 60, 62–63
    Transit, 60
Satellite relay, 27, 29–30, 92
    communication, 219–23

    Earth stations, 223–25
Satellite systems
    aeronautical mobile, 9–10
    broadcasting, 12–13
    cable requirements, 133
    Earth exploration, 12
    fixed, 19
    global positioning, 254–56
    intersatellite, 19
    locating/tracking, 46
    surveillance, 37
    mobile, 20, 25
    security, 255
    transmitter, 563–64
    uplink/downlink, 19–20, 30, 92, 223, 156
SAW. *See* Surface acoustic wave
Sawtek Inc., 489–90, 494
S-band radar, 38, 46, 78, 158, 282, 569
SBF method, 79
SCA. *See* Subsidiary communication
    authorization
Scale factor, 394, 396, 399–400
Scanning methods
    frequency, 416–18
    phase, 415–16
    space feed, 418–19
Scattering, 87–88
    backscatter, 89–90, 261
    coefficients, 83–84
    forward, 17
    radar, 261–62
Scattering parameter, 443–46, 522, 541–43
Schottky-barrier
    diode, 168, 361, 499, 501, 529–30
Screen grid, 550
Screw tuner, 356–57
Sea clutter, 49, 266–67
Sea level atmospheric attenuation, 77–78
Search/surveillance radar, 158
Second-order amplifier response
    curve, 148–50
Selective availability, 255
Selenium, 500
Self-biasing, 534, 537
Semiconductor infrared laser diode, 508–11
Semiconductor materials, 499–500
Semiloop wire, 486
Sequential gating, 286–87
Sequential lobbing, 273–74, 286
Series capacitance, 434, 475, 552
Series-fed array, 394

Series-mounted PIN diode, 502–3
Series RC circuit, 434
Series reactance, 470
Series resistance, 538
Series resonant mode, 160, 376, 475–77, 479
Series resonant RLC circuit, 435–36
Series to parallel circuit, 438–40
Series-tuned circuit, 164
Serpentine feed, 417
Setup channel, 242–43
Sheet-metal type antenna, 392–93, 398–400
SHF, 9, 12, 19–21, 25, 87
Shielding, amplifier, 150, 540
Shift oscillator, 219
Ship, tracking, 49
Ship-based radar, 38, 46, 49
Ship-to-air communication, 209
Ship-to-ship communication, 209
Ship-to-shore communication, 209, 222
Shore-based search/surveillance radar, 46
Short-circuited bandpass filter, 327–28
Short-circuited stub, 313–15, 356
Short-circuited termination, 315–16
Short-slot hybrid junction, 364–65
Shunt-mounted PIN diode, 502–3
Shunt stub DC return, 319
Shunt susceptance, 459–60, 469–70
Sidebands
    double, 118–19
    power, 117–18
    single, 120–21
    vestigial, 120
Sideband-suppression filter, 183–84
Side-coupled bandpass filter, 327–28
Sidelobe, 126–27, 115, 122, 369, 421
Sidelobe noise jammer, 64, 66–67
Side-looking radar, 47–48
Side-wall slot antenna, 403–5
Signal boosting, 123–24
Signal-to-jamming ratio, 113–14
Signal-to-noise ratio, 31, 87, 138, 189, 278, 282, 284, 287
    communication system link, 109–11
    improvement through integration, 105–7
    radar system link analysis, 111–12
    required, 107–9
Silence-interval detection, 251
Silicon, 146, 164, 256–57, 499–501, 503–4, 508, 511, 513–14, 530, 533
Silicon-point diode, 361
Simplex system, 229

Sine of sine, 122
Sine wave, 178, 383
Single-balanced diode mixer, 168–69
Single-channel tracking, 254
Single-channel transmitter, 131–33
Single-ended amplifier, 154
Single-ended mixer, 167–68, 173–74, 363
Single-hop propagation, 85
Single-mode fiber, 257
Single-mode propagation, 30
Single-pole double-throw switch, 331–32, 361
Single-sideband detection, 196, 198
Single-sideband modulation, 120–21, 183–85, 205–8, 223, 231
Single-sideband suppressed carrier modulation, 133
Single-span link, 17
Single-tuned transformer, 208, 462, 553, 557
Sinuous feed, 417
Sinusoidal current, 164
Sixteen amplitude-phase keying, 140
Sixteen-phase-shift keying, 192
Sixteen phase-state keying, 139
Sixteen-quadrature amplitude modulation, 192, 194
Skin effect, 343, 425–26
Sky temperature, 69
Sky wave propagation. *See* Ionospheric reflection propagation
Sleeve antenna, 386–87
Sleeve tuner, 315–16
Slot, waveguide wall, 343–44
Slot antennas
    cavity-backed, 402
    open-, 401–2
    waveguide-fed, 402–5, 415
Slot coupling, 366
SMA connector. *See* Subminiature A series connector
Smart bomb, 50
Smith chart, 304–15
    circuit analysis, 438, 441–43
    dipole impedance, 381–82
    impedance matching, 446–52, 543
    S-parameter conversion, 445
Snake feed, 417
Snap-off varactor, 164, 503
Snow, 78–79, 217, 266–67
Soft handoff, 253
Solar burst, 103
Solar noise, 95, 102–4

Solenoid, 427–28
Solid dielectric, 292
Solid-state radar, 158–59, 181, 224, 275, 277, 433
Sony cordless telephone, 229–30
Space-based radar, 46, 158–59
Space diversity, 17
Space feed, 418–19
Space operation service, 25
Spacer, 482–83
Space research service, 25
Spacing factor, 395–96
S-parameter. *See* Scattering parameter
Sparkgap switch, 206, 218, 334–35
Spatial frequency reuse, 232–35
SPDT. *See* Single-pole double-throw switch
Spectral lines envelope, 115–16
Spectrum envelope, 127–28
Specular scattering, 83
Speech coding, 190, 247–48, 251, 253
Speed measurement radar, 43, 159
Sphere, conducting, 261–62
Spherical resonator, 484
Spherical wavefront, 408–9, 420
Spinning electron, 484–85
Spiral antennas, 391
    Archimedean, 393–94
    conical, 394
    equiangular, 392–93
    helix, 254
Spiral inductor, 428–29
Spiroline cable, 294
Split inline hybrid divider, 322–23
Splitter, hybrid, 471–72
SPMT switch, 333
SPP-ID910 cordless telephone, 229–30
Spread spectrum modulation, 125–26, 232
Spurious amplitude modulation, 219
Spur, 133
Square-law transfer, 165
SRD, 165
SSB modulation. *See* Single-sideband modulation
Stability, amplifier, 539
Stable local oscillator, 279, 281
Stalo. *See* Stable local oscillator
Standard frequency, 26
Standing-wave ratio, 172, 296, 301–4, 315, 323, 331, 353, 381, 387, 393, 399, 402
Standing-wave ratio circle, 206, 306–7, 312
Standoff jammer, 66

Static radar cross-section, 49–51
Stealth bomber, 262
Step-index fiber, 30, 88, 257
Step-recovery diode, 164–66, 500, 504
Stripline conductor, 425
Stripline diode mixer, 168
Stripline filters
    bandpass, 327–29
    highpass, 326–27
    lowpass, 325–26
Stripline frequency multiplier, 165
Stripline transmission line, 298–300, 431, 541
Stubs
    open-circuit, 356
    short-circuited, 313–15, 356
    shunt, 319
Stub matching, 376
Stub tuner, 356
Styroflex cable, 294
Styrofoam, 408
Subcommutating, 125
Submarine cable, 31–32, 257
Subminiature A series connector, 295–96, 332, 482, 511
Subreflector, 421
Subsidiary communication authorization, 211, 214
Subtraction circuit, 283
Sulfur, 500
Superconducting magnet, 570
Supergroup, 186–87
Superheterodyne receiver, 152–53, 231
Supermastergroup, 186
Supervisory audio tone, 242, 244
Suppressed carrier modulation, 133
Suppressor grid, 551
Surface acoustic wave delay line, 473, 486–93
Surface acoustic wave filter, 154, 183–84, 208, 463
Surface-based fire-control radar, 159
Surface installation search, 158
Surface-mounted antenna, 402–403, 475
Surface reflection coefficient, 83
Surface wave propagation. *See* Ground wave propagation
Surveillance radar, 38, 40–41, 158
Switched frequency counter, 278
Switches
    coaxial electromechanical, 331–33
    directional coupler, 443–44
    in frequency synthesis, 162

Switches (continued)
    gas-tube, 361, 364–65
    multiposition, 333
    pin diode, 333–34
    single-pole double-throw, 361
    sparkgap, 206, 218, 334–35
    transfer, 332–33
Switching circuit, 168
Switching mixer, 170–71
SWR. *See* Standing-wave ratio
Symbol-timing recovery circuit, 202–3
Synchronization, 194, 202
Synchronous detector, 107
Synthetic aperture antenna, 47
System-initiated disconnect, 245

TACAN. *See* Tactical air navigation
Tactical air navigation, 55, 59
Talking procedure, 243–44
Tank circuit, 145, 454–55
Tansponder, 60
Tantalum, 426, 433
Tapered transition section, 353
Tapped delay line, 489
Target location/tracking, 38, 49, 278
TDA. *See* Tunnel diode amplifier
TDM. *See* Time division multiplexing
TDMA. *See* Time-division multiple access
TD system, 29
TE10 mode, 338–39, 344, 347–49
Teflon, 294–95, 297–98
Teflon fiberglass, 297–300
Telecommunication Industries
    Association, 190, 195, 251–54
Teledyne Microwave, 496
Telegraphy, 137–38
Telemetry, 124–25
Telephones, 26–27, 219
    cellular, 22–25, 488
    cordless, 227–30
Television, 11, 21, 121, 123, 153, 157, 219,
    222, 224, 291, 378, 488, 550, 561
Telex, 137–38
Tellurium, 500
TEM. *See* Transverse electromagnetic wave
TE mode. *See* Transverse electric mode
Temperature-compensated
    crystal oscillator, 475
Termination, coaxial, 315–16
Termination insensitive mixer, 171, 173
Terrain avoidance, 46–47
Terraine-following radar, 47

Test range, 49–50
Tetrode grid tube, 549–52, 557
Tetrode vacuum tube, 177–79, 181, 211
Texas Instruments, 533
T-fed slot antenna, 402–3
Thermal runaway, 534
Thermionic cathode, 549
Thermistor-resistor network, 475
Thick-film circuit, 426
Thin-film circuit, 426, 496
Thin-wire dipole antenna, 377–80
Thin-wire monopole antenna, 371–72
Third-order amplifier response, 148–50
Threaded coupling, 295–96, 464
Threaded mating coupling, 296
Three-hop propagation, 85
Three-screw tuner, 357
Threshold voltage, 531, 538
TIA. *See* Telecommunication Industries
    Association
TIM. *See* Termination insensitive mixer
Time-division multiple
    access, 22, 194–95, 224–25,
    232, 246
Time-division multiplexing, 27, 29, 124–25,
    135–37, 192–96, 219, 223
Time-domain filter, 281
Time signal service, 26
Time slot, digital cellular, 246
Time-slot repetition rate, 136
Timing circuit, 202–3
Titanate, 479
TIX3024 transistor, 533
T-junction, 353
TM mode. *See* Transverse magnetic mode
TNC cable connector, 295
Toll network, 27
Top-wall slot antenna, 403–5
Toroidal inductor, 429–30
Tracking radar, 273–74
    frequency-modulated CW, 278
    monopulse, 285–86
    sequential gating, 286–87
Transceiver, 7, 8, 14, 23, 205, 207–8, 240, 242
Transconductance, 520, 523, 527, 531
Transducer, 369–70, 486–87, 490,
    492–93, 496, 541
Transfer switch, 332–33
Transformer-coupled amplifier, 154
Transformers, 156, 169–71, 178, 180,
    199–200, 206, 208

common-emitter amplifier, 516–17
conventional, 457–60, 469–72
dipole antenna, 378
high-frequency wideband, 464–68
impedance, 318–19, 376, 539–41
magnetic core material, 460–62
power combiner/splitter, 471–72
quarter-wave, 310–13
single-tuned, 553, 557
transmission-line, 468–72
tuned, 462–64
Transistor amplifier, 154, 156, 159, 501, 539–41
Transistor feedback oscillator, 159–61
Transistor mixer, 173–75, 196
Transistor multiplier, 165
Transistor oscillator, 476–77
Transistors
  biasing, 426
  data sheets, 543, 545
  heterojunction bipolar, 535
  high electron mobility, 534–35
  transfer curve, 148–149, 150
  See also Bipolar junction transistor; Field-effect transistor
Transit navigational system, 60
Transit time, tube amplifiers, 558
Transmission-line transformer, 464, 468–472
Transmission media, 27
  balun types, 316–17
  communications satellites, 29–30
  fiber optics cable, 27, 30–31, 88–89, 256–59
  highpass structure, 171
  microwave relay, 28–29
  open-wire line, 28
  reflection coefficients, 301
  slot antenna feeding, 402
  Smith chart, 304–15
  standing-wave ratio, 301–4
  stripline, 298–300, 431, 541
  submarine cable, 31–32
  twisted-pair line, 28
  two-wire, 291–92, 318–19
  wave velocity, 300–301
  See also Coaxial cable; Microstrip transmission line
Transmit antenna, 369–70, 387, 420
Transmit/receive switch, 285, 361, 364–65
Transmitter systems, 50
  amplifiers, 143–44, 154–57
  amplitude modulation, 179
  antennas, 71–72, 89

double-sideband, 180–81
frequency division multiplex, 133–34
frequency modulation, 240
mixers, 219
power oscillator, 280
shutdown, 245
single-channel, 131–33
time division multiplex, 135–37
traveling wave tube, 563–64
Transparent relay system, 221
Transponder, 56, 68, 223
Transverse electric mode, 337–39, 344–47, 351–52, 358, 366
Transverse electromagnetic wave, 73–74, 129, 295, 297, 349, 358
Transverse magnetic mode, 338, 346, 366
TRAPATT. *See* Trapped plasma avalanche transit time
Trapezoidal pulse, 127
Trapezoidal-toothed log-periodic antenna, 398–400
Trapped plasma avalanche transit time, 162
Traveling wave tube, 156–59, 181, 219, 221, 287
Traveling wave tube amplifiers
  circuit configurations, 571–72
  coupled-cavity, 566
  helix-type, 562–65
  velocity modulation, 558
Traveling wave tube/crossed-field amplifier, 157–59
TRC-485 CB radio, 231
Triangular pulse, 128
Triangular-toothed log-periodic antenna, 400–401
Trimmer capacitor, 160, 476
Triode grid tube, 549–50, 557
Triode power amplifier, 181
Triode vacuum tube, 177–79
Tropospheric scatter propagation, 17–18, 87–88
TR switch. *See* Transmit/receive switch
Truncated paraboloid, 421
Trunk line, 26, 239–40, 242
T-section impedance matching, 446–47, 449, 451
Tube-type power amplifier, 154, 156, 158
Tubular capacitor, 433
Tuned-input tuned-output oscillator, 160, 476
Tuning
  antenna, 87

Tuning (continued)
  cavity resonator, 366
  circuit, 145
  dielectric resonator, 482
  magnetron, 568
  transformer, 462–64, 516
  YIG resonator, 486
  waveguide, 356–57, 361
Tuning inductor, monopole antenna, 375–76
Tuning screw, 168
Tunnel diode, 499
Tunnel diode amplifier, 505–6
Tunnel diode oscillator, 161, 504
Turnstile antenna, 211, 384–85
Twisted-pair line, 26, 28
Two-hop propagation, 85
Two-plate capacitor, 432
Two-screw tuner, 356–57
Two-state amplitude keying, 137
Two-state modulation, 137–38
Two-way atmospheric attenuation, 76–77, 80
Two-way communication, 7, 14, 23, 217, 224, 232, 270
Two-way free-space path loss, 76
Two-winding transformer, 457–58
Two-wire transmission line, 291–92, 318–19, 377–78, 386–87, 399
TWT. See Traveling wave tube
Twystron, 158, 279, 561–62

UHF. See Ultra high frequency
Ultra-high frequency, 7–9, 11–12, 14, 17, 19–21, 25, 38, 49, 80, 87, 157–58, 162, 181, 232, 276, 297, 337, 339, 380, 382, 399, 515, 522, 532, 539, 550–51, 554, 561, 566
Ultra-high frequency cable connector, 295–96
Ultra-high frequency communication system, 208–10
Ultra-wideband microwave generator, 128–29
Umbrella-type loading, 375–76
Undersea cable, 27, 133
Unity conductance, 452
Up-converter, 119, 133, 154, 181, 183, 186–87, 206, 225, 237, 279
Uplink, satellite, 19–20, 30, 92, 156, 223

Vacuum-tube amplifiers, 156, 177–79
  cavity resonator, 557–58
  grid-type, 549–57
  power, 177–79
  triode/tetrode/pentode, 549–53, 557–58

Vane-type magnetron, 567–68
Varactor, 160, 162, 164–66, 188, 211, 433, 475, 482, 500, 165, 502–4
Varian Associates, 561–65
V-band radar, 40
VCO. See Voltage-controlled oscillator
Vector addition/subtraction, 75
Vector sum-excited linear predictive, 247–48
Vegetative absorption coefficient, 83–84
Velocity, transmission line wave, 300–301
Velocity-gate pulloff, 68
Velocity measurement
  continuous wave radar, 270–76
  pulse radar, 272–73
Velocity modulation
  helix-type amplifier, 562–63
  microwave tube amplifier, 558
Vertical polarization, 11, 30, 73, 83, 371
Very high frequency, 7–9, 11, 14–15, 20–21, 25, 37–38, 49, 80, 87, 145, 147, 162, 181, 208–10, 219, 232, 339, 372, 399, 429, 464, 522–24, 549–52, 554, 557
Very high frequency propagation, 91–92
Very high frequency omnidirectional range, 55–56
Very high frequency transformer, 464, 467–68
Very large array, 68
Very low frequency, 12, 14, 20, 50, 84, 86, 95, 375
Very small antenna terminal, 221
Vestigial sideband modulation, 120, 181–83
VGPO. See Velocity-gate pulloff
VHF. See Very high frequency
Video amplifier, 285
Visited-location register, 252
VLA. See Very large array
VLF. See Very low frequency
Voice communication, 122
Voltage breakdown, 500–1, 512
Voltage-controlled amplifier, 146
Voltage-controlled oscillator, 160, 163, 189, 198, 492
Voltage-controlled resistor, 526
Voltage probe, 347–48
Voltage ratio, 6
Voltage reference, 501
Voltage reflection coefficient, 306, 443
Voltage standing-wave ratio, 172, 296, 312, 315, 323, 331, 381, 387, 393, 399, 402

Voltage standing-wave ratio circle, 306–7, 315
Voltage-tuned capacitor, 160
Voltage-variable capacitor, 433, 503
VOR. *See* Very high frequency omnidirectional range
VSAT. *See* Very small antenna terminal
VSB modulation. *See* Vestigial sideband modulation
VSELP. *See* Vector sum-excited linear predictive
VSWR. *See* Voltage standing-wave ratio
V-type dipole antenna, 377

Walkie-talkie, 231
Water cooling, 551
Water vapor, 76
Waveguide
   cavity resonator, 365–67
   characteristic impedance, 349
   characteristics, 337–38
   circular, 350–52
   detector/mixer, 168, 361
   directional coupler, 357–58
   duplexer, 362–65
   ferrite isolator, 358–61
   flexible section, 346
   gas-tube switch, 361–62
   hardware, 352–54
   hybrid junction, 354–56
   impedance matching, 356–57
   rectangular, 338–47, 349
   resistive load, 357
   ridged, 350
   TE_{10} mode, 347–48, 347–49
Waveguide-fed slot antenna, 402–5
Waveguide frequency multiplier, 165
Waveguide mode suppression, 348–349

Wavelength, 3–4, 73
Wave velocity, 300–1
W-band radar, 40
Weather, 78–79, 214, 217, 266–67
Weather-avoidance radar, 39, 44, 159
Weather-detection radar, 38–39, 46
Weather echo, 38
Wedge angle, 398–400
Whip antenna, 211, 236
White Sands Missile Range, 50
Wideband amplifier, 145, 426, 516, 564
Wideband antenna, 129, 372–74
Wideband balun, 318
Wideband barrage noise jammer, 64
Wideband communications, 14, 29, 31
Wideband frequency modulation, 123
Wideband hybrid combiner, 472
Wideband transformer, 318–19, 460, 464–68
Wiltron Corporation, 541
Winding, transformer, 156, 180, 199, 464–67, 469–71
Wire-type log-periodic antenna, 398–400

X-band airborne multiple-function radar, 94
X-band radar, 39, 43, 361

Yagi-Uda antenna, 211, 385–86, 421
YIG resonator oscillator. *See* Yttrium iron garnet resonator oscillator
Y-junction ferrite circulator, 329–30, 361
Yttrium iron garnet resonator oscillator, 160–63, 484–86

Zener diode, 218, 500–1
Zero-crossing detector, 201–2
Zinc, 500

The Artech House Microwave Library

Acoustic Charge Transport: Device Technology and Applications, R. Miller, C. Nothnick, and D. Bailey

Advanced Automated Smith Chart Software and User's Manual, Version 2.0, Leonard M. Schwab

Analysis, Design, and Applications of Fin Lines, Bharathi Bhat and Shiban K. Koul

Analysis Methods for Electromagnetic Wave Problems, vol. 2, Eikichi Yamashita, editor

C/NL2 for Windows: Linear and Nonlinear Microwave Circuit Analysis and Optimization, Software and User's Manual, Stephen A. Maas and Arthur Nichols

C/NL2 for Windows® 95, NT, and 3.1: Version 1.2-Linear and Nonlinear Microwave Circuit Analysis and Optimization, Software and User's Manual, Stephen A. Maas

Computer-Aided Analysis, Modeling, and Design of Microwave Networks: The Wave Approach, Janusz A. Dobrowolski

Designing Microwave Circuits by Exact Synthesis, Brian J. Minnis

Dielectric Materials and Applications, Arthur von Hippel, editor

Dielectrics and Waves, Arthur von Hippel

Electrical and Thermal Characterization of MESFETs, HEMTs, and HBTs, Robert Anholt

Frequency Synthesizer Design Handbook, James Crawford

Frequency Synthesizer Design Toolkit Software and User's Manual, Version 1.0, James A. Crawford

Fundamentals of Distributed Amplification, Thomas T. Y. Wong

GSPICE for Windows, Sigcad Ltd.

HELENA: HEMT Electrical Properties and Noise Analysis Software and User's Manual, Henri Happy and Alain Cappy

HEMTs and HBTs: Devices, Fabrication, and Circuits, Fazal Ali, Aditya Gupta, and Inder Bahl, editors

High-Power Microwaves, James Benford and John Swegle

LINPAR for Windows: Matrix Parameters for Multiconductor Transmission Lines, Software and User's Manual, Antonije Djordjevic, Miodrag B. Bazdar, Tapan K. Sarkar, Roger F. Harrington

Low-Angle Microwave Propagation: Physics and Modeling, Adolf Giger

MATCHNET: Microwave Matching Networks Synthesis, Stephen V. Sussman-Fort

Matrix Parameters for Multiconductor Transmission Lines: Software and User's Manual, A. R. Djordjevic et al.

Microelectronic Reliability, Volume I: Reliability, Test, and Diagnostics, Edward B. Hakim, editor

Microstrip Lines and Slotlines, Second Edition, K.C. Gupta, Ramesh Garg, Inder Bahl, and Prakash Bhartia

Microwave and Millimeter-Wave Diode Frequency Multipliers, Marek T. Faber, Jerzy Chramiec, Miroslaw E. Adamski

Microwave Engineers' Handbook, Two Volumes, Theodore Saad, editor

Microwave and Millimeter Wave Heterostructure Transistors and Applicatons, F. Ali, editor

Microwave and Millimeter Wave Phase Shifters, Volume I: Dielectric and Ferrite Phase Shifters, S. Koul and B. Bhat

Microwave and Millimeter Wave Phase Shifters, Volume II: Semiconductor and Delay Line Phase Shifters, S. Koul and B. Bhat

Microwave Mixers, Second Edition, Stephen Maas

Microwave Tubes, A. S. Gilmour, Jr.

Microwaves: Industrial, Scientific, and Medical Applications, J. Thuery

Modern GaAs Processing Techniques, Ralph Williams

Nonuniform Line Microstrip Directional Couplers, Sener Uysal

PC Filter: Electronic Filter Design Software and User's Guide, Michael G. Ellis, Sr.

PLL: Linear Phase-Locked Loop Control Systems Analysis Software and User's Manual, Eric L. Unruh

RF Design Guide: Systems, Circuits, and Equations, Peter Vizmuller

RF Systems, Components, and Circuits Handbook, Ferril Losee

Scattering Parameters of Microwave Networks with Multiconductor Transmission Lines: Software and User's Manual, A. R. Djordjevic et al.

Transmission Line Design Handbook, Brian C. Waddell

TRAVIS Pro: Transmission Line Visualization Software and User's Manual, Professional Version, Robert G. Kaires and Barton T. Hickman

TRAVIS Student: Transmission Line Visualization Software and User's Manual, Student Version, Robert G. Kaires and Barton T. Hickman

Yield and Reliability in Microwave Circuit and System Design, Michael Meehan and John Purviance

For further information on these and other Artech House titles, contact:

Artech House
685 Canton Street
Norwood, MA 02062
617-769-9750*
Fax: 617-769-6334*
Telex: 951-659
e-mail: artech@artech-house.com

Artech House
Portland House, Stag Place
London SW1E 5XA England
+44 (0) 171-973-8077
Fax: +44 (0) 171-630-0166
Telex: 951-659
e-mail: artech-uk@artech-house.com

WWW: http://www.artech-house.com

* As of September 1, 1997, new area code is (781)